Electrochemistry in Ionic Liquids

Angel A.J. Torriero
Editor

Electrochemistry in Ionic Liquids

Volume 1: Fundamentals

Springer

Editor
Angel A.J. Torriero
Centre for Chemistry and Biotechnology
School of Life and Environmental Sciences
Faculty of Science, Engineering and Built Environment
Deakin University, Melbourne Burwood Campus
Burwood, VIC, Australia

ISBN 978-3-319-34999-2 ISBN 978-3-319-13485-7 (eBook)
DOI 10.1007/978-3-319-13485-7

Springer Cham Heidelberg New York Dordrecht London

Softcover reprint of the hardcover 1st edition 2015

Printed on acid-free paper

Springer International Publishing AG Switzerland is part of Springer Science+Business Media (www.springer.com)

Preface for First Volume

Electrochemistry is dominated by the study of species dissolved in solution. The use of a solvent as the reaction medium helps electrochemists to control important reaction conditions such as pH, rate of mass transfer, concentration of reactant, solubility, solvation, etc. Water and organic solvents are the most popular media. However, by using appropriate ionic liquids, reactants and products that are unstable in those media remain stable, and redox reactions that are impossible in water and organic solvents become possible. The reaction environments are markedly wider in some ionic liquids than in other solvent systems. In spite of this, some fundamental electrochemical concepts generally used in conventional solvent systems are not always valid in ionic liquids.

The aim of this book is to present the insights of experts on emerging experimental techniques and theoretical concepts that are or will be in the vanguard of the field of electrochemistry in ionic liquids. The two volumes of this book provides the reader with a broad and self-contained account of electrochemical techniques available to work in ionic liquids. It also gathers and critically discusses the important properties of protic ionic liquids, deep-eutectic solvents, task-specific ionic liquids, polymeric ion gels, and lithium-ion solvation, relevant to electrochemistry.

The two volumes of this book will be of interest for researchers in relevant industries as well as for academics, for scientists, and for the general chemical and electrochemical audience. It also can be given to graduate students and postdoctoral fellows looking for guidance to start working in this research field.

Burwood, VIC, Australia — Angel A.J. Torriero

Preface for Second Volume

The field of electrochemistry in ionic liquids has blossomed in the last decade, since some of these ionic liquids tend to be good electrochemical solvents, exhibit interesting conductive properties, and offer the unique possibility to tune the physicochemical properties. These exclusive properties make them very useful as both the electrolyte and the solvent in electrochemical applications. There are several comprehensive reviews of this field, but none of them have covered the electrochemical applications of ionic liquids to the widest extent. Thanks to the contribution of scientists from different countries with strong expertise in the field, this new book collates the formerly dispersed knowledge into two handy volumes. Readers will find ten chapters gathered in three sections. The book commences with the latest understanding of electrodeposition, and electroless deposition, followed by a critical discussion of voltammetry of adhered microparticles, electrochemistry of organic and organometallic compounds, electrocatalytic reactions, oxygen reduction reaction, and ionic liquids in surface protection and lubrication. The book finishes with an overview on the current industrial application of ionic liquids and the challenges, issues, and recycling methods used in industrial developments.

The two volumes of this book will be of interest to researchers in relevant industries as well as academics, scientists, and anyone interested in general chemistry and electrochemistry. The book can also be used as a guide for graduate students and postdoctoral fellows who wish to start working in this field.

Burwood, VIC, Australia — Angel A.J. Torriero

Contents

Volume I

Volume II

Contributors

Volume I

Mérièm Anouti Laboratoire PCM2E (EA 6296), Université François Rabelais de Tours, Parc de Grandmont, Tours, France

Cameron L. Bentley School of Chemistry, Monash University, Clayton South, VIC, Australia

Energy Flagship, Commonwealth Scientific and Industrial Research Organisation (CSIRO), Clayton South, VIC, Australia

Alan M. Bond School of Chemistry, Monash University, Clayton South, VIC, Australia

Hiroyuki Doi Graduate School of Science and Technology, Niigata University, Nishi-ku, Niigata City, Japan

Kenta Fujii Graduate School of Science and Engineering, Yamaguchi University, Ube, Yamaguchi, Japan

Joshua A. Hammons X-ray Science Division, Advanced Photon Source, Argonne National Laboratory, Argonne, IL, USA

Anthony F. Hollenkamp Energy Flagship, Commonwealth Scientific and Industrial Research Organisation (CSIRO), Clayton South, VIC, Australia

Jan Ilavsky X-ray Science Division, Advanced Photon Source, Argonne National Laboratory, Argonne, IL, USA

Peter J. Mahon Faculty of Science, Engineering and Technology, Swinburne University of Technology, Hawthorn, VIC, Australia

David Mecerreyes POLYMAT, University of the Basque Country UPV/EHU, San Sebastian, Spain

Abdul Rehman Department of Chemistry, Oakland University, Rochester, MI, USA

Ikerbasque, Basque Foundation for Science, Bilbao, Spain

Laurent Rubatat EPCP, UMR IPREM 5256, Université de Pau et des Pays de l'Adour, Pau, France

Maitane Salsamendi POLYMAT, University of the Basque Country UPV/EHU, San Sebastian, Spain

Shiro Seki Materials Science Research Laboratory, Central Research Institute of Electric Power Industry, Komae, Tokyo, Japan

Angel A.J. Torriero Centre for Chemistry and Biotechnology, School of Life and Environmental Sciences, Faculty of Science, Engineering and Built Environment, Deakin University, Burwood, VIC, Australia

Yasuhiro Umebayashi Graduate School of Science and Technology, Niigata University, Nishi-ku, Niigata City, Japan

Darren A. Walsh School of Chemistry, University of Nottingham, Nottingham, UK

Zhe Wang Department of Chemistry, Oakland University, Rochester, MI, USA

Xiangqun Zeng Department of Chemistry, Oakland University, Rochester, MI, USA

Fan Zhang Materials Measurement Science Division, Material Measurement Laboratory, National Institute of Standards and Technology, Gaithersburg, MD, USA

Jie Zhang School of Chemistry, Monash University, Clayton South, VIC, Australia

Hua Zhao Department of Chemistry and Forensic Science, Savannah State University, Savannah, GA, USA

Volume II

Joaquín Arias-Pardilla Grupo de Ciencia de Materiales e Ingeniería Metalúrgica, Departamento de Ingeniería de Materiales y Fabricación, Universidad Politécnica de Cartagena, Cartagena, Spain

Cameron L. Bentley School of Chemistry, Monash University, Clayton, VIC, Australia

Energy Flagship, CSIRO, Clayton South, VIC, Australia

María Dolores Bermúdez Grupo de Ciencia de Materiales e Ingeniería Metalúrgica, Departamento de Ingeniería de Materiales y Fabricación, Universidad Politécnica de Cartagena, Cartagena, Spain

Alan M. Bond School of Chemistry, Monash University, Clayton, VIC, Australia

Natalia Borisenko Institute of Electrochemistry, Clausthal University of Technology, Clausthal-Zellerfeld, Germany

Andinet Ejigu School of Chemistry, University of Nottingham, Nottingham, UK

Tulia Espinosa Grupo de Ciencia de Materiales e Ingeniería Metalúrgica, Departamento de Ingeniería de Materiales y Fabricación, Universidad Politécnica de Cartagena, Cartagena, Spain

Michael D. Horne Mineral Resources Flagship, CSIRO, Clayton South, VIC, Australia

Yasushi Katayama Department of Applied Chemistry, Faculty of Science and Technology, Keio University, Kanagawa, Japan

Douglas R. MacFarlane School of Chemistry, Monash University, Clayton, VIC, Australia

Takeo Ohsaka Department of Electronic Chemistry, Interdisciplinary Graduate School of Science and Engineering, Tokyo Institute of Technology, Yokohama, Japan

Cristina Pozo-Gonzalo ARC Centre of Excellence for Electromaterials Science, IFM-Institute for Frontier Materials, Deakin University, Burwood, VIC, Australia

Amal Siriwardana TECNALIA, Parque Tecnológico de San Sebastián, Mikeletegi Pasealekua, Donostia-San Sebastián, Gipuzkoa, Spain

Angel A.J. Torriero Centre for Chemistry and Biotechnology, School of Life and Environmental Sciences, Faculty of Science, Engineering and Built Environment, Deakin University, Burwood, VIC, Australia

Darren A. Walsh School of Chemistry, University of Nottingham, Nottingham, UK

Da Zhang School of Environmental and Materials Engineering, Shanghai Second Polytechnic University, Shanghai, China

Department of Electronic Chemistry, Interdisciplinary Graduate School of Science and Engineering, Tokyo Institute of Technology, Yokohama, Japan

Jie Zhang School of Chemistry, Monash University, Clayton, VIC, Australia

Shu-Feng Zhao School of Chemistry, Monash University, Clayton, VIC, Australia

Mineral Resources Flagship, CSIRO, Clayton South, VIC, Australia

About the Editor

Dr. Angel A.J. Torriero

Dr. Angel A.J. Torriero is a Lecturer of Chemistry and Electrochemistry at Deakin University, Melbourne, Australia. He has published more than 50-refereed papers (h-index = 18; Scopus, November 2014), six book chapters, several patents, and one book, *Electrochemical Properties and Applications of Ionic Liquids* in 2011.

Dr. Torriero has a broad interest in both fundamental and applied electrochemistry and has made significant contributions in a number of fields, including analytical electrochemistry, biosensor, bioelectrochemistry, organic and organometallic electrochemistry, and most recently internal reference systems for ionic liquids.

Chapter 1
Introduction

Angel A.J. Torriero

As liquid media at a temperature less than 100 °C that possess some level of ionic conductivity, it is immediately of interest to consider the use of ionic liquids (ILs) as electrolytes to carry out electrochemical processes. This has of course the origins of the modern era of interest in ionic liquids via the work of Wilkes and coworkers in the 1990s [1]. Applications in electrowinning and electrodeposition have developed including processes for a range of metals from copper and zinc to lithium and aluminium [2]. Some metals such as titanium remain, however, stubbornly difficult to electrodeposit [3]. A range of applications in electrochemical devices, including batteries, fuel cells, and solar cells have also emerged and are being thoroughly discussed in Volume 2 (*Electrochemistry in ionic liquids. Applications*). In parallel to this, there has emerged the need to understand more in detail some important fundamental concepts of electrochemistry as well as the interest on fundamental electrochemical process taking place in an ionic liquid medium and in identifying the ways in which the processes differ, or not, from conventional solvent systems as a result of the highly charged medium [4–6]. Thereby, in this book, special emphasis is placed on showing which aspects of electrochemistry in ionic liquids are different from electrochemistry in conventional solutions. Furthermore, new electrochemical concepts and theories are presented. The book commences with a deep and comprehensive discussion on electrode/electrolyte interface reactions, interface structure, and its critical properties for all electrochemical applications. Chapter 2 discusses these fundamental concepts along with some in situ techniques, such as electrochemical impedance and Fourier transform infrared spectroscopy, cyclic voltammetry, and electrochemical quartz crystal microbalance, suitable for the characterization of electrode/IL interfaces.

A.A.J. Torriero (✉)
Centre for Chemistry and Biotechnology, School of Life
and Environmental Science, Faculty of Science Engineering
and Built Environment, Deakin University, Burwood, VIC 3125, Australia
e-mail: angel.torriero@deakin.edu.au

A.A.J. Torriero (ed.), *Electrochemistry in Ionic Liquids*,
DOI 10.1007/978-3-319-13485-7_1

One of the open problems of electrochemistry in ionic liquids nowadays is the choice of a reference electrode. Chapter 3 discusses the latest understanding and practical approaches to the application of reference electrodes and internal reference redox scales in ionic liquids. Problems and limitations of reference systems are presented with the help of experimental examples. Comparison of the behavior of internal reference redox systems in organic solvents with added supporting electrolyte and ionic liquids is provided and new observations critically discussed. This chapter is very important for any experimentalist starting work in the electrochemical and IL fields. Further information about internal reference redox scales can also be found in Volume 2 (Chap. 14).

Chapter 4 describes the specific challenges and opportunities for electrochemists attempting to perform scanning electrochemical microscopic and voltammetric measurements with ultramicroelectrodes in ionic liquids. Meanwhile, Chap. 5 focuses its attention on the electroanalytical application of semiintegral and convolution voltammetry in ionic liquids. This robust and versatile analytical method involves transforming a transient peak-shaped voltammogram into a form that closely resembles a steady-state voltammogram. This transformation is highly advantageous to determine the diffusion coefficient of redox active species, along with other thermodynamic parameters such as the bulk concentration and the stoichiometric number of electrons transferred. Other techniques to quantify these three parameters are summarized in Volume 2 (Chap. 18).

Small-angle X-ray scattering is also a powerful in situ technique that could be used together with electrochemistry to study surfaces and resolve the size and structure of phases on the nanometer scale. Chapter 6 discusses details of the small-angle X-ray scattering technique with special emphasis on the experimental setup and analysis.

Chapter 7 summarizes the synthesis and physical properties of protic ionic liquids and deep eutectic solvents and draws the connection to real-world applications. Further applications of protic ionic liquids can be found in Volume 2 (Chaps. 17 and 20). The "designable" or "tunable" nature of ILs, their physicochemical properties, and high applicability in the electrochemical field is discussed in detail in Chap. 8. The application of deep eutectic solvents presented in Chap. 7, is also discussed in this chapter. Further applications of task-specific ILs can as well be found in Chap. 2.

Chapter 9 summarizes the preparation methods, properties, and applications of ion gels, which are a new type of gels where the liquid phase, percolating throughout the solid phase, is an ionic liquid. The application of these new materials in the development of energy storage systems, in addition to sensors, optoelectronics, catalytic membranes, separation membranes, and in drug release is discussed.

Lithium ions are generally solvated by IL ions to form metal-ion solvation complexes. Electrochemical and physicochemical properties of the lithium-containing IL solutions, such as ionic conductivity, redox reaction behavior, viscosity, and density, strongly depend on the static and dynamic structures of the mentioned solvation complexes. Chapter 10 discusses recent progress in lithium solvation by imidazolium-containing ionic liquids, a topic that is of both fundamental and practical significance.

This book brings to the scientists, engineers, and students new electrochemical concepts and summarized results in the field of ionic liquid. Of course, as the mentioned research areas are in a constant state of development, this book can hardly be comprehensive, but certainly they will stimulate further the continuing development of *Electrochemistry in Ionic Liquids*.

References

1. Wilkes JS, Zaworotko MJ (1992) Air and water stable 1-ethyl-3-methylimidazolium based ionic liquids. J Chem Soc Chem Commun 13:965–967
2. Endres F, MacFarlane DR, Abbott AP (eds) (2008) Electrodeposition from ionic liquids. Wiley-VCH, Weinheim
3. Endres F, El Abedin SZ, Saad AY, Moustafa EM, Borissenko N, Price WE, Wallace GG, MacFarlane DR, Newman PJ, Bund A (2008) On the electrodeposition of titanium in ionic liquids. Phys Chem Chem Phys 10(16):2189–2199
4. Torriero AAJ (2014) Characterization of decamethylferrocene and ferrocene in ionic liquids: argon and vacuum effect on their electrochemical properties. Electrochim Acta 137:235–244. doi:10.1016/j.electacta.2014.06.005
5. Torriero AAJ, Sunarso J, Howlett PC (2012) Critical evaluation of reference systems for voltammetric measurements in ionic liquids. Electrochim Acta 82:60–68. doi:10.1016/j.electacta.2012.01.115
6. Torriero AAJ, Muhammad S (eds) (2011) Electrochemcal properties and application of ionic liqudis. Nova Science, New York, NY

Part I
Fundamental Concepts of Electrochemistry in Ionic Liquids

Chapter 2
Electrode–Electrolyte Interfacial Processes in Ionic Liquids and Sensor Applications

Xiangqun Zeng, Zhe Wang, and Abdul Rehman

2.1 Introduction

Electroanalytical methods or electrochemical sensors involve a group of quantitative and qualitative methods in which an electric excitation function is applied to an electrochemical cell and a response function is measured. Based on the properties of the applied excitation function, they are categorized as potentiometry, voltammetry, impedance spectroscopy, etc. All these methods require the use of electrodes where the electron-transfer reactions occur and an electrolyte, an ionically conducting medium, to transport charge within the electrochemical cells, to establish the electrochemical contact of all electrodes effectively, and to solubilize the reactants and products for efficient mass transport. Ionic liquids (ILs) are a class of compounds containing organic cations or anions that melt at or near room temperature. They represent a new type of nonaqueous and biphasic electrolytes that combine the benefits of both solid and liquid systems [1–7]. They are shown to have negligible vapor pressure, high ionic conductivity, wide potential window (up to 5.5 V), high heat capacity, and good chemical and electrochemical stability and have been demonstrated as media in electrochemical devices including super-capacitors, fuel cells, lithium batteries, photovoltaic cells, electrochemical and mechanical actuators, and in electroplating [8–10]. The liquid crystalline structure as well as the solvation properties of ILs has their own significance in many of these applications. In the past, ILs are regarded as completely nonvolatile, but the work of Earle et al. has shown that some ILs may be evaporated and recondensed [11]. However, generally speaking, most ILs have zero vapor pressure at room temperature; thus, there is no drying out of the electrolytes presenting many potential benefits for analytical methods and technique developments in comparison to traditional aqueous and nonaqueous electrolytes/solvents.

X. Zeng (✉) • Z. Wang • A. Rehman
Department of Chemistry, Oakland University, Rochester, MI 48309, USA
e-mail: zeng@oakland.edu

A.A.J. Torriero (ed.), *Electrochemistry in Ionic Liquids*,
DOI 10.1007/978-3-319-13485-7_2

It is well known that the maximum efficiency of electrochemical devices depends upon electrochemical thermodynamics, whereas real efficiency depends upon the electrode kinetics. To understand and control electrode reactions and the related parameters at an electrode and solution interface, a systematic study of the kinetics of electrode reactions is required. When ILs are used as solvents and electrolytes, many of the electrochemical processes will be different and some new electrochemical processes may also occur. For example, the properties of the electrode/electrolyte interface often dictate the sensitivity, specificity, stability, and response time, and thus the success or failure of the electrochemical detection technologies. The IL/electrode interface properties will determine many analytical parameters for sensor applications. Thus, the fundamentals of electrochemical processes in ILs need to be studied in order to have sensor developments as well as many other applications such as electrocatalysis, energy storage, and so on. Based on these insights, this chapter has been arranged into three parts: (1) Fundamentals of electrode/electrolyte interfacial processes in ILs; (2) Experimental techniques for the characterization of dynamic processes at the interface of electrodes and IL electrolytes; and (3) Sensors based on these unique electrode/IL interface properties. And in the end, we will summarize the future directions in fundamental and applied study of IL–electrode interface properties for sensor applications.

2.2 Fundamentals of Electrode/Electrolyte Interfacial Processes

Electroanalytical methods and electrochemical sensors are based on changes of electrode–electrolyte interface or the electrochemical reactions occurring at electrode/electrolyte interface for the detection of a specific analyte. The nature of electrode/electrolytic solution interfaces plays a central role in electrochemical surface science. A large number of important reactions take place at the interface between a metal and a solution: metal deposition and dissolution, electron- and proton-transfer reactions, corrosion, film formation, and electro-organic synthesis. If we want to understand and control these processes using ILs, it requires a detailed knowledge of the properties of the electrode/IL interface such as the adsorption of the ILs on electrode surfaces upon the variation of the electric potential, which acts as a driving force for all charge transfer reactions. As shown in Fig. 2.1, a study of electrochemical reaction at an electrode/solution interface is performed in an electrochemical cell. Typically, an electrochemical cell consists of three electrodes. The reference electrode establishes a constant electrical potential and the working electrode is where electrode reaction takes place. The electrical current is expected to flow between the working electrode and the auxiliary electrode, often also called the counter electrode. By holding certain variables of an electrochemical cell constant and applying an external electrical excitation function to the working electrode, information about the analyte can be obtained based on the measurement of how other variables change with the controlled variables. The common electrical

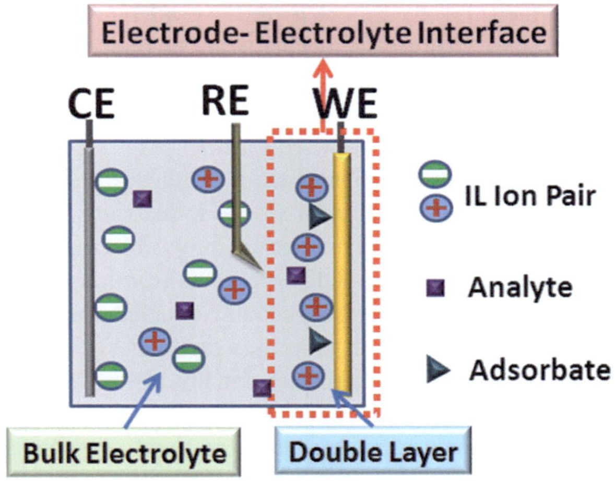

Fig. 2.1 Schematic of electrochemical cell containing the bulk electrolyte, interface and double-layer regions. *CE* counter electrode, *RE* reference electrode, *WE* working electrode

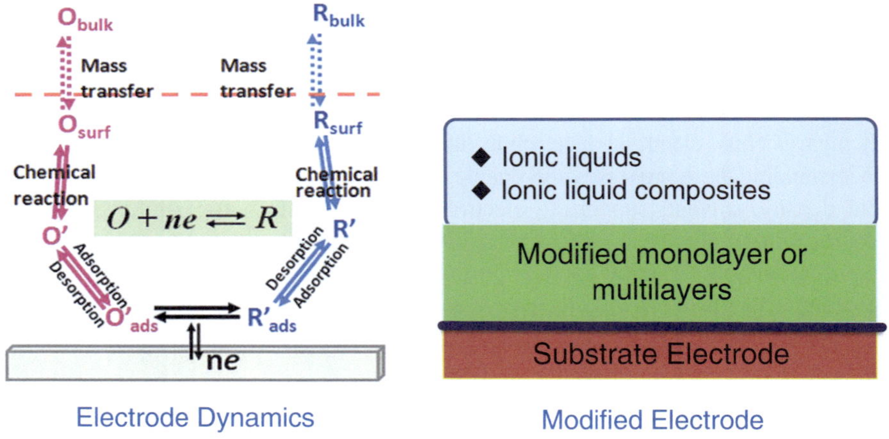

Fig. 2.2 Schematic of electrode processes at IL/electrode interface (*left*) and interfacial structure (*right*) for the electrochemical sensors at IL modified electrodes

excitation parameters are potential, current, and charge. Electrode kinetics for any electrochemical reactions can be controlled by the electrical excitation parameters (e.g., potential). As shown in Fig. 2.2, the electrode reactions typically involve three major steps: (1) Mass transport step (diffusion, diffusion-convection, and migration); (2) Heterogeneous step (charge transfer, adsorption and desorption, nucleation, crystal growth, and surface diffusion); and (3) Homogeneous step (chemical reactions in solution). The rate of an electron-transfer reaction occurring at the electrode/solution interface mainly depends on the electrical potential. However, other physical processes may also contribute to the overall kinetics of any particular reaction creating complex electrolysis mechanism. The overall reaction rate will be limited by the slowest step. In particular, the mass transport process (i.e., the transport of the reactant molecules to the electrode surface and the removal of the reaction products from electrode interface when formed) plays significant roles for the kinetics of electrode reactions. Chemical reactions at the electrode or in the solution also affect the mechanisms and kinetics of electrode reactions. The chemical reactions involved in electrolysis mechanisms are not only occurring in the homogenous solution phase (e.g., if the products of the electrode reactions are not chemically stable), but a significant range of heterogeneously coupled chemical reactions can also occur. Furthermore, the chemical processes that occur at electrode/electrolyte interface can also be coupled to mass transport. Finally, electrode materials and properties influence electrochemical reactions. For example, the surface area and the geometry of an electrode could influence the current and may also affect the mass transfer process, as in the case of a rotating disk electrode [12] and an ultramicro electrode [13]. Another area of electrode/electrolyte interface is that an electrode can be modified by the immobilization of a layer of material on to its surface so that the surface character of the immobilized material can control the electrode's performance and can be explored for a wide range of applications. The modified surfaces can offer novel properties such as the ability to hold charge (as in battery), induce novel specific reactions, and act as organic semiconductors or chemical/biosensors. The modification of the electrode greatly expands the application of electrochemistry [14–16] such as the enzyme electrodes that are widely used in biosensors and bio-electrochemistry. Thus, electrode–electrolyte interface processes can be complicated, and understanding these processes will enable the selection of the electrode potential, current density, solvent, supporting electrolyte, solution pH, temperature, deliberate modification of electrode surface or addition of other reagents, etc., based on the requirements of a specific application including sensors. Below we will detail the most important processes at electrode/electrolyte interface.

2.2.1 Adsorption

Electrode surface conditions such as roughness and modification are related with the electrochemical active sites and the properties of the electrode/electrolyte interface. Electrode surface conditions also include their single crystal faces. ILs contain organic cations or anions that could adsorb specifically or nonspecifically on the electrode surfaces. The adsorption of organic substances at an electrode/solution interface is of fundamental importance for many chemical and electrochemical processes. For example, adsorption can affect and change the electrode reaction kinetics. Molecules that exhibit structure-sensitive chemisorption are expected to undergo structure-sensitive electrocatalytic reactions. The adsorption of electrochemically active materials could change the peak shape, peak potential, and peak current of a cyclic voltammogram [17]. Depending upon the site of adsorption, electrochemically active species can be oxidized at different potentials. More complicated adsorption mechanisms exist when surface adsorbed species are not stable in either its charged or uncharged forms, or if they are desorbed during potential cycling. An IL's tremendous diversity in structural and chemical properties will allow various electrode–IL interfaces to be designed to unravel the IL–IL and IL–electrode substrate interaction. Typically, in the case of chemisorptions, the dominant IL–electrode interaction will be due to strong covalent or ionic chemical forces; in the case of physisorption, one often finds that the van der Waals force acting directly between the adsorbates dominates. In this case, the optimum adsorbate–substrate bonding geometry can be overridden by the lateral adsorbate–adsorbate interactions, yielding, for example, incommensurate structures in which the overlayer and the substrate have independent lattices. The IL–IL interactions will include electrostatic interactions, hydrogen bond interactions, orbital-overlapping interactions, and van der Waals interactions. These interactions may be attractive or repulsive depending on the properties of IL and the substrate electrode. Adsorption phenomena in IL–electrode interface are important for understanding the structure of the electric double layer (EDL) and for their various electrochemical applications.

2.2.2 Electrical Double Layer

The double-layer model is used to visualize the ionic environment in the vicinity of a charged surface which can be either a metal under potential control or a dielectric surface with ionic groups. It is easier to understand this model as a sequence of steps that would take place near the surface if its neutralizing ions were suddenly stripped away. The double layer formed, in turn, causes an electrokinetic potential between the surface and any point in the mass of the suspending liquid. This voltage difference is on the order of millivolts and is referred to as the often surface potential. The magnitude of the surface potential is related to the surface charge and the thickness of the double layer (Table 2.1).

Table 2.1 The molecular interaction forces present in IL solvents

Type of interaction	The energy equation
Ion–dipole	$U_{\text{ion-solvent}} = -(ze\mu \cos\theta)/(4\pi\varepsilon_0)r^2$
Dipole–dipole	$U_{\text{dipole-dipole}} = -(2\mu_1^2\mu_2^2)/(4\pi\varepsilon_0)3K_BTr^6$
Dipole-induced dipole	$U_{\text{dipole-induced dipole}} = -(\alpha_1\mu_2^2 + \alpha_2\mu_1^2)/(4\pi\varepsilon_0)r^6$
Ion-induced dipole	$U_{\text{ion-induced dipole}} = -(z^2e^2\alpha)/(4\pi\varepsilon_0)2r^4$
Dispersion	$U_{\text{dispersion}} = -(3\alpha_1\alpha_2)(I_1I_2)/(I_1 + I_2)2r^6$

Symbols: z—the charge of the ion; θ—the dipole angle relative the line joining the ion to the center of dipole; e—electric charge of -1.602×10^{-19} coulomb; I—the ionization potential; ε_0—the vacuum permittivity; r is the separation distance between the charges; q—the charge (assuming equal value of charge on each side of the molecule); α—polarizability of the solvents; K_B—the Boltzmann constant; T—the absolute temperature; μ_i—permanent dipole moment of solvent

ILs are free of solvents; thus, electrostatic interaction is the predominating force between the electrode and IL electrolyte. The electrode may have interactions with dipole moments of the anion and cation functional groups due to the bulky anions or cations of ILs. Even though the IL–electrode interface structure has some similarity to traditional electrode–electrolyte double layer, the IL–electrified metal interface behaves differently from those described by Gouy–Chapman–Sterns double-layer model. The study shows a well-defined structural organization of IL at Au(111) electrode interface with three structurally distinct regions: the interfacial (innermost) layer composed of ions in direct contact with the electrode; the transition region over which the pronounced interfacial layer structure decays to the bulk morphology; and the bulk liquid region where structure depends on the degree of ion amphiphilicity [18]. The innermost layer which templates ion arrangements in the transition zone is much more structured where the ions form a single layer at the surface to screen the electrode charge [19]. AFM results on gold and graphite in various ILs showed that the double layer at IL–electrified metal interface is not one layer thick, thereby extending the transition zone up to five ion pair diameters from the interface [20, 21]. In addition to the Coulombic short range electrostatic interaction, the ions in ILs are often large in size, flexible, highly polarizable, and chemically complex with a number of interionic forces (such as dispersion forces, dipole–dipole interaction, hydrogen bonding, and pi-stacking forces). Steric effects will contribute to the IL–electrode interface structure as well. Moreover, adsorptions of anions and/or cations are likely to occur at the IL–electrode interfaces. The specific adsorption, ion structures and properties, and the applied potential all influence the structure and thickness of the double layer of ILs near metal surface. For example, multiple sterically hindered allylic functional groups could be incorporated to minimize IL–electrode interactions and maximize compressibility of the solvation layers [22–24]. It was reported that the adsorption and dissolution of gas molecules in the IL–electrified metal interface can lead to the change of the double-layer structure and thickness and the change of IL viscosity and the dielectric constant (i.e., less organized arrangement of ILs) [25, 26]. The molecular structure of ILs such as alkyl chain length influences the relative permittivity [27]. Capacitance at a fixed potential will depend on double-layer thickness and permittivity of the liquid at the interface. Consequently, varying the structure of IL anions or cations and applied potential can result in the rearrangement of ions and/or adsorbates that enables the modification of the double-layer structure at the IL–electrified metal interface (e.g., thickness and permittivity) [28–30] and subsequently the modification of the IL double-layer capacitance.

2.2.3 Chemical Reactions

An electron-transfer process at electrode/electrolyte interface can often have two types of chemical reactions. One is the electron-transfer reaction coupled with heterogeneous chemical reactions including those surface adsorption processes that occur at electrode/electrolyte interface. Another is the electron-transfer reaction-coupled homogeneous reactions such as EC reaction: ($A+ne \leftrightarrow B$; $B \rightarrow$ product), ECE reaction: ($A+ne \leftrightarrow B$; $B \rightarrow C$, $C+ne \leftrightarrow D$), EC′ mechanism ($A+ne \leftrightarrow B$; $B+Y \rightarrow A+$product), where the prime (′) represents a catalytic process. Among these coupled chemical reactions, catalytic reactions are widely sought after since they often represent a clean and efficient method of enhancing chemical reactivity. The reaction mechanisms change with the solvents. In ILs, the electrophilic (Lewis acid) organic ions can be reduced electrochemically at the cathode or chemically by reducing agents ($Q+e \rightarrow Q^{\bullet-}$). On the other hand, many of the nucleophilic (Lewis base) organic ions can be oxidized electrochemically at the anode or chemically by oxidizing agents ($Q \rightarrow Q^{\bullet+}+e$). Here, $Q^{\bullet-}$ and $Q^{\bullet+}$ are radical anion and cation, respectively, and are extremely reactive if their charges are localized. However, if the charges are more or less delocalized, the radical ions are less reactive. The dynamic properties of ILs can have remarkable influences on the potentials of various types of redox couples and on the mechanisms of the electron-transfer processes that occur either heterogeneously at the electrode or homogeneously in the solution. For example, in high-permittive ILs, ILs can easily dissociate into ions. The anion or cation of ILs can form ion-pair complex with the radical cation or anion, respectively. With the decrease in permittivity of ILs, however, complete dissociation becomes difficult. Some part of the IL electrolyte remains undissociated and forms an ion pair ($A^{+}+B^{-} \leftrightarrow A^{+}B^{-}$). Ion pairs contribute neither ionic strength nor electric conductivity to the solution. Since there is high concentration of anion and cations in ILs, the formation of triple ions ($A^{+}+A^{+}B^{-} \leftrightarrow A^{+}B^{-}A^{+}$) and quadruples can also occur. ILs could also play roles as proton donors to lower the energy needed to oxidize the electroactive materials. In summary, the study of redox processes in ILs can be very complex requiring the adaptation of a diverse range of electrochemical and spectroscopic tools to characterize their properties and mechanisms during electron-transfer processes at electrode/electrolyte interface.

2.3 Techniques for Studying Interfacial Properties in ILs

The interfacial properties of IL/electrode interfaces are different from other media (i.e., aqueous or traditional nonaqueous media) because of the unique properties of ILs, especially the electrochemical properties. To understand the electrode/electrolyte interface chemistry for sensor research, the mechanisms of the electrochemical reactions, and the essential performance-limiting factors, both in the bulk and at the surface of the electrode materials need to be investigated, preferably in situ. In situ analysis is much desired due to the fact that ex situ measurements are usually not able to follow the fast kinetics at electrode interfaces. The past decades have been characterized by a spectacular development of in situ techniques for studying interfacial processes at metal electrodes. Radioactive tracer [31, 32], pulse potentiodynamic [33, 34], and galvanostatic methods [35] have been applied quantitatively to study the adsorption of organic compounds at solid metals. In the study of complex

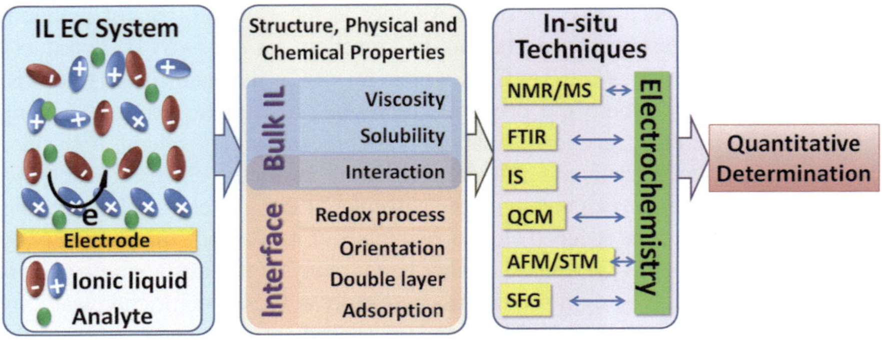

Fig. 2.3 Methods for characterization of IL–electrode interface

electrode reactions, vibrational spectroscopies, such as Raman [36–40], the infrared (IR) [41–43], and UV–vis spectroscopies [44, 45], have been used to identify adsorbed intermediates and the nature of the chemisorption bond [46]. And the electroreflectance spectroscopies [47] have provided information about the orientation of molecules in adsorbed overlayers. X-ray absorption spectroscopies [48, 49] were employed to determine the coordination numbers of adsorbed species and the substrate–adsorbate and adsorbate–adsorbate bond lengths. Ab initio-based computations on the property of electrode materials and continuum modeling have been incorporated to provide theoretical insights and predictions [50, 51]. Finally, the images of adsorbate-covered electrode surfaces have been obtained using scanning tunneling [52–54] and atomic force microscopies [53, 55].

As shown in Fig. 2.3, many of these techniques have also been used to study the IL/electrode interface properties. Baldelli et al. studied the surface structure at the IL–electrified electrode interfaces by SFG [19, 56–58]. And Endres' group use ATM and SPM to investigate the IL interface morphology and interaction force [59, 60]. As expected, their studies show that the IL/Pt electrode interface has a very thin double-layer structure where the ions form a single layer at the surface to screen the electrode charge. However, it is difficult to obtain high-resolution AFM and SFG and to obtain the dynamics of the interfacial structure of IL real time. Difficulties also exist in how to interpret the AFM, electrochemical impedance spectroscopy (EIS), and SFG results. Additionally, these studies were limited to the non-Faradaic region of the electrode where there are no electron-transfer reactions to occur. Below we will illustrate the principles of electrochemical techniques as well as spectroelectrochemical and microscopic methods that have been used for in situ characterization of the electrochemical processes at IL–electrode interfaces. Combining complementary characterization methods in situ can provide mechanistic understanding of interfacial structure, interfacial dynamics, interfacial chemical composition, etc., and enable the development of new analytical methods based on IL chemistry for the identification and quantitative determination of analytes.

2.3.1 *Cyclic Voltammetry*

Cyclic voltammetry (CV) is a widely used electrochemical technique to study electrochemical reaction mechanisms and the rates of oxidation and reduction processes. In CV, the current at the working electrode is plotted versus the applied potential to give the cyclic voltammogram traces of an analyte in solution. For studying an electrochemical process using CV, a stable supporting electrolyte and good solubility of analyte in the electrolyte is essential. ILs can offer benefits in both these aspects for CV characterizations. The solubility of the redox active materials in ILs would not be a problem due to the availability of a wide range of ILs with desired functionalities to choose from. ILs have high thermal stability. Most of ILs are stable up to 300 °C [61]. However, the electrochemical stability of an IL electrolyte is important for various electrochemical applications.

2.3.1.1 Electrochemical Stability (Electrochemical Potential Window)

The electrochemical stability of nonaqueous electrolytes is crucial for any electrochemical application. The electrolyte must be electrochemically stable up to the high voltage cutoff (the highest voltage at which cell is allowed to operate), and stable down to the lowest voltage cutoff. Typically, electrochemical potential window lies at a voltage range in which the electrolyte is not oxidized or reduced. The electrochemical potential window is a measure of the electrochemical stability of an electrolyte against oxidation and reduction processes. This value, on one hand, characterizes the electrochemical stability of ILs, i.e., the limits of the potential window correspond to the oxidation limits and the reduction limits of the electrochemical decomposition of the involved ions. On the other hand, the width of the electrochemical window governs the range of potentials available for studying the electrochemical processes that will not be affected by the electrolytes and/or the solvents. One of the important considerations in electrochemical characterization of the potential window is the cell design and reference electrode. In order to avoid contamination to the IL electrolytes due to the reference electrode filling solution (e.g., KCl), a quasi-reference electrodes often used are rather than the saturated Ag/AgCl or calomel electrode. The quasi-reference electrode often used is Ag, Au, or Pt thin film or wire. The quasi-reference electrode and the auxiliary electrode are placed far apart so that the product at the counter electrode will not diffuse to the reference electrode which can affect its stability. Compared with different reference electrodes in the same IL system, the width of potential window should be consistent. The change of reference electrode materials only affects the absolute value of potential in most of IL systems. And all of them can be calibrated by using stable redox couple, such as Fc^+/Fc. Thus, we can compare the potential windows reported using different quasi-reference electrodes.

The structures of ILs make them more electrochemically inert than other solvents such as H_2O and acetonitrile [62].The electrochemical window in ILs is much larger than that of most conventional aqueous and nonaqueous electrolytes [63].

Table 2.2 Potential windows of common RTILs

RTIL	Potential window (vs. Fc^+/Fc, ±0.2 V)
$[C_2mim][NTf_2]$	−2.6 to 3.0 V
$[C_4mim][CF_3SO_3]$	−2.6 to 2.2 V
$[C_4mim][BF_4]$	−2.8 to 3.2 V
$[C_4mim][PF_6]$	−2.5 to 3.0 V
$[C_4mim][HSO_4]$	−2.5 to 3.0 V
$[C_4mim][Cl]$	−2.5 to 1.5 V
$[C_4mim][NO_3]$	−2.5 to 2.0 V
$[P_{4,4,4,4}][NO_3]$	−2.3 to 2.0 V
$[N_{4,4,4,4}][NO_3]$	−3.5 to 2.0 V
$[N_{6,2,2,2}][NTf_2]$	−3.5 to 3.0 V
$[P_{14,6,6,6}][NTf_2]$	−3.4 to 3.0 V
$[P_{14,6,6,6}][FAP]$	−3.5 to 3.0 V
$[C_4mPy][NTf_2]$	−3.2 to 3.0 V

Table 2.2 lists the potential windows of some commercial ILs, obtained from references [64–67] and also from our own investigations. Here, the purities of all ILs are higher than 95 %. The electrochemical stability of ILs with cations of either quaternary ammonium structures or pyrrolidinium is better than that of imidazolium. However, anion stability is an issue. Halides can be easily oxidized; thus, halides-based ILs have lower electrochemical stabilities. Additionally, the high chemical activity of halides may result in side reactions during the analyte redox process. The fluorine-containing anions (BF_4^- and PF_6^-) are not as stable and can be decomposed in the presence of trace amounts of water producing HF under electrochemical process, although this reaction usually isn't observable during electrochemical study for short time testing, since both B–F and P–F bonds are weaker than C–F bond [68]. Additionally, for most organic anions, the negative charge is delocalized which increases their stability. It denotes that the electrochemical potential window is sensitive to impurities. Schröder and coworkers have reported that ILs such as $[C_4mim][BF_4]$ can rapidly absorb atmospheric water and that the presence of water has a profound effect on cyclic voltammetric experiments with platinum electrodes [69–71]. Any other impurities such as moisture residues from preparation could give an electrochemical signal during CV characterization. Since the purity of ILs affects the accuracy of the potential window measurement, the potential window obtained experimentally can be regarded to have some uncertainty. Theoretically potential window of ILs was basically dominated by the stability of ions. A general trend for the magnitude of the electrochemical oxidative limit is observed as follows: $[P_{14,6,6,6}]^+ > [N_{6,2,2,2}]^+ > [C_nmim]^+ \approx [C_4mPy]^+$ and $[FAP]^- \approx [NTf_2]^- > [BF_4]^- > [PF_6]^- > [MeSO_4]^- > [NO_3]^- > [CF_3SO_3]^-$ [61, 62]. The anodic "breakdown" is presumed to be due to the oxidation of the anionic component of the ILs. In most cases, the stability of the anion determines the anodic potential limits, and stability of the cation determines the cathodic potential limits. But in the practical measurement, the predicted values depend also on the electrode material and current density.

2.3.1.2 Mass Transport

The mass transport of electroactive species plays significant roles in electrode kinetics. The electroactive species can move by diffusion, convection, and migration. The measured mass transport is a combination of these three processes:

1. Migration: movement of charged species under the influence of an electric field.
2. Diffusion: movement of species because of their concentration gradient.
3. Convection: movement of species under the influence of a physical disturbance.

Typically, the contribution of the charged electroactive species which is often in very low concentration to migration process is much smaller than the contribution of inert electrolyte. In conventional electrolyte solutions, ions exist in the form of solvent complexes, and consequently, the solvent molecules are parts of moving ions. As a result, in addition to the flux of ions, there is also a flux of the solvent. This flux leads to the formation of a concentration gradient between the cathodic and anodic compartments during the electrochemical reaction. In the ILs, there is no solvent. ILs can be considered as an excess of inert ions carrying the current; thus, migration of electroactive species can be neglected. In an IL electrolyte, the ionic conduction is related to both the anion and the cation mobility and their transference number. The total cation and anion transference numbers is equal to unity.

$$t_+ + t_- = 1 \tag{2.1}$$

where t_+ and t_- are the transference numbers, depicting the portion of charge carried by cations or anions, respectively. Since ILs have much higher viscosities than common electrochemical supporting electrolytes, they have an effect on the diffusion coefficients of species. A previously reported study compares the diffusion coefficients of a neutral molecule and the radical cation produced after an electrochemical reaction in an IL and acetonitrile. The diffusion coefficient of the cation radical was consistently about half that of the value of the neutral molecule. In contrast, in acetonitrile, the ratio of the diffusion coefficients was nearly 60 % or higher [72], providing an indication that viscosity and charge have considerable effects on the transport of diffusing species in ILs [73–75]. We can consider that mass transport is mainly governed by the diffusion process. In this case, the diffusion coefficients, for both the reduced and oxidized forms of the redox species, can be calculated from the slope of the dependence of the peak current i_p on the square root of the scan rate, according to the Randles–Sevcik equation [76]:

$$i_p = 0.446 nFA \left(\frac{nF}{RT} \right)^{1/2} C D^{1/2} v^{1/2} \tag{2.2}$$

The Stokes–Einstein equation [77] is generally applicable for molecular liquids and ILs, at least semi-quantitatively, for analyzing the diffusion coefficient D. Water is ubiquitous in the environment. It was shown that water induced accelerated ion diffusion compared to its effect on neutral species in ionic liquids [78]. We also observed that IL conductivity is significantly increased when [Bmim][BF_4] is exposed to normal atmosphere for 24 h with 0.25 % wt moisture since fewer cycles are needed to reach a fully conductive polyvinyl ferrocene film [79]. Because of its high molecular weight, the diffusion coefficient of an IL is typically smaller than that of analytes regardless of the medium. However, despite the presence of an additional interaction between the functional group in IL and analytes, this behavior contrasts with the case of a small anion or radical anion; ion-pair formation leads to a marked decrease in the diffusion coefficient.

2.3.1.3 Charging Current in CV

As mentioned before, ILs have a unique double-layer structure, and the double-layer capacitance can be measured by both cyclic voltammetry and EIS. Our studies and others have shown that the capacitance of IL/electrode interface is potential dependent, and the adsorption of gas on the IL/electrode interface will decrease the double-layer capacitance under a specific DC bias potential [80]. In contrast to the EDL typically observed in dilute aqueous or nonaqueous electrolytes, the ions of ILs are strongly oriented near the electrode into an ordered layer structure with local ion density at maximum possible value [81, 82]. Multiple ion pair layers form at the interface of IL and metal electrode. Due to this heterogeneity, the lattice formation at the interface and the subsequent capacitance variations are not monotonic as described by the Gouy–Chapman–Stern's model [83–85]. Additionally, the hysteresis effects of an IL, where the double-layer capacitance depends on the scan direction of the DC potential, contribute to the slow pseudocapacitive process at the IL–electrode interface [86–88]. Since the change of IL orientation due to applied potential is a very slow process, the resulted charging currents only can be observed at a slow scan rate. Figure 2.4 shows that the small CV peaks related to heterogeneous IL orientations were observed in CV at slow scan rates (e.g., 20 mV/s) in the most of studied ILs. These small peaks will contribute to some uncertainty for the current measured, especially for the Faradaic process, under slow scan rates. Consequently, the double-layer charging current will increase the complexity of CV results. In order to obtain accurate Faradaic current measurement, it is recommended that the peak current of electrochemical processes should be measured under the high scan rate in CV study. The double-layer capacitance measured under this condition can be considered the average capacitance instead of the potential-dependent value.

2.3.1.4 Scan Rate Effects on the IL Double-Layer Charging Currents

In an EC redox process, the Faradaic current depends on the concentration of the analyte and increases monotonically with analyte concentrations. Numerous studies by others and our lab show that the structure and dynamics of the electrochemical double layer in room temperature of ILs are very different from the traditional

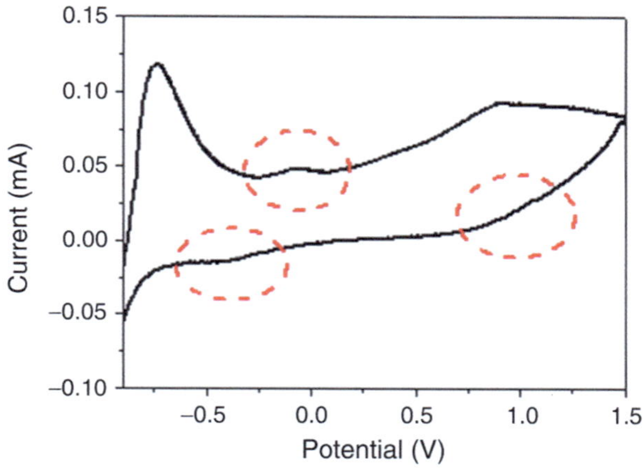

Fig. 2.4 CV of [C_4mPy][NTf_2] at 20 mv/s on Pt under air condition. The small peaks (*red circles*) are presenting charging current by the IL orientation during the potential scanning

aqueous and nonaqueous electrolytes. For example, in two separated studies, Gore and Druschler et al. [86, 89] have reported the hysteresis effects in the potential-dependent double-layer capacitance of room temperature ionic liquids at either a Pt electrode or a gold electrode. They explained this hysteresis effect due to the slow pseudocapacitive processes in a frequency range below 10 Hz. We have also observed a similar behavior in our study [90]. It is well known that the first potential cycle is critical to target analyte concentration profile for a Faradaic process in aqueous solutions. However, it is unpredictable in ionic liquid since its pseudocapacitance is scan rate related, and thus the value is highly dependent on the potential and the history of electrode conditions in the IL [91–93]. Currently, there is no method that can obtain the accurate value of pseudocapacitance in an IL consistently. As we know, increasing the scan rate in cyclic voltammetry allows one to decrease the thickness of the diffusion layer, and the pseudocapacitive process caused by the slow orientation of the IL will be negligible. Thus, the hysteresis of the IL double layer will be minimized using higher scan rate. This can help to obtain the reproducible and accurate current values for sensor application. High scan rate could facilitate a steady state of the IL–electrode double layer to be established during the multiple potential cycling. As shown in Fig. 2.5, for the methane-oxygen system [94], the difference of peak currents of methane oxidation (0.9 V) and oxygen redox (−1.2 V/0.9 V) processes in the 6th cycle and 20th cycle at 500 mV/s, is less than 0.2 μA, confirming that the methane and oxygen concentration profile in ionic liquid is reproducible and the double-layer charging current is relatively constant. Thus, electrochemical measurements based on redox processes in IL should be characterized at the high scan rate, and peak current of first cycle should not be used for the analyte quantification.

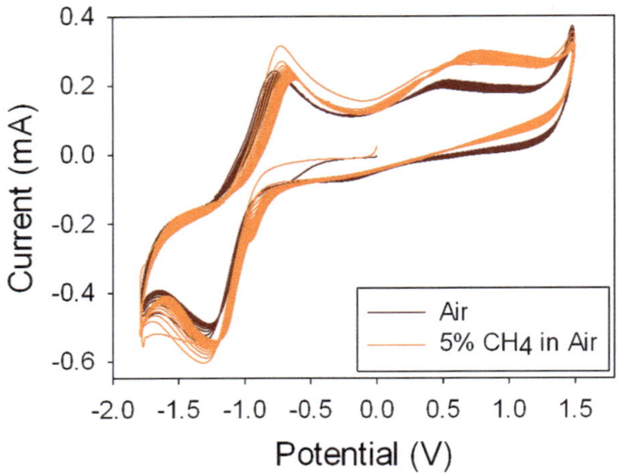

Fig. 2.5 Cyclic voltammograms of 5 % methane and dry air in [C_4mPy][NTf_2] (500 mV/s scan rate)

2.3.1.5 Electrolyte Conductivity

Compared to the conventional aqueous electrolyte, the conductivity of the ILs is at least one order of magnitude smaller. In contrast to the conventional aqueous electrolyte, the higher the concentration of the ILs, the smaller the conductivity of the solution due to their high viscosity. This is expected. At high salt concentrations, all solvent molecules are called "solvent in salt solution," in which the ions primary shell is more complex, and the resulting system shows different properties at the low concentration salt solution [95–97]. In such a case, the conductivity increases with the increasing amount of salt, goes through a maximum, and decreases with a further increase of the salt concentration. However, most of the molecular liquids have relatively low viscosity, and therefore, the dilution of the viscous neat IL with molecular diluents decreases the viscosity of the mixture [98]. Table 2.3 lists the dynamic viscosity (η) and specific conductivity (κ) values of common solvents and RTILs [66, 99]. In an IL system, the solution resistance (R_u) is high due to higher viscosity and slow ion mobility of ILs. Typically, the solution resistance was calculated from conductivity of the ILs and electrode geometry. In most ILs, the dynamic voltage drop (IR-drop) would not be negligible and can affect the accuracy of the applied electrode potential. The IR-drop is the voltage loss (drop) when current flows through a resistor. IR-drop occurs in most of the electrical circuits and electrochemical systems. In ILs, the solution resistance R_u could be measured by both AC and DC techniques. In most electrochemical systems, IR-drop could be estimated to be lower than 10 mV. However, in some high capacitance systems and high Faradaic current systems, such as supercapacitor and fuel cell, IR-drop is significant. Although many commercial electrochemical instruments have the capability to compensate IR-drop, it still has to be considered in some analysis of an electrochemical process [100, 101].

Table 2.3 A comparison of dynamic viscosity (η) and specific conductivity (κ) for a representative selection of molecular solvents and nonhaloaluminate RTILs

Solvents	Viscosity (cP)	Conductivity (mS/cm)
N,N-Dimethylformamide	0.794	4.0[a]
Acetonitrile	0.345	7.6[a]
Ethanol	1.074	0.6[a]
Dimethylsulfoxide	1.987	2.7[a]
$[C_2mim][NTf_2]$	28	8.4
$[C_4mim][NTf_2]$	44	3.9[b]
$[C_6mim][NTf_2]$	59	
$[C_8mim][NTf_2]$	74	
$[C_2mim][BF_4]$	43	13.0
$[C_2mim][PF_6]$		5.2
$[C_4mim][PF_6]$	275	1.5
$[N_{6,2,2,2}][NTf_2]$	167	0.67
$[N_{6,4,4,4}][NTf_2]$	595	0.16
$[C_3mPy][NTf_2]$	63	1.4
$[C_4mPy][NTf_2]$	85	2.2

Unless otherwise stated, $T = 25$ °C

[a]Conductivity for organic solvent containing 0.1 M tetrabutylammonium perchlorate at 22 °C

[b]20 °C

2.3.2 *Potential-Step Methods (Chronoamperometry)*

Potential-step method is an electrochemical technique in which the potential of the working electrode is either held at constant or stepped to a predetermined value, and the resulting current due to Faradaic processes and double-layer charging processes occurring at the electrode (caused by the potential step) is monitored as a function of time. Especially for practical electrochemical sensor in real-world application, chronoamperometry is preferred due to its simplicity and low cost of instrumentation. Besides its wide use in most electrochemical sensor systems, chronoamperometry has been used in the understanding of the kinetics of the electrochemical processes as well.

2.3.2.1 Kinetics Study for Electrochemical Process

Chronoamperometry was used to determine the electron-transfer kinetics and it could provide information regarding the dynamic model of an electrochemical process. In chronoamperometry, the current is integrated over relatively long time intervals; thus, it gives a better signal to noise ratio in comparison to other voltammetric techniques. The Faradaic current, which is due to electron-transfer events and is most often the current component of interest, decays as described by the Cottrell equation [77] in planar electrodes.

$$i(t) = id(t) = \frac{nFAD_O^{1/2}C_O^*}{\pi^{1/2}t^{1/2}} \tag{2.3}$$

In most electrochemical cells, the decay of Faradaic current vs. time is much slower than that of double-layer charging current; however, electrochemical systems with no supporting electrolytes are notable exceptions. ILs are more viscous than traditional organic solvents, thus slowing the diffusion of the electroactive species leads to transient behaviors. While the current behavior in the ILs system could be interpreted using the continuum model, it was recognized that the mechanisms by which an IL–electrode interface responds to an applied electrical field are quite different from a traditional, polar solvent. In the latter, solvent molecules re-orientate, whereas in an IL the molecules must move relative to each other. These differences must impact upon the overall observed rate of electron transfer, even if the free energies of activation are similar. That is, while inner-sphere electron-transfer rates might be expected to be very similar in ILs compared to polar organic solvents, outer-sphere electron transfers will be very different due to the screening response mechanisms [102, 103]. In order to possibly minimize the effect of "IR drop" and "capacitive charging" on the estimation of the heterogeneous electron-transfer rate constant, the microelectrode, scanning electrochemical microscopy (SECM), and other techniques have been used in many investigations by Mirkin, Bard, and Compton [104–107].

2.3.2.2 Faradaic Current Determination

Chronoamperometry (or potential step) can be regarded as an extreme fast potential sweep of CV which is especially beneficial for the practical IL-based sensor development due to its unique double-layer structure as discussed in the earlier section. However, as with all pulsed voltammetric techniques, chronoamperometry generates high charging currents, which decay exponentially with time as in any Randles circuit. As shown in our early work [90], the time-dependent oxygen reduction currents, $i(t)$, in ILs, can be described as the sum of the Faradaic current for EC reaction, i_f, and the double-layer charging current, i_c, on the electrode as shown in Eq. (2.4):

$$i(t) = i_f + i_c = i_\infty + \frac{C_{dl}dV_{dl}}{dt} + \frac{C_{bulk}dV_{bulk}}{dt} \tag{2.4}$$

In this expression, C and V represent capacitance and electrode potential, with "dl" and "bulk" denoting double layer and bulk solution, respectively. The Faradaic current (e.g., due to oxygen reduction) ordinarily decays much more slowly than the charging current (cells with no supporting electrolyte are notable exceptions). Typically the cell time constant $\tau = R_uC_d$, where R_u is the uncompensated resistance and C_d is the double-layer capacitance, is very useful to determine the timescale of

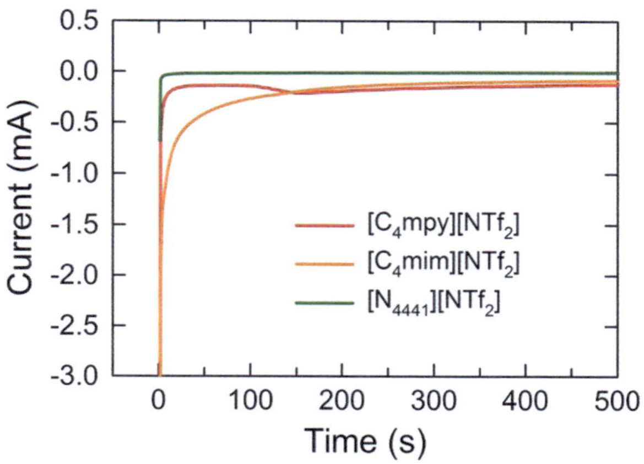

Fig. 2.6 Potential step experiment from 0 V to −1.2 V at 100 % air in three ILs. Time constant $\tau = 6$ s and 50τ is about 300 s

a potential step method. As such, it is important that analysis is applied over a sufficiently broad time interval in order to assure the reliability of the results. The charging current can be considered insignificant when current is sampled at times exceeding $5R_uC_d$. As shown in Fig. 2.6, the time constant obtained from potential step experiment (step from 0.0 V to −1.2 V in 100 % air) is $\tau = 6$ seconds (the time required for current decays to 37 % maximum current). Thus, in order to minimize the large double-layer charging current of IL, we measured the current at the 300th second which is 50τ. The current can be mainly attributed to the Faradaic current of oxygen reduction, and the double-layer charging current is negligible after 300 s, which is much longer than that in aqueous electrolytes (lower than few seconds). And the double-layer capacitance is also potential-dependent [80], because under different potentials the orientations of IL are different and the values of relative capacitances vary as well. Thus, cell time constant τ is not a constant and varies in different systems. The value of τ is related to the IL structure. Thus, chronoamperometry is not an efficient method to study the dynamic interfacial process in an IL system. For practical electrochemical applications, the significant conditioning process or preconditioning step (>50τ) is required to establish a constant IL–electrode interfacial boundary layers. Obviously, in different ionic liquids, the interface relaxation processes due to potential step vary and these processes have not been fully investigated so far.

2.3.3 Electrochemical Impedance Spectroscopy

In CV and potential step, the electrode surface reaction is studied through large perturbation on the system to drive the electrode to a condition far from equilibrium and observe the response. Typically in EIS, the electrochemical reaction is perturbed with an alternating signal (e.g., potential) of small magnitude and the way in which the system follows the perturbation in steady states is measured. The use of low amplitude perturbation in EIS allows the application of a linear model to interpret the results which can yield useful information about the physicochemical properties of the system. Analysis is generally carried out in the frequency domain, although measurements are sometimes made in the time domain and then Fourier transformed to the frequency domain. For the analytical application, EIS is a powerful, informative, and noninvasive method, which can be used to study the interfacial events and trace the blocking behaviors (as complexation and adsorption) or diffusion effects at modified electrodes and serve as a transducer [108], especially in the electrochemical system. EIS has been used to measure the resistance of an electrolyte solution, grain boundary resistance, and interfacial capacitance. The Nyquist plot (Z_{im} vs. Z_{Re}) at high frequency (near 100 kHz) is determined by the solution resistance (R_s) and the electrode capacitance. However, for a specific electrochemical process and physical status, EIS is by no means limited to the measurement and the analysis of data at the impedance level (e.g., impedance vs. frequency) but may involve any of the four basic immittance levels; thus, most generally, IS stands for immittance spectroscopy. The signal of the whole system can be equivalent to a combination of certain electric components described by an "equivalent circuit."

2.3.3.1 Randles Circuit

An ideal electrode–electrolyte interface with an electron-transfer process can be described using Randle equivalent circuit shown in Fig. 2.7. The Faradaic electron-transfer reaction is represented by a charge transfer resistance R_{ct}, and the mass transfer of the electroactive species is described by Warburg element (W). The electrolyte resistance R_s is in series with the parallel combination of the double-layer capacitance C_{dl} and an impedance of a Faradaic reaction. However, in practical application, the impedance results for a solid electrode/electrolyte interface often reveal a frequency dispersion that cannot be described by simple Randle circuit and simple electronic components. The interaction of each component in an electrochemical system contributes to the complexity of final impedance spectroscopy results. The EIS results often consist of resistive, capacitive, and inductive components, and all of them can be influenced by analytes and their local environment, corresponding to solvent, electrolyte, electrode condition, and other possible electrochemically active species. It is important to characterize the electrode/electrolyte interface properties by EIS for their real-world applications in sensors and energy storage applications.

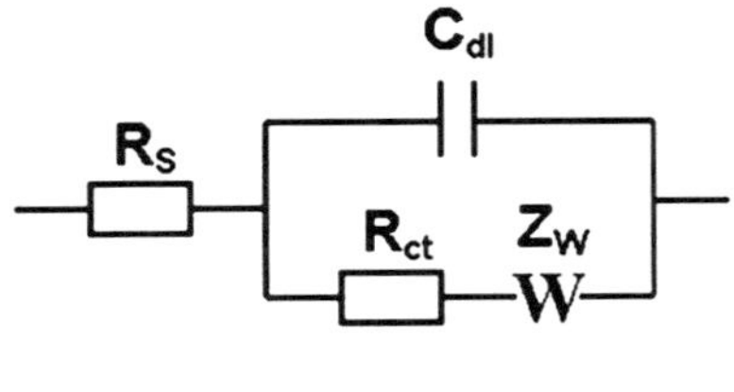

Fig. 2.7 The typical Randle's circuit

2.3.3.2 Proposed Equivalent Circuit in ILs

The typical IL system could be considered as a solvent-free system, in which it can simplify the EIS analysis significantly which spurs its wide use in the characterization of the IL–electrode interface. However, due to low mobility of ions in an IL and multiple molecular interactions present in an IL, more time is needed to reach to a steady state of IL–electrode interface structure and arrangement, when a potential is applied. Furthermore, the electron-transfer process in ILs is different from that in traditional solvents containing electrolytes. Thus, the interfacial structures of IL are more complex than other systems. Even the electrode geometry could affect the EIS results of IL systems. It is noted that the bulk ILs could not be simply described by a resistor (R_s) as in classic electrochemical systems. And the electrode double layer in IL electrolyte couldn't be simply depicted as a capacitor. So the Randle equivalent circuit is not sufficient to describe an IL system. Significant efforts have been made to illustrate the properties of diffusion layer and the bulk ILs with equivalent circuits. However, currently there is no general equivalent circuit model to describe the interface of an IL system.

2.3.3.3 EIS Measurement in IL

For practical EIS characterization of IL system, the purity of IL used, the hysteresis effect of IL, and the different experimental measurement conditions can all contribute to the variability of EIS results. Thus, for the EIS measurement: (1) A significant conditioning time for IL–electrode interface is required to minimize the hysteresis effect so that the interfacial boundary layer of IL is at a steady state for each measurement; (2) The electric field applied to the working electrode should be parallel. For sensor applications, the field effect transistor electrode structure in which the voltage applied across the gate and source terminals or the interdigited electrode structure is considered as better electrode designs for IL system using EIS testing. Additionally, EIS has the ability to monitor the ion movement and orientation in real time. However, the presence of the Faradaic process increases the complexity of results; because of the large capacitance of double layer, it is hard to analyze the contributions due to electron-transfer process or the charge transfer processes based only on impedance result. Combining impedance with spectroscopic methods is often used to obtain conclusive results.

2.3.4 Infrared Spectroscopy

Spectrometric methods are based on the interactions of electromagnetic radiations with the atomic and molecular species. Among them, IR spectroscopy (with wavenumbers ranging from about 12,800 to 10 cm^{-1} or wavelengths from 0.78 to 100 μm) is mostly used to obtain structural information about the molecular species.

IR absorption, emission, and reflection spectra for molecular species either in solid, liquid, or gas phases arise mostly from various changes in energy due to transitions of molecules from one vibrational or rotational energy state to another. The frequency or wavelength of this energy transition is characteristic of the specific chemical bond vibration and/or rotation in the molecule which are determined by the molecular structure, the masses of the atoms, and the associated vibrational energy coupling. Attenuated total reflectance (ATR) and reflection-mode of IR in conjunction with electrochemical methods allow samples to be examined directly in the solid or liquid state without further preparation and are widely used in the characterization of electrode–electrolyte interface properties. Most of ILs are IR-active molecules. Since ILs are stable and chemically inert, the IR characterization can be easily performed on the IL-based system directly.

2.3.4.1 IR Signal from IL

ILs have negligible vapor pressure making them not volatile, and this property permits in situ IR characterization of IL systems at various timescales without considering solvent loss generated uncertainties. High solubility of analytes in the ILs enables accurate characterization of their IR spectra in ILs. On one hand, the usage of ILs can eliminate the effect of water, and on the other hand, the signal from IL functional group can complicate the interpretation of Fourier transform infrared spectroscopy (FTIR) results. For example, hydrogen bonding is barely present in highly hydrophobic ILs, and as a result, the relative band shifts caused by the hydrogen bonding are negligible when analyzing the IR results in ILs [100, 109–111]. For those high purity IL systems, the IR spectra could be considered as the sum of the spectra of anion and cation and other species. Most of the methods used for studying the traditional ionic compounds such as ionic melts can be applied to the study of ILs. It is also noted that the IR response in liquid states of IL is different from the one obtained in a crystalline IL, predominantly because of lower energy conformations of the molecules in liquid states. For example, the IR spectra of $[NTf_2]^-$ in crystalline $[C_4mim][NTf_2]$ shows intensive bands at 649 and ~600 cm^{-1} (Fig. 2.8). Therefore, one may conclude that the anion is in *cis* conformation in this crystal. Furthermore, the intensity of both bands gradually decreases in the premelting region due to the predominance of the *trans* conformation in the liquid state. But they are still detectable [112–115]. Those conformation differences are present in the fingerprint region of IR (under 1,000 cm^{-1}), whereas most variability comes from anions. $[NTf_2]^-$ has O=S=O (1,100 cm^{-1} and 1,400 cm^{-1}) and C–F bonding (1,000 cm^{-1}) [112–115]. B–F of BF_4 is located at 924 cm^{-1} and 1,312 cm^{-1}, and P–F of PF_6 is about 843 cm^{-1} [116–119]. For cations, the bands are located in the rest of the regions and address the IL structure, such as C–C (around 1,450 cm^{-1}), C=C (around 1,600 cm^{-1}), and C–N bonding (around 1,450 cm^{-1}). Only C–H bonds around 3,000 cm^{-1} and 1,200 cm^{-1} are highly sensitive to their chemical environment. Typically, the addition of an analyte will change the C–H vibration and hydrogen bonding as well (Fig. 2.8).

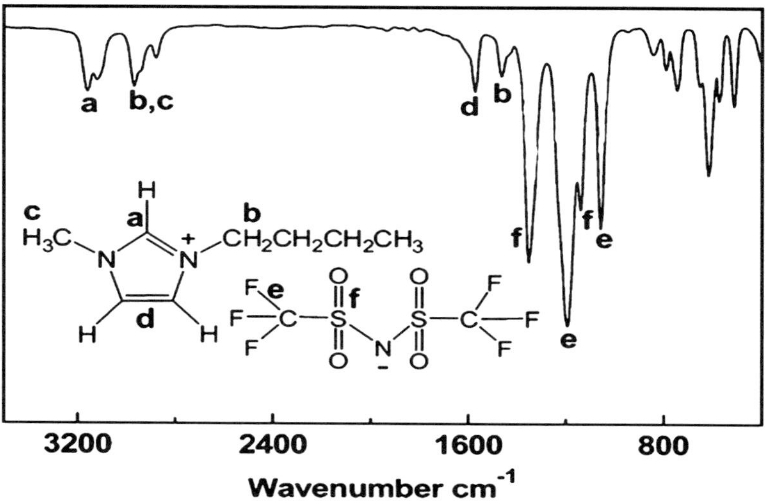

Fig. 2.8 FTIR spectra of $[C_4mim][NTf_2]$ under a nitrogen atmosphere

Table 2.4 Main IR bands (cm^{-1}) assigned C–H in ILs

Wavelength of the band (cm^{-1})	Vibration
3171, 3124	ν(C–H) aromatic, strong
2933, 2939, 2878	ν(C–H) aliphatic, strong
1575, 1467	ν(ring), strong, sym
1431, 1386	MeC–H, asym
1170	Ring, strong, sym

A comparison of IR spectra of acidic and basic ILs reveals only a small distortion of the aromatic ring, which is not of such extent as in the case of a salt with a smaller cation. This means that the hydrogen bond between the hydrogen on the carbon atom and a chloride ion is very weak or absent. This kind of bond is seen in the case of $[C_4mim][Cl]$, where there is a band at 3,088 cm^{-1}, and the peaks at 3,100 cm^{-1} are considerably smaller [120–123].The weak absorption from 2,500 cm^{-1} to 2,850 cm^{-1} may be attributed to the formation of hydrogen bonds between the aromatic protons and the halide ions. In basic ILs, the extent of hydrogen bond acceptor Cl^- is limited. Hydrogen bonds are weak or even not present. Other anions such as $[PF_6]^-$, $[BF_4]^-$, and $[NTf_2]^-$ are weakly complexing anions and not expected to participate in strong hydrogen bonding. Indeed, even $[C_4mim][PF_6]$ shows no hydrogen bonding bands in the region 3,000–3,100 cm^{-1} where C-H…Cl^- interactions on imidazolium chloride were observed previously. Therefore, IR results are consistent with a lack of hydrogen bonding within weakly complexing anions [124]. This means that cation–anion coulombic attraction is driving the overall structure, with local steric effects influencing the final orientation of ions. Table 2.4 lists the IR response of C–H bond in ILs [112–119].

2.3.4.2 Effects on Dissolved Analytes (Peak Position and Width)

ILs could provide a highly purified environment to monitor the interface reactions, and the species of interests can be localized near the interface because of the high viscosity of the ILs. Both these effects can enhance the observed IR signals. The p-type polarized IR is considered best as the light source for measuring IR-active species through bulk IL. However, typically the strength of signals is attenuated due to the long optical path in ILs.

Inorganic ILs such as the halogen salts of alkali metals provides strong ionic environment. The strong interaction between the analyte and the salt is manifested by peak shift from its position in gaseous state or liquid state of itself. For example, ν_3 of CO_2 was found with about 150 cm^{-1} shift from its position in gaseous state. The shift of ν_2 of CO_2 is about 40 cm^{-1}confirming this fact [125–128]. In an organic-based IL, this effect is not that strong. The position of this band for CO_2 dissolved in IL is very close to that of CO_2 dissolved in glassy polymers, such as poly(methyl methacrylate) (PMMA). The blue-shift is lower than 30 cm^{-1} [129–132]. However for H_2O the polar molecule structure and hydrogen bonding make this effect amplified. The presence of two bands corresponding to the antisymmetric and symmetric stretching modes of water indicated that molecules of water are present in the "free" (not self-aggregated) state bound to the basic anion via H-bonding [133–135]. The blue-shift of the antisymmetric stretching band has been used to correlate the relative strength of the interaction between water and RTILs with different anions. The strength of H-bonding increases in the order $[PF_6]<[SbF_6]<[BF_4]<[ClO_4]<[CF_3SO_3]<[NO_3]<[CF_3CO_2]$ [110, 111, 136–139].

Another effect is on the IR peak width of the analyte. Since most of the anions in IL are Lewis acids, the IR signal of dissolved analyte could be increased by the interaction. For example, the anions could act as weak Lewis bases, and CO_2 appears to be a useful probe to sense the extent of their basicity. The width of the ν_2 mode, measured as an effective average width of the split band (the average width represents the width of the doublet at the half maximum of the absorbance), is an estimate of the strength of the interaction of CO_2 with the Lewis bases: the width increases with the increase of the strength of the interaction [113, 130, 140]. This could also be a result of the strength of interaction decreasing with the increase of anion size. Thus, the strength of these interactions cannot be solely responsible for the solubility of CO_2 in these ILs, and, presumably, a free volume contribution in the IL plays a significant role. The strength of anion–cation interaction in the IL affects the available free volume, and one would anticipate that a weaker interacting anion leads to more free volume being available. It makes the result more complex.

2.3.4.3 Interface Signal

For the study of electrochemical processes, the interfacial information is even more important. However, due to the strong bulk absorption of IL, the transmission mode was limited to the electrode design. Using ATR mode may partially eliminate interference from ILs, ATR–IR spectroscopy overcomes some of the limitations inherent to other techniques when the objective is to gain spectra of molecules dissolved in highly absorbing media. And also the interface information could be collected using the reflection mode as well. But the IR signal is limited by the thickness of the liquid film. The signal/noise ratio is another issue that should be considered. Typically, these methods are limited to verify the products during the electrochemical processes and are not able to provide the information regarding interfacial structure and dynamics. To understand the dynamical electrochemical process in IL and to obtain detailed molecular level information on the interface, the surface-enhanced spectroscopic techniques have to be applied since the bulk adsorption of IL is relatively strong. Among many surface enhanced spectroscopy techniques, SFG has been explored to characterize the electrode–IL interface. The nonlinear laser spectroscopy SFG method has long been recognized as an important tool for deducing the composition, orientation distributions, and some structural information of molecules at gas–solid, gas–liquid, and liquid–solid interfaces without interference from the bulk electrolyte. Thus, the interface is probed in situ, while the electrode potential is changed. Currently, SFG is the most efficient method to directly obtain the interfacial structure from the electrode surface.

SFG measurements of metal/aqueous electrode surfaces have been conducted on Au, Ag, and Pt electrode surfaces. The most extensive studies have been conducted with platinum single-crystalline and polycrystalline surfaces [141–143]. Using a free electron laser, Guyot-Sionnest and Tadjeddine demonstrated the first use of Vibrational Sum Frequency Spectroscopy (VSFS) to study ionic adsorption at an electrified metal/aqueous electrolyte interface [144–146]. Conboy et al. used it to determine the orientation of the imidazolium cation at the RTIL/SiO_2 interface, as well as the structure of the interfacial water molecules [147–149]. Baldelli et al. extended it to investigate the metal/IL interface. It provides solid evidences on the potential-dependant double layer of IL [19, 150]. SFG could determine the ion's location and orientations on the metal surface under variable applied potential via measuring the tilt of the terminal alkyl $-CH_3$ group on the cation. The results can be compared with those in situ EIS results to obtain an in-depth understanding of IL/solid interface [151, 152].

2.3.5 *In Situ EQCM Methods*

Electrochemical quartz crystal microbalance (EQCM) is a powerful analytical technique for probing electrode–electrolyte interfacial processes by using a quartz crystal electrode. Figure 2.9 shows an integrated EQCM electrode fabricated on a monolithic quartz wafer. AT-cut quartz crystals oscillating in a shear thickness mode exhibit extraordinary high sensitivity to potential-induced changes in the mass at electrode–electrolyte interface via the change of its resonance frequency. In addition to that, the monitoring of the mechanical state of the viscoelastic coatings at a quartz crystal electrode via a change in their shear storage and loss modulus is

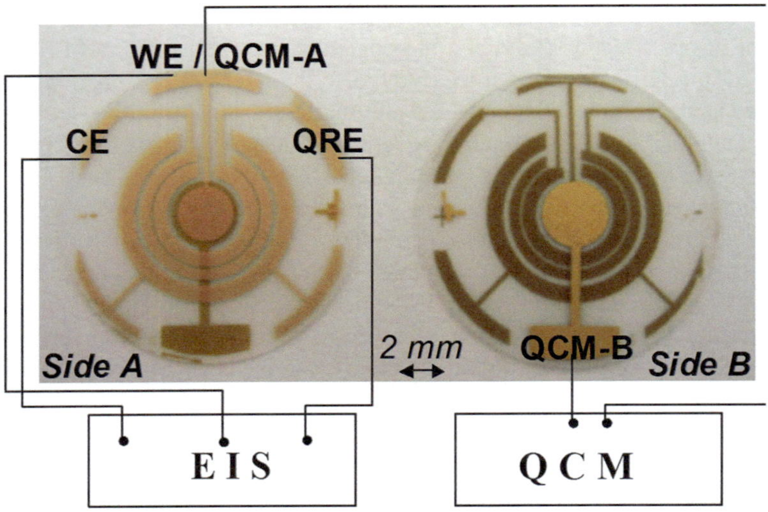

Fig. 2.9 Demonstration of the microfabrication of EIS–QCM electrode chip that was made on a 10 MHz AT cut quartz crystal formed EIS. It has a 2.5 mm disk-working electrode on both sides of 10 MHz AT cut quartz. Ring-shaped counterelectrodes (CE) and quasi-reference electrodes (QRE) are patterned around the circular working electrodes on one side and narrow gaps between disk and ring electrodes to support EIS and QCM

possible. When the coating on the quartz crystal electrode remains rigid, the major component of the QCM depends on the hydrodynamic interactions between the film on the electrode and the contacting electrolyte solution due to potential-induced volume and surface roughness changes in the coating. However, in the case of viscoelastic coatings, there are a variety of ways in which oscillation energy of the coated crystal may dissipate, and a large number of factors affect the change of the resonance frequency which are the indicators of the nature of the physicochemical processes behind the charging processes in the electrode coatings, requiring a careful analysis of the potential dependencies of both resonance frequency shift and resonance bandwidth. In some modern instruments, this damping of the oscillation is expressed through a quantity called dissipation. For a detailed understanding of these quantities and their effects, the readers can be directed to a recent review [153] presenting the main concepts of EQCM in context of its historical development, focusing on the exceptional sensitivity of the method. As described therein, its dynamic capability allows for the real-time monitoring of minute mass changes during electrochemical reactions in electroactive polymers and the accompanying viscoelastic changes [153]. Thus, a QCM used with electrochemistry methods such as EIS and CV allows real-time in situ measurement of the mass changes in the nanogram range during the electrochemical perturbation of the electrode solution interfaces.

ILs are highly viscous media thereby having large viscoelastic effects. Therefore, whether they are used as thin films for absorption–desorption-based phenomena or

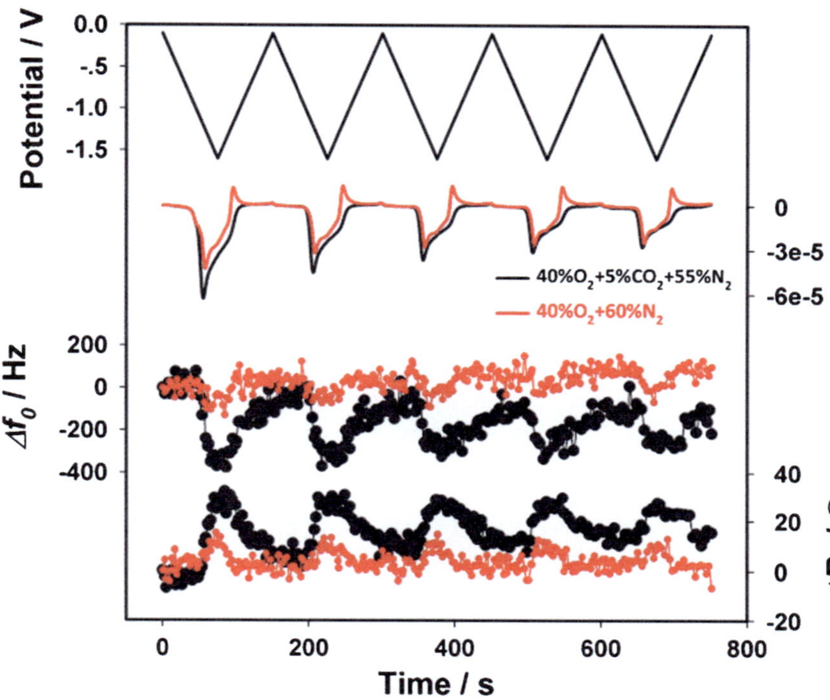

Fig. 2.10 Comparison of five-cycle-CV EQCM responses for O_2 (vol. 40 %) electrochemical reduction in the absence and presence of CO_2 in [C_4mPy][NTf_2]. Scan rate: 20 mV/s

more conventionally used as electrolytes for various electrochemical applications, in situ EQCM technique is quite powerful for characterizing the interfacial phenomena in ILs. For example, we have shown that in situ EQCM enables to answer several very important questions regarding the IL and electrified metal interfaces that have not been addressed before: what are the mechanisms of electrode oxidation and reduction reactions in ILs? Is the adsorption of ILs potential dependent? Is the adsorption of IL specific or nonspecific under potential control? What are the double-layer structure changes when the electrode changes from a metal to its oxide? These capabilities of EQCM methods facilitate the understanding of the structure/property relationships of electrode reactions in IL electrolytes. For instance, the electrochemical reactions of O_2, CO_2, and their mixture were investigated in three structurally different ILs by in situ EQCM [154]. Compared with electrochemical method only, the QCM integrated with the electrochemical method was shown to be significantly more sensitive and powerful for the characterization of the subtle differences (mass change or viscoelastic change) at IL/electrode interface as depicted in Fig. 2.10 showing current, frequency, and resistance responses obtained simultaneously. This led to the conclusion that the CO_2 reduction in ILs is irreversible and forms $CO_2^{\bullet-}$ adsorbate at electrode interface. With increasing concentrations of CO_2, the reduction of oxygen is switched from a one electron process to the overall two-electron process, and forms adsorbed $CO_4^{\bullet-}$ intermediate species. Even though the mechanisms of electrochemical reaction between CO_2 and electrochemical-generated superoxide radical ($O_2^{\bullet-}$) in three structurally different ILs are found to be similar, the simultaneous EQCM experimental results show that the different cations in three ILs can modify the kinetics of the electrode reactions of O_2 and CO_2 due to a competition between the ILs' cation and CO_2 to react with $O_2^{\bullet-}$. The reactivity of $O_2^{\bullet-}$ toward CO_2 follows the order of the stability of the ILs' cation under $O_2^{\bullet-}$ attack. Understanding of such adsorption of redox intermediate at an electrode/electrolyte interface during CO_2 reduction is of fundamental importance for many chemical and electrochemical processes. In this way, EQCM technique can be used for the study of various gaseous molecules' redox processes in ILs in order to characterize the dynamics of the electrode reactions at an electrode and IL interface for a broad range of applications such as electrochemical synthesis, supercapacitor, Li-ion battery solid electrolyte interface, electrochemical conversion, fuel cells, and sensors.

EQCM has also been used to explore the ion–solvent coupling mechanisms in a redox polymer, polyvinylferrocene (PVF) in ILs, and their solutions. From the data of frequency and current values, the mass changes and the moles of electrons transferred can be calculated which can be plotted against each other to obtain very useful information as shown in Fig. 2.11. This study showed that the unique solvation and ionic properties of ILs significantly affected the break-in process and the ion–solvent transport mechanisms of PVF redox switching which is indicated in the form of difference for the significant difference of mass changes during first and fifth cycle of the redox reaction. Due to the existence of strong IL–polymer interaction, not only the anions but also the IL molecules interacted with the PVF matrix. The cations were later removed from the PVF matrix to balance the excessive positive charge in PVF oxidation. This also confirmed that IL was not only an electrolyte but also a

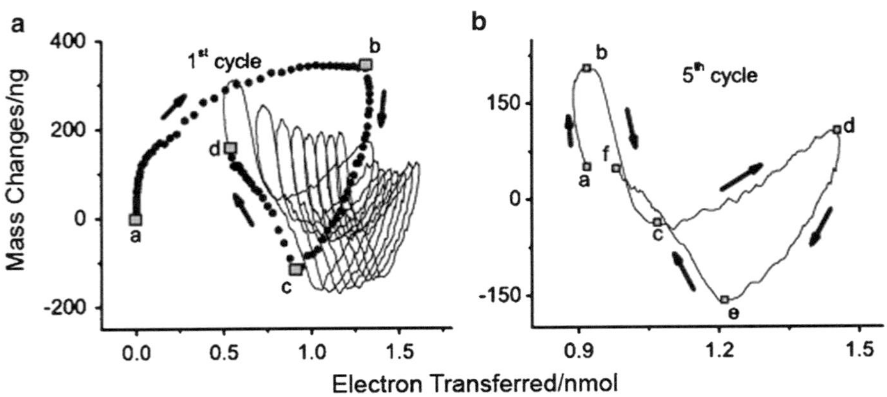

Fig. 2.11 Mass changes vs. electron transferred during PVF conditioning with [C_5mPyr][NTf_2] as measured by EQCM. Potential scan rate: 50 mV/s. Scan range: 0–0.6 V. (**A**) The *dots* and the *arrows* correspond to the first potential sweeping cycle and the *solid line* to the following cycles. (**B**) The mass changes during the fifth potential sweeping cycle

solvent in PVF redox switching processes. Various types of interactions between PVF and the IL, including dispersion, dipole induction, dipole orientation, hydrogen bonding, or ionic/charge-charge interactions, could significantly change the PVF redox dynamics. The strong IL–polymer interaction in certain ILs, e.g., methanesulfonate ILs, may change even the polymer sublattice structure. This property was being utilized by preconditioning the PVF films with suitable ILs, such as *N*-alkyl-*N*-methylpyrrolidinium bis(trifluoromethanesulfonyl)imides, which allows PVF film to relax to a state compatible with reversible glycylglycylglycine peptide (GGG^-) doping/undoping. In contrast, ILs with structures and properties highly divergent from the target GGG^- failed to properly condition PVF to a compatible state [155]. These examples further confirmed that IL's tremendous diversity in structural and chemical properties and their distinctive properties offer an excellent opportunity to explore IL–polymer interactions and to dynamically control the conductive polymer relaxation processes and their redox switching mechanism for various applications.

Another new direction of EQCM-based exploration of ILs is by using in situ admittance mode of EQCM [156] which can characterize the changes in the composition of the ionic part of the EDL during charge/discharge of high surface area electrodes, such as activated carbons. Reliable dynamic information on the rather complicated charging processes in meso- and microporous carbon electrodes can be obtained by the measurement and analysis of the change in the bandwidth of the resonance. It has been shown that EQCM is able to monitor ionic and solvent fluxes [157, 158], and at the same time can reflect the mechanical status of composite activated carbon electrodes, with high sensitivity. The use of this technique provides excellent opportunities for designing new porous electrodes, for the optimization of already existing supercapacitor devices, and for developing effective desalination cells for capacitive deionization (CDI) of water. In situ EQCM frequency–potential responses have also been used for understanding the mechanism of the redox reaction of $[Au(III)Cl_4]^-$ at the Pt electrode in $EmimBF_4$ [159]. The results show that a series of $2e^-$ and $1e^-$ reductions occurs at the Pt electrode. Once metallic Au was formed on the Pt electrode surface, the subsequent reduction of $[Au(III)Cl_4]^-$ could occur on the Au deposit with a lower over-potential for electrodeposition. In addition, in situ EQCM experiments proved that the Au deposited was produced by the disproportionation of $[Au(I)Cl_2]^-$ to $[Au(III)Cl_4]^-$ and Au metal. The dissolution of the Au deposit was also investigated in this work by in situ EQCM method using Pt-QCM, thereby proving it to be a very powerful method in elucidating electrochemical surface phenomena accompanying mass changes in ILs.

2.3.6 In Situ Electrochemical–Scanning Probe Microscopy

STM and AFM are powerful tools for the in situ studies of the structure and morphology of electrode interface and have been used to characterize IL electrode interface as well [60]. These techniques are known to routinely achieve atomic resolution images in high vacuum conditions [160] and even in air. However, in situ measurements in liquids tend to be less well resolved owing to inhibited cantilever resonance and a vast increase in noise [161]. When it comes to ILs, AFM imaging is quite a complicated process, not only in terms of cell designs [162] and arrangement but also in terms of image acquisition [18, 60]. Even for the aqueous and organic electrolytes, electro-viscous effect is quite evident. The viscosity of these solutions when in contact with the electrode increases more than twofold under the influence of an electric field [163]. Compared to other electrolytes, the excellent stability of ILs provides suitability for the long time operation. Moreover, their high purity can make it possible to investigate the molecular level information in IL at the interface. However, an IL represents a dense system of cations and anions with no solvent, so that the individual interactions between neighboring ions create an extraordinary viscosity effect [164]. In addition to this high viscosity, the adsorption of moisture by the ILs also creates challenges for the practical measurements. However, despite these challenges, some fundamental studies have been done in IL systems which can form the basis of new and more relevant discoveries. For instance, AFM measurements have been done to reveal that ILs are strongly adsorbed on solid surfaces and that several layers are present adjacent to the surface [124, 165]. This is consistent with X-ray reflectivity measurements [166] and the proposed theoretical models [167, 168]. But as expected from the diversity of features in ILs, the functional groups as well as the chain lengths can significantly affect the adsorption strength. Study of Atkin and Endres et al. [124] showed that [C_4mPy][NTf_2] and [C_2mim] [NTf_2] behave quite differently on Au(111). Both liquids are adsorbed at the open circuit potential, but the first one adsorbed approximately four times more strongly than the latter. This adsorption of ILs on a solid surface is not at all surprising as water and organic solvents can also be adsorbed [169]. However, the force to rupture IL layers is one order of magnitude higher than for aqueous/organic solvents. Such a strong adsorption must influence the image quality in an in situ STM experiment [170], especially under electrochemical control thereby affecting the electrochemical reactions. This work also emphasizes that ILs should not be regarded as neutral solvents which all have similar properties, particularly regarding the differences in interfacial structure (solvation layers) on metal electrodes which is a function of ionic liquid species. Conventional double layers do not form on metal electrodes in IL systems. Rather, a multilayer architecture is present. The number of these solvation layers is determined by the IL species and the properties of the surface and up to seven discrete interfacial solvent layers are present on electrode surfaces. In a recently published paper [171], the structure and dynamics of the interfacial layers between the air- and water-stable IL [C_4mPy][FAP] and Au(111) were investigated using STM, CV, EIS, and AFM measurements. In situ STM measurements reveal that the Au (111) surface also undergoes a reconstruction, forming a "herringbone" superstructure at −1.2 V. AFM force–distance profiles also showed that IL becomes more structured at higher cathodic potentials with an increase in both the solvation layers and the push-through effects. EIS measurements showed a capacitive process at the IL/Au (111) interface which is considerably slower than electrochemical double-layer formation and seems to be related to the reconstruction of Au(111). In another work from the same group, the anodic regime for the same IL system [C_4mPy][FAP] and Au(111) was studied. It was again shown that structures form on the nanoscale

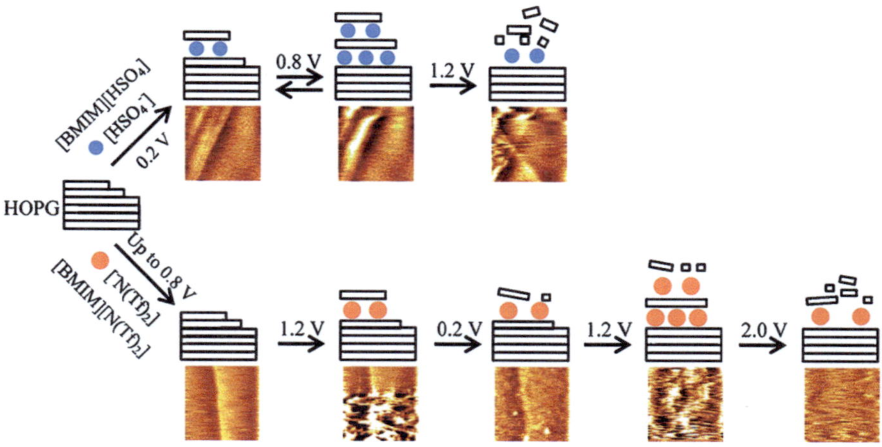

Fig. 2.12 Summary of the EC-AFM investigation of electrochemical intercalation–deintercalation of the anions into HOPG in pure [C_4mim][HSO_4] and [C_4mim][NTf_2]

for electrode potentials of up to +2 V which disappear when the electrode potential is returned to the open circuit potential, and the original surface is recovered. Thus, these observed processes on the surface are reversible. Interestingly, the work done in our group has indicated that this heterogeneity of the double-layer structure posed by different ILs and at different potentials is extremely beneficial for designing sensing interfaces. Therefore, the information gathered by the EC-AFM and sole AFM studies can significantly enhance the sensor development process.

Heterogeneity in IL's structure can also influence their interaction with substrate electrodes as well, during the electrochemical reactions and such interaction can also be studied by EC-AFM. Again by understanding these phenomena, more stable and innovative electrochemical systems can be designed both for sensors and fundamental electrochemistry. We have studied [172] such interaction of ILs with highly oriented pyrolytic graphite (HOPG) electrodes using EC-AFM, as summarized in Fig. 2.12. It was observed that electrochemical intercalation–deintercalation processes in pure ILs have caused the morphological changes in HOPG. Such changes due to electrochemical intercalation–deintercalation of ions during electrochemical processes are a common phenomenon occurring in graphite electrodes. HOPG has been widely studied as a model to understand the processes during electrochemical intercalation–deintercalation of ions into graphite. According to the model proposed for aqueous electrolytes, solvent (H_2O) molecules play an important role during the oxidative reactions on HOPG. Therefore, studying the intercalation–deintercalation process in ILs which are composed entirely of ions is of primary significance. Being composed entirely of ions, morphological changes in HOPG will be mostly due to the intercalation–deintercalation of ions in ILs. Thus, the ionic sizes can have strong influence on intercalation into HOPG. The hydrophobic character of the ILs can also affect the intercalation of the ILs with the hydrophobic HOPG surface. However, these effects may play smaller roles compared to the anionic size effect. Our data proved that these morphological changes are reversible for intercalation and deintercalation, which are strongly dependent on the anionic size of the ILs used. It was also possible to get the quantitative estimate of the amount of thickness changes on the HOPG surface during the intercalation–deintercalation process. And above a certain voltage, degradation of the steps on the HOPG surface occurs to form carbon nanoparticles. Furthermore, STM results can reveal the relationship of the electrochemical properties and the strength of this interaction. The results of these measurements are used to find out the symmetry and interatomic distances of metal atoms under potential control. Surface diffusion of metal atoms in contact with IL electrolytes has also been explored because of its close relation to the reactivity of the atoms on the surface. Such studies provide a key to understand many kinds of electrochemical processes at an electrode–electrolyte interface, such as anion desorption, electrodeposition, corrosion, and surface alloying with a deposit.

As discussed earlier, high viscosity of ILs leads to a complicated situation in terms of AFM measurements that result in much higher signal-to-noise ratios. In order to obtain better spatial resolution in such a scenario, a new EC-AFM technique has been developed which involves independent control of the AFM tip and working electrode substrate and is named as frequency-modulated AFM (FM-AFM) [173]. FM-AFM measures tip–sample interaction very sensitively by detecting changes in the resonant frequency of the oscillating cantilever. This allows the tip–sample interaction stiffness and tip deflection to be monitored simultaneously, providing the atomic resolution images. By utilizing this technique for ILs, not only the topography of the substrate can be accurately ascertained at very high resolution, but the energy dissipation curves for oscillating cantilever can be obtained. Thus, the microscopic information regarding the electrical double layer at the electrode surface can also be attained by the assessment of energy dissipation curves.

2.4 New Approaches for Sensor Development Using IL Interface

Following the 1980s PC revolution and the 1990s Internet revolutions, recent decades have experienced a revolution in sensor research which promises to have a significant impact on a broad range of applications including national security, health care, environment, energy, industry and food safety, and manufacturing. The parameters important in the evaluation of sensors can be thought of as the “five S’s” of sensor characterization, and include *Sensitivity, Selectivity, Speed of response, Stability*, and *Size/shape/cost*. The unique properties of ILs and IL interface enable a systematic design process across all design layers, with judicious choices in sensing materials, sensing chemistries, and transducers to uniquely overcome all performance challenges, especially those related to sensing capability and miniaturized sensor system implementation. As summarized in Fig. 2.13, (1) ILs’ remarkable dissolution properties for organic and inorganic substances enable analyte preconcentration and sensitivity enhancement for the detections; (2) various molecular noncovalent interactions between ILs and the analytes contribute in increasing reversibility; (3) synthetic flexibility of ILs enables selective sensing by

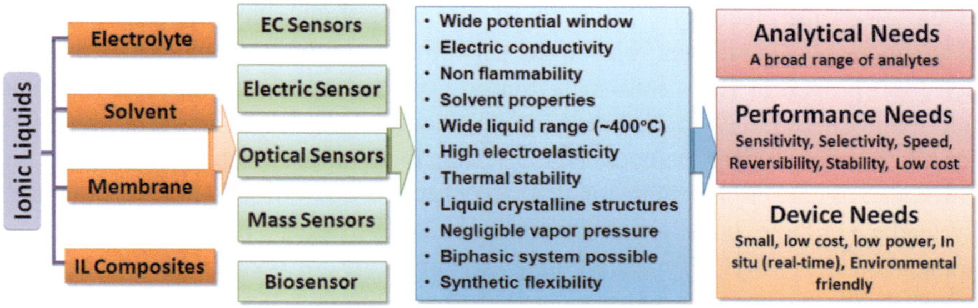

Fig. 2.13 Multifunctional character of ILs as related to different sensing platforms using their highly unique and diverse properties. All these properties help in improving one or more of the performance evaluation parameter of the sensors

constructing ILs' sensor arrays; (4) the high viscosity of ILs allows the formation of stable thin films for mass-based sensors; (5) high thermal stability of ILs enables high-temperature detection; (6) ILs' negligible vapor pressure and nonflammable nature provides stability to the sensing film; (7) ILs' negligible vapor pressure, wide potential windows, high ionic conductivity, and good solubility enable their uses as new electrolytes for many improved electrochemical sensor systems [90, 174–180]; (8) ILs can also serve as a medium or a solvent for dyes and fluorescent compounds to form stable optical matrices, which can be the basis of accurate optical detection systems; (9) ILs can be incorporated into conventional matrices, such as biopolymers, cellulose, carbon nanotubes (CNTs), metal nanoparticles, and sol–gel-based silica matrices [181, 182], to form stable composite materials. These composite materials amalgamate multifunctional properties of materials involved in preparing the composites, thereby enhancing both their utility and their performance for a variety of sensing applications. Moreover, the unique properties of ILs and their miscibility/immiscibility with other solvents can be easily controlled by changing the anions and cations, thereby providing unlimited options for their applications designs, and especially for sensors where cross functionalities are essentially required. In the following sections, we will discuss examples of ILs that have been successfully explored in multiple ways for their chemical and biological sensing capabilities. We will also illustrate the principles of some of the modifications in order to obtain good sensor interfaces which can then generate different types of sensor systems based on ILs.

2.4.1 ILs as Electrolytes for Electrochemical Sensors

Most electroanalytical methods are based on the measurements of electrochemical cell potential, current, charge, or impedance at a fixed or controlled external applied function to the electrochemical cell. There are three main types of electrochemical sensors based on either the measurement of a redox current (amperometry), the potential (potentiometry), or the impedance or capacitance (impedance spectroscopy) at the electrode/electrolyte interface of interests. Both electrode materials and electrolytes play significant roles in the performance of the interface. The analytical detection properties (e.g., sensitivity, selectivity, response time, drift, stability, and interference) are largely determined by the electrode/electrolyte interface properties for all electroanalytical methods. Notably, a majority of the electrochemical sensors developed to date employ aqueous-based electrolytes that suffer from solvent volatility leading to the exhaustion of solvent in a reaction batch. The other option is the nonaqueous electrolyte. Typically, there are three kinds of nonaqueous electrolytes: aqueous salts that could be dissolved in organic solvents (e.g., lithium perchlorate, and other perchlorate compounds, which have good solubility and ionization in particular organic solvents, such as CH_3CN), organic compounds with charged ion center (e.g., tetra-*n*-butylammonium perchlorate) or polyelectrolytes (e.g., perfluorosulfonate [183, 184] and Nafion ionomers [185, 186]), and ILs. However, some of the aqueous electrolytes could not be dissolved in the organic solvents or cannot be ionized. The dielectric constants of organic solvents are usually smaller than the aqueous solvents too, which affects the conductivity a lot. So for a certain organic solvent, there are only some particular electrolytes used for a certain kind of solvent. Because of their nonvolatility and much higher ionic conductivity, ILs have the ability to replace almost all conventional electrolytes in all types of electrochemical sensors [174, 175, 187, 188]. Here, we will discuss the principles and examples of the most important analytical applications of ILs where they are used as electrolytes and solvents in electrochemical sensing.

2.4.2 Sensory Design Based on the Potentiometry

Potentiometry is a method in which the electrochemical cell potential is measured at equilibrium at which the current is zero. The properties of the interface region differ from the bulk properties. A potential is established at the phase boundaries, e.g., between the solution and the electrode surface. The potential of electrochemical cells is the sum of all interface potentials including electrode/electrolyte interface and liquid/liquid interface (i.e., the two electrolyte solutions of different compositions that are in contact with each other). Ideally the measured potential should depend only on the potential between the interfaces of interest for analytical purpose. This is typically accomplished by keeping all other interfaces constant through a suitable electrode construction. Potentiometric sensors (e.g., ion selective electrodes) usually consist of a membrane that contains ion exchangers, lipophilic salts, and plasticizers, and the transmembrane potential gives the activity of the analyte ion in solution.

2.4.2.1 Sensing Based on the Liquid Junction Potentials

ILs have a broad range of polarity and can be used to design the interface between two immiscible ILs such as water/organic solvent interface to detect ions that are not redox active. The transfer of ions across the interface, as opposed to electron exchange at solid electrodes, generates a potential which is directly related to the concentrations of the ions of interest. ILs have been explored as an alternative water-immiscible solvent phase for sensing at the liquid/liquid interface by the detection of ions transferring across liquid/liquid interfaces as described in an excellent review by Samec and Kakiuchi [189]. The main advantage of this type of sensing compared to sensing in conventional electrolytes is that the IL is usually pre-saturated with water, so any differences in humidity in the environment will not affect the overall sensor response. The group of Kakiuchi in collaboration with Mirkin has recently reported [105] the kinetics of ion transfer at the IL/water nanointerface. The interface was formed at the tip of a nanopipette, and the transfer of tetrabutylammonium from water to the hydrophobic IL trihexyl(tetracdecyl) phosphonium bis(nonafluorobutylsulfonyl)$[P_{14,6,6,6}][C_4C_4N]$ was reported. Although this work was not specifically related to sensing, this could have implications in the field due to the use of very small (nano) interfaces that give rise to lower ohmic drop and higher current density, which is highly beneficial for sensing applications. The same group has also highlighted the ultraslow relaxation of the electrical double layer in ILs at millimeter-sized interfaces [190], which may lead to the self-inhibition of the charge-transfer step across the IL–water boundary, limiting the performance of a sensor that relates charge transfer to analyte concentration. More research in this area is needed to understand the impact of the ultraslow relaxation on sensor responses and behavior, using smaller sized interfaces.

2.4.2.2 Ion-Selective Electrodes with ILs

Ion selective electrodes (ISEs) are electrochemical sensors that allow the potentiometric determination of the activity of an analyte ion in the presence of other interfering ions. An ion-selective membrane consists of four major constituents: ionophore [ion carrier, e.g., valinomycin (selective to K^+ over Na^+)], lipophilic salt as ion exchanger [e.g., K^+R^- (where R^-=lipophilic anion)], plasticizer [e.g., 2-nitrophenyloctylether (NPOE)] as organic solvent, and polymeric substrate matrix such as Polyvinyl chloride (PVC). The different potential response has, as its principal component, the Gibbs energy change associated with permselective mass transfer across a phase boundary. The ionophore in membrane can selectively complex the target ions at membrane interface that causes the interfacial charge separation. The measured potential differences of ISEs are linearly dependent on the logarithm of the activity of a given ion in solution as described in Nicolsky-Eisenman equation (Eq. 2.5).

$$E_i = E_i^0 + s_i \log\left[a_i + \sum_{i \neq j} K_{i,j}^{\text{pot}} \left(a_j\right)^{z_i/z_j} \right] \tag{2.5}$$

E_i: electromotive force (potential) of the cell assembly; E_i^0: cell constant; a_i: activity of primary ion **I** in sample solution; a_j: activity of interfering ion **J** in sample solution; z_i, z_j: charges of **I** and **J**, respectively; and $K_{i,j}^{\text{pot}}$: potentiometric selectivity coefficient of primary ion $\mathbf{I}^{z_i}$ against interfering ion $\mathbf{J}^{z_j}$. An ideal selective electrode would show all $K_{i,j}^{\text{pot}} = 0$.

ILs possess high ion concentration, high conductivity, and good electrochemical stability. So they seem to be natural candidates for liquid ion exchanger membrane electrodes. For these reasons, they have been employed as alternatives to plasticizers and ion exchangers in membranes of ISEs. Most charged carriers applied in conventional ISEs are ion exchangers that do not exhibit specific interactions with the ions to be sensed. As a result, selectivity is determined by the free energy of hydration of the ions and corresponds to the so-called Hofmeister series [191]. However, ILs can be designed by modifying cations such as imidazolium or pyridinium and pendant alkyl groups with functionality; thus increased selectivity or even specificity for a given set of ions can be obtained. This new type of IL-based membrane electrode could be smaller, more portable, and less expensive, with the possibility of conducting simultaneous and continuous analysis of multiple analytes. It is also possible to provide real-time output and higher sensitivity and selectively. Since no sample preparation is required and no secondary labels are needed, these ISEs are very user friendly. For example, low melting ionic solids, i.e., the ILs that melt slightly above room temperature, can be used in ISEs for the potentiometric determination of salicylate, perchlorate, thiocyanate, and iodide ions in water. Three of such ILs [$N_{6,6,6,6}$][BSB], [$N_{8,8,8,8}$][BSB], and [$N_{8,8,8,1}$][BSB] were proved as suitable salts for both liquid-contact and solid-contact ISE membranes, and [$N_{8,8,8,8}$][BSB] was chosen as the most suitable because of its lowest melting point. A Nernstian response was observed over the range 10^{-5} M to 10^{-1} M for the four chosen analyte ions, with good reproducibility and reversibility and a fast response time of <10 s. Peng et al. [192] also reported a sulfate ion sensor based on a PVC membrane containing either one of the two ILs [C_8mim][Cl] or [$P_{14,6,6,6}$] [Cl] as both an anion exchanger and plasticizer, and sulfate ionophore I (1,3-[Bis(3-phenylthioureidomethyl)]benzene). Both membranes exhibited ideal Nernstian responses to sulfate over the range 10^{-5} M to 10^{-1} M and were successfully applied for the analysis of sulfate in drinking water samples. We used two imidazolium-based ILs, 1-butyl-3-methyl-imidazolium bis(trifluoromethanesulfonyl)imide [C_4mim][NTf_2] and 1,3-dibutylimidazolium bis(trifluoromethanesulfonyl)imide [C_4C_4im][NTf_2], to construct PVC liquid membrane electrodes for the detection of [NTf_2]$^-$. As shown in Fig. 2.14, the ISEs we made have improved selectivity and sensitivity as compared to anion-sensitive electrodes based on organic ammonium salts, but we also observed less than Nernstian response due to the partial solubility of the two imidazolium-based ILs in aqueous solutions. The uptake of water is expected for IL membranes being in contact with aqueous samples. The wide potential window is regarded as one of the most important advantages of ILs when they are used as pure electrolytes [189, 193]. This is not the case for IL–modified electrodes used as ISE, because they operate mostly in contact with aqueous solution. Similar to our observation, IL [$P_{14,6,6,6}$][Cl]/PVC membrane-based potentiometric sensor also suffered from severe potential instability due to the tendency of the IL to absorb water, but this was improved by the addition of carbon nanotubes (CNTs) to the membrane, resulting in a stable Nernstian response over the concentration range 10^{-5} M to 10^{-1} M [194]. Thus, hydrophobic ILs and low melting ionic solid are choices for IL-based new type of ISEs. The solubility of hydrophobic ILs in

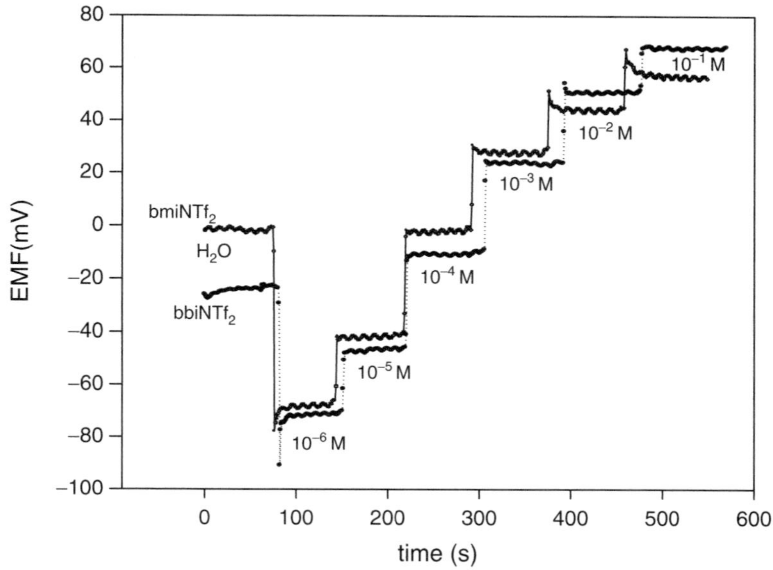

Fig. 2.14 EMF time profile of the [C_4mim][NTf_2] and [C_4C_4im][NTf_2] PVC membrane at various Li[NTf_2] concentrations

water is much smaller than the solubility of water in IL and in some cases is much below the fraction of percent. This property is crucial from the point of view of stability of IL membrane electrodes, because of the possibility of microliter-sized liquid deposit dissolution in few milliliters of the surrounding aqueous electrolyte. However, IL–membrane electrodes are reported to be stable at least in electrochemical timescale indicating slow dissolution in the case of more hydrophilic ILs. For example, Nishi et al. [195] used the hydrophobic IL methyltrioctylammonium bis(nonafluorobutylsulfonyl)imide [$N_{8,8,8,1}$][C_4C_4N] containing the ionophore dicyclohexano-18-crown-6 (DCH18C6) for the potentiometric detection of K^+ in water. Facilitated ion transfer was reported for K^+ and the two-phase system was used for the potentiometric determination of K^+. A Nernstian response to K^+ (59 mV per decade) was observed over the concentration range 10^{-5} M to 10^{-1} M with a response time of 20 min. It is clear from these reports that ILs have the ability to be used in membranes of ISEs. However, the long-term stability of such sensors may be restricted by the leaching of IL ions from the membrane into the water phase. At present, the ILs available may not be sufficiently lipophilic to serve as suitable replacements for components in conventional ISEs (e.g., ion exchangers in PVC membranes). This has prompted the researchers to use IL composites in ISEs. For example, polymer/IL/multiwalled CNT composite electrode was developed [194] as an all solid-state potentiometric sensor for various anions (SO_4^{2-}, Cl^-, NO_3^-, SCN^-, and ClO_4^-). Moreover, a potentiometric Er(III) sensor consisting of a multi-walled CNT/[C_4mim][BF_4] carbon paste electrode containing 5-(dimethylamino) naphthalene-1-sulfonyl-4-phenylsemicarbazide (NSP) as the sensing material was also constructed. A Nernstian response of 19.8 mV per decade was observed in the range 10^{-7} M to 10^{-1} M with a detection limit of 5×10^{-8} M for Er(III). This shows that the use of IL as a binder demonstrated improved performance compared to mineral oil (paraffin), and the combination of the three sensing materials gave better sensitivity, selectivity, response time, and stability compared to traditional Er(III) carbon paste sensors. Thus, this strategy is going to be more useful in preparing ISEs until new highly lipophilic ILs are synthesized.

2.4.2.3 Nonclassical Potentiometry

In ISE, the equilibrium potential change of the electrode is inversely proportional to the charge of the analyte being detected. Thus, it is not feasible for its use for detecting large, highly charged molecules, due to the very small sensor response to highly charged macromolecules. However, in 1990s, several research groups demonstrated a nonclassical potentiometry [196] for the detection of different macromolecular polyions, including heparin, protamine, carrageenan, and DNA. The response mechanism of nonclassical potentiometry is fundamentally different, employing nonequilibrium, steady-state conditions to generate a useful sensor response. Ionic liquids have also been used in the nonclassical potentiometric electrode fabrication. Langmaier and Samec [197] employed a thin microporous membrane impregnated with valinomycin and the hydrophobic IL $[N_{12,12,12,1}]$[TFPB] for the facilitated amperometric detection of K^+ and Na^+. The use of IL in the membrane in this case provided enhanced stability and selectivity compared to conventional K^+ ISEs. A similar strategy can be applied with the addition of a water-soluble crown ether (18-crown-6) to the aqueous phase, in order to make it possible to distinguish the voltammetric waves of various alkali metal cations (K^+, Na^+, Li^+, and Mg^{2+}), and the promising results suggest that the IL membrane may be suitable for application in an amperometric ISE for K^+. The work of Silvester and Arrigan [198] is quite important in this regard who reported the transfer of several common ions across the interface between water and a commercially available hydrophobic IL $[P_{14,6,6,6}]$ [FAP] with a melting point of <−50 °C. This was the first report of voltammetry at an *array* of water/IL microinterfaces, rather than at a single interface or porous polymer-supported interface. The interface array was formed within the micropores of a silicon chip membrane (30 pores and 23 μm diameters) and provided advantages of micron-sized interfaces such as radial diffusion, higher current densities, and lower iR drop, but with larger currents for sensing low concentrations of redox-inactive ions. This work has shown that sensing of various ions is possible at the water–IL interface, but it was limited to the ions that transfer at potentials less negative than the IL anion (e.g., $[FAP]^-$) transfers to water. As a result, the synthesis of ILs with extremely hydrophobic cations and anions which are not yet available in the liquid form is required in order to extend the potential windows to those similar to organic solvent/supporting electrolyte systems (e.g., 0.9 V). If this is made possible, this should allow for the observation of species that have been detected at the positive potential limit in water/organic solvent systems, e.g., important biomolecules such as lysozyme and hemoglobin.

It is already known that the contamination of IL with water shrinks potential window to that exhibited in aqueous electrolyte solution [78]. This is also the case of most hydrophobic ILs composed of large 1-alkyl-3-methyl-imidazolium cations and homologs of [NTf_2] anions. Therefore, the uptake of water is expected for IL deposit being in contact with aqueous electrolyte in the nonclassical potentiometry.

2.4.2.4 Sensing Based on Voltammetry

Voltammetry is a powerful suite of tools that enables qualitative and quantitative information about the analytes to be obtained by the measurement of the current as a function of applied potential. The technique can be applied to any species that can undergo redox transitions on solid electrodes, or any ion that can be transferred across interfaces (in the case of liquid/liquid interface). The analyte species diffuses through the electrolyte to be detected at the working electrode surface. It was also used to monitor the diffusion coefficients of species in solution and to understand their behaviors and electrochemical reaction mechanisms. Voltammetry typically takes place with two- or three electrodes that are connected through an electrolyte medium (e.g., water or organic solvents containing a background electrolyte). There has been a lot of work focused on understanding the reaction mechanisms, kinetics, and thermodynamics of electrochemical reactions in ILs. However, ILs have recently shown much promise as stable and nonvolatile electrolytes, especially in amperometric gas sensors. A typical amperometric gas sensor consists of three electrodes connected through an electrolyte, which is covered by a gas-permeable membrane. The gas passes through the membrane, diffuses through the electrolyte, and is detected at the working electrode. Most commercially available amperometric sensors of gases (e.g., O_2, CO, SO_2, H_2S, NO_2, and Cl_2) currently employ conventional solvents (e.g., H_2SO_4/H_2O mixtures, or organic solvents such as acetonitrile or propylene carbonate) that cannot survive drastic temperature changes or extremely dry or humid conditions. The "lifetime" of a sensor is often determined by how quickly the electrolyte dries up and the solvent often has to be replaced every few days/weeks in the most extreme conditions. ILs possess negligible volatility and high chemical stability, making them ideal electrolyte media in robust gas sensors for potential application in more extreme operating conditions (e.g., up to 300 °C), with no possibility of solvent evaporation or degradation. This, combined with the intrinsic conductivity (no need for supporting electrolyte), wide potential windows (to investigate compounds that may have been inaccessible otherwise), and, in some cases, increased gas solubility in ILs, makes them ideal electrolyte media for gas detection. Furthermore, ILs can stabilize the radical cation or radical anions which enables new redox chemistry of many analytes that are quasi reversible. However, the high viscosity of ILs often results in slow and less reproducible response [199] that have limited their direct use in practical sensors regarding the response time and sensitivity. To increase response speed and circumvent the high viscosity of ILs, we introduced an innovative electrochemical cell (Fig. 2.15) with a porous Teflon membrane as the support for ILs that facilitate diffusional transport through the pores and provide two levels of selectivity (i.e., the size of the pore and the IL membrane) [90]. Analytes entering from the back side

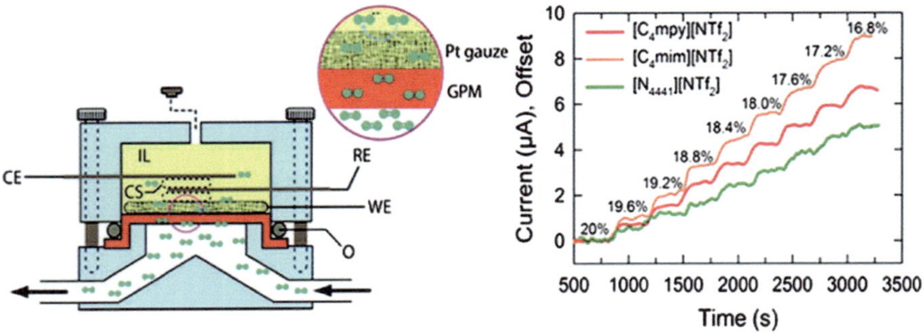

Fig. 2.15 Schematic diagram (not to scale) of the Clark-type oxygen sensing cell with the analyte flow from the back side to circumvent the high viscosity of ILs, and the current transients recorded at −1.2 V for different oxygen concentrations using three different ILs. Nitrogen was used as the carrier gas

were able to quickly reach the electrode/electrolyte interface without passing through the IL diffusion barrier, reducing response time to mere seconds. Figure 2.15 also shows the amperometric response of three [NTf_2]-based ILs to oxygen and illustrates a strong dependence on the choice of cation, as desired. High sensitivity, complete reversibility, long-term stability (less than 0.3 % potential shift over 90 days), rapid response, and strong selectivity (selectivity coefficient ~0.01 % for CO_2, NO, NO_2, SO_2) of this IL oxygen sensor were observed.

Recent advances in voltammetry-based gas sensing involving ILs have been reviewed by Rogers et al. [200] who describe the electrochemistry of various gases including oxygen, carbon dioxide, hydrogen, ammonia, hydrogen sulfide, sulfur dioxide, and nitrogen dioxide in ILs. From these studies, it appears that the most logical step forward in this area is to employ thin IL layers to overcome the slow diffusion often associated with viscous ILs. This may result in decreased response times, on the order of minutes or seconds, which is ideal for most gas sensor device requirements. Microelectrodes fabricated on a silicon chip can be an important advancement in this regard as shown by Compton's group [179]. The use of small (micron-sized) working electrodes can minimize any iR/Ohmic drop limitations that are often associated with voltammetry in highly resistive ILs on larger electrodes (larger currents). Also from these studies, the electrochemical reaction mechanisms of carbon dioxide, hydrogen sulfide, and chlorine gas in ILs can be explored. For example, the electrochemical reduction of CO_2 in the RTIL [C_4mim][Ac] was studied on a Pt microelectrode using cyclic voltammetry. CO_2 undergoes a chemically irreversible one-electron reduction to the radical anion $CO_2^{\bullet-}$ and subsequent followup chemistry including interaction/complexation with the IL. The behavior was found to be irreversible due to strong absorption of CO_2 in the IL, suggesting that this system may not be suitable for real-time sensing of CO_2; however, the high solubility (>1.5 M) of CO_2 in [C_4mim][Ac] suggests a method for sequestration of the greenhouse gas. We have also used voltammetry in ILs in combination with QCM technique [154] as described in in situ EQCM section proving that the CO_2 reduction in ILs is irreversible and forms $CO_2^{\bullet-}$ adsorbate at electrode interface. In the presence of O_2 and with increasing concentrations of CO_2, the reduction of oxygen is switched from a one-electron process to overall two-electron process, and forms adsorbed $CO_4^{\bullet-}$ intermediate species. Such information about the mechanisms of the reaction is highly beneficial for the development of the sensor systems.

2.4.2.5 IL Redox Chemistry Enabled Reliable Voltammetric Sensing

Study focusing on understanding the reaction mechanisms and comparing behavior to that in conventional electrolyte systems is an active research area and always establishes the foundation of reliable sensing before the actual sensing protocols are developed. In some cases, reactions and mechanisms have been found to be the similar in ILs as those in conventional solvents, however, in some cases, notable differences and unusual behavior have also been observed. It is extremely important to understand the atypical behavior in ILs before they can be employed as direct replacements for conventional molecular solvents (containing supporting electrolytes) in sensor devices. Compton's group [201], for example, has reported the two-electron reduction of 100 % chlorine gas to chloride on a Pt microelectrode in a range of ILs ($[C_2mim][NTf_2]$, $[C_4mim][NTf_2]$, $[C_4mPy][NTf_2]$, $[C_4mim][BF_4]$, $[C_4mim][PF_6]$, $[C_4mim][CF_3SO_3]$, $[N_{6,2,2,2}][NTf_2]$, and $[C_6mim][Cl]$). The behavior of the voltammetry at various scan rates was highly unusual, with limiting currents observed to decrease with increasing sweep rate. This intriguing observation was assigned to a mechanism of adsorption of chlorine gas on the Pt electrode surface. The adsorbed chlorine itself cannot be reduced, but must undergo desorption before the electron-transfer step. At slower scan rates with longer timescales, there is more desorption resulting in more surface available for electron transfer, thereby resulting in higher currents at slower scan rates. The large voltammetric currents observed suggest that Cl_2 has a very high solubility (1–10 M) in ILs, making these solvents attractive for the purposes of Cl_2 gas sensors. The authors did not report the effect of varying concentrations of Cl_2 on the current response, but this seems like the next logical step especially given the unusual adsorption mechanism. If the current scales linearly with concentration, ILs may prove to be a useful medium for robust Cl_2 gas sensing.

Although IL electrolytes provide partial selectivity, the primary selectivity of an IL–electrochemical sensor comes from the redox properties of the analyte observed using amperometric methods, wherein the electrical current generated by reaction of an analyte at an electrode at a fixed or variable potential is measured [22]. We have shown redox chemistry that occurs only in ILs and can be exploited to enhance sensor performance [202]. As shown in Fig. 2.16, we discovered that at platinum electrode in $[NTf_2]$-based ionic liquids (ILs), facile methane electro-oxidation is observed suggesting a unique catalytic Pt-$[NTf_2]$ interface for electron-transfer reaction of methane at room temperature. Little methane electro-oxidation signals are observed in ILs with other anions. In this experiment, an oxygen reduction process was included; as a result, the incomplete methane oxidation product CO was fully oxidized to CO_2 in the presence of superoxide ions, since superoxide can quickly adsorb on the electrode surface to form the active species, and [C_4mPy][NTf_2] is one of the best solvents to stabilize superoxide. In contrast to many methane oxidation systems reported, there are no incomplete oxidation products such as CO or COOH detected and the final methane oxidation products in [C_4mPy][NTf_2] are confirmed to be CO_2 and water by in situ infrared spectroscopy and electrochemistry. The electrochemical oxidation of methane at Pt/[C_4mPy][NTf_2] interface depends on the methane concentration linearly (from 1 % to 10 vol% methane concentration). As shown in Fig. 2.16, taking advantage of the unique coupled chemical reactivity of superoxide with the methane electrode oxidation product CO_2 that takes place in [C_4mPy][NTf_2] at room temperature, we introduce an innovative internally referenced electrochemical method for methane detection that increases the reliability of the measurements by determining these two ambient gases within a single sample matrix in real time, with cross-validation occurring for a single sensory element. The coupled reactions facilitate the complete oxidation of methane to CO_2 and water in [C_4mPy][NTf_2] and the in situ generated CO_2 was used as an internal standard for oxygen detection which addresses the fundamental limitations to accuracy and the long-term stability and reliability in chemical analysis. We have validated this ionic liquid-based methane sensor employing both conventional solid macroelectrodes and flexible microfabricated electrodes using single- and double-potential step chronoamperometry and the analytical parameters [202]. Figure 2.16 shows the amperometric response of a methane sensor with the IL [C_4mPy][NTf_2], as methane concentration is increased and oxygen concentration is decreased. Both analytes can be selectively quantified by alternating the bias potential. Oxygen can be reduced to superoxide radical which was used for oxygen sensor development (−1.2 V) as shown in Fig. 2.16. High sensitivity, complete reversibility, long-term stability (less than 0.3 % potential shift over 90 days), rapid response, and strong selectivity (selectivity coefficient ~0.01 % for CO_2, NO, NO_2, and SO_2) of this IL oxygen sensor were observed. The simplicity of detection using low-power and low-cost potential step chronoamperometry and microfabricated electrodes provides strong foundation for their easy integration with engineering advancements such as portable electronics, networked sensing, and next-generation monolithic implementation for autonomous, maintenance-free sensor operation with a long lifetime for the detection of methane with high sensitivity and specificity using smaller, inexpensive, and low-power transducers for their applications for monitoring the methane emissions.

2.4.3 Electrified IL/Electrode Interface and Related Sensor Based on Impedance Technique

EIS, which involves the application of a sinusoidal electrochemical perturbation (potential or current) over a wide range of frequencies, allows for the measurement of impedance changes in the forms of double-layer capacitance and the

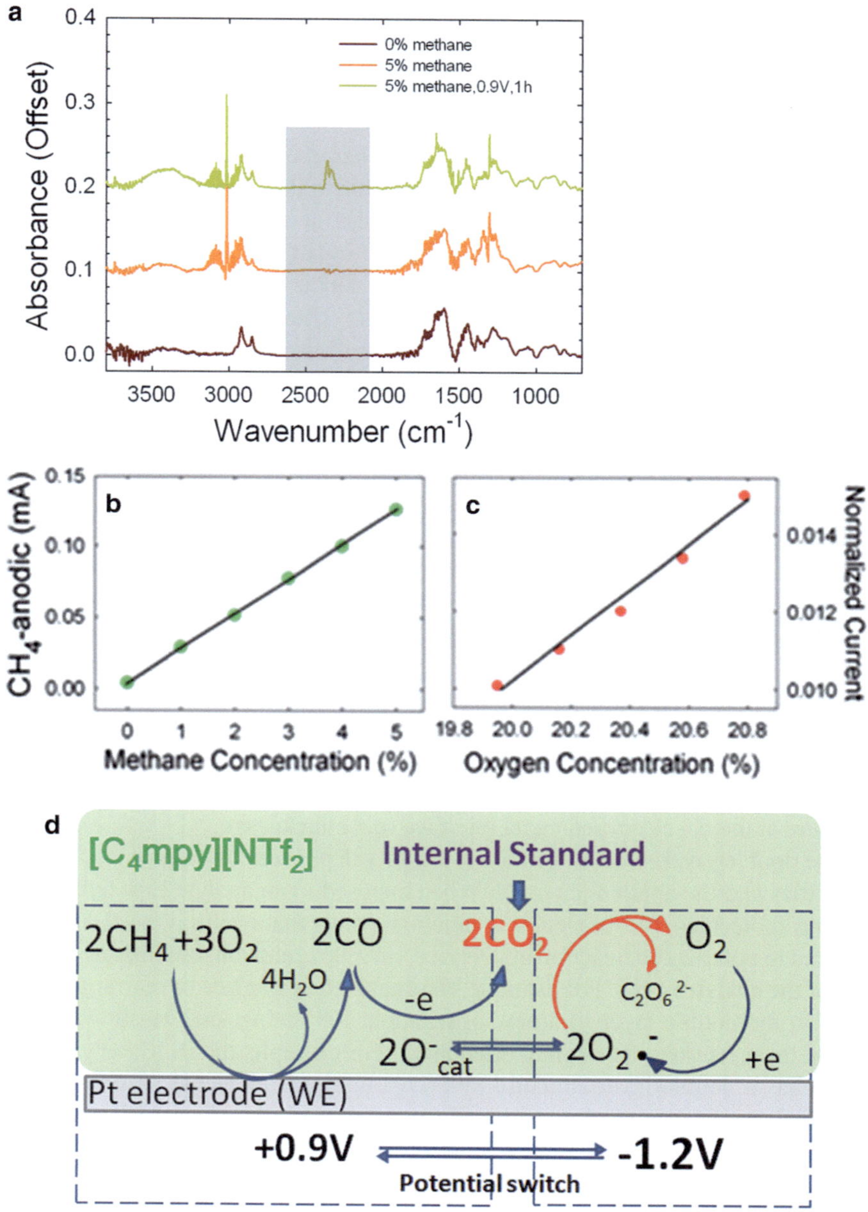

Fig. 2.16 (**a**) In situ FT-IR study of methane oxidation in [C$_4$mPy][NTf$_2$]; (**b**) plots of methane anodic currents at 0.9 V applied vs. methane concentration and (**c**) plots of normalized oxygen reduction currents at −1.2 V vs. oxygen concentration. (**d**) Scheme of methane oxidation and coupled oxygen reduction in [C$_4$mPy][NTf$_2$]

charge-transfer conductance that originate from the change of polarization of IL–electrode interface. The current signals and the impedance signals are all related to the active area of the electrodes. Thus, EIS-based sensors not only provide orthogonal detection modes to amperometric sensors but they also permit self-monitoring of the sensor's stability and automated calibration for drift mechanisms. We have demonstrated an innovative capacitance sensor based on the change of the surface charge at IL–electrified metal interfaces either due to the change of ion or electron transport in ILs resulted from the redox process of analyte or the polarization reaction due to adsorption/dissolution of analyte in ILs [80]. At room temperature, ILs have high viscosity, and the diffusion and conductivity of ions are normally lower in ILs than in aqueous electrolyte solutions. Thus, they can be regarded as a solid and liquid interface simultaneously. As a result, ILs are ideally suited for applications that require a thin or intensely concentrated layer of ionic charge, such as capacitance sensor development. Compared with molten salts at room temperature, the ions comprised of ILs are often large, flexible, highly polarizable, and chemically complex with a number of interionic forces (such as dispersion forces, dipole–dipole interaction, hydrogen bonding, and pi-stacking forces) in addition to coulombic forces present. Additionally, adsorption of anions and/or cations is likely to happen at the interfaces with ILs. Specific adsorption of IL on solid surface depends on the chemistry of the ions and applied potential. Varying the applied potential can result in the rearrangement of adsorbed ions which allows the modification of the IL double-layer capacitance. Capacitance at a fixed potential will depend on double-layer thickness and permittivity of the liquid at the interface. Ion size and applied potential influences the double-layer thickness, and alkyl chain length influences the relative permittivity. Consequently, varying the structure of IL anions or cations and applied potential enables the modification of the double-layer structure at the IL–electrified metal interface (e.g., thickness).

The double-layer structures in ILs are highly dependent on IL physicochemical properties and the applied potential. When exposed to an analyte especially small gaseous molecules, the molecular interaction events that occur at the IL interface can lead to polarization reaction at interface via redox reaction, binding, or dissolution of the analyte in ILs. For example, the analyte can displace or rearrange the IL order in the double layer to a new orientation. Removing analyte allows the IL double layer return to its original orientation. For example, the IL [C_4mPy][NTf_2], which has a double-layer structure favoring small gas molecule adsorption, can sensitively and selectively measure concentrations of CH_4 (Fig. 2.17) using EIS with a −0.3 V DC bias [80]. Figure 2.17c also clearly shows that the sensor returns to its baseline value when the analyte is removed, demonstrating the reaction is fully reversible with less than 0.1 % drift. The selectivity comes from the unique highly ordered arrangement of ions in the innermost layer of IL–electrode interface which is potential dependent. The degree of ordered structure at IL–electrode interface can be tuned by the applied bias (e.g., −0.3 V for CH_4) and by the unique molecular structure of the IL ions. The high viscosity of ILs that is usually considered a limitation to practical electrochemical applications due to slow rate of mass transport is an advantage in capacitance [203] measurement due to a more ordered

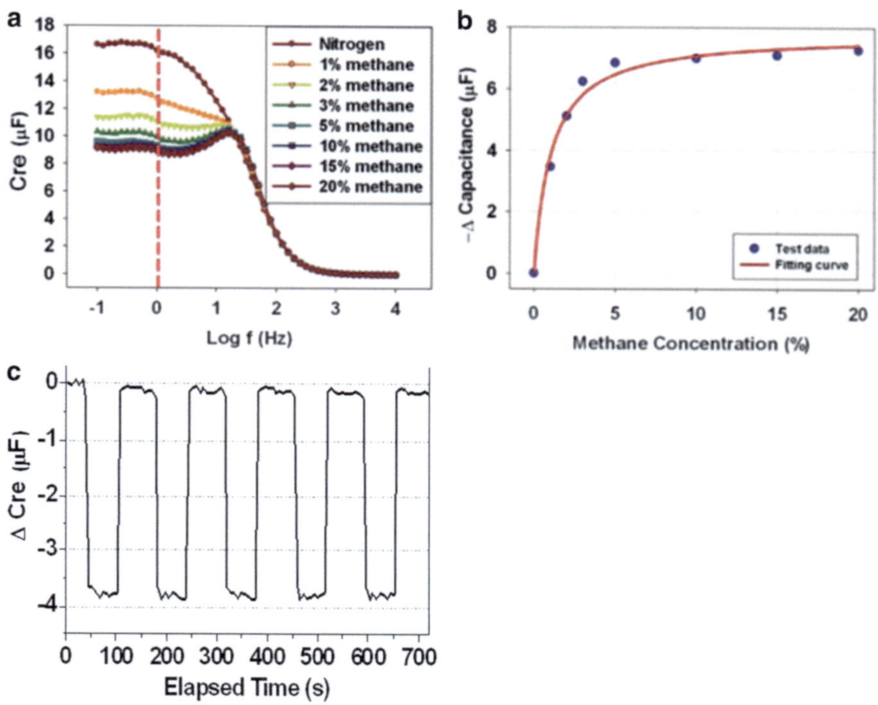

Fig. 2.17 (**a**) Plots of real part of complex capacitance at different methane concentrations. (**b**) Change of real part of complex capacitance (C_{re}) at 1 Hz vs. methane concentrations. [C_4mPy][NTf_2] is the IL and DC bias is −0.3 V; (**c**) C_{re} measured over five cycles of alternate exposure to 5 % methane (N_2 carrier gas at 200 sccm). DC potential is −0.3 V vs. Au quasi reference electrode. AC frequency is at 1 Hz

and concentrated double layer. Thus, impedance measurements provide an orthogonal detection mode to amperometric measurement, enhancing measurement sensitivity and improving sensor robustness.

2.4.4 IL Electrochemical Microarrays

For real-time multiple analyte monitoring, a group of sensors must be brought together, typically in an array format. We have demonstrated small arrays of IL sensors using both amperometric and impedance approaches. An array of four IL-coated glassy carbon electrodes was tested for the voltammetric detection of DNT and TNT, and a correlation (with 100 % classification accuracy) between the redox properties and the physiochemical parameters of the species involved was revealed (Fig. 2.18) [204]. Detection limits in liquid phase of 190 nM and 230 nM for TNT and DNT, respectively, and a linear range up to 100 μM were obtained.

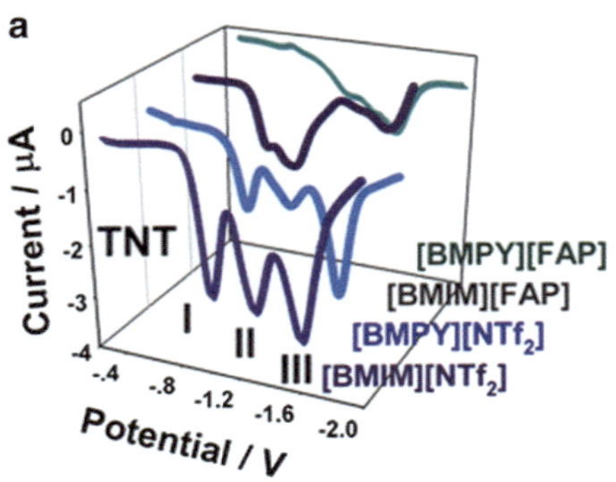

Fig. 2.18 Square wave voltammograms of 0.1 mM of TNT and DNT in four different ILs

Interference from water was eliminated by calibrating with a redox peak that was not proton related. Gas phase analysis showed strong redox signals for TNT and DNT and demonstrated that ILs serve as a preconcentrator to improve the sensitivity of very low vapor pressure analytes. These examples demonstrate that the combination of IL materials and electrochemical transducers overcomes many obstacles in forming an effective sensor system. IL–electrochemical sensors are also well suited for miniaturization and can be fabricated with very low cost. In contrast to nonspecific transducers such as a surface acoustic wave device, arrays of amperomertic and EIS transducers allow a secondary perturbation (e.g., potential) that enhances selectivity and increases analytical information content without increasing the number of physical sensor elements.

2.4.4.1 Microarray Fabrication

To improve current density and to overcome iR/Ohmic drop common in resistive ILs, small-sized working electrodes and eventually "lab-on-a-chip" type systems are being investigated. Microfabrication technologies, particularly thin film deposition of microelectrodes and formation of microfluidic channels, have been widely applied to analytical systems [205, 206]. Electrode arrays, both macro- and microscale, are now commonly fabricated using thin film metal deposition and photolithography [207–210]. For example, Fig. 2.19 shows an interdigitated electrode (IDE), which was fabricated using thin film deposition of gold on porous Teflon. Figure 2.19 also shows the data using this electrode for the measurement of oxygen concentrations. These miniaturized sensors yield similar sensitivity to results shown before. Similar fabrication techniques can be applied to optimize electrodes for the sensor interfaces developed within this project and design array structures. As fabrication of these electrodes relies on photolithography, the size and geometry of the electrode can easily be varied to meet the needs of the sensor interfaces, and different electrode configurations can be utilized within the same array to improve analytical performance.

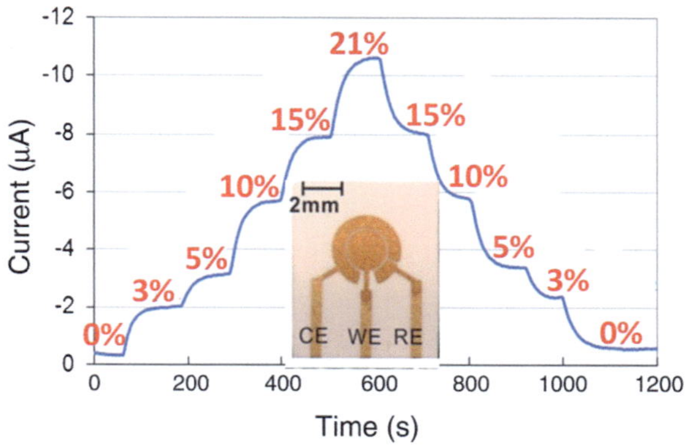

Fig. 2.19 Amperometric sensing of oxygen using (inset) micro-fabricated concentric ring disk electrode on porous Teflon

An ethylene gas sensor employing a thin layer of the IL [C_4mim][BF_4] on a "lab-on-a-chip" sensor was proposed by Zevenbergen et al. [177] The sensor consisted of a 1-mm diameter Au working electrode surrounded by a Pt ring-shaped quasi-reference electrode and a Pt rectangular counter electrode. CV curve of ethylene in the IL revealed an oxidation peak before the onset of gold oxidation. Interestingly, the ethylene oxidation peak was only visible when water was present in the IL; no response was observed with humidity levels less than 20 %. As a result, the authors studied the dependence on humidity and observed larger responses when the humidity level was higher. A detection limit of 760 ppb and a linear response (current vs. concentration) up to 10 ppm were achieved. This suggests that amperometric ethylene detection in ILs is possible; however, it is limited only to environments with sufficient humidity levels (more than 40 %), which limits the application of such a sensor in extremely dry environments.

2.4.5 *IL as Component for New Electrode Materials*

Besides using ILs as electrolytes for electrochemical sensor development, their low melting points and low tendency to crystallize, realized by the combination of large, usually asymmetric cation and smaller anion, make them good candidates for electrode modifications that can lead to a diverse set of applications in sensing fields. These modifications can be done in many different ways as shown in Fig. 2.20; however, the choice of modification is entirely based on the application requirements. For example, IL/CNT-based composite materials, consisting of highly electroactive carbon nanotubes and fluid electrolyte, can be utilized for a wide variety of electrochemical applications, such as sensors, capacitors, and actuators. In electrochemical biosensors, these composite materials can also be used as an immobilizing matrix to entrap proteins and enzymes, which provide a favorable microenvironment

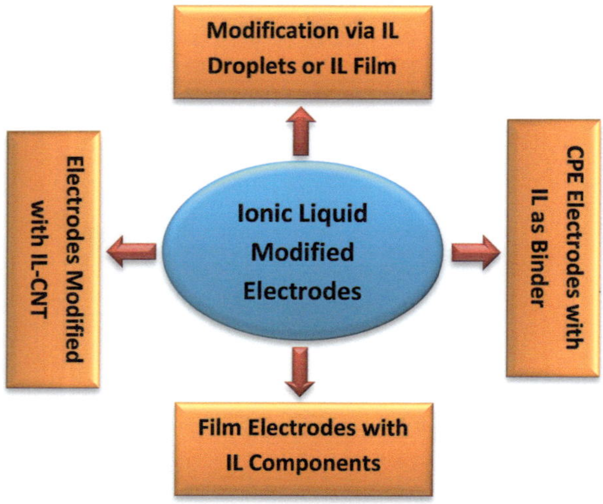

Fig. 2.20 Most common modes of the electrode modifications using ILs for different sensing applications

for redox proteins and enzymes to retain their bioactivity and perform direct electrochemistry and electrocatalysis. In the following sections, we will discuss some of the most common modifications for IL-based electrodes for sensing applications.

2.4.5.1 Electrodes Modified with IL Droplets or IL Films

Among many different types of electrode modifications possible for sensor applications, electrodes modified with IL droplets or films are the most commonly employed. In this method, electrodes are prepared by direct deposition of IL on the electrode surface [211] or from its diluted solution in volatile solvent [212]. These two procedures may provide different geometry of the deposit. Thin liquid film geometry is most important for sensors which can be enforced by surface coverage with porous Teflon membranes. The preparation of IL–modified electrode in situ by the adsorption of IL from aqueous solution has also been reported [213]. The appropriate selection of electrode substrate and IL is important in this case to obtain stable liquid deposit in contact with aqueous solution. Although research on IL–modified electrodes started from electrode substrate covered by droplets or liquid film, in the absence of other film components, now the focus has been shifted to film electrodes with organic or inorganic polymers, nanoparticles, nanotubes, and other micro- or nano-objects as other components. Their complexity ranges from simple ones as polymer film plasticized with IL to multicomponent films. The electrodes consisting of polymer membrane saturated with IL-based ionophore solution exhibit ion selectivity and were applied as ion-selective electrode. By using similar strategy with Ag/AgCl substrate, reference electrodes can be prepared. In particular, gel composed of poly(vinylidene fluoride-co-hexafluoropropylene) and viscous IL

(1-methyl-3-octylimidazolium bis-(trifluoromethyl-sulfonyl)imide) allows to obtain solid state and easy to miniaturize devices.

In many of these cases of direct modification, ILs, especially the imidazolium based, can also be appended to demonstrate functionalities which are related to their immobilization onto the electrodes to generate different types of modified electrodes. These functionalities include self-assemblies of IL on the electrode surface, covalent bonding of IL to the electrode surface, IL functions that can be used for the preparation of IL-based polymer (PILs) films, and IL used for the functionalization of conductive elements of the film. When the electrode is ready, the counter ions are electrostatically attracted to positively charged functionalities and they can be exchanged after immobilization on the electrode surface. Such small modifications can be highly suitable for incorporating the desired characteristics to the electrode materials. But most importantly, the electrodes modified by any of these methods can then be utilized for various sensing applications. For example, Ng et al. [214] employed a nanocomposite gel consisting of a three-dimensional graphene material and the IL $[C_4mim][PF_6]$ for the amperometric detection of nitric oxide (NO). A linear response of current vs. concentration over the range 1–16 μM NO was observed, with a fast response time of less than 4 s and a low detection limit of 16 nM. The improved response of this modified electrode is attributed to the porous graphene material that has a high specific surface area and superior conductivity which generated a 3-D graphene/IL nanocomposite to provide a novel platform for sensitive NO detection.

Although the generic combinations of IL with other electrode materials as shown in case of NO sensing have proven quite useful for sensing applications, the results are again generic in most cases with much less selectivity than required. Therefore, the focus is now shifting to use more task-specific ILs in the electrode modifications. Lu et al. [215] have shown that a task-specific IL in combination with bismuth oxide (Bi_2O_3) can be used for the electrochemical detection of heavy metal oxides including cadmium oxide (CdO), copper oxide (CuO), and lead oxide (PbO). The IL contained an $[NTf_2]^-$ anion and a tetraalkylammonium cation with one carbon chain functionalized with a carboxylic acid group. The presence of the acid group allowed for the solubilization of metal oxides into the IL. The IL was coated as a thin layer onto a surface containing three indium tin oxide (ITO) printed electrodes and acted as the selective solubilization medium and electrolyte. It is envisioned that more analyte species may begin to be detected by employing newly synthesized ILs with specific functional groups.

2.4.5.2 Carbon Paste Electrodes with IL as Binder

The high viscosity of ILs can also be employed to enhance the performance of the modified electrodes, for example, ILs as binders in carbon paste electrodes (CPE) which can then be used as sensing materials as well. In most cases, hydrophobic ILs are used for IL-based carbon paste electrodes (IL-CPE). Contrary to classic CPEs, their binder is composed of charged species and exhibits ionic conductivity which enhances the electrochemical sensing capabilities of these modified electrodes. Typically, IL-CPEs are prepared as classic CPEs by mixing or grinding the graphite particles with IL and placing the mixture in a cavity of the polymer or glass tube. After polishing, the electrode is ready to use. The IL/graphite particles ratio has to be optimized from the point of view of not only mechanical stability but also capacitive current, resistance, and specific electrode process, and 7:3 graphite-to-IL ratio has become a popular composition. Imidazolium-type ILs also tend to form physical gel when ground with SWCNTs by physical cross-linking of the nanotube bundles, mediated by local molecular ordering of ILs. Recently, ILs have been found to be efficient binders in the preparation of carbon composite/carbon paste electrodes. They are prepared by mixing or grinding graphite particles with the IL, followed by the transfer of this mixture into a cavity of a polymer. Higher currents (both Faradaic and capacitive) are often observed at IL-CPEs compared to traditional CPEs. This is believed to be due to the larger electroactive area for electron transfer in the IL-CPEs due to the conductive IL medium. In traditional oil-based CPEs, electron transfer can only take place at the carbon/aqueous electrolyte interface. Other reasons for the higher currents may also be due to the changes in paste morphology, better solubility of polar analytes in the IL (compared to the binder), or the presence of additional interface where transfer across the liquid/liquid interface can occur.

From all this discussion, it is quite clear that ILs can be easily employed as a modifying component of the electrodes, and in whatever way it is done, they come up with good responses for analyte detection in synthetic and real-world samples. In particular, the higher currents at IL-CPEs compared to oil-CPEs have huge advantages for sensing applications. Task specificity renders high selectivity, if employed in this modification. Otherwise, many of the sensing performance parameters are enhanced up to certain extents, even if the ILs are used randomly. The ease of preparation and the low cost of such sensing systems suggest that this could be a very active field for future sensing of species in the food/drink industry and for environmental monitoring of soil/water samples.

2.4.6 Adsorption and Absorption-Based Chemical Sensing Using IL Thin Films

In contrast to electrochemical sensor, the high viscosity of ILs is beneficial for adsorption based sensor since they can be cast into a thin film which makes them suitable for many chemical sensing applications. Being highly volatile, solvents are seldom used as sensing elements. In contrast, ILs have no significant volatility, allowing chemical processes to be carried out with essentially zero emission of toxic organic solvents and enabling their utilization as recognition elements. In these applications, ILs behave as both solid and liquid interfaces simultaneously, thus overcoming the issues associated with the interchangeable use of solid and liquid phases and the requirement of solvents to generate these phases. Depending on the physical phenomenon involved for the solvation of analytes in ILs, a coating of IL can swell, shrink, or undergo a viscosity change. The solvation in ILs is controlled by the variable contributions of adsorption-desorption and partition phenomena depending upon IL and analyte properties to furnish additional selectivity, thus opening up excellent opportunities to design different arrays of chemically selective IL films. Such characteristics make these ILs suitable for the detection of analytes in liquid or atmospheric environments via mass sensing approaches such as Quartz Crystal Microbalance (QCM) or MEMS.

2.4.6.1 IL-Based QCM Mass Sensor

QCM is a mass sensing transducer, the key component of which is quartz that functions for signal transduction and substrate for sensing materials. Applying alternating current to the gold electrodes coated on the quartz wafer induces mechanical oscillations modifiable by interfacial mass changes according to the Sauerbrey's equation:

$$\Delta f = -\frac{\Delta m f_0^2}{A\sqrt{\mu_q \rho_q}} \tag{2.6}$$

Here Δf=frequency shift, Δm=mass changes, f_0=fundamental frequency, A=electrode area, μ_q=quartz shear modulus, and ρ_q=quartz density. This principle assumes that the foreign mass is strongly coupled to the resonator making it quite straightforward for gas phase operations, where the added mass binds tightly to the surface, and the films are stiff and thin. However, density and viscosity strongly impact the nonrigid systems such as ILs, for which the following expression is used:

$$\Delta f = -f_0^{3/2}\left(\frac{\eta_L \rho_L}{\pi \mu_q \rho_q}\right)^{1/2} \tag{2.7}$$

Here η_L=viscosity and ρ_L=density of the liquid in contact with the crystal. ILs form stable layers just like the polymer matrices; however, being liquids, they bound firmly but not rigidly. QCM is capable of measuring the change of mass as well as the energy dissipation properties of thin films simultaneously, upon analyte adsorption/desorption or partition processes in thin films, thus providing rich information of the dynamic processes occurring during the adsorption–desorption events as depicted in Fig. 2.21. Dai and coworkers [216] first developed a sensor for organic vapors based on QCM using ILs as sensing materials. When analytes were dissolved in an IL, the viscosity of this IL changed rapidly which generated a frequency shift of the QCM device. The response of the QCM depended on the nature of both the analyte and the IL. Some of these ILs also have strong affinities for selected chemical species. Sensors comprising multiple of these sensors, each with a different IL coating, thus can not only detect target analytes but also help identify chemical speciation. In most of these cases, ILs are solely used as sensing films which interact with targets through ion and dipole forces, hydrogen bonding, and

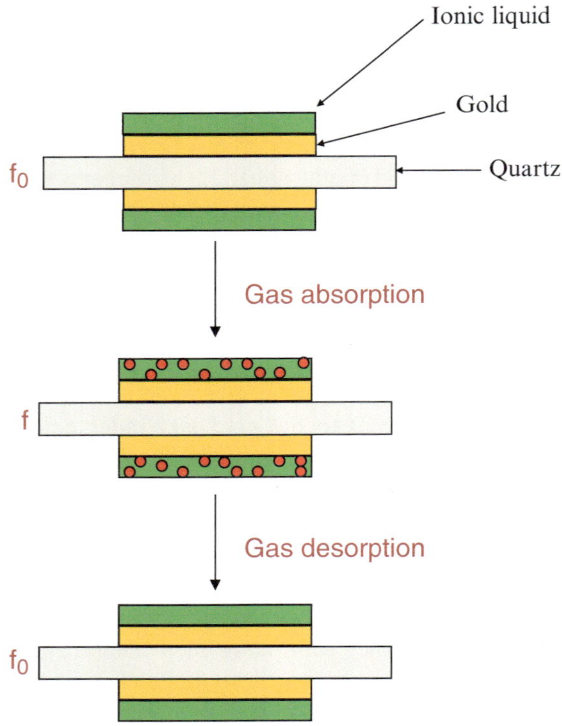

Fig. 2.21 Schematic illustration of the mechanism of frequency responses obtained in a QCM sensor

van der Waals interactions on various spatial and temporal scales. All these interactions depend upon variations in ionic composition and structure of ILs, thereby generating a criterion for selective incorporation and interaction of analytes. In addition, great solvation ability of ILs enables the rapid and reversible incorporation of gases, as shown in the work done by our group for detecting different explosive gases [176]. We have also shown that the very different solubilities of ILs for specific reactant gases create a membrane-like selectively to enhance or suppress (preconcentrate) the transport of specific reactants and provide a mechanism for selectivity [90]. Using this principle, we have successfully classified four volatile organic compounds (benzene, hexane, methylene chloride, and ethanol) via seven IL film-based mass sensors, where the unique selectivity of the ILs results from strong hydrogen bond basicity and significant capacity for dipole-type interactions with the analytes [217].

2.4.6.2 IL-Thin Film Formation

For the sorption-based sensor too, thin films on the electrodes are usually formed by employing the methods in the electrode modification sections. However, the effectiveness of most thin film-based sorption sensors relies on high sensitivity and specificity of the detection interface. The obvious approach to increase the sensitivity of

QCM sensors is to increase the thickness/amount of the sensing material. Pure ILs inherently possess certain limitations in this regard, especially for low molecular weight analytes, where sensitivity enhancement by thicker IL films can substantially influence the reproducibility through temperature-controlled variations in IL-layer thickness and the spreading out effects. Nonrigidity of thicker films makes the Sauerbrey's equation invalid in addition to slower responses, requiring a strategy to achieve higher IL loading while maintaining the IL-film integrity. For these situations, IL composite films are usually employed. Several approaches have been developed for fabricating IL-composite sensing films. In one such example, ILs are trapped as nanodroplets into the cylindrical cavities of solid alumina matrix, thus avoiding liquid wetting and softness [218]. Although this matrix can hold more IL than planar gold surface, the detection limits were not low enough for selected analytes. Contrarily, conducting polymers (CPs) and polyelectrolytes have charges, which make them ideal template materials to make IL-composite films. The proof of the concept was provided with the example of methane [219], a highly inflammable gas with very low molecular weight, and hence an evidential candidate for higher sensitivity in mass sensing devices requiring abundant total absorption/partition into the films. An ideal template for this purpose should be a porous and stable scaffold that can be modified to generate the required surface area and wettability for IL immobilization. Polyaniline (PAN) can provide this dimensionality. Four different oxidation states of PAN including the doped and undoped ones can be tested for the analytical response with an IL, e.g., 1-ethyl-3-methylimidazolium camphorsulfonate [C_2mim][CS], which can form hydrogen bonds through the sulfonate group; however, the doped PAN showed the highest sensitivity being highly charged, facilitating IL wettability through electrostatic interactions, in addition to hydrogen bonding. This increased IL surface area exposed to the analyte. The IL distribution into the nanosized channels of the PAN film helped to increase the response owing to the increase of IL film coverage. UV–vis and FT–IR data in this case confirmed [220] the formation of hydrogen bonds between camphorsulfonate and the nitrogen sites of protic acid-doped PAN. These bonds force the anions to align along the polymer backbone in a comb-like manner so as to enhance the long-range π-orbital conjugation. The interacting methane molecules can fit into these comb spaces and thus an enhanced sensitivity was observed. The enthalpy and entropy of the dissolution were shown to be higher than those in pure IL or PAN which further supported the existence of methane inside the composite generating a more ordered structure. Molecular mechanics simulation agreed with these results as well. This example and others indicate that CPs often have fairly rigid structures with tunable porosity and charge states, which can promote the rational development of the CP/IL interface alongside the IL-controlled parameters. The chemical selectivity can be provided by varying the oxidation states of CP and ionic structure of IL.

Besides this, many useful approaches are available for IL-composite formation both for their use in QCM sensors, biosensors, and even for their electrochemical sensing applications. The most widely used techniques include direct mixing, casting and rubbing, physical adsorption, electrodeposition, layer-by-layer assemblies, and sol–gel encapsulation. Physical adsorption of the IL onto a solid support is most often used to prepare the sensing films, which is based on binding forces including ionic interactions, hydrogen bonds, van der Waals forces, hydrophobic interactions, and so on. However, the more sophisticated methods for this immobilization are electrodeposition, sol–gel encapsulation, and layer-by-layer assembly. In electrodeposition, a clean electrode is immersed in a bath containing the supporting electroactive material and corresponding IL which are then electrochemically deposited onto the electrodes [221, 222]. Electrodeposition of metal nanoparticles onto IL/CNT is another well-studied technique to prepare AuNP/IL/CNT nanocomposite-based sensors [223]. In sol–gel encapsulation, an IL–silica sol is synthesized which is then mixed with different types of interesting molecules to form the films [224], whereas in layer-by-layer method, a sequential deposition of multiple layers is achieved [225], one being the IL and the other one that is supporting material, onto the electrode surface by electrostatic, van der Waals, hydrogen bonding, and charge transfer interactions. From all these approaches, merits and demerits can be identified for any of them; however, the selection is highly related to the particular application in which the approach is going to be applied.

2.4.6.3 IL High Temperature Sensor

We have also demonstrated that the IL sensors show great promise for high-temperature gas sensing [226]. For that purpose, a polar IL, trihexyl(tetracdecyl) phosphonium dodecylbenzenesulphonate $[P_{14,6,6,6}][DBS]$, was prepared via the alcohol-to-alkyl halide conversion method and coated onto QCM from its ethanol solution. This sensor was studied for the exposure of both polar (ethanol, dichloromethane) and nonpolar (heptane, benzene) vapors even at 200 °C showing linear response pattern and clear signals as shown in Fig. 2.22. As expected thermodynamically, the sensor signal decreased with increasing temperature, but still we could achieve a 5 % detection limit, which is encouraging because most solid surfaces are unable to adsorb vapors at temperatures that much higher than their boiling points. Additionally, there was an excellent reversibility for adsorption–desorption processes, requiring no experimental manipulation for sensor regeneration. The data for the damping resistance showed that physiochemical parameters such as Henry's constant can be more accurately determined at higher temperatures because of the lowered viscosity of the ILs.

2.4.6.4 IL Solvation and Other Sensing Platform

Although in many mass sensing applications, IL films play the role of solvents to dissolve the target analytes thereby having a partition coefficient, their roles as solvating species can be best described from the examples of optical sensing. Here, they dissolve many important optical substances to form stabilized optical matrices. For instance, Oter et al. reported [227] an optical CO_2 sensor using the ILs ($[C_4mim][BF_4]$ or $[C_4mim][Br]$) as the matrix with 8-hydroxypyrene-1,3,6-trisulfonic acid

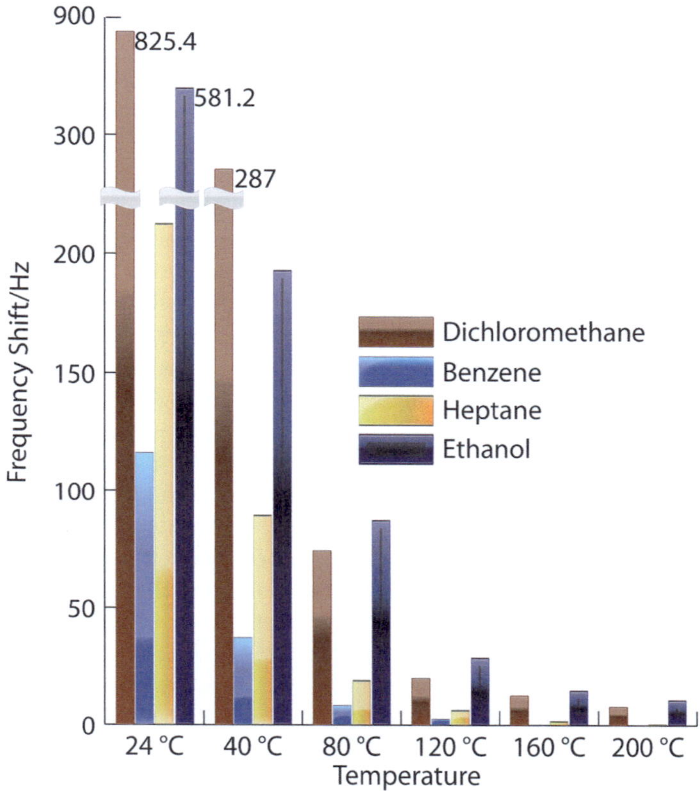

Fig. 2.22 Responses shown by IL-sensors from ambient to elevated temperatures for different volatile organic compounds (VOCs)

trisodium salt (HPTS). The detection of CO_2 is based on the fluorescence signal change of HPTS when pairing to CO_2. More recently, the same group reported [228] that IL modification of an ethyl cellulose matrix extended the detection range to 0–100 % pCO_2. In comparison to conventional solid matrix or liquid optical sensors, which have leaching problems, short life times because of evaporation of solvents, and poisoning issues from the ambient air and interfering gases, IL-based sensor showed much better performance due to IL's unique properties and better CO_2 adsorption than polymeric materials. Similar principles have been applied for temperature sensors, in which temperature indicators are dissolved in IL matrix to monitor temperatures in microregions, such as in case of microfluidics [229]. A hybrid electrochemical-colorimetric sensing platform for detecting explosives was also developed [230]. The product of the electrochemical reaction was detected by a colorimetric device. A thin layer of $[C_4mim][PF_6]$ played an important role in this platform: the IL coating selectively preconcentrated explosives by its solvation power and quickly transported them to the electrodes, and it facilitated the formation of reduction products.

2.4.6.5 Detection of Explosives and Chemical Warfare Agents

Although many of the new developments in this area can be covered by one of the previous headings, it seems appropriate to give this topic a separate section to highlight the importance of this area. With the increasing need of sensors for explosive materials at airports and security installations which may be at risk from improvised explosive devices, the interest in developing electrochemical and absorptive sensors for the detection of explosives is growing rapidly, and in the past few years, there have been several reports on the detection of explosive and toxic materials using ILs as solvent/supporting electrolytes. ILs are conductive and have low volatility, allowing them to be deployed in many types of extreme environments (e.g., hot/dry/arid conditions) where other sensor electrolytes would fail. In the prospects of such huge importance, the researchers have even used a combination of different sensing methodologies for these detections. For example, Forzani et al. proposed a hybrid electrochemical-colorimetric sensing platform for the detection of explosive trinitrotoluene (TNT) vapors [230]. A thin layer of the IL [C_4mim][PF_6] was found to selectively preconcentrate the explosive materials and quickly transport the analytes to the electrodes. The explosive vapors were detected by electrochemical (cyclic voltammetry) and colorimetric (absorbance change) methods at the working electrode. The observed currents and distinct color changes provide a fingerprint for the identification and quantification of TNT. The same group later employed a conducting polymer nanojunction that was sensitive only to the reduction products of TNT and was able to discriminate from other redox-active interferents found in ambient air. The sensor simultaneously measured the current for the reduction of TNT and the resulting conductance change of the polymer, such as poly(ethylenedioxythiophene) (PEDOT). A linear response (current vs. concentration) was observed at concentrations of 30 pM to 6 nM TNT and the sensor was capable of detecting extremely low levels of TNT within 1–2 min.

Carbon materials have got a real interest in this sense, as they possess high surface area which can enhance the IL distribution in the film as well as help in improving the sensitivity which is critically required for explosive detections. Graphene is even special and, in conjunction with ILs, has been employed for highly selective sensing of explosive TNT by Guo et al. [231]. The IL [C_4mim][PF_6] was combined with three-dimensional graphene to make an IL-graphene paste electrode with a large surface area, low background current, and pronounced mesoporosity. A linear relationship was observed between peak current using absorptive stripping voltammetry and concentration from 2 to 1,000 ppb, with a low detection limit of 0.5 ppb for TNT. This performance was superior to that demonstrated by IL-CNT and IL-graphite composites. We have also shown for many of explosive targets and employing electrochemical [204], piezoelectric [176], and EQCM platforms that by using IL materials, certain performance parameters can be achieved which are not usually possible with conventional materials.

These interesting works have shown that it is possible to selectively detect highly dangerous explosive and toxic species using an IL as an electrolyte and/or preconcentration medium. In many cases, the IL has been combined with other materials (e.g., nanomaterials), or the electrochemical technique has been combined with a complementary technique. It seems that the high viscosity of ILs compared to conventional solvent/electrolyte systems (resulting in slower mass transport, smaller diffusion coefficients, and therefore lower currents) may prevent the detection of trace quantities of explosives that are required for a viable real-world sensor. However, perhaps one of their most useful applications could be as electrolytes in sensors that can be deployed directly after an explosion in an enclosed area where temperature could still be high and other electrolytes would volatilize. This would allow for the fast identification of "hot," "warm," and "cold" zones, allowing forensic personnel to determine the areas that are safe to enter for postexplosion analysis.

2.5 Future Directions and Concluding Remarks

This chapter gives a brief highlight of the current state of knowledge of IL-electrode interface study and the notable contributions to the sensor fields using IL materials from our laboratories and others. ILs bring in several benefits as nonvolatile and stable electrolytes and solvents for electrochemical sensor as well as sensing materials for adsorption-based mass sensor developments. By synergistically utilizing IL interface electrochemistry as well as molecular design and control of IL composites, ILs-based electrochemical and mass sensors are shown to enable multiple gas detections with multiple modes detection principles that can address many gas sensors challenges, especially for sensors that can operate in unconventional and hostile environments (i.e., elevated temperature, low pressure) or in the presence of typically interfering substances (e.g., high humidity, pyrophorics) as well as for sensor miniaturization and robustness. However, in order to further the developments of ILs for the sensor applications discussed here, many aspects of the behavior of ILs need urgent investigation as well: the fundamental electrochemical issues such as the structure of the double layer and diffusion layer at electrode/ionic liquid interface; the speciation of solute ions such as metal ions and polymerized ions; transference numbers and how these are influenced by speciation; dielectric properties and how these merge with conduction properties as a function of surface chemistry and the wetting phenomena of ILs at gas/liquid/solid interfaces; the thermodynamic quantities such as chemical potentials of solutes in ILs, and how these influence redox potentials and redox reversibility; ion association and its effect on thermodynamic and transport properties; and interactions of ILs with solutes and interfaces. Additionally, ILs have been shown as promising alternative solvents in a variety of applications. However, only a few processes are commercialized due to the relatively high cost of ILs. The recovery and reuse of ILs play an important role in the commercialization of processes employing ILs. The understanding of recyclability of ILs based on the available literature in various application fields is still in its infancy stage which was recognized by ILs' research communities. With ever flourishing research and applications of ILs in the future, the study toward thermal stability of ionic liquids for engineering applications should analyze the ILs with thoughts on evaporation or temperatures at which evaporation kinetics are known. The temperature-dependent properties and the high dependency of the results on the experimental factors need to be investigated in the comparison of different conditions. In summary, many fundamental aspects of the physical chemistry and electrochemistry of ILs need to be thoroughly investigated, and they promise to further improve the potential of their various electrochemical and chemical applications.

Acknowledgement We like to thank the support from NIOSH R21, NIOSH R01, ONR, and MIIE. We also like to thank Professor Naifei Hu for helpful comments and suggestions.

References

1. Lu W, Fadeev AG, Qi BH, Smela E, Mattes BR, Ding J, Spinks GM, Mazurkiewicz J, Zhou DZ, Wallace GG, MacFarlane DR, Forsyth SA, Forsyth M (2002) Use of ionic liquids for pi-conjugated polymer electrochemical devices. Science 297(5583):983–987
2. Plechkova NV, Seddon KR (2008) Applications of ionic liquids in the chemical industry. Chem Soc Rev 37(1):123–150. doi:10.1039/B006677j
3. Blanchard LA, Brennecke JF (2001) Recovery of organic products from ionic liquids using supercritical carbon dioxide (vol 40, pg 289, 2001). Ind Eng Chem Res 40(11):2550
4. Endres F (2001) Electrodeposition of a thin germanium film on gold from a room temperature ionic liquid. Phys Chem Chem Phys 3(15):3165–3174
5. Handy ST (2003) Greener solvents: room temperature ionic liquids from biorenewable sources. Chemistry 9(13):2938–2944
6. Jensen MP, Neuefeind J, Beitz JV, Skanthakumar S, Soderholm L (2003) Mechanisms of metal ion transfer into room-temperature ionic liquids: the role of anion exchange. J Am Chem Soc 125(50):15466–15473
7. Yang CH, Sun QJ, Qiao J, Li YF (2003) Ionic liquid doped polymer light-emitting electrochemical cells. J Phys Chem B 107(47):12981–12988
8. Buzzeo MC, Evans RG, Compton RG (2004) Non-haloaluminate room-temperature ionic liquids in electrochemistry – a review. ChemPhysChem 5(8):1106–1120
9. Wei D, Ivaska A (2008) Applications of ionic liquids in electrochemical sensors. Anal Chim Acta 607(2):126–135
10. Rogers EI, Sljukic B, Hardacre C, Compton RG (2009) Electrochemistry in room-temperature ionic liquids: potential windows at mercury electrodes. J Chem Eng Data 54(7):2049–2053. doi:10.1021/Je800898z
11. Earle MJ, Esperança JMSS, Gilea MA, Lopes JNC, Rebelo LPN, Magee JW, Seddon KR, Widegren JA (2006) The distillation and volatility of ionic liquids. Nature 439:831. doi:10.1038/nature04451
12. Mays DE, Hussam A (2009) Voltammetric methods for determination and speciation of inorganic arsenic in the environment–a review. Anal Chim Acta 646(1–2):6–16. doi:10.1016/j.aca.2009.05.006
13. Macpherson JV, Marcar S, Unwin PR (1994) Microjet electrode – a hydrodynamic ultramicroelectrode with high mass-transfer rates. Anal Chem 66(13):2175–2179
14. Drummond TG, Hill MG, Barton JK (2003) Electrochemical DNA sensors. Nat Biotechnol 21(10):1192–1199. doi:10.1038/nbt873
15. Gallagher JJ, Svenson RH, Kasell JH, German LD, Bardy GH, Broughton A, Critelli G (1982) Catheter technique for closed-chest ablation of the atrioventricular-conduction system – a therapeutic alternative for the treatment of refractory supra-ventricular tachycardia. N Engl J Med 306(4):194–200

16. Wang J (2005) Carbon-nanotube based electrochemical biosensors: a review. Electroanal 17(1):7–14. doi:10.1002/elan.200403113
17. Laviron E (1974) Adsorption, autoinhibition and autocatalysis in polarography and in linear potential sweep voltammetry. Electroanal Chem Interracial Electrochem 52:355–393
18. Hayes R, Borisenko N, Tam MK, Howlett PC, Endres F, Atkin R (2011) Double layer structure of ionic liquids at the Au(111) electrode interface: an atomic force microscopy investigation. J Phys Chem C 115(14):6855–6863. doi:10.1021/Jp200544b
19. Baldelli S (2008) Surface structure at the ionic liquid-electrified metal interface. Acc Chem Res 41(3):421–431. doi:10.1021/ar700185h
20. Lockett V, Horne M, Sedev R, Rodopoulos T, Ralston J (2010) Differential capacitance of the double layer at the electrode/ionic liquids interface. Phys Chem Chem Phys 12(39):12499–12512. doi:10.1039/c0cp00170h
21. Hayes R, El Abedin SZ, Atkin R (2009) Pronounced structure in confined aprotic room-temperature ionic liquids. J Phys Chem B 113(20):7049–7052. doi:10.1021/Jp902837s
22. Stetter JR, Li J (2008) Amperometric gas sensors – a review. Chem Rev 108(2):352–366. doi:10.1021/Cr0681039
23. Tuantranont A, Wisitsora-at A, Sritongkham P, Jaruwongrungsee K (2011) A review of monolithic multichannel quartz crystal microbalance: a review. Anal Chim Acta 687(2):114–128. doi:10.1016/j.aca.2010.12.022
24. Rheaume JM, Pisano AP (2011) A review of recent progress in sensing of gas concentration by impedance change. Ionics 17(2):99–108. doi:10.1007/s11581-010-0515-1
25. Santos PS, Ando RA, Siqueira LJA, Bazito FC, Torresi RM (2007) The sulfur dioxide-1-butyl-3-methylimidazolium bromide interaction: drastic changes in structural and physical properties. J Phys Chem B 111(30):8717–8719. doi:10.1021/jp0743572
26. Sanchez LMG, Meindersma GW, de Haan AB (2007) Solvent properties of functionalized ionic liquids for CO2 absorption. Chem Eng Res Des 85(A1):31–39. doi:10.1205/cherd06124
27. Feng GA, Qiao R, Huang JS, Dai S, Sumpter BG, Meunier V (2011) The importance of ion size and electrode curvature on electrical double layers in ionic liquids. Phys Chem Chem Phys 13(3):1152–1161. doi:10.1039/C0cp02077j
28. Yuan HT, Shimotani H, Tsukazaki A, Ohtomo A, Kawasaki M, Iwasa Y (2010) Hydrogenation-induced surface polarity recognition and proton memory behavior at protic-ionic-liquid/oxide electric-double-layer interfaces. J Am Chem Soc 132(19):6672–6678. doi:10.1021/ja909110s
29. Sun J, Forsyth M, MacFarlane DR (1998) Room-temperature molten salts based on the quaternary ammonium ion. J Phys Chem B 102(44):8858–8864. doi:10.1021/jp981159p
30. MacFarlane DR, Meakin P, Sun J, Amini N, Forsyth M (1999) Pyrrolidinium imides: a new family of molten salts and conductive plastic crystal phases. J Phys Chem B 103(20):4164–4170. doi:10.1021/jp984145s
31. Bockris JOM, Swinkels DAJ (1964) Adsorption of n-decylamine on solid metal electrodes. J Electrochem Soc 111(6):736–743. doi:10.1149/1.2426222
32. Ryan JP, Kunz RJ, Shepard JW (1960) A radioactive tracer study of the adsorption of fluorinated compounds on solid planar surfaces. II. C8F17SO2N(C2H5)CH2COOH. J Phys Chem 64(5):525–529. doi:10.1021/j100834a003
33. Granese SL (1988) Study of the inhibitory-action of nitrogen-containing compounds. Corrosion 44(6):322–327
34. Vázquez CI, Lacconi GI (2013) Nucleation and growth of silver nanostructures onto HOPG electrodes in the presence of picolinic acid. J Electroanal Chem 691:42–50. doi:10.1016/j.jelechem.2012.12.017
35. Capon A, Parson R (1973) The oxidation of formic acid at noble metal electrodes: I. Review of previous work. J Electroanal Chem Interfacial Electrochem 44(1):1–7. doi:10.1016/S0022-0728(73)80508-X
36. Jeanmaire DL, Van Duyne RP (1977) Surface Raman spectroelectrochemistry: Part I. Heterocyclic, aromatic, and aliphatic amines adsorbed on the anodized silver electrode. J Electroanal Chem Interfacial Electrochem 84(1):1–20. doi:10.1016/S0022-0728(77)80224-6

37. Sandroff CJ, Herschbach DR (1982) Surface-enhanced Raman study of organic sulfides adsorbed on silver: facile cleavage of sulfur-sulfur and carbon-sulfur bonds. J Phys Chem 86(17):3277–3279. doi:10.1021/j100214a002
38. Rupprechter G, Dellwig T, Unterhalt H, Freund HJ (2001) High-pressure carbon monoxide adsorption on Pt(111) revisited: a sum frequency generation study. J Phys Chem B 105(18):3797–3802. doi:10.1021/jp003585s
39. Liu D, Ma G, Allen HC (2005) Adsorption of 4-picoline and piperidine to the hydrated SiO2 surface: probing the surface acidity with vibrational sum frequency generation spectroscopy. Environ Sci Technol 39(7):2025–2032. doi:10.1021/es0482280
40. Wang J, Buck SM, Chen Z (2002) Sum frequency generation vibrational spectroscopy studies on protein adsorption. J Phys Chem B 106(44):11666–11672. doi:10.1021/jp021363j
41. Bockris JOM, Jeng KT (1992) In-situ studies of adsorption of organic compounds on platinum electrodes. J Electroanal Chem 330(1–2):541–581. doi:10.1016/0022-0728(92)80330-7
42. Carlos-Cuellar S, Li P, Christensen AP, Krueger BJ, Burrichter C, Grassian VH (2003) Heterogeneous uptake kinetics of volatile organic compounds on oxide surfaces using a Knudsen cell reactor: adsorption of acetic acid, formaldehyde, and methanol on α-Fe2O3, α-Al2O3, and SiO2. J Phys Chem A 107(21):4250–4261. doi:10.1021/jp0267609
43. Bewick A, Kunimatsu K (1980) Infra red spectroscopy of the electrode-electrolyte interphase. Surf Sci 101(1–3):131–138. doi:10.1016/0039-6028(80)90604-4
44. Péré E, Cardy H, Cairon O, Simon M, Lacombe S (2001) Quantitative assessment of organic compounds adsorbed on silica gel by FTIR and UV–VIS spectroscopies: the contribution of diffuse reflectance spectroscopy. Vib Spectrosc 25(2):163–175. doi:10.1016/S0924-2031(00)00113-2
45. Park H, Choi W (2005) Photocatalytic reactivities of nafion-coated TiO2 for the degradation of charged organic compounds under UV or visible light. J Phys Chem B 109(23):11667–11674. doi:10.1021/jp051222s
46. Heinz TF, Chen CK, Ricard D, Shen YR (1982) Spectroscopy of molecular monolayers by resonant second-harmonic generation. Phys Rev Lett 48:478–481
47. Sagara T, Kawamura H, Nakashima N (1996) Electrode reaction of methylene blue at an alkanethiol-modified gold electrode as characterized by electroreflectance spectroscopy. Langmuir 12(17):4253–4259. doi:10.1021/la951530p
48. Chang J-K, Lee M-T, Tsai W-T, Deng M-J, Sun IW (2009) X-ray photoelectron spectroscopy and in situ X-ray absorption spectroscopy studies on reversible insertion/desertion of dicyanamide anions into/from manganese oxide in ionic liquid. Chem Mater 21(13):2688–2695. doi:10.1021/cm9000569
49. Gaillard C, Chaumont A, Billard I, Hennig C, Ouadi A, Wipff G (2007) Uranyl coordination in ionic liquids: the competition between ionic liquid anions, uranyl counterions, and Cl- anions investigated by extended X-ray absorption fine structure and UV–visible spectroscopies and molecular dynamics simulations. Inorg Chem 46(12):4815–4826. doi:10.1021/ic061864+
50. Del Pópolo MG, Lynden-Bell RM, Kohanoff J (2005) Ab initio molecular dynamics simulation of a room temperature ionic liquid. J Phys Chem B 109(12):5895–5902. doi:10.1021/jp044414g
51. Emel'yanenko VN, Verevkin SP, Heintz A (2007) The gaseous enthalpy of formation of the ionic liquid 1-butyl-3-methylimidazolium dicyanamide from combustion calorimetry, vapor pressure measurements, and ab initio calculations. J Am Chem Soc 129(13):3930–3937. doi:10.1021/ja0679174
52. De Feyter S, De Schryver FC (2003) Two-dimensional supramolecular self-assembly probed by scanning tunneling microscopy. Chem Soc Rev 32(3):139–150. doi:10.1039/b206566p
53. Hansma P, Elings V, Marti O, Bracker C (1988) Scanning tunneling microscopy and atomic force microscopy: application to biology and technology. Science 242(4876):209–216. doi:10.1126/science.3051380
54. De Feyter S, Gesquière A, Abdel-Mottaleb MM, Grim PCM, De Schryver FC, Meiners C, Sieffert M, Valiyaveettil S, Müllen K (2000) Scanning tunneling microscopy: a unique tool in

the study of chirality, dynamics, and reactivity in physisorbed organic monolayers. Acc Chem Res 33(8):520–531. doi:10.1021/ar970040g
55. Wilkinson KJ, Balnois E, Leppard GG, Buffle J (1999) Characteristic features of the major components of freshwater colloidal organic matter revealed by transmission electron and atomic force microscopy. Colloids Surf A Physicochem Eng Asp 155(2–3):287–310. doi:10.1016/S0927-7757(98)00874-7
56. Baldelli S, Bao JM, Wu W, Pei SS (2011) Sum frequency generation study on the orientation of room-temperature ionic liquid at the graphene-ionic liquid interface. Chem Phys Lett 516(4–6):171–173. doi:10.1016/j.cplett.2011.09.084
57. Santos CS, Baldelli S (2010) Gas-liquid interface of room-temperature ionic liquids. Chem Soc Rev 39(6):2136–2145. doi:10.1039/b921580h
58. Aliaga C, Baldelli S (2008) A sum frequency generation study of the room-temperature ionic liquid-titanium dioxide interface. J Phys Chem C 112(8):3064–3072. doi:10.1021/jp709753r
59. Endres F (2012) Interfaces of ionic liquids. Phys Chem Chem Phys 14(15):5008–5009. doi:10.1039/C2cp90031a
60. Endres F, Borisenko N, El Abedin SZ, Hayes R, Atkin R (2012) The interface ionic liquid(s)/electrode(s): in situ STM and AFM measurements. Faraday Discuss 154:221–233. doi:10.1039/C1fd00050k
61. Zhang SJ, Sun N, He XZ, Lu XM, Zhang XP (2006) Physical properties of ionic liquids: database and evaluation. J Phys Chem Ref Data 35(4):1475–1517. doi:10.1063/1.2204959
62. Rooney D, Jacquemin J, Gardas R (2010) Thermophysical properties of ionic liquids. In: Kirchner B (ed) Ionic liquids, vol 290, Topics in current chemistry. Springer, Berlin, pp 185–212. doi:10.1007/128_2008_32
63. Endres F, El Abedin SZ (2006) Air and water stable ionic liquids in physical chemistry. Phys Chem Chem Phys 8(18):2101–2116. doi:10.1039/B600519p
64. O'Mahony AM, Silvester DS, Aldous L, Hardacre C, Compton RG (2008) Effect of water on the electrochemical window and potential limits of room-temperature ionic liquids. J Chem Eng Data 53(12):2884–2891. doi:10.1021/je800678e
65. Dzyuba SV, Bartsch RA (2002) Influence of structural variations in 1-alkyl(aralkyl)-3-methylimidazolium hexafluorophosphates and bis(trifluorormethyl-sulfonyl)imides on physical properties of the ionic liquids. ChemPhysChem 3(2):161. doi:10.1002/1439-7641(20020215)3:2<161::Aid-Cphc161>3.0.Co;2-3
66. McEwen AB, Ngo HL, LeCompte K, Goldman JL (1999) Electrochemical properties of imidazolium salt electrolytes for electrochemical capacitor applications. J Electrochem Soc 146(5):1687–1695. doi:10.1149/1.1391827
67. Suarez PAZ, Einloft S, Dullius JEL, de Souza RF, Dupont J (1998) Synthesis and physical-chemical properties of ionic liquids based on 1-n-butyl-3-methylimidazolium cation. J Chim Phys Pcb 95(7):1626–1639
68. Wang HP, Liu WW, Cheng LY, Zhang YM, Yu MF (2008) The physical properties of aqueous solution of room-temperature ionic liquids based on imidazolium: database and evaluation. J Mol Liq 140(1–3):68–72. doi:10.1016/j.molliq.2008.01.008
69. Schroder C, Neumayr G, Steinhauser O (2009) On the collective network of ionic liquid/water mixtures. III. Structural analysis of ionic liquids on the basis of Voronoi decomposition. J Chem Phys 130(19), 194503. doi:10.1063/1.3127782
70. Schroder C, Rudas T, Neumayr G, Benkner S, Steinhauser O (2007) On the collective network of ionic liquid/water mixtures. I. Orientational structure. J Chem Phys 127(23), 234503. doi:10.1063/1.2805074
71. Schroder C, Rudas T, Neumayr G, Gansterer W, Steinhauser O (2007) Impact of anisotropy on the structure and dynamics of ionic liquids: a computational study of 1-butyl-3-methylimidazolium trifluoroacetate. J Chem Phys 127(4), 044505. doi:10.1063/1.2754690
72. Silvester DS, Compton RG (2006) Electrochemistry in room temperature ionic liquids: a review and some possible applications. Z Phys Chem 220(10–11):1247–1274. doi:10.1524/zpch.2006.220.10.1247

73. Ho TD, Zhang C, Hantao LW, Anderson JL (2014) Ionic Liquids in Analytical Chemistry: Fundamentals, Advances, and Perspectives. Analytical Chemistry 86 (1):262-285. doi:10.1021/Ac4035554
74. Soukup-Hein RJ, Warnke MM, Armstrong DW (2009) Ionic liquids in analytical chemistry. Annu Rev Anal Chem 2:145–168. doi:10.1146/annurev-anchem-060908-155150
75. Anderson JL, Armstrong DW, Wei GT (2006) Ionic liquids in analytical chemistry. Anal Chem 78(9):2892–2902. doi:10.1021/Ac069394o
76. Delahay P, Turner DR (1955) New instrumental methods in electrochemistry. J Electrochem Soc 102(2):46C–47C. doi:10.1149/1.2429993
77. Bard A, Faulkner L (2001) Electrochemical methods: fundamentals and applications. Wiley. citeulike-article-id:8415288. New York
78. Schroder U, Wadhawan JD, Compton RG, Marken F, Suarez PAZ, Consorti CS, de Souza RF, Dupont J (2000) Water-induced accelerated ion diffusion: voltammetric studies in 1-methyl-3-[2,6-(S)-dimethylocten-2-yl]imidazolium tetrafluoroborate, 1-butyl-3-methylimidazolium tetrafluoroborate and hexafluorophosphate ionic liquids. New J Chem 24(12):1009–1015. doi:10.1039/B007172M
79. Tang YJ, Zeng XQ (2008) Poly(vinyl ferrocene) redox behavior in ionic liquids. J Electrochem Soc 155(5):F82–F90. doi:10.1149/1.2868797
80. Wang Z, Mu X, Guo M, Huang Y, Mason AJ, Zeng X (2013) Methane recognition and quantification by differential capacitance at the hydrophobic ionic liquid-electrified metal electrode interface. J Electrochem Soc 160(6):B83–B89. doi:10.1149/2.138306jes
81. Taylor AW, Qiu FL, Hu JP, Licence P, Walsh DA (2008) Heterogeneous electron transfer kinetics at the ionic liquid/metal interface studied using cyclic voltammetry and scanning electrochemical microscopy. J Phys Chem B 112(42):13292–13299. doi:10.1021/Jp8024717
82. Muller EA, Strader ML, Johns JE, Yang A, Caplins BW, Shearer AJ, Suich DE, Harris CB (2013) Femtosecond electron solvation at the ionic liquid/metal electrode interface. J Am Chem Soc 135(29):10646–10653. doi:10.1021/ja3108593
83. Cremer T (2013) Ionic liquid/metal interfaces. Ionic liquid bulk and interface properties. Springer Theses. Springer International Publishing, pp 69–122. doi:10.1007/978-3-319-00380-1_4
84. Wu P, Huang JS, Meunier V, Sumpter BG, Qiao R (2011) Complex capacitance scaling in ionic liquids-filled nanopores. ACS Nano 5(11):9044–9051. doi:10.1021/Nn203260w
85. Kondrat S, Georgi N, Fedorov MV, Kornyshev AA (2011) A superionic state in nano-porous double-layer capacitors: insights from Monte Carlo simulations. Phys Chem Chem Phys 13(23):11359–11366. doi:10.1039/c1cp20798a
86. Druschler M, Huber B, Passerini S, Roling B (2010) Hysteresis effects in the potential-dependent double layer capacitance of room temperature ionic liquids at a polycrystalline platinum interface. J Phys Chem C 114(8):3614–3617. doi:10.1021/Jp911513k
87. Lau VWH, Masters AF, Bond AM, Maschmeyer T (2012) Ionic-liquid-mediated active-site control of MoS2 for the electrocatalytic hydrogen evolution reaction. Chemistry 18(26):8230–8239. doi:10.1002/chem.201200255
88. Kilic MS, Bazant MZ, Ajdari A (2007) Steric effects in the dynamics of electrolytes at large applied voltages. I. Double-layer charging. Phys Rev E 75(2), 021502. doi:10.1103/Physreve.75.021502
89. Gore TR, Bond T, Zhang WB, Scott RWJ, Burgess IJ (2010) Hysteresis in the measurement of double-layer capacitance at the gold-ionic liquid interface. Electrochem Commun 12(10):1340–1343. doi:10.1016/j.elecom.2010.07.015
90. Wang Z, Lin P, Baker GA, Stetter J, Zeng X (2011) Ionic liquids as electrolytes for the development of a robust amperometric oxygen sensor. Anal Chem 83(18):7066–7073. doi:10.1021/ac201235w
91. Galinski M, Lewandowski A, Stepniak I (2006) Ionic liquids as electrolytes. Electrochim Acta 51(26):5567–5580. doi:10.1016/j.electacta.2006.03.016
92. Sun P, Armstrong DW (2010) Ionic liquids in analytical chemistry. Anal Chim Acta 661(1):1–16. doi:10.1016/j.aca.2009.12.007
93. Hagiwara R, Lee JS (2007) Ionic liquids for electrochemical devices. Electrochemistry 75(1):23–34

94. Wang Z, Zeng X (2013) Bis(trifluoromethylsulfonyl)imide (NTf2)-based ionic liquids for facile methane electro-oxidation on Pt. J Electrochem Soc 160(9):H604–H611. doi:10.1149/2.039309jes
95. Seo DM, Borodin O, Balogh D, O'Connell M, Ly Q, Han SD, Passerini S, Henderson WA (2013) Electrolyte solvation and ionic association III. Acetonitrile-lithium salt mixtures-transport properties. J Electrochem Soc 160(8):A1061–A1070. doi:10.1149/2.018308jes
96. Frank HS, Wen W-Y (1957) Ion-solvent interaction. Structural aspects of ion-solvent interaction in aqueous solutions: a suggested picture of water structure. Discuss Faraday Soc 24:133–140. doi:10.1039/df9572400133
97. Kaminsky M (1957) Ion-solvent interaction and the viscosity of strong-electrolyte solutions. Discuss Faraday Soc 24:171–179. doi:10.1039/df9572400171
98. Visser AE, Rogers RD (2003) Room-temperature ionic liquids: new solvents for f-element separations and associated solution chemistry. J Solid State Chem 171(1–2):109–113. doi:10.1016/S0022-4596(02)00193-7
99. Bonhôte P, Dias A-P, Papageorgiou N, Kalyanasundaram K, Grätzel M (1996) Hydrophobic, highly conductive ambient-temperature molten salts. Inorg Chem 35(5):1168–1178. doi:10.1021/ic951325x
100. Lee PC, Han TH, Hwang T, Oh JS, Kim SJ, Kim BW, Lee Y, Choi HR, Jeoung SK, Yoo SE, Nam JD (2012) Electrochemical double layer capacitor performance of electrospun polymer fiber-electrolyte membrane fabricated by solvent-assisted and thermally induced compression molding processes. J Membr Sci 409:365–370. doi:10.1016/j.memsci.2012.04.007
101. Han XX, Zhou LX (2011) Optimization of process variables in the synthesis of butyl butyrate using acid ionic liquid as catalyst. Chem Eng J 172(1):459–466. doi:10.1016/j.cej.2011.06.025
102. Belding SR, Rees NV, Aldous L, Hardacre C, Compton RG (2008) Behavior of the heterogeneous electron-transfer rate constants of arenes and substituted anthracenes in room-temperature ionic liquids. J Phys Chem C 112(5):1650–1657. doi:10.1021/jp7103598
103. Law CS, Nodder SD, Mountjoy JJ, Marriner A, Orpin A, Pilditch CA, Franz P, Thompson K (2010) Geological, hydrodynamic and biogeochemical variability of a New Zealand deep-water methane cold seep during an integrated three-year time-series study. Mar Geol 272(1–4): 189–208. doi:10.1016/j.margeo.2009.06.018
104. Mirkin MV, Bard AJ (1992) Simple analysis of quasi-reversible steady-state voltammograms. Anal Chem 64(19):2293–2302. doi:10.1021/Ac00043a020
105. Wang Y, Kakiuchi T, Yasui Y, Mirkin MV (2010) Kinetics of ion transfer at the ionic liquid/water nanointerface. J Am Chem Soc 132(47):16945–16952. doi:10.1021/ja1066948
106. Barnes AS, Rogers EI, Streeter I, Aldous L, Hardacre C, Compton RG (2008) Extraction of electrode kinetic parameters from microdisc voltammetric data measured under transport conditions intermediate between steady-state convergent and transient linear diffusion as typically applies to room temperature ionic liquids. J Phys Chem B 112(25):7560–7565. doi:10.1021/jp711897b
107. Rogers EI, Huang X-J, Dickinson EJF, Hardacre C, Compton RG (2009) Investigating the mechanism and electrode kinetics of the oxygen|superoxide (O2|O2•–) couple in various room-temperature ionic liquids at gold and platinum electrodes in the temperature range 298–318 K. J Phys Chem C 113(41):17811–17823. doi:10.1021/jp9064054
108. Janek RP, Fawcett WR, Ulman A (1997) Impedance spectroscopy of self-assembled monolayers on Au(111): evidence for complex double-layer structure in aqueous NaClO4 at the potential of zero charge. J Phys Chem B 101(42):8550–8558. doi:10.1021/jp971698e
109. Wang ZW, Wu PY (2011) Spectral insights into gelation microdynamics of PNIPAM in an ionic liquid. J Phys Chem B 115(36):10604–10614. doi:10.1021/Jp205650h
110. Pasilis SP, Blumenfeld A (2011) Effect of nitrate, perchlorate, and water on uranyl(VI) speciation in a room-temperature ionic liquid: a spectroscopic investigation. Inorg Chem 50(17):8302–8307. doi:10.1021/Ic2008232
111. Gao Y, Li N, Zheng LQ, Zhao XY, Zhang J, Cao Q, Zhao MW, Li Z, Zhang GY (2007) The effect of water on the microstructure of 1-butyl-3-methylimidazolium tetrafluoroborate/TX-100/benzene ionic liquid microemulsions. Chemistry 13(9):2661–2670. doi:10.1002/chem.200600939

112. Roth C, Appelhagen A, Jobst N, Ludwig R (2012) Microheterogeneities in ionic-liquid-methanol solutions studied by FTIR spectroscopy, DFT calculations and molecular dynamics simulations. ChemPhysChem 13(7):1708–1717. doi:10.1002/cphc.201101022
113. Kortenbruck K, Pohrer B, Schluecker E, Friedel F, Ivanovic-Burmazovic I (2012) Determination of the diffusion coefficient of CO2 in the ionic liquid EMIM NTf2 using online FTIR measurements. J Chem Thermodyn 47:76–80. doi:10.1016/j.jct.2011.09.025
114. Zhang QH, Liu SM, Li ZP, Li J, Chen ZJ, Wang RF, Lu LJ, Deng YQ (2009) Novel cyclic sulfonium-based ionic liquids: synthesis, characterization, and physicochemical properties. Chemistry 15(3):765–778. doi:10.1002/chem.200800610
115. Tao GH, Zou M, Wang XH, Chen ZY, Evans DG, Kou Y (2005) Comparison of polarities of room-temperature ionic liquids using FT-IR spectroscopic probes. Aust J Chem 58(5):327–331. doi:10.1071/Ch05025
116. Shalu, Chaurasia SK, Singh RK, Chandra S (2013) Thermal stability, complexing behavior, and ionic transport of polymeric gel membranes based on polymer PVdF-HFP and ionic liquid, [BMIM][BF4]. J Phys Chem B 117(3):897–906. doi:10.1021/Jp307694q
117. Zhou D, Zhou R, Chen CX, Yee WA, Kong JH, Ding GQ, Lu XH (2013) Non-volatile polymer electrolyte based on poly(propylene carbonate), ionic liquid, and lithium perchlorate for electrochromic devices. J Phys Chem B 117(25):7783–7789. doi:10.1021/Jp4021678
118. Dobbelin M, Pozo-Gonzalo C, Marcilla R, Blanco R, Segura JL, Pomposo JA, Mecerreyes D (2009) Electrochemical synthesis of PEDOT derivatives bearing imidazolium-ionic liquid moieties. J Polym Sci Pol Chem 47(12):3010–3021. doi:10.1002/Pola.23384
119. Pieniazek PA, Stangret J (2005) Hydration of tetraethylammonium tetrafluoroborate derived from FTIR spectroscopy. Vib Spectrosc 39(1):81–87. doi:10.1016/j.vibspec.2004.11.004
120. Shamsuri AA, Daik R (2013) Utilization of an ionic liquid/urea mixture as a physical coupling agent for agarose/talc composite films. Materials 6(2):682–698. doi:10.3390/Ma6020682
121. Xie WL, Wang YB (2011) Synthesis of high fatty acid starch esters with 1-butyl-3-methylimidazolium chloride as a reaction medium. Starch-Starke 63 (4):190-197. doi:10.1002/star.201000126
122. Zadegan S, Hossainalipour M, Ghassai H, Rezaie HR, Naimi-Jamal MR (2010) Synthesis of cellulose-nanohydroxyapatite composite in 1-n-butyl-3-methylimidazolium chloride. Ceram Int 36(8):2375–2381. doi:10.1016/j.ceramint.2010.07.019
123. Du Z, Yu YL, Wang JH (2007) Extraction of proteins from biological fluids by use of an ionic liquid/aqueous two-phase system. Chemistry 13(7):2130–2137. doi:10.1002/chem.200601234
124. Atkin R, Abedin SZE, Hayes R, Gasparotto LHS, Borisenko N, Endres F (2009) AFM and STM studies on the surface interaction of [BMP]TFSA and [EMIm]TFSA ionic liquids with Au(111). J Phys Chem C 113(30):13266–13272. doi:10.1021/jp9026755
125. Hasib-ur-Rahman M, Larachi F (2012) CO2 Capture in alkanolamine-RTIL blends via carbamate crystallization: route to efficient regeneration. Environ Sci Technol 46(20):11443–11450. doi:10.1021/Es302513j
126. Rai R, Pandey S, Baker SN, Vora S, Behera K, Baker GA, Pandey S (2012) Ethanol-Assisted, Few Nanometer, Water-In-Ionic-Liquid Reverse Micelle Formation by a Zwitterionic Surfactant. Chemistry-a European Journal 18 (39):12213-12217. doi:10.1002/chem.201200682
127. Andanson JM, Jutz F, Baiker A (2009) Supercritical CO2/ionic liquid systems: what can we extract from infrared and Raman spectra? J Phys Chem B 113(30):10249–10254. doi:10.1021/Jp904440x
128. Chambreau SD, Schneider S, Rosander M, Hawkins T, Gallegos CJ, Pastewait MF, Vaghjiani GL (2008) Fourier transform infrared studies in hypergolic ignition of ionic liquids. J Phys Chem A 112(34):7816–7824. doi:10.1021/Jp8038175
129. Xue ZM, Zhang ZF, Han J, Chen Y, Mu TC (2011) Carbon dioxide capture by a dual amino ionic liquid with amino-functionalized imidazolium cation and taurine anion. Int J Greenh Gas Con 5(4):628–633. doi:10.1016/j.ijggc.2011.05.014
130. Sakellarlos NI, Kazarian SG (2005) Solute partitioning between an ionic liquid and high-pressure CO2 studied with in situ FTIR spectroscopy. J Chem Thermodyn 37(6):621–626. doi:10.1016/j.jct.2005.03.022

131. Du XZ, Miao W, Liang YQ (2005) IRRAS studies on chain orientation in the monolayers of amino acid amphiphiles at the air-water interface depending on metal complex and hydrogen bond formation with the headgroups. J Phys Chem B 109(15):7428–7434. doi:10.1021/Jp0441700
132. Andanson JM, Jutz F, Baiker A (2010) Purification of ionic liquids by supercritical CO2 monitored by infrared spectroscopy. J Supercrit Fluid 55(1):395–400. doi:10.1016/j.supflu.2010.08.012
133. Du XZ, Liang YQ (2000) Roles of metal complex and hydrogen bond in molecular structures and phase behaviors of metal N-octadecanoyl-L-alaninate Langmuir-Blodgett films. J Phys Chem B 104(43):10047–10052. doi:10.1021/Jp002016h
134. Xing DY, Peng N, Chung TS (2011) Investigation of unique interactions between cellulose acetate and ionic liquid [EMIM]SCN, and their influences on hollow fiber ultrafiltration membranes. J Membr Sci 380(1–2):87–97. doi:10.1016/j.memsci.2011.06.032
135. Dahl K, Sando GM, Fox DM, Sutto TE, Owrutsky JC (2005) Vibrational spectroscopy and dynamics of small anions in ionic liquid solutions. J Chem Phys 123(8), 084504. doi:10.1063/1.2000229
136. Siddiqui KA, Mehrotra GK, Mrozinski J, Butcher RJ (2008) H-bonded porous supramolecular network of a Cu-II complex assisted by assembled 2D sheet of chair form hexameric water cluster. Eur J Inorg Chem 26:4166–4172. doi:10.1002/ejic.200800463
137. Zhang MJ, Zhao PP, Leng Y, Chen GJ, Wang J, Huang J (2012) Schiff base structured acid-base cooperative dual sites in an ionic solid catalyst lead to efficient heterogeneous knoevenagel condensations. Chemistry 18 (40):12773-12782. doi:10.1002/chem.201201338
138. He HY, Chen H, Zheng YZ, Zhang XC, Yao XQ, Yu ZW, Zhang SJ (2013) The hydrogen-bonding interactions between 1-ethyl-3-methylimidazolium lactate ionic liquid and methanol. Aust J Chem 66(1):50–59. doi:10.1071/Ch12308
139. Liu M, Li LS, Da SL, Feng YQ (2004) Preparation of p-tert-butyl-calix[8]arene bonded silica stationary phase and separation for steroid hormone medicines. Anal Lett 37(14):3017–3031. doi:10.1081/Al-200035899
140. Soll S, Zhao Q, Weber J, Yuan JY (2013) Activated CO2 sorption in mesoporous imidazolium-type poly(ionic liquid)-based polyampholytes. Chem Mater 25(15):3003–3010. doi:10.1021/Cm4009128
141. Tadjeddine A, Peremans A, Guyotsionnest P (1995) Vibrational spectroscopy of the electrochemical interface by visible-infrared sum-frequency generation. Surf Sci 335(1–3):210–220. doi:10.1016/0039-6028(95)00419-X
142. Busson B, Tadjeddine A (2010) Chiral specificity of doubly resonant sum-frequency generation in an anisotropic thin film (vol 112C, pg 11813, 2008). J Phys Chem C 114(49):21891. doi:10.1021/Jp110099k
143. Bozzini B, Busson B, Humbert C, Mele C, Raffa P, Tadjeddine A (2011) Investigation of Au electrodeposition from [BMP][TFSA] room-temperature ionic liquid containing K[Au(CN)(2)] by in situ two-dimensional sum frequency generation spectroscopy. J Electroanal Chem 661(1):20–24. doi:10.1016/j.jelechem.2011.07.004
144. Guyotsionnest P, Tadjeddine A (1990) Spectroscopic investigations of adsorbates at the metal electrolyte interface using sum frequency generation. Chem Phys Lett 172(5):341–345. doi:10.1016/S0009-2614(90)87124-A
145. Tadjeddine A, Guyotsionnest P (1991) Spectroscopic investigation of adsorbed cyanide and thiocyanate on platinum using sum frequency generation. Electrochim Acta 36(11–12):1839–1847
146. Tadjeddine A, Guyotsionnest P (1993) Spectroscopic investigation of adsorbed cyanide and thiocyanate ions on platinum by sum-frequency generation. Sov Electrochem 29(1):94–101
147. Rollins JB, Fitchett BD, Conboy JC (2007) Structure and orientation of the imidazolium cation at the room-temperature ionic liquid/SiO2 interface measured by sum-frequency vibrational spectroscopy. J Phys Chem B 111(18):4990–4999. doi:10.1021/Jp0671906
148. Fitchett BA, Conboy JC (2004) Structure of the room-temperature ionic liquid/SiO2 interface studied by sum-frequency vibrational spectroscopy. J Phys Chem B 108(52):20255–20262. doi:10.1021/Jp0471251

149. Conboy JC, Messmer MC, Richmond GL (1997) Dependence of alkyl chain conformation of simple ionic surfactants on head group functionality as studied by vibrational sum-frequency spectroscopy. J Phys Chem B 101(34):6724–6733. doi:10.1021/Jp971867v
150. Baldelli S (2003) Influence of water on the orientation of cations at the surface of a room-temperature ionic liquid: a sum frequency generation vibrational spectroscopic study. J Phys Chem B 107(25):6148–6152. doi:10.1021/jp027753n
151. Roling B, Druschler M, Huber B (2012) Slow and fast capacitive process taking place at the ionic liquid/electrode interface. Faraday Discuss 154:303–311. doi:10.1039/C1fd00088h
152. Baldelli S (2005) Probing electric fields at the ionic liquid-electrode interface using sum frequency generation spectroscopy and electrochemistry. J Phys Chem B 109(27):13049–13051. doi:10.1021/Jp052913r
153. Hillman AR (2011) The EQCM: electrogravimetry with a light touch. J Solid State Electrochem 15(7–8):1647–1660. doi:10.1007/s10008-011-1371-2
154. Xiao CH, Zeng XQ (2013) In situ EQCM evaluation of the reaction between carbon dioxide and electrogenerated superoxide in ionic liquids. J Electrochem Soc 160(10):H749–H756. doi:10.1149/2.073310jes
155. Tang Y, Baker GA, Zeng X (2010) Ionic liquid conditioning of poly(vinylferrocene) for the doping/undoping of glycylglycylglycine tripeptide. J Phys Chem C 114(32):13709–13715. doi:10.1021/jp1030202
156. Levi MD, Sigalov S, Aurbach D, Daikhin L (2013) In situ electrochemical quartz crystal admittance methodology for tracking compositional and mechanical changes in porous carbon electrodes. J Phys Chem C 117(29):14876–14889. doi:10.1021/jp403065y
157. Levi MD, Salitra G, Levy N, Aurbach D, Maier J (2009) Application of a quartz-crystal microbalance to measure ionic fluxes in microporous carbons for energy storage. Nat Mater 8(11):872–875. doi:10.1038/nmat2559
158. Sigalov S, Levi MD, Salitra G, Aurbach D, Maier J (2010) EQCM as a unique tool for determination of ionic fluxes in microporous carbons as a function of surface charge distribution. Electrochem Commun 12(12):1718–1721. doi:10.1016/j.elecom.2010.10.005
159. Oyama T, Yamaguchi S, Rahman MR, Okajima T, Ohsaka T, Oyama N (2010) EQCM study of the [Au(III)Cl4](-)-[Au(I)Cl2](-)-Au(0) redox system in 1-ethyl-3-methylimidazolium tetrafluoroborate room-temperature ionic liquid. Langmuir 26(11):9069–9075. doi:10.1021/la904483y
160. Giessibl FJ (2003) Advances in atomic force microscopy. Rev Mod Phys 75(3):949–983
161. Batchelor-McAuley C, Dickinson EJF, Rees NV, Toghill KE, Compton RG (2011) New electrochemical methods. Anal Chem 84(2):669–684. doi:10.1021/ac2026767
162. Valtiner M, Ankah GN, Bashir A, Renner FU (2011) Atomic force microscope imaging and force measurements at electrified and actively corroding interfaces: challenges and novel cell design. Rev Sci Instrum 82(2):023703. doi:10.1063/1.3541650
163. Guriyanova S, Mairanovsky VG, Bonaccurso E (2011) Superviscosity and electroviscous effects at an electrode/aqueous electrolyte interface: an atomic force microscope study. J Colloid Interface Sci 360(2):800–804. doi:10.1016/j.jcis.2011.04.072
164. Fedorov MV, Kornyshev AA (2008) Ionic liquid near a charged wall: structure and capacitance of electrical double layer. J Phys Chem B 112(38):11868–11872. doi:10.1021/jp803440q
165. Atkin R, Warr GG (2007) Structure in confined room-temperature ionic liquids. J Phys Chem C 111(13):5162–5168. doi:10.1021/jp067420g
166. Mezger M, Schroder H, Reichert H, Schramm S, Okasinski JS, Schoder S, Honkimaki V, Deutsch M, Ocko BM, Ralston J, Rohwerder M, Stratmann M, Dosch H (2008) Molecular layering of fluorinated ionic liquids at a charged sapphire (0001) surface. Science 322(5900):424–428. doi:10.1126/science.1164502
167. Fedorov MV, Georgi N, Kornyshev AA (2010) Double layer in ionic liquids: the nature of the camel shape of capacitance. Electrochem Commun 12(2):296–299. doi:10.1016/j.elecom.2009.12.019
168. Bazant MZ, Storey BD, Kornyshev AA (2011) Double layer in ionic liquids: overscreening versus crowding. Phys Rev Lett 106(4):046102

169. Lim RYH, O'Shea SJ (2004) Discrete solvation layering in confined binary liquids. Langmuir 20(12):4916–4919. doi:10.1021/la036200g
170. Endres F, Hofft O, Borisenko N, Gasparotto LH, Prowald A, Al-Salman R, Carstens T, Atkin R, Bund A, Zein El Abedin S (2010) Do solvation layers of ionic liquids influence electrochemical reactions? Phys Chem Chem Phys 12(8):1724–1732. doi:10.1039/b923527m
171. Atkin R, Borisenko N, Druschler M, el-Abedin SZ, Endres F, Hayes R, Huber B, Roling B (2011) An in situ STM/AFM and impedance spectroscopy study of the extremely pure 1-butyl-1-methylpyrrolidinium tris(pentafluoroethyl)trifluorophosphate/Au(111) interface: potential dependent solvation layers and the herringbone reconstruction. Phys Chem Chem Phys 13(15):6849–6857. doi:10.1039/c0cp02846k
172. Singh PR, Zeng X (2011) Size-dependent intercalation of ions into highly oriented pyrolytic graphite in ionic liquids: an electrochemical atomic force microscopy study. J Phys Chem C 115(35):17429–17439. doi:10.1021/jp203833v
173. K-I U, K-I F (2010) Observation of redox-state-dependent reversible local structural change of ferrocenyl-terminated molecular island by electrochemical frequency modulation AFM. Langmuir 26(11):9104–9110. doi:10.1021/la904797h
174. Rehman A, Zeng X (2012) Ionic liquids as green solvents and electrolytes for robust chemical sensor development. Acc Chem Res 45(10):1667–1677. doi:10.1021/ar200330v
175. Shiddiky MJA, Torriero AAJ (2011) Application of ionic liquids in electrochemical sensing systems. Biosens Bioelectron 26(5):1775–1787. doi:10.1016/j.bios.2010.08.064
176. Rehman A, Hamilton A, Chung A, Baker GA, Wang Z, Zeng XQ (2011) Differential solute gas response in ionic-liquid-based QCM arrays: elucidating design factors responsible for discriminative explosive gas sensing. Anal Chem 83(20):7823–7833. doi:10.1021/ac201583c
177. Zevenbergen MAG, Wouters D, Dam V-AT, Brongersma SH, Crego-Calama M (2011) Electrochemical sensing of ethylene employing a thin ionic-liquid layer. Anal Chem 83(16):6300–6307. doi:10.1021/ac2009756
178. Aguilar AD, Forzani ES, Leright M, Tsow F, Cagan A, Iglesias RA, Nagahara LA, Amlani I, Tsui R, Tao NJ (2010) A hybrid nanosensor for TNT vapor detection. Nano Lett 10(2):380–384. doi:10.1021/nl902382s
179. Huang X-J, Aldous L, O'Mahony AM, Javier del Campo F, Compton RG (2010) Toward membrane-free amperometric gas sensors: a microelectrode array approach. Anal Chem 82(12):5238–5245. doi:10.1021/ac1006359
180. Torriero AAJ, Bond AM (2009) Critical evaluation of dynamic electrochemistry in ionic liquids. In: Electroanalytical chemistry research trend. Nova Science, New York, NY, pp 1–63
181. Turner MB, Spear SK, Holbrey JD, Rogers RD (2004) Production of bioactive cellulose films reconstituted from ionic liquids. Biomacromolecules 5(4):1379–1384. doi:10.1021/bm049748q
182. Liu Y, Wang M, Li J, Li Z, He P, Liu H, Li J (2005) Highly active horseradish peroxidase immobilized in 1-butyl-3-methylimidazolium tetrafluoroborate room-temperature ionic liquid based sol-gel host materials. Chem Commun 13:1778–1780. doi:10.1039/b417680d
183. Bontempelli G, Comisso N, Toniolo R, Schiavon G (1997) Electroanalytical sensors for non-conducting media based on electrodes supported on perfluorinated ion-exchange membranes. Electroanalysis 9(6):433–443
184. Cox JA, Alber KS, Brockway CA, Tess ME, Gorski W (1995) Solid-phase extraction in conjunction with solution or solid-state voltammetry as a strategy for the determination of neutral organic-compounds. Anal Chem 67(5):993–998
185. Barroso-Fernandez B, Lee-Alvarez MT, Seliskar CJ, Heineman WR (1998) Electrochemical behavior of methyl viologen at graphite electrodes modified with Nafion sol-gel composite. Anal Chim Acta 370(2–3):221–230
186. Otsuki S, Adachi K (1995) Characterization of Nafion solutions and films and observation of the casting process using basic-dyes as optical probes. J Appl Polym Sci 56(6):697–705
187. Rehman A, Xiao C, Zeng X (2012) Dynamics of redox process in ionic liquids and their interplay for discriminative electrochemical sensing. Anal Chem 84(3):1416–1424
188. Stetter JR, Korotcenkov G, Zeng X, Tang Y, Liu Y (2011) Electrochemical gas sensors: fundamentals, fabrication, and parameters. In: Korotcenkov G (ed) Chemical sensors comprehensive

sensor technologies, vol 5, Sensors technology series. Momentum Press, LLC, New York, NY, pp 1–89
189. Samec Z, Langmaier J, Kakiuchi T (2009) Charge-transfer processes at the interface between hydrophobic ionic liquid and water. Pure Appl Chem 81(8):1473–1488. doi:10.1351/Pac-Con-08-08-36
190. Yasui Y, Kitazumi Y, Ishimatsu R, Nishi N, Kakiuchi T (2009) Ultraslow response of interfacial tension to the change in the phase-boundary potential at the interface between water and a room-temperature ionic liquid, trioctylmethylammonium bis(nonafluorobutanesulfonyl) amide. J Phys Chem B 113(11):3273–3276. doi:10.1021/jp9006312
191. Morf WE (1981) The principles of ion-selective electrodes and of membrane transport. Elsevier, Amsterdam
192. Peng B, Zhu J, Liu X, Qin Y (2008) Potentiometric response of ion-selective membranes with ionic liquids as ion-exchanger and plasticizer. Sensors Actuators B Chem 133(1):308–314. doi:10.1016/j.snb.2008.02.027
193. Liu H, Liu Y, Li J (2010) Ionic liquids in surface electrochemistry. Phys Chem Chem Phys 12(8):1685–1697
194. Zhu J, Qin Y, Zhang Y (2009) Preparation of all solid-state potentiometric ion sensors with polymer-CNT composites. Electrochem Commun 11(8):1684–1687. doi:10.1016/j.elecom.2009.06.025
195. Nishi N, Murakami H, Yasui Y, Kakiuchi T (2008) Use of highly hydrophobic ionic liquids for ion-selective electrodes of the liquid membrane type. Anal Sci 24(10):1315–1320. doi:10.2116/analsci.24.1315
196. Bakker E, Pretsch E (2007) Modern potentiometry. Angew Chem Int Ed 46(30):5660–5668. doi:10.1002/anie.200605068
197. Langmaier J, Samec Z (2009) Voltammetry of ion transfer across a polarized room-temperature ionic liquid membrane facilitated by valinomycin: theoretical aspects and application. Anal Chem 81(15):6382–6389. doi:10.1021/ac9008258
198. Silvester DS, Arrigan DWM (2011) Array of water vertical bar room temperature ionic liquid micro-interfaces. Electrochem Commun 13(5):477–479. doi:10.1016/j.elecom.2011.02.025
199. Buzzeo MC, Hardacre C, Compton RG (2004) Use of room temperature ionic liquids in gas sensor design. Anal Chem 76(15):4583–4588. doi:10.1021/ac040042w
200. Barrosse-Antle LE, Bond AM, Compton RG, O'Mahony AM, Rogers EI, Silvester DS (2010) Voltammetry in room temperature ionic liquids: comparisons and contrasts with conventional electrochemical solvents. Chemistry 5(2):202–230. doi:10.1002/asia.200900191
201. Huang X-J, Silvester DS, Streeter I, Aldous L, Hardacre C, Compton RG (2008) Electroreduction of chlorine gas at platinum electrodes in several room temperature ionic liquids: evidence of strong adsorption on the electrode surface revealed by unusual voltammetry in which currents decrease with increasing voltage scan rates. J Phys Chem C 112(49):19477–19483. doi:10.1021/jp8082437
202. Wang Z, Guo M, Baker GA, Stetter J, Lin L, Mason AJ, Zeng X (2014) Methane–oxygen electrochemical coupling in an ionic liquid: a robust sensor for simultaneous quantification. Analyst, 139, 5140–5147
203. Ohsaka T, Alam MT, Islam MM, Okajima T (2007) Measurements of differential capacitance in room temperature ionic liquid at mercury, glassy carbon and gold electrode interfaces. Electrochem Commun 9(9):2370–2374. doi:10.1016/j.elecom.2007.07.009
204. Xiao CH, Rehman A, Zeng XQ (2012) Dynamics of redox processes in ionic liquids and their interplay for discriminative electrochemical sensing. Anal Chem 84(3):1416–1424. doi:10.1021/ac2024798
205. Schienle M, Paulus C, Frey A, Hofmann F, Holzapfl B, Schindler-Bauer P, Thewes R (2004) A fully electronic DNA sensor with 128 positions and in-pixel A/D conversion. J Solid-State Circuits 39(12):2438–2445
206. Zhu X, Ahn CH (2006) On-chip electrochemical analysis system using nanoelectrodes and bioelectronic CMOS chip. IEEE Sensors J 6(5):1280–1286

207. Jin X, Huang Y, Mason A, Zeng X (2009) Multichannel monolithic quartz crystal microbalance gas sensor array. Anal Chem 81(2):595–603
208. Li L, Liu X, Qureshi WA, Mason AJ (2011) CMOS amperometric instrumentation and packaging for biosensor array applications. IEEE Trans Biomed Circuits Syst 5(5):439–448
209. Trombly N, Mason A (2008) Post-CMOS electrode formation and isolation for on-chip temperature-controlled electrochemical sensors. IET Electron Lett 44(1):29–30
210. Yang C, Huang Y, Hassler BL, Worden RM, Mason AJ (2009) Amperometric electrochemical microsystem for a miniaturized protein biosensor array. IEEE Trans Biomed Circuits Syst 3(3):160–168
211. Niedziolka J, Rozniecka E, Stafiej J, Sirieix-Plenet J, Gaillon L, di Caprio D, Opallo M (2005) Ion transfer processes at ionic liquid based redox active drop deposited on an electrode surface. Chem Commun 23:2954–2956. doi:10.1039/b502194d
212. Wadhawan JD, Schröder U, Neudeck A, Wilkins SJ, Compton RG, Marken F, Consorti CS, de Souza RF, Dupont J (2000) Ionic liquid modified electrodes. Unusual partitioning and diffusion effects of Fe(CN)64–/3– in droplet and thin layer deposits of 1-methyl-3-(2,6-(S)-dimethylocten-2-yl)-imidazolium tetrafluoroborate. J Electroanal Chem 493(1–2):75–83. doi:10.1016/S0022-0728(00)00308-9
213. Yu P, Lin Y, Xiang L, Su L, Zhang J, Mao L (2005) Molecular films of water-miscible ionic liquids formed on glassy carbon electrodes: characterization and electrochemical applications. Langmuir 21(20):9000–9006. doi:10.1021/la051089v
214. Ng SR, Guo CX, Li CM (2011) Highly sensitive nitric oxide sensing using three-dimensional graphene/ionic liquid nanocomposite. Electroanalysis 23(2):442–448. doi:10.1002/elan.201000344
215. Lu D, Shomali N, Shen A (2010) Task specific ionic liquid for direct electrochemistry of metal oxides. Electrochem Commun 12(9):1214–1217. doi:10.1016/j.elecom.2010.06.022
216. Liang C, Yuan C-Y, Warmack RJ, Barnes CE, Dai S (2002) Ionic liquids: a new class of sensing materials for detection of organic vapors based on the use of a quartz crystal microbalance. Anal Chem 74(9):2172–2176. doi:10.1021/ac011007h
217. Yu L, Garcia D, Rex RB, Zeng XQ (2005) Ionic liquid high temperature gas sensors. Chem Commun 17:2277–2279
218. Goubaidoulline I, Vidrich G, Johannsmann D (2004) Organic vapor sensing with ionic liquids entrapped in alumina nanopores on quartz crystal resonators. Anal Chem 77(2):615–619. doi:10.1021/ac048436a
219. Jin X, Yu L, Zeng X (2008) Enhancing the sensitivity of ionic liquid sensors for methane detection with polyaniline template. Sens Actuators B 133(2):526–532. doi:10.1016/j.snb.2008.03.022
220. Yu L, Jin X, Zeng X (2008) Methane interactions with polyaniline/butylmethylimidazolium camphorsulfonate ionic liquid composite. Langmuir 24(20):11631–11636. doi:10.1021/la8018327
221. Zeng X, Li X, Xing L, Liu X, Luo S, Wei W, Kong B, Li Y (2009) Electrodeposition of chitosan–ionic liquid–glucose oxidase biocomposite onto nano-gold electrode for amperometric glucose sensing. Biosens Bioelectron 24(9):2898–2903. doi:10.1016/j.bios.2009.02.027
222. Xi F, Liu L, Wu Q, Lin X (2008) One-step construction of biosensor based on chitosan–ionic liquid–horseradish peroxidase biocomposite formed by electrodeposition. Biosensors and Bioelectronics 24 (1):29-34. doi:10.1016/j.bios.2008.03.023
223. Gao R, Zheng J (2009) Amine-terminated ionic liquid functionalized carbon nanotube-gold nanoparticles for investigating the direct electron transfer of glucose oxidase. Electrochem Commun 11(3):608–611. doi:10.1016/j.elecom.2008.12.060
224. Liu Y, Shi L, Wang M, Li Z, Liu H, Li J (2005) A novel room temperature ionic liquid sol-gel matrix for amperometric biosensor application. Green Chem 7(9):655–658. doi:10.1039/b504689k
225. Wu X, Zhao B, Wu P, Zhang H, Cai C (2009) Effects of ionic liquids on enzymatic catalysis of the glucose oxidase toward the oxidation of glucose. J Phys Chem B 113(40):13365–13373. doi:10.1021/jp905632k

226. Jin X, Yu L, Garcia D, Ren RX, Zeng X (2006) Ionic liquid high-temperature gas sensor array. Anal Chem 78(19):6980–6989. doi:10.1021/ac0608669
227. Oter O, Ertekin K, Topkaya D, Alp S (2006) Emission-based optical carbon dioxide sensing with HPTS in green chemistry reagents: room-temperature ionic liquids. Anal Bioanal Chem 386(5):1225–1234. doi:10.1007/s00216-006-0659-z
228. Oter O, Ertekin K, Derinkuyu S (2008) Ratiometric sensing of CO2 in ionic liquid modified ethyl cellulose matrix. Talanta 76(3):557–563. doi:10.1016/j.talanta.2008.03.047
229. Kan T, Aoki H, Binh-Khiem N, Matsumoto K, Shimoyama I (2013) Ratiometric optical temperature sensor using two fluorescent dyes dissolved in an ionic liquid encapsulated by parylene film. Sensors 13(4):4138–4145
230. Forzani ES, Lu D, Leright MJ, Aguilar AD, Tsow F, Iglesias RA, Zhang Q, Lu J, Li J, Tao N (2009) A hybrid electrochemical – colorimetric sensing platform for detection of explosives. J Am Chem Soc 131(4):1390–1391. doi:10.1021/ja809104h
231. Guo CX, Lu ZS, Lei Y, Li CM (2010) Ionic liquid-graphene composite for ultratrace explosive trinitrotoluene detection. Electrochem Commun 12(9):1237–1240. doi:10.1016/j.elecom.2010.06.028

Chapter 3
Reference Systems for Voltammetric Measurements in Ionic Liquids

Angel A.J. Torriero

3.1 Introduction

One of the open problems of electrochemistry in ionic liquids (ILs) nowadays is the choice of a reference electrode. In order to carry out reliable electrochemical measurements in ILs, a robust and stable reference electrode is required. Unfortunately, many literature reports describe the use of poorly chosen, unreliable reference electrodes or the use of quasi-reference electrodes of unknown potential versus any standard potential scale, so they could not be reproduced, being therefore, of little or no value [1]. In addition, the potential differences of reliable reference electrodes in ILs have rarely been measured, preventing the comparison of potential data measured by different research groups in the same IL. A similar situation was reported by IUPAC almost 30 years ago for nonaqueous systems [2]. As a consequence, it is necessary to agree for ILs systems on some procedures and limitations to measure and report electrode potentials versus generally accepted standards. There are in general two possible ways: (a) to use a reference electrode or (b) to refer all potentials versus internal reference redox systems.

3.2 Reference Electrodes

The use of quasi-reference (or pseudo-reference) electrodes, QRE, is a common practice in electrochemical studies in nonaqueous solvents, where, as is the case with ionic liquid investigations, well-defined reference electrodes are difficult to

A.A.J. Torriero (✉)
Centre for Chemistry and Biotechnology, School of Life and Environmental Science, Faculty of Science Engineering and Built Environment, Deakin University, Burwood, VIC 3125, Australia
e-mail: angel.torriero@deakin.edu.au

A.A.J. Torriero (ed.), *Electrochemistry in Ionic Liquids*,
DOI 10.1007/978-3-319-13485-7_3

prepare and stabilize. In most ionic liquids, volatility is not an issue compared to the situation prevailing in some organic solvents in which evaporation can vary the composition of the reference electrode half cell. However, the small volume of IL typically used, in general several tenth or hundred of microliters, in order to perform simple and cost-efficient research, justifies the use of this kind of reference system. The majority of reports that employ QRE in ILs are based on silver [3–12] or platinum wire [13–16] immersed directly into the working solution. Lithium, zinc, and magnesium [17, 18] are less commonly used.

As discussed by Snook and Bond, the function of silver or platinum wire as a QRE is based on the presence of various compounds (most likely oxides) on the metal surface, where the exact identity of those "redox couples" is never known with any confidence [1]. In addition, the amount of the oxidized component(s), which is very small when compared with properly constructed reference electrodes, is also unknown [19]. Therefore, the potential of the QRE is likely to change greatly in the event of (i) reaction with the IL components or impurities present in the media; (ii) dissolution of the oxidized compounds (present onto the metal surface) in the IL; (iii) polarization, due to a lack of potentiostatic control, or (iv) its recent history [20]. These changes can occur dynamically throughout the time period when the wire is immersed in the IL system during, or even as a consequence of, the electrochemical experiment. Therefore, the assumption that the potential(s) of the electrochemical processes of interest recorded against a QRE can be calibrated, at some stage (or before and after the experiment), against a suitable reference electrode or a standard redox compound, cannot always be relied upon to provide an accurate potential value over the entire range of potentials measured.

To illustrate this, the electrochemistry of cobaltocenium, Cc^+, and ferrocene, Fc, in the protic ionic liquid triethylammonium acetate, $[HN_{222}][CH_3CO_2]$, was studied on a glassy carbon electrode (GC, Fig. 3.1) under low water content (<100 ppm) by using an argon-filled glove box. The QRE used in this setup was a silver wire, which together with the platinum wire used as a counter electrode was immersed directly into the working solution. For cobaltocenium (Fig. 3.1a), the mid-point potential, E_m {where E_m was calculated from the average of the oxidation (E_{peak}^{anodic}) and reduction ($E_{peak}^{cathodic}$) peak potentials $\left[E_m = \left(E_{peak}^{anodic} + E_{peak}^{cathodic}\right)/2\right]$ } is −0.763 V vs. Ag QRE at 0.1 V s^{-1}, which systematically shifts to more positive potentials during consecutive scans with increased scan rates, up to a value of −0.623 V vs. Ag QRE at 1.0 V s^{-1}. Immediately after this experiment, a solution of 1.9 mM Fc in $[HN_{222}]$ $[CH_3CO_2]$ was evaluated under identical experimental conditions, and the result is shown in Fig. 3.1b. Here, although there was no systematic shift in the E_m, a 0.007 V change was observed when the scan rate was systematically increased from 0.1 to 1.0 V s^{-1} (0.463 V at 0.1 V s^{-1}, and 0.470 V at 0.1 V s^{-1}).

The electrochemistry of Fc and Cc^+ was also studied in 1-butyl-3-methylimidazolium hexafluorophosphate, $[C_4mim][PF_6]$, 1-butyl-3-methylimidazolium trifluoromethanesulfonate, $[C_4mim][CF_3SO_3]$, and 1-ethyl-3-methylimidazolium tetracyanoborate, $[C_2mim][B(CN)_4]$, using an argon-filled glove box [20]. The QRE used in this experiment was a platinum wire, which together with a second platinum wire used as a

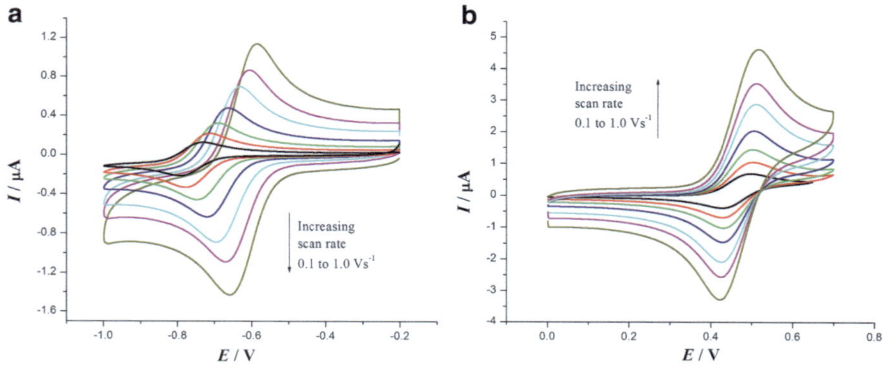

Fig. 3.1 Cyclic voltammograms for (**a**) the reduction of 2.1 mM $CcPF_6$ and (**b**) oxidation of 1.9 mM Fc in $[HN_{222}][CH_3CO_2]$ on a GC electrode (diameter = 1 mm) at scan rates of 0.1, 0.2, 0.3, 0.4, 0.5, 0.7, 1.0 V s^{-1} and using Ag wire as a QRE. $T = 21 \pm 1$ °C

counter electrode was immersed directly into the working solution. For cobaltocenium (Fig. 3.2a), the E_m measured after the Pt wire was immersed for 7 min in $[C_4mim][PF_6]$ was −1.082 V at 0.1 V s^{-1} (curve *a*). This value systematically shifted to more positive potentials with time, up to a value of –0.829 V after 45 min (curve *c*). This indicates a drift in the QRE potential of around 0.250 V during this timeframe. Figure 3.2b shows voltammograms of 10.3 mM Cc^+ in $[C_4mim][CF_3SO_3]$, where a drift in the E_m of 0.034 V between two consecutive potential scans is observed. Similar observations were previously reported by Huber and Roling, where a drift of 0.330 V in the Pt wire potential was reported after the QRE was immersed for 10 min in 1-ethyl-3-methylimidazolium bis(trifluoromethylsulfonyl)imide, $[C_2mim][N(Tf)_2]$ [21].

For reversible processes, such as Fc and Cc^+ couples, the separation between the potentials of the forward and reverse peaks (called peak-to-peak potential separation), ΔE_p, at 25 °C, is expected to be 0.059 V. A departure of 20–30 mV from the theoretical value (especially at relatively high scan rates) does not compromise the criterion of *reversibility* in ILs, as both the slower heterogeneous electron transfer rate constant observed for redox compounds in ILs with respect to organic solvents [22] and the eventual presence of solution resistances not compensated by the electrochemical instrumentation tend to simultaneously "spread" the forward and reverse peaks of Fc and Cc^+, thus increasing the relative ΔE_p value. It is worth noting that this is strictly true only if a stable reference system is being used. Figure 3.2c shows voltammograms obtained under the same experimental conditions (including a constant distance between the QRE and the working electrode) in order to avoid changes in the solution resistance. The peak-to-peak potential separation changes from 0.078 V (*a*) to 0.066 (*b*) in two consecutive potential scans. As observed in this figure, the ferrocenium reduction peak (Ic) drifts from 0.152 (curve *a*) to 0.164 V (curve *b*) vs. Pt QRE, while the oxidation peak (Ia) is unaltered. Therefore, the around 0.006 V error in the mid-point potential determination in Fig. 3.2c, curve *a*, is directly related to a drift in the Pt QRE potential during the potential scan instead

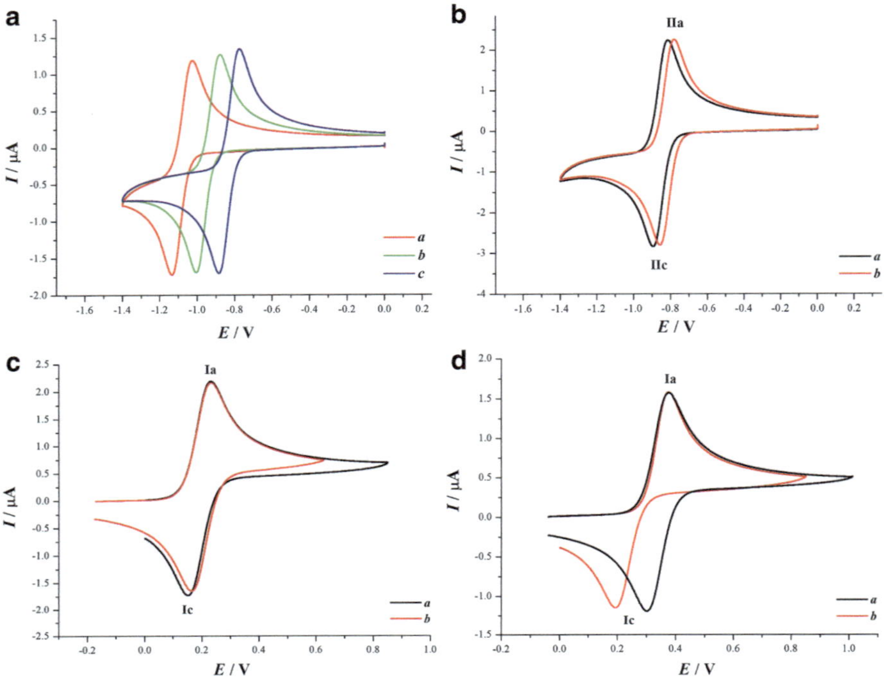

Fig. 3.2 Cyclic voltammograms of 10.3 mM Cc^+ (**a** and **b**) and 10.3 mM Fc (**c** and **d**) on a glassy carbon electrode (diameter 1 mm) at scan rate of 0.1 (**a**, **b**, and **d**) and 0.05 V s^{-1} (**c**) and using Pt wire as a QRE. (**a**) voltammograms after cleaning the Pt QRE and immersion for (*a*) 7 min, (*b*) 10 min, and (*c*) 45 min in $[C_4mim][PF_6]$. (**b–d**) Two consecutive voltammetric scans after Pt QRE cleaning and its immersion (**b**) in $[C_4mim][CF_3SO_3]$, and (**c**, **d**) in $[C_4mim][PF_6]$. Voltammograms were recorded in a homemade argon glove box ($H_2O < 100$ ppm). Reproduced from [20] with permission from Elsevier

of modification in the uncompensated resistance magnitude. Figure 3.2d shows an extreme experimental situation observed in the same solution when the potential sweep rate was 0.05 V s^{-1}. Voltammogram *a* in this figure shows a ΔE_p of 0.074 V, however, an important drift in the Pt QRE potential was produced during the consecutive potential sweep scan (curve *b*) affecting the potential of peak Ic, which introduced an error of 0.055 V in the calculated E_m value.

The most common drift in the QRE potential is shown in Fig. 3.2b, where both cathodic (IIc) and anodic peak (IIa) potentials drift similarly in two consecutive scans, leading to the ΔE_p value of around 0.077 V in both voltammograms. Examples of Pt QRE potential drift could also be observed in Fig. 3.2a, where a $\Delta E_p = 0.104$, 0.127, and 0.112 V is observed in voltammograms *a*, *b*, and *c*, respectively [20].

This series of experiments clearly illustrates the problems associated with the use of QREs, where a variety of influences, including the nature of the IL cation and anion as well as time-dependent changes (most likely related to changes at the surface of the QRE), lead to a variety of different outcomes in terms of the absolute position of the QRE on the potential scale.

Figure 3.3a illustrates cyclic voltammograms obtained at a GC electrode for ferrocene (peaks Ia and Ic) and cobaltocenium (peaks IIc and IIa) redox processes in $[C_2mim][B(CN)_4]$ using a Ag/AgCl, $[C_2mim][B(CN)_4]$ reference system separated from the test solution via a glass frit. Using this setup, the potential of the reference system was stable enough to confirm that the distance between the E_m of Fc and Cc^+ (ΔE_m) in this IL is 1.325 ± 0.002 V and independent of the direction of the scan potential [20]. For comparison, voltammograms obtained under similar conditions in $[C_2mim][B(CN)_4]$ using a platinum wire QRE showed a ΔE_m of 1.293 V and 1.263 V (Fig. 3.3b, curves *a* and *b*, respectively). This indicates an error in the potential measurement of 0.032 V and 0.062 V, respectively. To easily visualize the Pt wire QRE drifts during (and not just between) the voltammetric experiments, all voltammograms in Fig. 3.3 were calibrated against the cobaltocenium/cobaltocene, $Cc^{+/0}$, potential scale. Figures 3.3c and 3.3d display different error magnitudes in

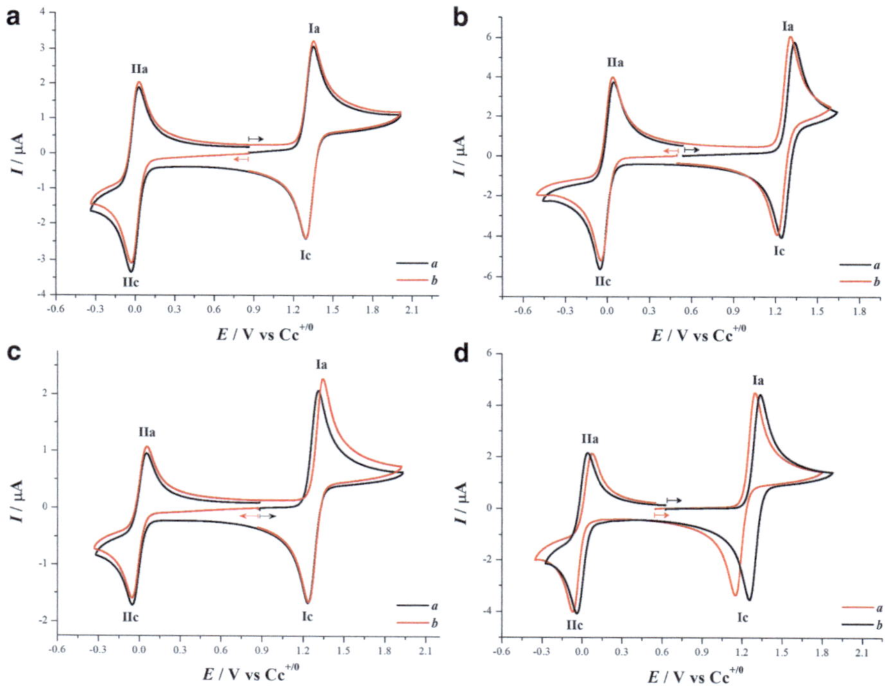

Fig. 3.3 Two consecutive cyclic voltammograms obtained for the oxidation of Fc and reduction of Cc^+ at a GC electrode (diameter = 1 mm) at 25 ± 1 °C. (**a**) 5.70 mM Cc^+ and 5.55 mM Fc in $[C_2mim][B(CN)_4]$ using a Ag|AgCl, $[C_2mim][B(CN)_4]$ reference electrode; (**b**) 10.1 mM Cc^+ and 10.3 mM Fc in $[C_2mim][B(CN)_4]$ using Pt wire as a QRE; (**c**) 10.3 mM Cc^+ and 10.3 mM Fc $[C_4mim][PF_6]$ using Pt wire as a QRE; and (**d**) 10.3 mM Cc^+ and 10.5 mM Fc in $[C_4mim][N(Tf)_2]$ using Pt wire as a QRE. Voltammograms were recorded in a homemade argon glove box (H_2O < 100 ppm) at a scan rate = 0.1 V s^{-1}. Reproduced from [20] with permission from Elsevier

the ΔE_m determination in $[C_4mim][PF_6]$ and 1-butyl-3-methylimidazolium bis(trifluoro-methylsulfonyl)imide, $[C_4mim][N(Tf)_2]$, respectively [20].

Silvester et al. also reported results related with the stability of silver and platinum QREs [23]. The QRE was inserted into a glass tube and separated from the test solution via a vycor tip. The reference electrode was, prior to recording the respective voltammogram, preoxidized for different times by holding the potential at a predetermined value. From this experiment, platinum QRE showed a shift of around 90 mV when the preoxidizing time was increased from 5 to 40 min. Meanwhile, the shift in potential from experiment to experiment in the silver wire QRE was minimized by introducing the wire into a compartment filled with the same IL present in the working solution and separated from this solution by a glass frit [6, 19, 23]. However, even when the behavior of this system may appear to be stable, comparison of day-to-day potential stability indicates a relatively large variability [19].

Activated carbon (AC) has been proposed recently as an alternative QRE for use in ionic liquids [24]. This electrode was tested in 1-ethyl-3-methylimidazolium tetrafluoroborate, $[C_2mim][BF_4]$, 1-ethyl-3-methylimidazolium bis(trifluoromethylsulfonyl)imide, $[C_2mim][N(Tf)_2]$, 1-ethyl-3-methylimidazolium trifluorotris(pentafluoroethyl)phosphate, $[C_2mim][FAP]$, and 1-methyl-1-propylpyrrolidinium bis(trifluoromethylsulfonyl)imide, $[C_3mPyr][N(Tf)_2]$, using ferrocene (50 mM) as the redox probe. Although the AC QRE is stable for more than 40 h and it is not affected by impurities, the electrode showed a potential of 0.190 ± 0.010 V vs. $Fc^{0/+}$ independently of the IL used. This last point is a real concern, because recent studies have proven that $Fc^{0/+}$ redox process is sensible to the solvation effect of ILs, and a difference in the E_m of Fc of approximately 0.092 V was observed when the IL was changed from $[C_2mim][FAP]$ to 1-butyl-3-methylimidazolium tetrafluoroborate, $[C_4mim][BF_4]$ (see below) [25]. Consequently, a variation on the potential of $Fc^{0/+}$ vs. AC QRE is expected by changing the nature of the IL. It is worth noting that the last-mentioned ILs are similar to those used to validate the AC QRE. Consequently, more studies are required to prove the applicability of this new QRE to IL solvent systems.

A highly stable reference system in IL was described by Snook et al., which was formed by immersing a silver wire in a fixed concentration of silver trifluoromethanesulfonate ($Ag[CF_3SO_3]$) dissolved in $[C_4mPyr][N(Tf)_2]$ and separated by a porous junction from the test solution [19]. This reference electrode, when stored in an argon-filled glove-box away from light, was found to be stable for 3 weeks, but it drifts significantly at longer times. The reproducibility of this electrode was studied by repeated preparations of the $Ag|Ag^+$(10 mM $Ag[CF_3SO_3]$, $[C_4mPyr][N(Tf)_2]$) electrode, which gave a variation in potential of ±2 mV. Once prepared, the reproducibility intra- and interday was of ±0.001 V, for up to about 3 weeks [19].

Even when this electrode was proven to be useful in different ILs, in situations where a large number of experiments are to be performed in the same IL, it is recommended to prepare the $Ag|Ag^+$ reference electrode using the IL under study as the solvent in the reference system and evaluate its potential against a standard redox compound.

Table 3.1 shows that the potential of an Ag|Ag^+ system is highly dependent on the ionic liquid used in the reference electrode compartment and varies by more than 0.600 V vs. $Fc^{0/+}$. In order to overcome this issue, Lewandowski et al. have proposed the use of cryptand reference electrodes, which are prepared by immersing a silver wire in a solution of 0.01 M silver perchlorate and 0.1 M cryptand 222 ligand (4,7,13,16,21,24-hexaoxa-1,10-diazobicyclo[8,8,8]hexacosane) in an IL or with the use of an organic solvent ([Ag|Ag^+ 222, R] where R represents the IL or the organic solvent used) [26, 30, 31]. The E_m of $Fc^{0/+}$ vs. [Ag|Ag^+ 222, CH_3CN] varies by around 0.110 V in different ILs, and between 0.240 and 1.000 V when compared with a Ag|Ag^+(10 mM Ag[CF_3SO_3], IL) system in the same IL (Table 3.1).

Table 3.1 Mid-point potentials of the ferrocene/ferrocenium redox process against different Ag-based reference electrodes

Reference electrolyte	E_m (V)	Reference electrode	Ref
[C_4mPyr][N(Tf)$_2$]	−0.39	Ag|10 mM Ag[CF_3SO_3]	[19]
[C_4mPyr][N(Tf)$_2$]	−0.45	Ag|100 mM Ag[CF_3SO_3]	[19]
	0.55	Ag|Ag^+ 222, CH_3CN[a]	[26]
[C_4mim][PF_6]	0.04	Ag|10 mM Ag[CF_3SO_3]	[20]
	0.40	Ag|AgCl, [Bu_4N]Cl(0.1 M), CH_3CN[b]	[27]
[C_4mim][BF_4]	−0.58	Ag|10 mM Ag[CF_3SO_3]	[20]
	0.46	Ag|AgCl, [Bu_4N]Cl(0.1 M), CH_3CN[b]	[27]
[C_2mim][BF_4]	−0.51	Ag|Ag[BF_4]$_{(sat)}$	[28]
[C_2mim][CF_3SO_3]	−0.25	Ag|10 mM Ag[CF_3SO_3]	[20]
[C_2mim][CH_3SO_3]	0.26	Ag|10 mM Ag[CF_3SO_3]	[20]
	0.50	Ag|Ag^+ 222, CH_3CN[a]	[26]
[C_2mim][N(Tf)$_2$]	−0.37	Ag|10 mM Ag[CF_3SO_3]	[20]
	0.20	Ag|10 mM Ag[N(Tf)$_2$]	[21]
	0.52	Ag|Ag^+ 222, CH_3CN[a]	[26]
[C_4mim][N(Tf)$_2$]	0.46	Ag|Ag^+ Kryptofix22, CH_3CN[c]	[29]
	0.45	Ag|AgCl, [Bu_4N]Cl(0.1 M), CH_3CN[b]	[27]
[$N_{1\,1\,1\,6}$][N(Tf)$_2$]	−0.49	[d]	[28]
[C_4m_βpy][BF_4]	0.57	Ag|Ag^+ 222, CH_3CN[a]	[26]
[$S_{2\,2\,1}$][N(Tf)$_2$]	0.52	Ag|Ag^+ 222, CH_3CN[a]	[26]
[$S_{2\,2\,2}$][N(Tf)$_2$]	0.53	Ag|Ag^+ 222, CH_3CN[a]	[26]
[C_3mpip][N(Tf)$_2$]	0.55	Ag|Ag^+ 222, CH_3CN[a]	[26]
[C_8py][BF_4]	0.56	Ag|Ag^+ 222, CH_3CN[a]	[26]
[C_1HPyr][BF_4]	0.48	Ag|Ag^+ 222, CH_3CN[a]	[26]
[Hmim][BF_4]	0.44	Ag|Ag^+ 222, CH_3CN[a]	[26]
[C_2mim][$EtSO_4$]	0.39	Ag|Ag^+ Kryptofix22, CH_3CN[c]	[29]

[a] Ag|$AgClO_4$(0.01 M) 222(0.1 M), acetonitrile

[b] 50 μL of a 1 M [Bu_4N]Cl acetonitrile solution was mixed with 450 μL of the respective IL to give a final concentration of 0.1 M [Bu_4N]Cl in the reference compartment

[c] Ag|$AgNO_3$(0.01 M) Kryptofix®22(0.1 M) (Kryptofix®22 = 1,7,10,16-tetraoxa-4,13-diazacyclooctadecane)

[d] Ag|Ag[N(Tf)$_2$] reference electrode, where the Ag[N(Tf)$_2$] concentration was not reported

Consequently, different factors such as (i) silver ion coordination or solvation by the IL anions and/or cryptand 222 ligand, (ii) junction potential, and (iii) effect of IL nature on the E_m of Fc (discussed below) could simultaneously be operative and responsible for the observed potential variation [20]. The stable lifetime of the [Ag|Ag^+ 222, CH_3CN] reference electrode is about 4 weeks and, thereby, similar to that reported by Snook et al. [19], even when the silver(I) stability is increased by the formation of the Ag^+-222 inclusion complex [30].

Rogers et al. have studied the kinetics of the Ag|Ag^+ system in [C_4mPyr][$N(Tf)_2$] and found that the interconversion of Ag and Ag^+ is slow, with a charge transfer rate constant ($k^{0\prime}$) of 2.0 cm × 10^{-4} s^{-1} when using the Ag[CF_3SO_3] salt [32]. It is even slower when $[N(Tf)_2]^-$ is used as the counter ion (1.5×10^{-4} cm s^{-1}) and yet again slower with nitrate as the counter ion (1.0×10^{-5} cm s^{-1}). This means fast measurements such as a fast scan rate cyclic voltammetry if a two-electrode system is utilized and high-frequency electrochemical impedance spectroscopy (EIS) if a three-electrode system is utilized could be limited [32]. Even in aqueous systems using a Ag|AgCl reference electrode in a three-electrode setup, the use of EIS is limited to frequencies below 10 kHz. Above this range, artifacts (e.g., inductive loops) due to the sluggish kinetics of the reference half-cell reaction will begin to dominate. To minimize this kinetic limitation with the reference electrode, Rogers et al. [32] have introduced the TMPD|$TMPD^{+\bullet}$ couple (TMPD = N,N,N′,N′-tetramethyl-*p*-phenylenediamine) as a reference electrode of the third kind (Type III or reference electrodes of the third kind involve immersing an inert metal (e.g., platinum) in a solution containing both halves of a redox couple). In this case, $k^{0\prime}$ is an order of magnitude higher ($2.6–2.8 \times 10^{-3}$ cm s^{-1}) but does depend on the counter ion used. However, a true TMPD|$TMPD^{+\bullet}$ reference electrode was not actually constructed. Rather, the voltammetry of a solution of TMPD and of $TMPD^{+\bullet}$ was measured using a quasi-reference electrode in the form of a silver wire, in order to carry out charge transfer rate constant calculations. The authors recognize that the availability, solubility, and stability of the TMPD radical cation salt may limit the use of TMPD|$TMPD^{+\bullet}$ in the construction of a reference electrode of the third kind [32].

In a work by Andriyko et al., an AgCl-coated Ag wire (or Ag|AgCl reference electrode) was immersed in [C_4dmim][BF_4] containing a saturated solution of AgCl [33]. The reference electrode was contained in a Vycor fritted separate compartment for studies in $[C_4dmim]^+$-based ionic liquids. A similar reference electrode configuration was employed by Ikeda and colleagues [34]. However, even when this Ag|AgCl system could, in principle, be used to construct reference electrodes in many ionic liquids, the AgCl coating is prone to being stripped off from the Ag wire, as it is often partly soluble or sparingly soluble in many ionic liquids [35], thereby introducing instability in the reference electrode system, which in addition acts as a source of impurities in the general electrochemical setup.

Saheb et al. described an Ag|Ag^+ reference electrode (for use in an imidazolium-based ionic liquid) in which the concentration of silver ion was established by addition of silver(I) nitrate in acetonitrile [27]. A problem noted with the use of this system is that the potential of the reference half cell varies significantly with the volume fraction of acetonitrile, which is likely to vary after extended periods of use [27]. Lordermeyer et al. also used an acetonitrile-based reference electrode to study the electroplating of dysprosium from an ionic liquid [36]. Other studies have described the use of the saturated calomel electrode (aqueous based) in electrochemical investigations in ionic liquids [37–40]. However, a problem with these and related approaches is that large junction potentials may exist between the two very different media (molecular solvent in the reference compartment and IL in the working solution compartment) and a contamination of the IL with the molecular solvent part of the reference system is unavoidable [20].

3.3 Internal Reference Redox Scales

Internal reference redox scales or internal reference redox systems, IRRS, are reversible or nearly reversible redox systems used to provide a known and stable reference point in nonaqueous solvents in conditions where reliable reference electrodes are difficult to be established and/or stabilized [20]. In general, IRRS is used in conjunction with a QRE [18, 41–44].

The establishment of a solvent-independent reference electrode for the comparison of redox potentials in nonaqueous systems has a long story. It began with the concept of $Rb|Rb^+$ or $Rb(Hg)|Rb^+$, followed by the proposal of using organometallic redox couples [2]. In order to limit the number of redox systems used and then make the comparison easier, IUPAC recommended that the systems ferrocene|ferrocenium, $Fc^{0/+}$, and bis(biphenyl)chromium(0)|bis(biphenyl)chromium(I), $BCr^{0/+}$, need to be used as internal reference redox systems in nonaqueous media [2]. These two complexes were selected arbitrarily from several published redox systems [2].

An effective IRRS must possess a number of properties, many of which were suggested by Gritzner and Kuta in their IUPAC recommendation, and most of which can be deduced by common sense [2]. We recently suggested the following to be essential properties of an IRRS [45]:

1. The OX_{IRRS}/RED_{IRRS} redox couple ($OX_{IRRS} + e \leftrightarrows RED_{IRRS}$; being OX_{IRRS} and RED_{IRRS} the oxidized and reduced form of IRRS, respectively) must be reversible or nearly reversible under the operative measurement conditions. It is advantageous for the IRRS electron transfer to be a simple one-electron outer-sphere electron transfer.
2. It is very important that the IRRS redox moieties do not specifically interact with the working electrode (e.g., they do not adsorb on the working electrode).
3. The mid-point potential for IRRS should not overlap the active potential zone of the analyte (ANS) under study.
4. The initially present IRRS redox component, OX_{IRRS} or RED_{IRRS}, must be stable over the time period of experimentation; it should not react with the solvent, supporting electrolyte or ANS.
5. IRRS redox moiety, OX_{IRRS} or RED_{IRRS}, which is not initially present, must be stable on the timescale required to execute the slowest cyclic voltammetry (CV).

6. The molecule or salt that is the source of the selected IRRS component should be pure, easily synthesized, or otherwise, readily available and should have a long shelf life.

There are also some constraints and requirements for ANS [45]:

1. ANS is initially present as either an oxidizable moiety (RED_{ANS}) or reducible moiety (OX_{ANS}). It is worth noting that RED_{ANS} and OX_{ANS} species may or may not be redox partners, and the oxidation of RED_{ANS} and reduction of OX_{ANS} may be complicated.
2. The initially present ANS moiety should be stable and should not react with IRRS.

It is necessary to notice that IUPAC recommendation has been proposed for organic solvents and their mixtures. By extrapolation from those systems, the $Fc^{0/+}$ redox couple is also being used as an IRRS in ionic liquids [18, 26, 46, 47].

Nowadays, it is well known that Fc is not an "ideal" redox process as mentioned in the IUPAC recommendation, as its formal potential is highly dependent on the solvation effects of the organic solvent and supporting electrolyte used [48, 49]. There are two types of solvation around the Fc molecule, which includes the electrostatic interaction of solvent/electrolyte with (i) the iron center and (ii) the cyclopentadienyl ring [48]. Yang et al. has recently computed the molecular electrostatic potential (MEP) for ferrocene and ferrocenium ion (Fig. 3.4) [50]. As seen in

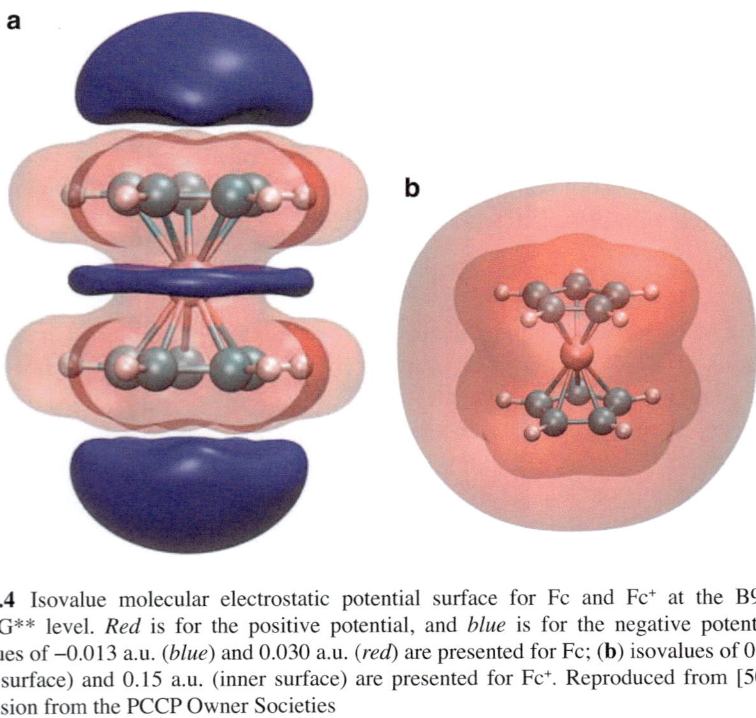

Fig. 3.4 Isovalue molecular electrostatic potential surface for Fc and Fc^+ at the B97-D/6-311++G** level. *Red* is for the positive potential, and *blue* is for the negative potential. (**a**) Isovalues of –0.013 a.u. (*blue*) and 0.030 a.u. (*red*) are presented for Fc; (**b**) isovalues of 0.10 a.u. (outer surface) and 0.15 a.u. (inner surface) are presented for Fc^+. Reproduced from [50] with permission from the PCCP Owner Societies

Fig. 3.4a, negative electrostatic potentials are observed on the top of the Fc cyclopentadienyl rings and as a planetary distribution on a plane going through the central iron atom, which favor the electrostatic interaction between Fc and positively charged molecules and ions (including electron–acceptor moieties). Meanwhile, an interaction with electron–donor moieties from the solvent and/or electrolyte is also possible from the side of the cyclopentadienyl rings, due to the presence of a positive electrostatic potential. A different situation can be observed for the ferrocenium ion MEP (Fig. 3.4b), where only positive electrostatic potentials are observed. In this last situation, the electrostatic interaction between Fc^+ and either negatively charged ions or molecules with electron–donor moieties is favored. The MEP difference between Fc and Fc^+ is very important to understand the dissimilarities in the mass transport properties of Fc and Fc^+ observed experimentally in different ionic liquids (see below).

Decamethylferrocene (DmFc or $[Fe^{II}(\eta^5\text{-}C_5(CH_3)_5)_2]$) shows at least an order of magnitude weaker solvent–solute interaction (in organic solvent systems) compared to ferrocene, as the methyl-substituent groups in the cyclopentadienyl rings prevent both specific (previously mentioned for Fc) and nonspecific interactions by hindering the access of organic solvent and supporting electrolyte molecules to the metal center and to the cyclopentadienyl ring [48, 51–54]. This observation was also confirmed after analysis of the X-ray diffraction patterns of decamethylferrocene, which indicate that the inter-ring methyl groups of DmFc are within Van der Waals distances [55]. Consequently, DmFc is better suited than Fc as an IRRS in organic solvents.

Decamethylferrocene is oxidized in organic solvents in a reversible, one-electron transfer process to decamethylferrocenium ($DmFc^+$ or $[Fe^{III}(\eta^5\text{-}C_5(CH_3)_5)_2]^+$) and reduced back to DmFc according to Eq. (3.1). The E_m value of this redox couple is more negative than that of Fc as a consequence of the electron-donating effect of the methyl groups (inductive effect) which, by pushing electron density toward the metal ion, facilitates electron removal by the electrode. As given in Table 3.2, the difference in the E_m values between $Fc^{0/+}$ and $DmFc^{0/+}$ varies from a low of 0.413 ± 0.005 V in tetrahydrofuran/0.1 M $[Bu_4N][BF_4]$ to a high of 0.614 ± 0.005 V in dichloromethane/0.1 M $[Bu_4N]$[TFAB] ([TFAB] = $[B(C_6F_5)_4]^-$), which represents a difference of 0.201 V. Furthermore, a variation of about 0.152 V can be observed by changing the organic solvent form tetrahydrofuran to 2,2,2-trifluoroethanol and keeping constant the supporting electrolyte nature and concentration (0.1 M $[Bu_4N][ClO_4]$). To minimize uncertainties in the comparison of mid-point potentials in Table 3.2, all potentials quoted are relative to the $DmFc^{0/+}$ potential scale. Consequently, since the values presented in this table represent the difference in E_m, the variations in the absolute values of E_m of the $Fc^{0/+}$ and $DmFc^{0/+}$ IRRS may probably be larger than those indicated here.

$$\left[Fe^{II}\left(\eta^5 - C_5\left(CH_3\right)_5\right)_2\right] \leftrightarrows \left[Fe^{III}\left(\eta^5 - C_5\left(CH_3\right)_5\right)_2\right]^+ + e^- \tag{3.1}$$

The solvent–Fc interaction is also present in ILs. This issue was addressed by studying the effect of ionic liquid structure on the mid-point potentials of Fc and DmFc in 11 different ILs as well as in dichloromethane with added IL as the

Table 3.2 Redox potentials of ferrocene versus decamethylferrocene$^{0/+}$ redox couple in different organic media

Solvent	Electrolyte	$Fc^{0/+}$ vs. $DmFc^{0/+}$ (V)	Ref
Diethyl ether	0.1 M $[Bu_4N][BArF_{24}]$[a]	0.550±0.005	[49]
	0.1 M $Na[BArF_{24}]$[a]	0.583±0.005	[49]
Anisole	0.1 M $[Bu_4N][PF_6]$	0.518±0.005	[49]
	0.1 M $[Bu_4N]$[TFAB][b]	0.607±0.005	[49]
Methanol	0.1 M $[Bu_4N][ClO_4]$	0.497±0.002	[48]
Ethanol	0.1 M $[Bu_4N][ClO_4]$	0.473±0.005	[48]
2,2,2-Trifluoroethanol	0.1 M $[Bu_4N][ClO_4]$	0.575±0.004	[48]
2-Propanol	0.1 M $[Bu_4N][CF_3SO_3]$	0.455±0.003	[48]
Tetrahydrofuran	0.1 M $[Bu_4N][BF_4]$	0.413±0.005	[49]
	0.1 M $[Bu_4N][CF_3SO_3]$	0.438±0.005	[49]
	0.1 M $[Bu_4N][ClO_4]$	0.423±0.005	[49]
		0.427±0.002	[48]
	0.1 M $[Bu_4N][PF_6]$	0.446±0.005	[49]
	0.1 M $[Bu_4N][BPh_4]$	0.485±0.005	[49]
	0.1 M $Na[BArF_{24}]$[a]	0.502±0.005	[49]
	0.1 M $[Bu_4N]$[TFAB][b]	0.484±0.005	[49]
	0.1 M $[Bu_4N][BArF_{24}]$[a]	0.521±0.005	[49]
Pyridine	0.1 M $[Bu_4N][ClO_4]$	0.517±0.004	[48]
Dichloromethane	0.1 M $[Bu_4N]Cl$	0.534±0.005	[49]
	0.1 M $[Bu_4N][ClO_4]$	0.532±0.002	[48]
	0.1 M $[Bu_4N][PF_6]$	0.548±0.003	[56]
	0.1 M $[Et_4N][BF_4]$	0.541±0.003	[57]
	0.1 M $[Bu_4N]$[TFAB][b]	0.614±0.005	[49]
	0.1 M $[C_4mPyr]$[FAP]	0.589±0.003	[57]
	0.1 M $[C_2mim]$[FAP]	0.590±0.003	[57]
	0.1 M $[C_2mim][B(CN)_4]$	0.588±0.003	[57]
	0.1 M $[C_4mim][N(Tf)_2]$	0.570±0.003	[57]
	0.1 M $[C_4mPyr][N(Tf)_2]$	0.568±0.003	[57]
	0.1 M $[C_2mim]$[FSI]	0.569±0.003	[57]
	0.1 M $[C_3mim]$[FSI]	0.568±0.003	[57]
	0.1 M $[C_4mPyr][N(CN)_2]$	0.564±0.003	[57]
	0.1 M $[C_4mim][PF_6]$	0.556±0.003	[57]
	0.1 M $[C_4mim][BF_4]$	0.557±0.003	[57]
	0.1 M $[C_4mim][CF_3SO_3]$	0.556±0.003	[57]
1,2-Dichloroethane	0.1 M $[Bu_4N][ClO_4]$	0.532±0.001	[48]
1,2-Dibromoethane	0.1 M $[Bu_4N][ClO_4]$	0.475±0.007	[48]
Benzonitrile	0.1 M $[Bu_4N]Cl$	0.524±0.005	[49]
	0.1 M $[Bu_4N][ClO_4]$	0.523±0.001	[48]
	0.1 M $[Bu_4N][PF_6]$	0.530±0.005	[49]
	0.1 M $[Bu_4N]$[TFAB][b]	0.543±0.005	[49]
Bromobenzene	0.1 M $[Bu_4N][ClO_4]$	0.489±0.005	[48]

(continued)

Table 3.2 (continued)

Solvent	Electrolyte	$Fc^{0/+}$ vs. $DmFc^{0/+}$ (V)	Ref
Chlorobenzene	0.1 M $[Bu_4N][ClO_4]$	0.497 ± 0.001	[48]
1,2-Dichlorobenzene	0.1 M $[Bu_4N][ClO_4]$	0.535 ± 0.001	[48]
Benzyl alcohol	0.1 M $[Bu_4N][ClO_4]$	0.508 ± 0.003	[48]
Nitrobenzene	0.1 M $[Bu_4N][ClO_4]$	0.514 ± 0.002	[48]
Aniline	0.1 M $[Bu_4N][ClO_4]$	0.527 ± 0.004	[48]
Toluene	$[Bu_4N][BF_4]$[c]	0.430 ± 0.005	[49]
Acetone	0.1 M $[Bu_4N]Cl$	0.451 ± 0.005	[49]
	0.1 M $[Bu_4N][ClO_4]$	0.479 ± 0.004	[48]
	0.1 M $[Bu_4N][PF_6]$	0.487 ± 0.005	[49]
	0.1 M $[Bu_4N]$[TFAB][b]	0.504 ± 0.005	[49]
Propylene carbonate	0.1 M $[Bu_4N][ClO_4]$	0.495 ± 0.002	[48]
Acetonitrile	0.1 M $[Bu_4N]Cl$	0.501 ± 0.005	[49]
	0.1 M $[Bu_4N][ClO_4]$	0.505 ± 0.002	[48]
	0.1 M $[Bu_4N][PF_6]$	0.509 ± 0.003	[56]
	0.1 M $[Bu_4N]$[TFAB][b]	0.517 ± 0.005	[49]
Nitromethane	0.1 M $[Bu_4N]Cl$	0.505 ± 0.005	[49]
	0.1 M $[Bu_4N][ClO_4]$	0.516 ± 0.004	[48]
	0.1 M $[Bu_4N][PF_6]$	0.510 ± 0.005	[49]
	0.1 M $[Bu_4N]$[TFAB][b]	0.516 ± 0.005	[49]
Formamide	0.1 M $[Bu_4N][ClO_4]$	0.510 ± 0.003	[48]
N-methylformamide	0.1 M $[Bu_4N][ClO_4]$	0.510 ± 0.002	[48]
N,N-dimethylformamide	0.1 M $[Bu_4N]Cl$	0.475 ± 0.005	[49]
	0.1 M $[Bu_4N][ClO_4]$	0.458 ± 0.003	[48]
	0.1 M $[Bu_4N][PF_6]$	0.478 ± 0.005	[49]
	0.1 M $[Bu_4N]$[TFAB][b]	0.493 ± 0.005	[49]
N,N-dimethylacetamide	0.1 M $[Bu_4N][ClO_4]$	0.455 ± 0.008	[48]
Dimethyl sulfoxide	0.1 M $[Bu_4N][PF_6]$	0.486 ± 0.005	[49]
	0.1 M $[Bu_4N]$[TFAB][b]	0.493 ± 0.005	[49]
	0.1 M $[Bu_4N][ClO_4]$	0.468 ± 0.001	[48]
Chloroform	0.1 M $[Bu_4N][ClO_4]$	0.483 ± 0.001	[48]

[a] $[BArF_{24}] = [B(C_6H_3(CF_3)_2)_4]^-$
[b] $[TFAB] = [B(C_6F_5)_4]^-$
[c] The toluene:$[Bu_4N][BF_4]$ electrolyte is of the 3:1 stoichiometry

supporting electrolyte [57]. As evidenced in Table 3.3, a variation in the E_m of Fc vs. $DmFc^{0/+}$ of about 0.100 V is observed when the IL was changed from $[C_4mim][CF_3SO_3]$ to $[C_2mim]$[FAP] under neat conditions (water content <100 ppm). Similarly, when DmFc and Fc were simultaneously present in dichloromethane with added IL as the supporting electrolyte—conditions under which DmFc is less sensitive to solvation effects—a variation in the E_m of Fc vs. $DmFc^{0/+}$ of about 0.050 V was observed when the supporting electrolyte was changed from 0.1 M $[Et_4N][BF_4]$ to 0.1 M $[C_2mim]$[FAP] (Table 3.2). A variable potential difference

Table 3.3 Redox potentials of transition-metal sandwich complexes obtained by cyclic voltammetry in different ionic liquids

		E_m vs. $DmFc^{0/+}$ (V)		E_m vs. $Fc^{0/+}$ (V)	
IL	Electrode	Fc	Cc^+	Cc^+	Ref
$[C_4mPyr][FAP]$	GC	0.575[a]	−0.747[a]	−1.322[a]	[20, 57]
$[C_2mim][FAP]$	GC	0.574[a]	−0.749[a]	−1.323[a]	[20, 57]
$[C_2mim][B(CN)_4]$	GC	0.526[a]	−0.799[a]	−1.325[a]	[20, 57]
$[C_2mim][N(Tf)_2]$	Au, Pt, GC	0.520[a]	−0.820[a]	−1.330[a]	[21, 58]
$[C_4mim][N(Tf)_2]$	GC	0.514[a]	−0.813[a]	−1.327[a]	[20, 57]
$[C_3mPyr][N(Tf)_2]$	Pt	–	–	−1.325[b]	[59]
	GC	–	–	−1.327[b]	[59]
$[C_4mPyr][N(Tf)_2]$	GC	0.511[a]	−0.816[a]	−1.327[a]	[20, 57]
	Au	–	–	−1.363[b]	[59]
	Pt	–	–	−1.333[c]	[19]
$[S_{2\,2\,1}][N(Tf)_2]$	Au	–	–	−1.348[b]	[59]
$[S_{2\,2\,2}][N(Tf)_2]$	Au	–	–	−1.354[b]	[59]
$[C_3mpip][N(Tf)_2]$	Au	–	–	−1.362[b]	[59]
$[HmimSC_4][N(Tf)_2]$	Au, Pt, GC	0.510[a]	−0.830[a]	−1.330[a]	[58]
$[C_2mim][FSI]$	GC	0.512[a]	−0.820[a]	−1.332[a]	[20, 57]
$[C_3mPyr][FSI]$	GC	0.510[a]	−0.822[a]	−1.332[a]	[20, 57]
$[C_2mim][N(CN)_2]$	GC	–	–	−1.332[b]	[59]
	Pt	–	–	−1.355[b]	[59]
$[C_4mPyr][N(CN)_2]$	GC	0.505[a]	–	–	[20, 57]
$[C_4mim][PF_6]$	GC	0.478[a]	−0.851[a]	−1.329[a]	[20, 57]
$[C_2mim][BF_4]$	GC	–	–	−1.336[d]	[60]
$[C_4mim][BF_4]$	GC	0.478[a]	−0.848[a]	−1.326[a]	[20, 57]
$[C_4mim][CF_3SO_3]$	GC	0.474[a]	−0.854[a]	−1.328[a]	[20, 57]
	Au	–	–	−1.362[b]	[59]
$[C_4mPyr][CF_3SO_3]$	GC	–	–	−1.347[b]	[59]

[a]Potentials were obtained using a scan rate of 0.1 V s^{-1}. E_m values were obtained with an accuracy of ±0.003 V ($n=5$)
[b]No experimental error was reported
[c]E_m value was obtained with an accuracy of ±0.001 V
[d]E_m values were obtained with an accuracy of ±0.002 V

which increases from 0.014 V in the case of $[C_4mPyr][FAP]$ to 0.082 V in the case of $[C_4mim][CF_3SO_3]$ is noticed from the comparison of the E_m of Fc vs. $DmFc^{0/+}$ between neat ILs and diluted conditions (dichloromethane with added IL as the supporting electrolyte) [57].

This interaction between Fc and IL ions has been theoretically studied by Yang et al. [50]. According to their MEP predictions, the possible interactions of ferrocene with 1,3-dimethylimidazolium tetrafluoroborate ($[C_1mim][BF_4]$, used as an IL model system) are shown in Figs. 3.5 and 3.6. In Fig. 3.5, it is possible to observe that the IL cation binds to Fc both on the top and from the side. Meanwhile, the IL anion is only able to interact with Fc from the side (Fig. 3.6).

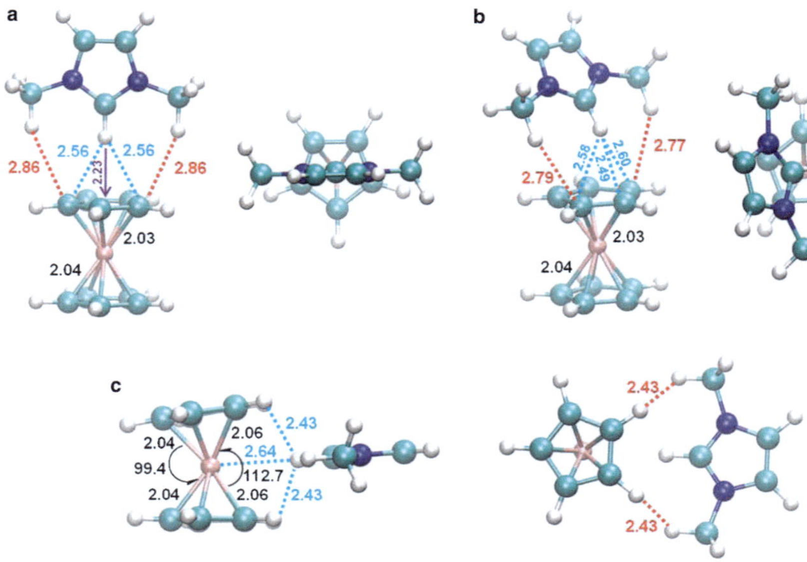

Fig. 3.5 Optimized geometries for Fc–$[C_1mim]^+$ complexes. Distances are in Å, and angles are in degrees. Molecules are rendered in CPK style, with carbon in *cyan*, nitrogen in *blue*, hydrogen in *white*, and iron in *pink*. The sphere for ion atom is lightly shrunk for clearer geometrical parameters labeling. (**a**) $[C_1mim]^+$ is perpendicular to the Cp plane of Fc (*left*: side view; *right*: top view). The *purple arrow* shows the H2–Cp distance; (**b**) $[C_1mim]^+$ is tilted on top of the Cp plane (*left*: side view; *right*: top view); (**c**) $[C_1mim]^+$ "orbits" around Fc, with H2–Fe distance of 2.64 Å (*left*: side view; *right*: top view). Reproduced from [50] with permission from the PCCP Owner Societies

As shown in Fig. 3.4b, ferrocenium ion only has positive electrostatic potentials. Thereby, a strong and exclusive interaction between the +1 charged ferrocenium ion with the IL anion is expected [50]. This interaction could form a stable and tight first solvation shell of anions, around which the IL cation could occupy the second solvation shell by strong electrostatic interactions with the anionic shell [50]. Consequently, the hydrodynamic radius of the diffusing Fc^+ species will be largely affected (increased), decreasing the mass transport rate of this oxidized species with respect to that of Fc. This implies the appearance of an inequality in the values of the diffusion coefficient for the reduced (D_R) and oxidized (D_O) component of the $Fc^{0/+}$ redox couple (Table 3.4) and a net effect in the E_m (Eq. 3.2) and in the half-wave potential, $E_{1/2}$ (Eq. 3.3), experimentally obtained [25]. For example, the $E_{1/2}$ obtained with the use of ultramicroelectrodes is larger by $(RT/nF)\ln(D_R/D_O)$ than the $E^{\theta'}$, the value of which is between 0.007 ± 0.001 and 0.017 ± 0.002 V depending on the IL used (Table 3.4).

$$E_m = E^{\theta'} + \frac{RT}{nF}\ln\sqrt{\frac{D_R}{D_O}} \tag{3.2}$$

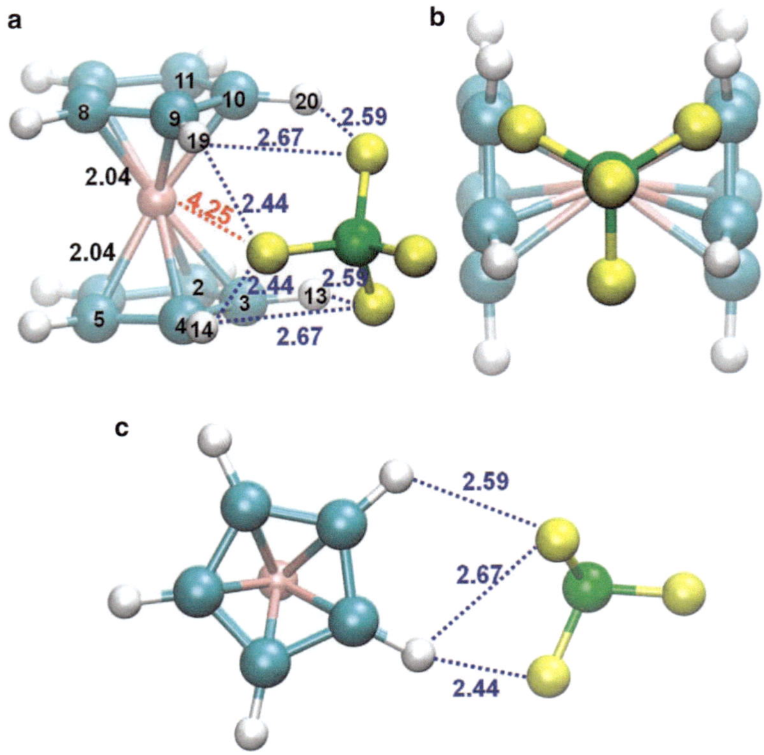

Fig. 3.6 Optimized geometries for the Fc–$[BF_4]^-$ complex. Distances are in Å, and angles are in degrees. Molecules are rendered in the same scheme as in Fig. 3.5. The complex structure is shown from the different perspectives in (**a**)–(**c**). Reproduced from [50] with permission from the PCCP Owner Societies

Table 3.4 Effect of diffusion coefficient inequality on the mid-point potential observed between $DmFc^{0/+}$ and $Fc^{0/+}$[a]

IL	$\frac{D_{Fc}}{D_{Fc+}}$	$\frac{D_{DmFc}}{D_{DmFc+}}$	$\frac{RT}{nF}\ln\frac{D_{Fc}}{D_{Fc+}}$ (10^{-3} V)	$\frac{RT}{nF}\ln\frac{D_{DmFc}}{D_{DmFc+}}$ (10^{-3} V)	$E_{m(Fc+)}$ vs. $DmFc^{0/+}$ (V)	$E^{\theta\prime}_{(Fc+)}$ vs. $DmFc^{0/+}$ (V)
$[C_4mim][PF_6]$	1.64 ± 0.16	1.56 ± 0.20	13 ± 2	11 ± 3	0.478	0.478
$[C_2mim][FAP]$	1.73 ± 0.06	1.00 ± 0.10	14 ± 1	0 ± 2	0.574	0.567
$[C_4mim][CF_3SO_3]$	1.31 ± 0.08	1.36 ± 0.14	7 ± 1	8 ± 2	0.474	0.475
$[C_4mim][N(Tf)_2]$	1.35 ± 0.07	1.09 ± 0.05	8 ± 1	2 ± 1	0.513	0.510
$[C_4mPyr][FAP]$	1.98 ± 0.29	1.07 ± 0.16	17 ± 3	2 ± 2	0.575	0.567
$[C_4mim][BF_4]$	2.01 ± 0.34	1.59 ± 0.10	18 ± 3	12 ± 2	0.478	0.475

[a] Optimal diffusion coefficient values were determined by numerical and digital fit of the experimental chronoamperometric and voltammetric data ($n = 5$). Values obtained from [25]

$$E_{1/2} = E^{\theta\prime} + \frac{RT}{nF}\ln\frac{D_R}{D_O} \tag{3.3}$$

It is worth noting that when the ratio of the diffusion coefficient for D_R and D_O components of a particular redox couple is close to unity, the difference between $E_{1/2}$, E_m, and $E^{\theta\prime}$ is not experimentally significant. However, this condition is not always true in the case of ILs. Thereby, in those ILs where $D_R/D_O \neq 1$, the $E_{1/2}$ obtained with the use of an ultramicroelectrode may differ from the E_m obtained with a macroelectrode. This is because the second term in Eq. (3.3) contains the first power of the diffusion coefficient, rather than the square root as in Eq. (3.2). Consequently, the experimentally obtained $E_{1/2}$ is larger by $(RT/nF)\ln(D_R/D_O)$ with respect to E_m. For example, the experimental $E_{1/2(Fc)} = 0.582 \pm 0.003$ V vs. $DmFc^{0/+}$ obtained in [C_4mPyr][FAP] and with the use of a carbon fiber ultramicroelectrode (6.2 μm diameter) is larger with respect to $E_m = 0.575 \pm 0.003$ V vs. $DmFc^{0/+}$ obtained with a glassy carbon macroelectrode (1 mm diameter) under the same experimental conditions. This difference is consistent, within the experimental error, to the calculated $(RT/nF)\ln(D_R/D_O)$ value of 0.008 V in this IL.

The E_m of Fc vs. $DmFc^{0/+}$, as well as the $E^{\theta\prime}$ obtained after correction for inequality in the respective diffusion coefficients of the charged and neutral species by using Eq. (3.2), is presented in Table 3.4. From this table, it is possible to observe that the inequality in the diffusion coefficients introduces negligible differences between the experimental E_m and $E^{\theta\prime}$ values when Fc and DmFc have a similar inequality ratio (e.g., in [C_4mim][PF_6] and [C_4mim][CF_3SO_3]). Meanwhile, the mentioned difference becomes larger than 0.003 V (consider a typical experimental error) as the inequality ratio difference increases. For example, a difference of 0.007 V and 0.008 V is observed for [C_2mim][FAP] and [C_4mPyr][FAP], respectively. For comparison, a difference of about 0.016 V between $E_{1/2}$ and $E^{\theta\prime}$ is observed in these ILs [25].

Figure 3.7 shows typical cyclic voltammograms for the oxidation of saturated solution of DmFc in [C_4mim][$N(Tf)_2$] and [C_4mPyr][FAP] on a 6.2 μm diameter carbon fiber ultramicroelectrode at a scan rate of 10 mV s^{-1} and under an argon atmosphere. Decamethylferrocene is oxidized in a reversible, one-electron transfer process to $DmFc^+$ and reduced back to DmFc according to Eq. (3.1). The $E_{1/2}$ for this redox reaction was arbitrarily set at 0.0 V.

As expected for a reversible process, a well-defined response with $|E_{3/4}-E_{1/4}|$ values of 56 and 52 was observed in [C_4mim][$N(Tf)_2$] and [C_2mim][FAP], respectively, which are close to the theoretical value of 56 mV at 21 °C. Values below 50 mV were observed in [C_4mim][PF_6], [C_4mim][CF_3SO_3], [C_4mPyr][FAP], and [C_4mim][BF_4] ILs. This departure from ideality could be attributed to a voltam metric behavior intermediate to that observed at micro- and macroelectrodes due to the very low diffusivity of DmFc and $DmFc^+$ in these ILs and the value of potential scan rate selected for this experiment. For example, Fig. 3.7a shows the voltammogram

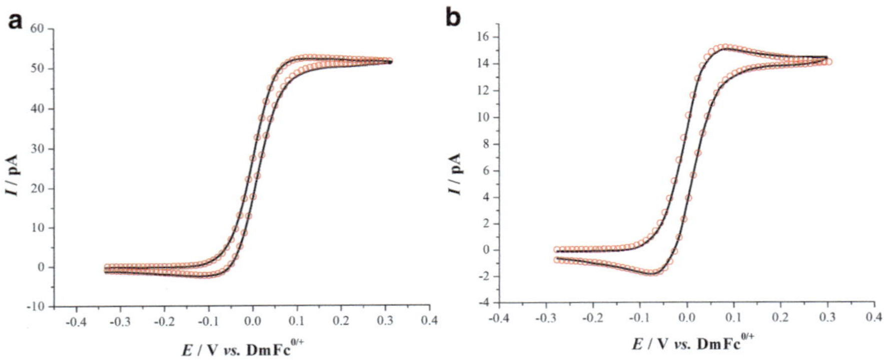

Fig. 3.7 Comparison of experimental (*black solid line*) and digitally simulated (*open red circles*) cyclic voltammograms obtained for the oxidation of saturated solution of DmFc in (**a**) [C_4mim][$N(Tf)_2$] and (**b**) [C_4mPyr][FAP], respectively, under argon atmosphere and using a carbon fiber ultramicroelectrode ($d = 6.2$ μm) as the working electrode. Scan rate = 10 mV s^{-1}; $T = 21 \pm 1$ °C. Reproduced from [25] with permission from Elsevier

of DmFc in [C_4mim][$N(Tf)_2$]. The potential scan rate of 10 mV s^{-1} used experimentally is smaller than the 20 and 18 mV s^{-1} predicted by the following inequality

$$v \ll \frac{RTD}{nFr^2} \tag{3.4}$$

and required to observe a near-steady-state voltammetric behavior for the oxidation of DmFc and the following reduction of $DmFc^+$, respectively ($D_{DmFc} = 0.76 \times 10^{-7}$ and the $D_{DmFc+} = 0.70 \times 10^{-7}$ $cm^2\ s^{-1}$). Meanwhile, Fig. 3.7b shows the cyclic voltammogram of DmFc in [C_4mPyr][FAP]. According to Eq. (3.4) and assuming $D_{DmFc} = 0.30 \times 10^{-7}$ and $D_{DmFc+} = 0.28 \times 10^{-7}$ $cm^2\ s^{-1}$ in this IL, near-steady-state behavior at the microdisk electrode is expected to be achieved at scan rates of ≪8 mV s^{-1} and 7 mV s^{-1} for the oxidation of DmFc and the following reduction of $DmFc^+$, respectively. Consequently, the scan rate of 10 mV s^{-1} used experimentally allows observing peak-shaped voltammograms, which is characteristic for more planar diffusion behaviors [25].

To further confirm the reversibility of the $DmFc^{0/+}$ redox process, a comparison of digital simulations of the cyclic voltammograms and experimental data was performed (Fig. 3.7). Simulations based on the reversible mechanism outlined in Eq. (3.1) give excellent agreement to experimental data for DmFc in all the ILs evaluated [25].

Table 3.4 shows the D_{DmFc}/D_{DmFc+} ratio obtained in different ILs. In contrast with Fc, where the D ratio is within the range of 1.31–2.01, the D_{DmFc}/D_{DmFc+} ratio is ≈1, except for [C_4mim][CF_3SO_3], [C_4mim][PF_6], and [C_4mim][BF_4], which have a higher ratio of 1.36 ± 0.14, 1.56 ± 0.20, and 1.59 ± 0.10, respectively. This result suggests that neutral DmFc species diffuse at the same rate than positively charged decamethylferrocenium species through the [C_4mim][FAP], [C_4mim][$N(Tf)_2$], and [C_4mPyr][FAP] media. Thereof, the $E_{1/2} \approx E^{\theta\prime}$ (within the experimental error) for DmFc in these ILs. Although this is not a ubiquitous property, the behavior observed in the latest mentioned ILs is consistent with that observed in traditional organic solvents with added supporting electrolyte and where DmFc is considered to be a less solvent-dependent internal reference redox system with respect to Fc [48].

It has been previously discussed that the formal potential of redox couples obtained in ILs could be affected by at least two factors [20]: (i) the different IL solvation effects on DmFc/$DmFc^+$ and Fc/Fc^+ couples and (ii) the inequality ratio between the D_R and D_O of these redox couples. An analysis of Table 3.4 shows that an increase in experimental E_m of Fc vs. $DmFc^{0/+}$ of about 0.101 V is observed when the IL is changed from [C_4mim][CF_3SO_3] to [C_4mPyr][FAP]. The increase in the $E^{\theta\prime}$ of Fc vs. $DmFc^{0/+}$ is equal to 0.092 V after the correction for inequality in the diffusion coefficients of both redox couples in the mentioned ILs. It is instructive to mention that the mentioned increase in the experimental $E_{1/2}$ of Fc vs. $DmFc^{0/+}$ is of 0.110 V by changing from [C_4mim][CF_3SO_3] to [C_4mPyr][FAP]. These differences between E_m, $E_{1/2}$, and $E^{\theta\prime}$ could easily be explained by considering the differences between Eqs. (3.2) and (3.3). Furthermore, from these values it is possible to see that even when both previously mentioned factors (i and ii) are present in the obtained potential difference, the solvation property of the ILs as a function of the chemical structure is the predominant reason for the observed variations.

The cobaltoceniumlcobaltocene, $Cc^{+/0}$ (also known as $[Co^{III}(\eta^5\text{-}C_5H_5)_2]^+/[Co^{II}(\eta^5\text{-}C_5H_5)_2]$), redox couple is used in organic solvent and ILs as an alternative IRRS. It was suggested by Strehlok et al. in 1960 as an IRRS for organic solvent systems [61]. Although it was not considered by IUPAC as an alternative to the $Fc^{0/+}$ couple, it shows negligible volatility and high stability and solubility in some organic solvents [62] and ILs [6, 47, 58, 63]. If the cathodic potential window of the solvent is large enough, two chemically reversible, one electron reduction processes can be observed (Eqs. 3.5 and 3.6).

$$\left[Co^{III}\left(\eta^5-C_5H_5\right)_2\right]^+ + e^- \leftrightarrows \left[Co^{II}\left(\eta^5-C_5H_5\right)_2\right] \quad (3.5)$$

$$\left[Co^{II}\left(\eta^5-C_5H_5\right)_2\right] + e^- \leftrightarrows \left[Co^{I}\left(\eta^5-C_5H_5\right)_2\right]^- \quad (3.6)$$

The mid-point potential for the $Cc^{0/-}$ couple (Eq. 3.6) was found to be considerably solvent dependent and can not always be determined accurately at room temperature, mainly due to the overlap and interaction of this process with the electrochemical response of the organic solvent used [62]. Thereby, only the $Cc^{+/0}$ redox couple (Eq. 3.5) has been extensively used in organic solvents to provide a known and stable reference point. As an extrapolation of the concept, it is also used in IL systems. Surprisingly, little is known about the solvation effect of either different organic solvents with added supporting electrolytes or IL structure on the $Cc^{+/0}$ formal potential.

Table 3.3 reports the IL effect on the $Cc^{+/0}$ process using both $Fc^{0/+}$ and $DmFc^{0/+}$ as IRRS. In the first case, the E_m of $Cc^{+/0}$ is almost constant and could be represented by a value of 1.327±0.005 V vs. $Fc^{0/+}$ in 14 different ILs and using glassy carbon as the working electrode (exceptions are [C_2mim][BF_4] and [C_4mPyr][CF_3SO_3], where

Table 3.5 Redox potentials of cobaltocenium hexafluorophosphate vs. $Fc^{0/+}$ in different organic media

Solvent	Electrolyte	$Cc^{+/0}$ vs. $Fc^{0/+}$ (V)	Ref
Acetonitrile	0.1 M $[Et_4N][ClO_4]$	−1.350 ± 0.003	[62]
	0.1 M $[Bu_4N][BF_4]$	−1.348 ± 0.003	[62]
	0.1 M $[Bu_4N][PF_6]$	−1.337 ± 0.005	
Dichloromethane	0.1 M $[Et_4N][ClO_4]$	−1.359 ± 0.007	[62]
	0.1 M $[Bu_4N][BF_4]$	−1.355 ± 0.007	[62]
Ethanol	0.1 M $[Et_4N][ClO_4]$	−1.351 ± 0.004	[62]
	0.1 M $[Bu_4N][BF_4]$	−1.356 ± 0.004	[62]
Water	0.1 M $Li[ClO_4]$	−1.367 ± 0.002	[62]
Toluene	0.4 M $[P_{6\,6\,6\,14}][FAP]$	−1.340 ± 0.005	[64]

$[P_{6\,6\,6\,14}][FAP]$ = trihexyl(tetradecyl)phosphonium tris(pentafluoroethyl)trifluorophosphate

a value of 1.336 V and 1.347 V was reported, respectively). Meanwhile, when $DmFc^{0/+}$ is used as IRRS, an increase of 0.107 V in the potential separation between $Cc^{+/0}$ and $DmFc^{0/+}$ couples is observed upon changing the IL from $[C_4mPyr][FAP]$ to $[C_4mim][CF_3SO_3]$ and using GC as the working electrode.

To further understand this observation, we need to compare the variation in the E_m of $Cc^{+/0}$ vs. $Fc^{0/+}$ in organic solvents with added supporting electrolyte. It is well known that Fc is highly sensitive to solvation effects under these conditions (see above). As seen in Table 3.5, the E_m of $Cc^{+/0}$ in nine different solvent/supporting electrolyte combinations can be represented by an almost constant value of about 1.352 ± 0.015 V vs. $Fc^{0/+}$. As given in Table 3.2, the E_m value of the $Fc^{0/+}$ redox process varies from 0.437 ± 0.005 V in ethanol/0.1 M $[Bu_4N][ClO_4]$ to a value of 0.532 ± 0.002 V in dichloromethane/0.1 M $[Bu_4N][ClO_4]$, which represents a difference of 0.095 V. Meanwhile, the obtained Cc^+ E_m values vs. $Fc^{0/+}$ in the same two organic solvent systems (with 0.1 M $[Bu_4N][ClO_4]$ as the supporting electrolyte) show a variation of just 0.008 V (Table 3.5).

The comparison of data reported in Tables 3.2, 3.3, and 3.5 is of significant value as they suggest that the mid-point potentials of both $Cc^{+/0}$ and $Fc^{0/+}$ couples are dependent on the solvation properties of organic solvents (with added supporting electrolyte) as well as the solvation properties of ILs, particularly under the assumption that $DmFc^{0/+}$ redox couple is a less solvent dependent process [57]. Consequently, potentials reported versus $Cc^{+/0}$ and $Fc^{0/+}$ might need to be corrected using the reported values in case of comparison of experimental data obtained in different organic solvents (Tables 3.2 and 3.5) and/or ILs (Table 3.3).

Decamethylcobaltocenium ($DmCc^+$ or $[Co^{III}(\eta^5\text{-}C_5(CH_3)_5)_2]^+$) was studied to throw light further on this matter [65]. Figure 3.8 shows typical cyclic voltammograms for the reduction of a 4.36 mmol/kg $DmCc^+$ solution in $[C_4mPyr][N(Tf)_2]$ on a 1.0 mm diameter GC electrode at varying scan rates. Decamethylcobaltocenium is reduced by one electron to decamethylcobaltocene, DmCc, at a peak potential of *ca.* −1.424 V vs. $DmFc^{0/+}$, and oxidized back to $DmCc^+$ at a potential of *ca.* −1.359 V vs. $DmFc^{0/+}$, according to Eq. (3.7). The second reduction process, related to the one-electron reduction of decamethylcobaltocene (Eq. 3.8), however, was not observed in all ILs studied [65].

$$\left[\mathrm{Co^{III}}\left(\eta^5-\mathrm{C_5}\left(\mathrm{CH_3}\right)_5\right)_2\right]^+ + \mathrm{e}^- \leftrightarrows \left[\mathrm{Co^{II}}\left(\eta^5-\mathrm{C_5}\left(\mathrm{CH_3}\right)_5\right)_2\right] \quad (3.7)$$

$$\left[\mathrm{Co^{II}}\left(\eta^5-\mathrm{C_5}\left(\mathrm{CH_3}\right)_5\right)_2\right] + \mathrm{e}^- \leftrightarrows \left[\mathrm{Co^{I}}\left(\eta^5-\mathrm{C_5}\left(\mathrm{CH_3}\right)_5\right)_2\right]^- \quad (3.8)$$

As expected for reversible processes, E_m values obtained from the cyclic voltammograms are independent of the scan rates (0.1–0.7 V s^{-1}, Fig. 3.8) and concentrations (3–4.5 mmol/kg) studied. The peak current for the reduction of $DmCc^+$ varies linearly with the square root of the scan rate (0.1–0.7 V s^{-1}), characteristic of diffusion-controlled electrode processes [65].

To examine the influence of IL constituents on the redox potential of $DmCc^+$, ionic liquids containing the same anion but different cations, e.g., $[C_4mim][N(Tf)_2]$ and $[C_4mPyr][N(Tf)_2]$, and the same cation but different anions, e.g., $[C_4mim][N(Tf)_2]$ and $[C_4mim][CF_3SO_3]$, were selected. From Fig. 3.9 and Table 3.6, it is noted that the substitution of the ten hydrogen atoms of the cyclopentadienyl ring of cobaltocenium with ten methyl groups decreases the redox potential of the reduction process by 0.579 V, 0.583 V, and 0.541 V in $[C_4mPyr][N(Tf)_2]$, $[C_4mim][N(Tf)_2]$, and $[C_4mim][CF_3SO_3]$, respectively [65]. This effect may be attributed to the

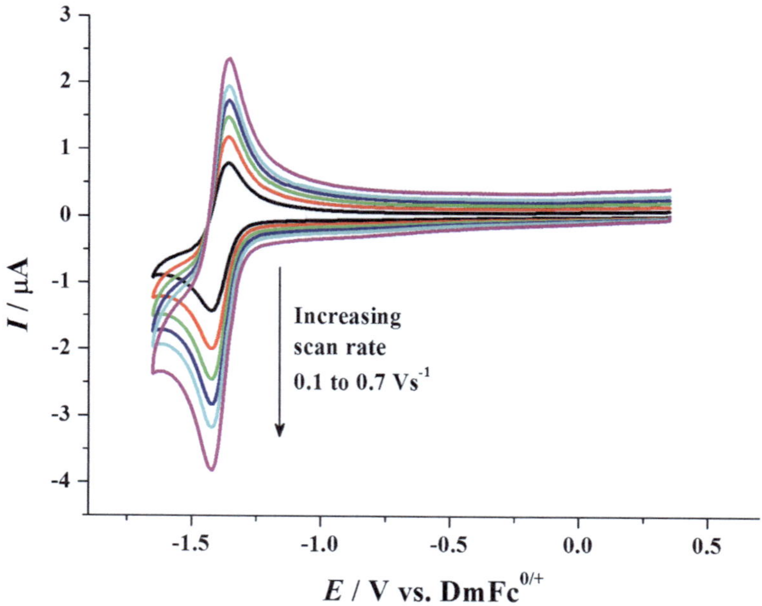

Fig. 3.8 Cyclic voltammetry for the reduction of 4.36 mmol/kg $DmCc^+$ in $[C_4mPyr][N(Tf)_2]$ on a 1 mm diameter GC electrode at varying scan rates (0.1, 0.2, 0.3, 0.4, 0.5, 0.7 V s^{-1}). Reproduced from [65] with permission from the PCCP Owner Societies

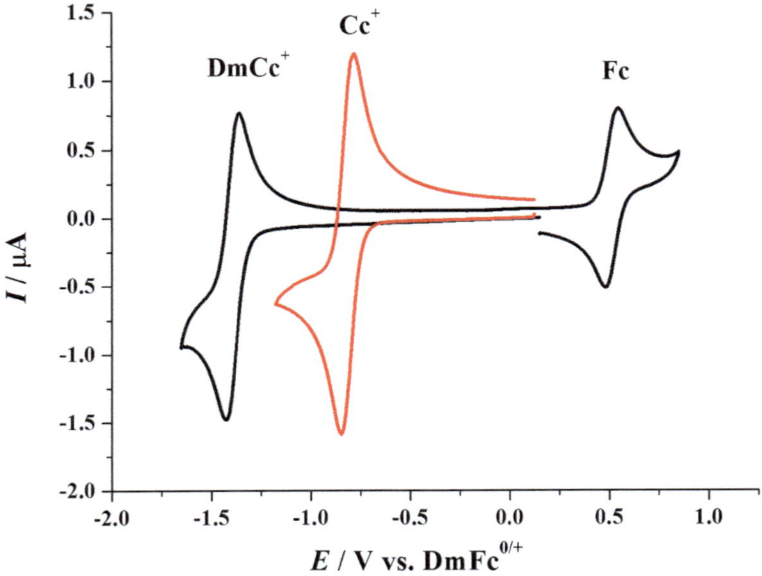

Fig. 3.9 Voltammetric responses for the oxidation of 1.39 mmol/kg Fc and reduction of 4.33 mmol/kg Cc^+ and 4.36 mmol/kg $DmCc^+$ in $[C_4mPyr][N(Tf)_2]$ on a GC electrode (diameter = 1 mm) at a scan rate of 0.1 V s^{-1}. Reproduced from [65] with permission from the PCCP Owner Societies

electron-donating effect of the methyl groups, which leads to the pushing of the electron density toward the metal center and, thereby, obstructing the electron addition to the $DmCc^+$ by the electrode.

Table 3.6 shows a subtle variation of the E_m of $DmCc^+$, Cc^+, and Fc vs. $DmFc^{0/+}$ upon changing the IL from $[C_4mim][N(Tf)_2]$ to $[C_4mPyr][N(Tf)_2]$. Magnitude changes of around 0.040 V, nonetheless, are observed for Cc^+, and Fc by keeping the $[C_4mim]$ cation constant and changing the anion from $[N(Tf)_2]^-$ to $[CF_3SO_3]^-$. The observed variation in the Cc^+ and Fc E_m values vs. $DmFc^{0/+}$ when the weakly coordinating $[N(Tf)_2]$ anion is replaced with $[CF_3SO_3]^-$can be rationalized in terms of the ion-pairing tendencies of small versus large anions and the differences in their donor abilities [66, 67]. This tendency also justifies the Cc^+ and Fc E_m values vs. $DmFc^{0/+}$ observed with the bulky $[FAP]^-$-containing IL, $[C_2mim][FAP]$, and in comparison with the previously mentioned ILs, where for example, the Cc^+ E_m value vs. $DmFc^{0/+}$ shifts in the cathodic direction by *ca.* 0.100 V by changing the IL from $[C_2mim][FAP]$ to $[C_4mim][CF_3SO_3]$. Surprisingly, the difference in the E_m between $DmCc^+$ and $DmFc^{0/+}$ remains almost constant in all four ILs.

Other alternatives, such as 1,1′-dimethylferrocene, 1,1-diacetylferrocene, ferrocenecarboxaldehyde, etc., could be used to provide a stable redox potential [6, 58, 68]. However, these systems have not been fully characterized for this role in ILs as yet.

Table 3.6 Redox potentials of transition-metal sandwich complexes obtained by cyclic voltammetry in different ionic liquids[a]

		E_m vs. $DmFc^{0/+}$ (V)			E_m vs. $Fc^{0/+}$ (V)	
IL	Electrode	Fc	Cc^+	$DmCc^+$	Cc^+	Ref
[C_2mim][FAP]	GC	0.574	−0.749	−1.392	−1.323	[65]
[C_4mPyr][$N(Tf)_2$]	GC	0.511	−0.816	−1.395	−1.327	[65]
[C_4mim][$N(Tf)_2$]	GC	0.514	−0.813	−1.396	−1.327	[65]
[C_4mim][CF_3SO_3]	GC	0.474	−0.854	−1.395	−1.328	[65]

[a]Potentials were obtained by cyclic voltammetry and using a scan rate of 0.1 V s^{-1}. E_m values were obtained with an accuracy of ±0.003 V ($n=5$)

3.3.1 Limitations in the Application of IRRS in Ionic Liquids

- *Poor solubility of Fc in some ionic liquids*, mostly in those with high viscosity [63],
- *Moderate volatility of Fc* [25, 69, 70],
- *Reactivity of* Fc^+ *with the IL components:* generally observed when the anodic potential window of the IL is close to the formal redox potential of $Fc^{0/+}$ [58, 63].
- *The* E_m *of* $Fc^{0/+}$ *redox process differs from the* $E^{0\prime}$ *:* this effect is related to the inequality in the diffusion coefficients of Fc and Fc^+ in almost all known ILs [25, 46].
- *Poor solubility of decamethylcobaltocene in some ILs:* Fig. 3.10 shows a voltammogram of $DmCc^{+/0}$ in [C_4mim][PF_6] and [C_4mim][BF_4] ILs. An almost symmetrical, bell-shaped anodic peak is observed in the positive going potential scan at −1.310 V vs. $DmFc^{0/+}$ in [C_4mim][PF_6] (Fig. 3.10a). This is a consequence of the poor solubility of the reduction product, DmCc, which translates to important electrode adsorption effects. Meanwhile, stronger electrode adsorption processes of $DmCc^+$ and its oxidation product can be observed in [C_4mim][BF_4] (Fig. 3.10b), where the adsorption process can be observed both in the reduction and oxidation voltammetric scan direction. This adsorption process prevents the accurate determination of the E_m, and consequently, the application of this redox system as a reliable IRRS in these particular ILs.
- *Limited solubility of decamethylferrocene in ILs:* Table 3.7 shows the solubility of DmFc and Fc in different ILs. As possible to be seen, the solubility of DmFc is much smaller (0.5–5.6 mM) than that of ferrocene in the same ILs.
- *Reactivity of decamethylferrocene in the presence of oxygen*: Fig. 3.11a-*a* shows a picture of a 2.19 mmol/kg DmFc solution in [C_4mPyr][$N(Tf)_2$] under argon atmosphere. Under this experimental condition, the DmFc solution was yellow and stable over time, even with the addition of different water concentrations [57, 65]. However, after the exposure to an O_2 and H_2O saturated atmosphere (atmospheric condition), the IL solution turns green (Fig. 3.11a-*b*). This reaction was followed by UV–Vis spectroscopy, where an absorption band due to the

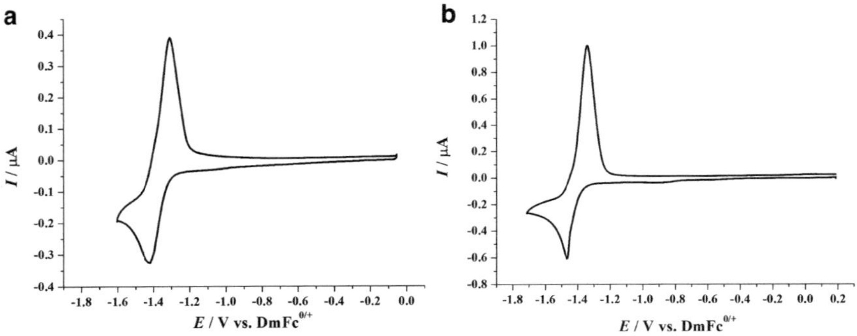

Fig. 3.10 Cyclic voltammetry for the reduction of (**a**) 4.27 mmol/kg $DmCc^+$ in $[C_4mim][PF_6]$ and (**b**) 4.71 mmol/kg $DmCc^+$ in $[C_4mim][BF_4]$ on a 1 mm diameter GC electrode at scan rate of 0.02 V s^{-1}. Reproduced from [65] with permission from the PCCP Owner Societies

Table 3.7 Solubilities of DmFc and Fc in various ILs

IL	DmFc solubility (mM)[a]	Fc solubility (mM)[b]
$[C_4mim][PF_6]$	1.10±0.21	38.2±1.5
$[C_2mim][FAP]$	2.80±0.12	–
$[C_4mim][CF_3SO_3]$	1.90±0.09	58.3±0.1
$[C_4mim][N(Tf)_2]$	5.26±0.14	77.4±0.5
$[C_4mPyr][FAP]$	3.50±0.17	–
$[C_4mim][BF_4]$	0.52±0.09	27.5±0.2

[a]Data obtained from [25]
[b]Data obtained from [46]

$DmFc^+$ centered at 776 nm develops, whereas the absorption peak for DmFc centered at 428 nm is replaced by a very large absorbance in the UV range with peaks at 242, 278, and 315 nm and the appearance of an isosbestic point, IP, at 258 nm. An almost identical spectrum can be observed from a 1.15 mM DmFc solution in acetonitrile (0.1 M $[Bu_4N][PF_6]$) during its oxidative electrolysis at 0.2 V vs. $DmFc^{0/+}$ [65]. These observations confirm the occurrence of DmFc oxidation to $DmFc^+$ in the IL solution.

The spontaneous reaction observed between DmFc and oxygen brings important consequences in the establishment of conditions under which DmFc should be used. Indeed, it is a suitable IRRS to work under inert atmospheres, such as those present inside argon or nitrogen-filled glove boxes. However, the application of this redox system as IRRS should be avoided when the experiments are being conducted under atmospheric conditions. Moreover, DmFc sample preparation in ILs should be also performed under an inert atmosphere.

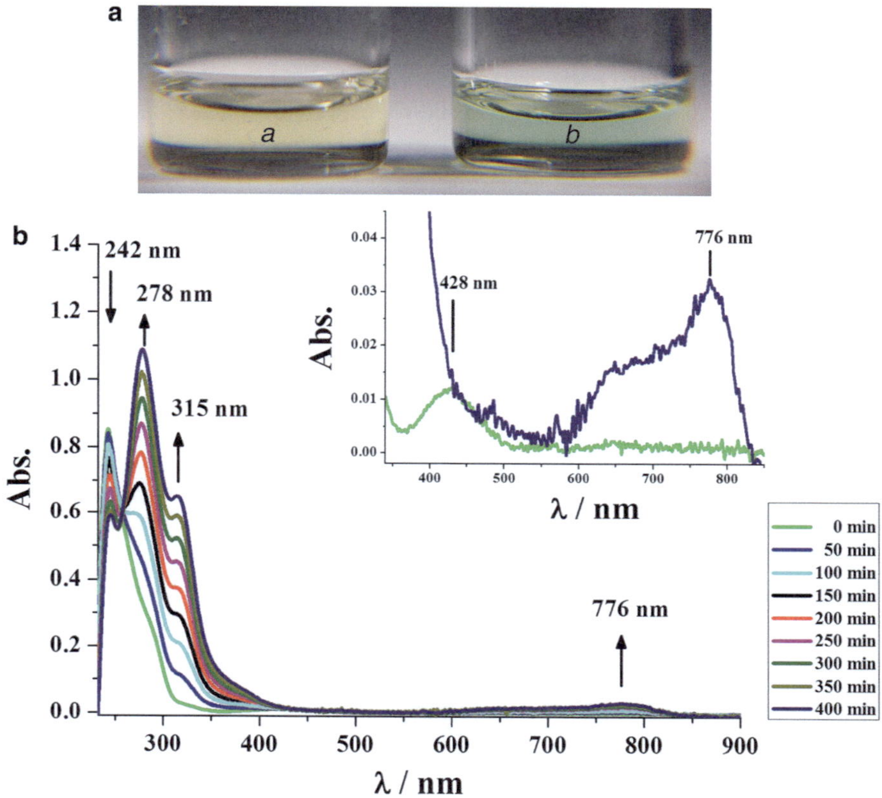

Fig. 3.11 (**a**) Comparison of a 2.19 mmol/kg DmFc in $[C_4mPyr][N(Tf)_2]$ solution under argon (a) and atmospheric (b) conditions. (**b**) UV–Vis spectra of the $[C_4mPyr][N(Tf)_2]$ solution shown in A-a exposed to atmospheric condition and recorded in function of time. Inset: magnification of (**b**) to show the differences in the spectra before (*green*) and after (*blue*) 400 min exposition to atmospheric condition. Adapted from [65] with permission from the PCCP Owner Societies

3.3.2 *Selection of an Internal Reference Redox Scale for Voltammetric Measurements*

The appropriated selection of IRRS to provide a known and stable reference point in voltammetry is very important both in organic and IL solvent systems. The objective is to choose IRRS (either the OX_{IRRS} or RED_{IRRS} component of a reversible redox couple) so that the cyclic voltammetric response for the simultaneously present electroactive ANS can be observed independently of the IRRS response. This means that if ANS to be studied is in the oxidized form, then the reduced form of IRRS must be added to the system. Conversely, if ANS is in the reduced from, then the oxidized form of IRRS must be added to the system. However, it is possible to see a large number of published works, which do not follow this basic rule.

This matter was recently addressed and three fundamental paradigms describing the relative positioning of IRRS and ANS on the potential scale, the operative redox components for IRRS and ANS, and the starting potential (E_{start}), reversing potential (E_{rev}), and ending potential (E_{end}) for a cyclic voltammetric scan have been identified [45]. These paradigms are as follows:

Paradigm A *an optimal paradigm, which can produce completely independent voltammetric responses for ANS and for IRRS.*

Paradigm B *a less-than-optimal paradigm which can produce an independent CV response for ANS or a mixed response for IRRS with that response on top of ANS response.*

Paradigm C *a problematic paradigm that can produce an independent CV response for IRRS or a mixed response for ANS with that response on top of the IRRS response.*

Any mixed response produces a thermodynamically favored redox-cross-reaction which couples the IRRS and ANS systems and which can complicate the analysis of ANS and IRRS responses [45].

The various paradigmatic arrangements along a potential axis are shown in Table 3.8. For the purposes of illustration of the concepts, we will assume that cyclic voltammetry is the operative experimental protocol.

The paradigms A-1a and A-1b have identical arrangement of the OX_{ANS} and RED_{IRRS} along the potential scale. They only differ in the initial scan direction and E_{rev}. The paradigm A-1a produces the independent CV response for ANS (Eq. 3.9). Meanwhile, the paradigm A-1b produces the independent CV response for IRRS (Eq. 3.10). A cross-reaction (Eq. 3.11) is not being established, as the necessary reactants are never simultaneously present in the diffusion layer. An analogous situation can be seen with the paradigms A-2a and A-2b, the oxidation state of the electroactive species (RED_{ANS} and OX_{IRRS}) being the major difference between A-1 and A-2 paradigm [45].

$$\mathbf{A-1a}: \quad OX_{ANS} + e \Leftrightarrow RED_{ANS} \quad (OX_{ANS}\ \text{initially present}) \tag{3.9}$$

$$\mathbf{A-1b}: \quad OX_{IRRS} + e \Leftrightarrow RED_{IRRS} \quad (RED_{IRRS}\ \text{initially present}) \tag{3.10}$$

$$RED_{ANS} + OX_{IRRS} \Leftrightarrow OX_{ANS} + RED_{IRRS} \quad (K_{cross} = 2.82 \times 10^{8}) \tag{3.11}$$

K_{cross} being the equilibrium constant for a simple outer-sphere second-order cross-reaction.

$$K_{cross} = \exp\left[\frac{F}{RT}\left|E^{\theta\prime}_{ANS} - E^{\theta\prime}_{IRRS}\right|\right] \tag{3.12}$$

Figure 3.12 shows simulated voltammograms related to the paradigms A-1a and A-1b, and following the mechanism depicted in Eqs. (3.9)–(3.11). Setting

Table 3.8 The three fundamental paradigms (A, B, and C) characterizing the cyclic voltammetric responses for IRRS and ANS compounds [45]

Paradigm	Potential zone	CV response for	Cross-reaction
A-1a	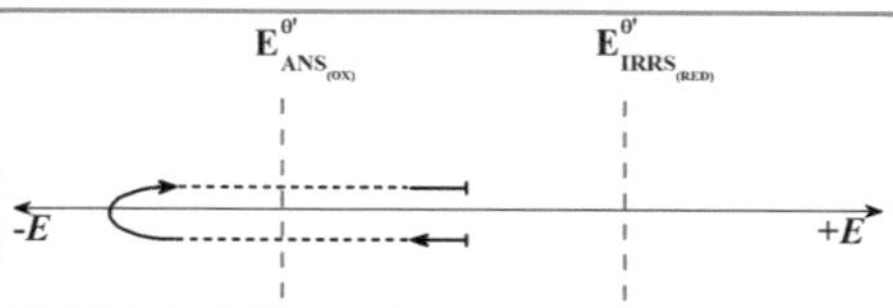	ANS	NO
A-1b	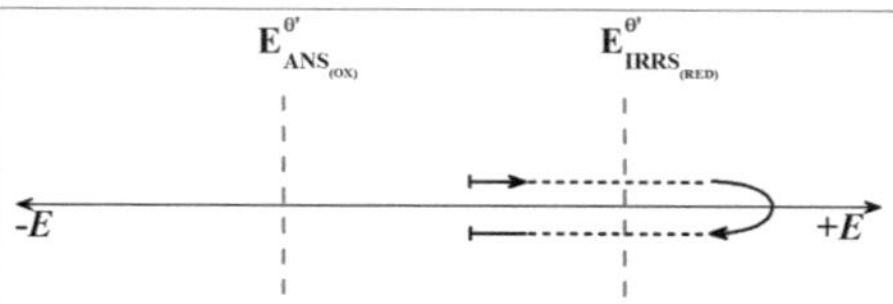	IRRS	NO
A-2a	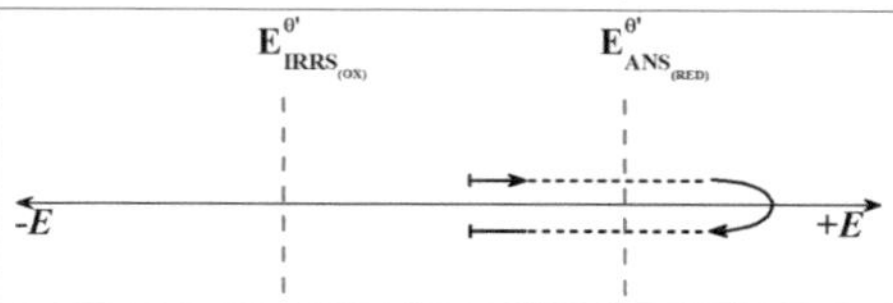	ANS	NO
A-2b	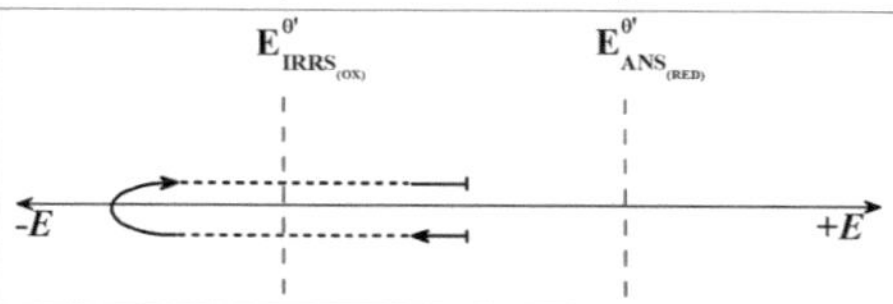	IRRS	NO
B-1a	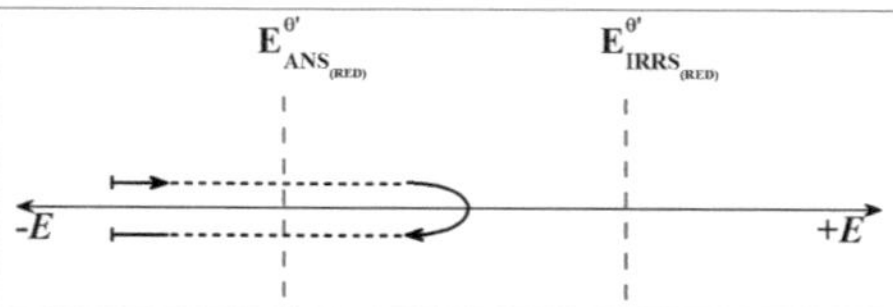	ANS	NO
B-1b	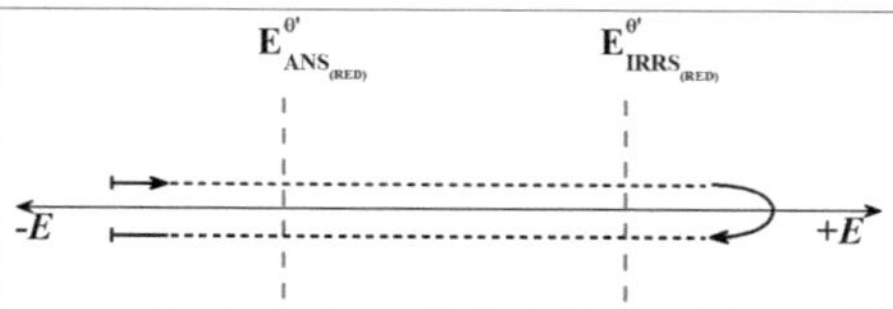	IRRS + ANS	YES
B-2a	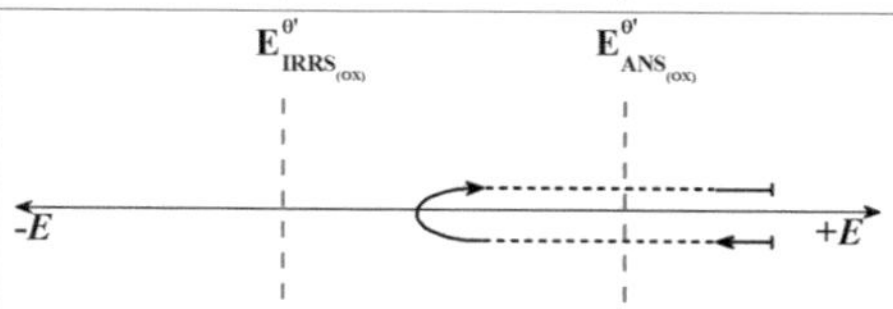	ANS	NO
B-2b	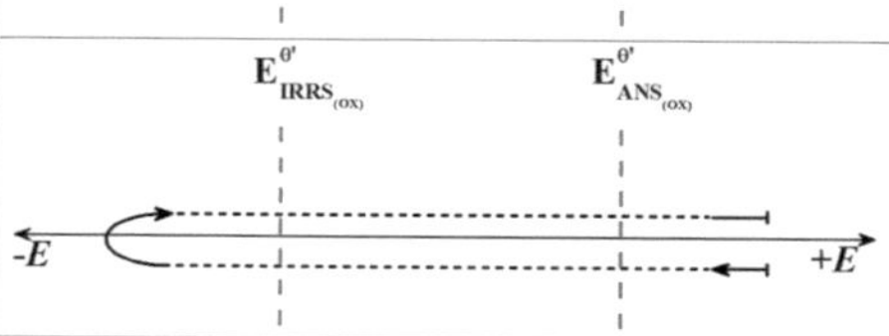	IRRS + ANS	YES

(continued)

Table 3.8 (continued)

Paradigm	Potential zone	CV response for	Cross-reaction
C-1a	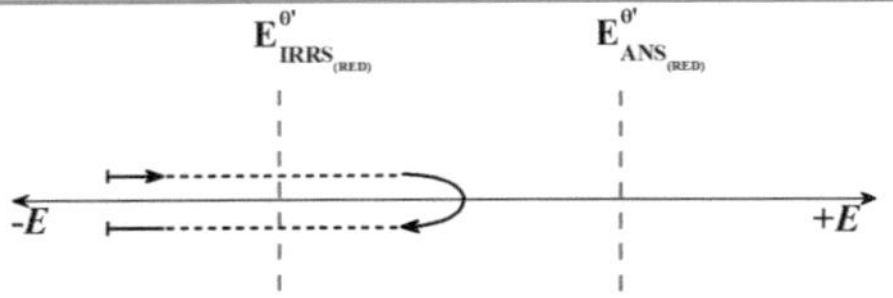	IRRS	NO
C-1b	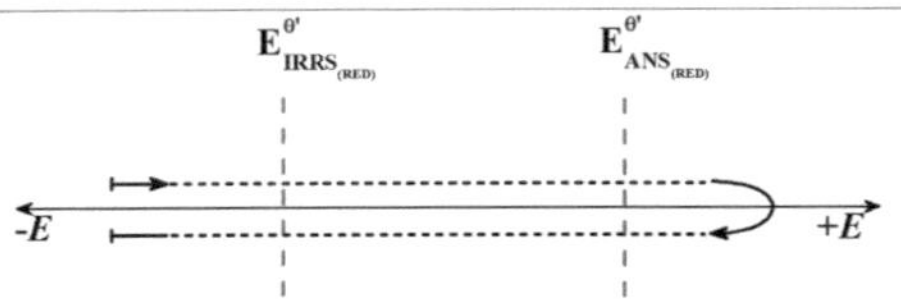	ANS + IRRS	YES
C-2a	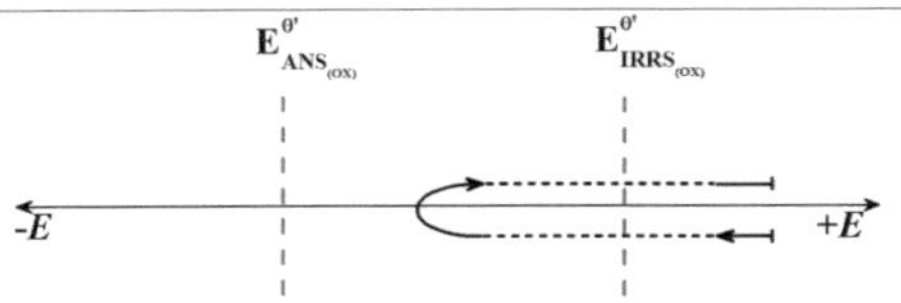	IRRS	NO
C-2b	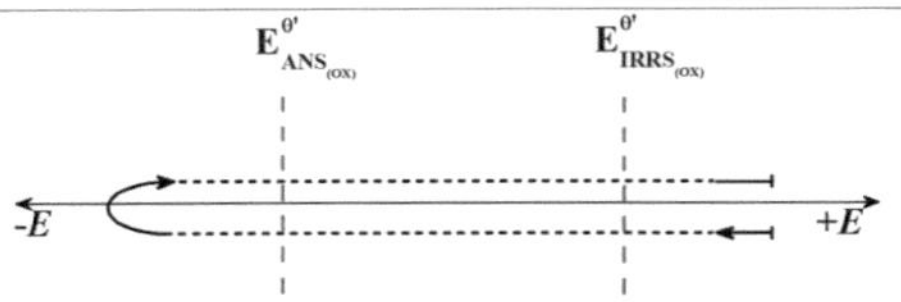	ANS + IRRS	YES

⟵ or ⟼: The potential zone where a given CV scan is initiated (E_{start}) and the initial direction of the scan (following the IUPAC sign convention: negative scan: ⟵; positive scan: ⟼). This potential zone should be chosen so that the current is initially zero before the scan commences. ↻ or ↺: The potential zone where a scan is reversed (E_{rev}). ⊢— or —⊣: The potential zone where a given CV scan is finished (E_{end}). $E^{\theta'}_{IRRS}$: The formal potential zone where IRRS is electroactive. $E^{\theta'}_{ANS}$: The formal potential zone where ANS is electroactive. Adapted from [45] with permission from RSC

$E_{start} = -0.25$ V, $E_{rev} = -0.75$ V, and $E_{end} = E_{start} = -0.25$ V produces the independent voltammogram for ANS. Meanwhile, by setting $E_{start} = -0.25$ V, $E_{rev} = 0.25$ V, and $E_{end} = E_{start}$, the independent voltammogram for IRRS is obtained. Note that there are no conditions where there will be significant concentrations of both RED_{ANS} and OX_{IRRS} generated within the diffusion layer. Of course, one could execute multiple CV cycles over the entire potential range (e.g., between −0.75 and 0.25 V) and generate the full menu of species and a cross-reaction [45].

Paradigms B-1a (Eq. 3.13) and B-1b (Eq. 3.14) have identical arrangements of the RED_{IRRS} and RED_{ANS} but differ in the choice of the reversal zone. The B-2a and B-2b paradigms are analogously related, the oxidation state of the electroactive species (OX_{ANS} and OX_{IRRS}) being the major difference. Depending upon the initial scan direction and reversing potential the B-paradigm will produce independent CV responses for ANS and will also produce the mixed CV response for ANS and IRRS (Eqs. 3.13–3.15). In this particular case scenario, the CV response for IRRS is on top of the CV response for ANS [45].

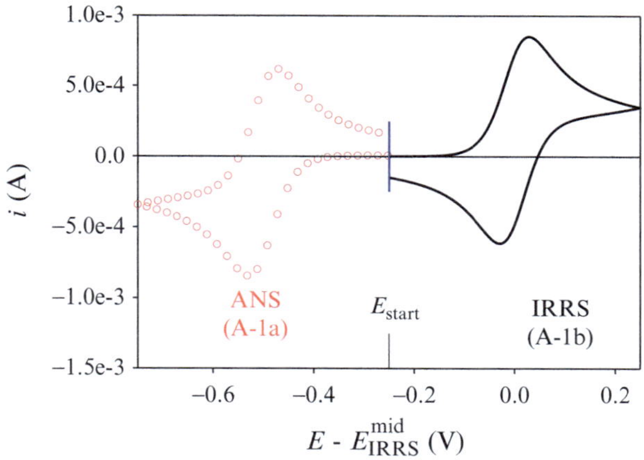

Fig. 3.12 Simulations of the CV responses for ANS (*open red circles*, paradigm A-1a) and IRRS (*black solid line*, Paradigm A-1b) with electrode area = 1.0 cm^2, scan rate = 1.0 V s^{-1}, T = 298.2 K, $c_{OX_{ANS}} = 1\times10^{-3}$ M, $D_{OX_{ANS}} = D_{RED_{ANS}} = D_{OX_{IRRS}} = D_{RED_{IRRS}} = 1.0\times10^{-5}\,\text{cm}^2\text{s}^{-1}$, $E^{\theta\prime}_{ANS} = -0.5$ V, $c_{RED_{IRRS}} = 1\times10^{-3}$ M, $E^{\theta\prime}_{IRRS} = 0$ V, $k_{s,ANS} = k_{s,IRRS} = 1.0\times10^4$ cm s^{-1}, $K_{cross} = 2.82\times10^8$, rate constant for a simple outer-sphere second-order cross-reaction ($k_{cross} = 1.0\times10^7$ M^{-1} s^{-1}, $E_{start} = -0.25$ V). Reproduced from [45] with permission from the RSC

$$\mathbf{B-1a}: \text{OX}_{ANS} + \text{e} \Leftrightarrow \text{RED}_{ANS} \quad (\text{RED}_{ANS}\ \text{initially present}) \tag{3.13}$$

$$\mathbf{B-1b}: \text{OX}_{IRRS} + \text{e} \Leftrightarrow \text{RED}_{IRRS} \quad (\text{RED}_{IRRS}\ \text{initially present}) \tag{3.14}$$

$$\text{RED}_{ANS} + \text{OX}_{IRRS} \Leftrightarrow \text{OX}_{ANS} + \text{RED}_{IRRS} \quad (K_{cross} = 2.82\times10^8) \tag{3.15}$$

Simulated examples of the B-1a and B-1b voltammograms are shown in Fig. 3.13. Setting E_{start} = −0.75 V, E_{rev} = −0.25 V, and $E_{end} = E_{start}$ = −0.75 V produces the independent voltammogram for ANS; setting E_{start} = −0.75 V, E_{rev} = 0.25 V, and $E_{end} = E_{start}$ produces the mixed response with IRRS response on top of ANS response. Note that as soon as an increasingly positive potential begins to drive Eq. (3.14) to the left, thereby producing OX_{IRRS}, Eq. (3.15) will proceed to the right creating an additional path for the generation of OX_{ANS} [45].

Paradigm C will produce an independent CV response for IRRS (Eq. 3.16) and will also produce the mixed CV response for IRRS and ANS (Eqs. 3.16–3.18). In this case, the CV response for ANS is on top of the CV response for IRRS. Please note that paradigm C cannot produce an independent CV response for ANS.

$$\mathbf{C-1a}: \text{OX}_{IRRS} + \text{e} \Leftrightarrow \text{RED}_{IRRS} \quad (\text{RED}_{IRRS}\ \text{initially present}) \tag{3.16}$$

$$\mathbf{C-1b}: \text{OX}_{ANS} + \text{e} \Leftrightarrow \text{RED}_{ANS} \quad (\text{RED}_{ANS}\ \text{initially present}) \tag{3.17}$$

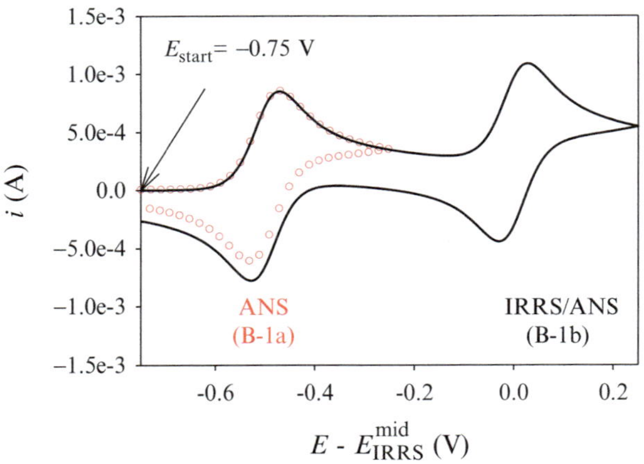

Fig. 3.13 Simulations of the CV responses for ANS (*open red circles*, paradigms B-1a) and IRRS/ANS (*solid black line*, paradigms B-1b) with electrode area = 1.0 cm^2, scan rate = 1.0 V s^{-1}, T = 298.2 K, $c_{RED_{ANS}} = 1\times10^{-3}$ M, $D_{OX_{ANS}} = D_{RED_{ANS}} = D_{OX_{IRRS}} = D_{RED_{IRRS}} = 1.0\times10^{-5}\,\text{cm}^2\text{s}^{-1}$, $E^{\theta\prime}_{ANS} = -0.5$ V, $c_{RED_{IRRS}} = 1\times10^{-3}$ M, $E^{\theta\prime}_{IRRS} = 0$ V, $k_{s,ANS} = k_{s,IRRS} = 1.0\times10^4$ cm s^{-1}, $K_{cross} = 2.82\times10^8$, $k_{cross} = 1.0\times10^7\,\text{M}^{-1}\text{s}^{-1}$, $E_{start} = -0.75$ V. Reproduced from [45] with permission from the RSC

$$\text{OX}_{\text{ANS}} + \text{RED}_{\text{IRRS}} \Leftrightarrow \text{RED}_{\text{ANS}} + \text{OX}_{\text{IRRS}} \qquad \left(K_{\text{cross}} = 2.82\times10^{8}\right) \tag{3.18}$$

$$\text{OX}_{\text{ANS}} \Leftrightarrow \text{P} \qquad \left(K_{\text{d}} = 1.0\times10^{5};\ \ k_{\text{d}} = 10.0\,\text{s}^{-1}\right) \tag{3.19}$$

Figure 3.14 shows simulated examples of the C-1a and C-1b voltammograms. Setting $E_{start} = -0.25$ V, $E_{rev} = 0.25$ V, and $E_{end} = E_{start} = -0.25$ V produces the independent voltammogram for IRRS; setting $E_{start} = -0.25$ V, $E_{rev} = 0.75$ V, and $E_{end} = E_{start}$ produces the mixed response with the ANS response on top of the IRRS response.

Any ANS/IRRS mixed response (see Table 3.8, paradigms B-1b, B-2b, C-1b, and C-2b) is potentially problematic in that the CV response is a combination of the responses for the two electroactive systems [45]. Extracting an accurate estimate of that potential from a mixed response may not be that difficult (see Figs. 3.13 and 3.14) even when the ANS response is more complicated than a simple outer-sphere one-electron transfer. There is, however, an additional complication: every mixed response produces a thermodynamically favored cross-reaction (see Eqs. 3.15 and 3.18). Under certain conditions (both IRRS and ANS are perfectly reversible-electron transfers, a simple second-order cross-reaction and identical diffusion coefficients for all species) the cyclic voltammetric response is completely independent of the value of k_{cross} (see the analysis of Andrieux and Savéant [71] for the closely related sequential electron transfer, EE, mechanism [41]). Of course, it is

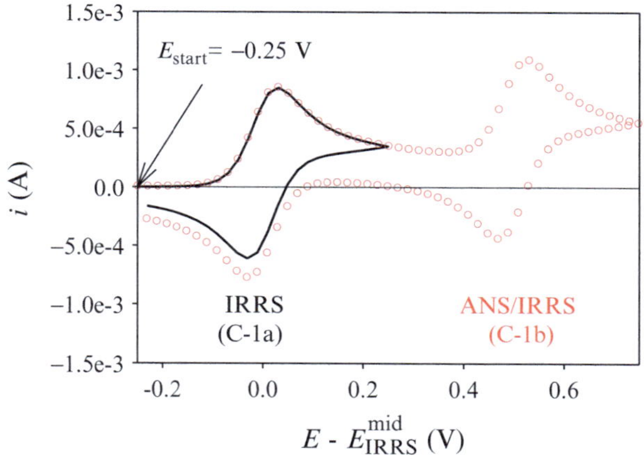

Fig. 3.14 Simulations of the CV responses for IRRS (*solid black line*, paradigm C-1a) and ANS/IRRS (*open red circles*, paradigm C-1b) with electrode area = 1.0 cm^2, scan rate = 1.0 V s^{-1}, T = 298.2 K, $c_{\mathrm{RED}_{\mathrm{ANS}}} = 1 \times 10^{-3}\,\mathrm{M}$, $D_{\mathrm{OX}_{\mathrm{ANS}}} = D_{\mathrm{RED}_{\mathrm{ANS}}} = D_{\mathrm{OX}_{\mathrm{IRRS}}} = D_{\mathrm{RED}_{\mathrm{iRRS}}} = 1.0 \times 10^{-5}\,\mathrm{cm^2 s^{-1}}$, $E^{\theta\prime}_{\mathrm{ANS}} = -0.5\,\mathrm{V}$, $c_{\mathrm{RED}_{\mathrm{IRRS}}} = 1 \times 10^{-3}\,\mathrm{M}$, $E^{\theta\prime}_{\mathrm{IRRS}} = 0\,\mathrm{V}$, $k_{\mathrm{s,ANS}} = k_{\mathrm{s,IRRS}} = 1.0 \times 10^4$ cm s^{-1}, $\mathrm{K_{cross}} = 2.82 \times 10^8$, $k_{\mathrm{cross}} = 1.0 \times 10^7\,\mathrm{M^{-1}s^{-1}}$, $\mathrm{E_{start}} = -0.25$ V. Reproduced from [45] with permission from the RSC

unlikely that the preconditions will be precisely met. Furthermore, when oxidation of $\mathrm{RED_{ANS}}$ or reduction of $\mathrm{OX_{ANS}}$ is not a mechanistically simple redox process, the cyclic voltammetric response can be quite sensitive to the magnitude of the rate constant for the cross-reaction, k_{cross} [45].

Figure 3.15 shows an example of this situation, where a C-paradigm system comprising the previously discussed ANS and IRRS reactions (Eqs. 3.16–3.18) and the added homogeneous reaction (Eq. 3.19) are present [45]. In this figure, voltammograms with $k_{\mathrm{cross}} = 0$ and $k_{\mathrm{cross}} = 1.0 \times 10^7$ M^{-1} s^{-1} and $k_{\mathrm{d}} = 10.0$ s^{-1} are compared. The presence of the first-order reaction (Eq. 3.19) amplifies the effect of the cross-reaction. Based on these results, it could be suggested that a conservative but wise experimental protocol is to use the paradigm A as the first option. If this is impossible, then paradigm B may be used. Both can produce independent responses for ANS.

We noted earlier that under certain, precisely defined conditions the cyclic voltammetric response is perfectly independent of the value of k_{cross} when paradigms B and C are operative [45]. However, the insensitivity of the electrochemical response to the value of k_{cross} does not mean that nothing is happening: even when those preconditions prevail (and when $K_{\mathrm{cross}} \gg 1$), the composition of the diffusion layer is dramatically sensitive to the value of k_{cross} and that can be seen spectroelectrochemically [72]. If k_{cross} is large enough the highly favored cross-reaction can produce what Amatore et al. refer to as an acute reaction front [73]. Thus, certainly for spectroelectrochemical measurements the analysis of ANS spectroelectrochemical response should be limited to paradigms A and B.

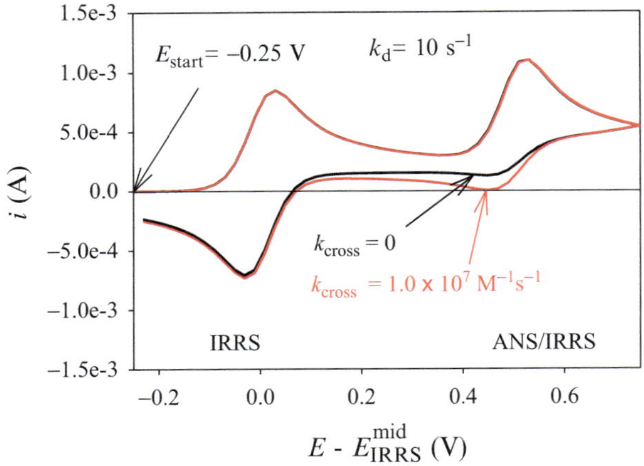

Fig. 3.15 Simulations of the CV responses for $k_{cross} = 0$ (*solid black line*) and $k_{cross} = 1.0 \times 10^7\ M^{-1}\ s^{-1}$ (*solid red line*); electrode area = 1.0 cm^2, scan rate = 1.0 V s^{-1}, $T = 298.2$ K, $c_{RED_{ANS}} = 1 \times 10^{-3}$ M, $D_{OX_{ANS}} = D_{RED_{ANS}} = D_{OX_{IRRS}} = D_{RED_{IRRS}} = 1.0 \times 10^{-5}\,cm^2 s^{-1}$, $E^{\theta\prime}_{ANS} = -0.5$ V, $c_{RED_{IRRS}} = 1 \times 10^{-3}$ M, $E^{\theta\prime}_{IRRS} = 0$ V, $k_{s,ANS} = k_{s,IRRS} = 1.0 \times 10^4$ cm s^{-1}, $K_{cross} = 2.82 \times 10^8$, $K_d = 10^5$, $E_{start} = -0.25$ V. Reproduced from [45] with permission from the RSC

3.4 Recommendations

As a conclusion, good practice for electrochemistry in ILs dictates that

- If anything other than qualitative data is sought, the use of QRE needs to be avoided. If impossible, the potential of the QRE must be calibrated against an IRRS. The key requirement under this situation is that the voltammetric response for ANS must be independent of the IRRS response (paradigms A or B). As we noted above, the selection of the optimal IRRS is based, to a great extent, on common sense. Errors associated with the QRE drift during the voltammetric experiments must be considered.
- In situations where a large number of experiments are to be performed in the same IL, it is recommended to prepare a Ag|0.01 M Ag[CF_3SO_3] reference electrode using the IL under study as the solvent in the reference system. If the solubility of Ag[CF_3SO_3] in the IL becomes an issue, this can be substituted by a silver salt that contains the same anion as the IL (e.g., in an IL in which the anion is [BF_4]$^-$, silver(I) could be introduced as Ag[BF_4]). Cryptane-based reference electrodes (Ag|Ag^+ 222, R) could be useful if the electrochemical experiment is exposed to a source of light. However, a precaution should be taken if R = acetonitrile, as it could be a source of impurity in long time experiments. Irrespective of which of these reference electrodes is used, it must be calibrated against an IRRS before and after the experiment, in order to evaluate possible drifts of the reference system during the electrochemical experiment.

- The redox potentials of Fc and Cc^{+} are sensitive to the solvation properties of the IL components. Consequently, the use of $Fc^{+/0}$ and $Cc^{+/0}$ couples as IRRS is expected to be adequate only when a single IL system is being studied; however, corrections may be required in conditions that require comparison between different IL systems.
- Even though DmFc is not freely soluble in ILs, the results presented throughout this chapter suggest that the $DmFc^{0/+}$ redox process is a suitable IRRS for voltammetric studies in ionic liquids, as it is less dependent on the IL nature in comparison to what is observed for ferrocene and cobaltocenium.
- If DmFc is not soluble in the IL under study, or if the $Fc^{+/0}$ and/or $Cc^{+/0}$ couples are preferred, the experimentally obtained E_m needs to be corrected to eliminate the errors associated with the inequality in the diffusion coefficients as well as the solvation effect of ILs ions.
- The spontaneous reaction observed between DmFc and oxygen brings important consequences in the establishment of conditions under which DmFc should be used. Indeed, it is a suitable IRRS to work under inert atmospheres, such as those present inside argon- or nitrogen-filled glove boxes. However, the application of this redox system as IRRS should be avoided when the experiments are being conducted under atmospheric conditions. Moreover, DmFc sample preparation in ILs should be also performed under an inert atmosphere.

References

1. Snook GA, Bond AM (2010) Reference electrodes for electrochemical studies in room temperature ionic liquids. In: Torriero AAJ, Shiddiky MJA (eds) Electrochemcial properties and applications of ionic liquids. Nova Science, New York, NY
2. Gritzner G, Kuta J (1983) Recommendations on reporting electrode potentials in nonaqueous solvents. Pure Appl Chem 56(4):461–466
3. AlNashef IM, Leonard ML, Matthews MA, Weidner JW (2002) Superoxide electrochemistry in an ionic liquid. Ind Eng Chem Res 41(18):4475–4478. doi:10.1021/ie010787h
4. Dickinson E, Williams ME, Hendrickson SM, Masui H, Murray RW (1999) Hybrid redox polyether melts based on polyether-tailed counterions. J Am Chem Soc 121(4):613–616
5. Endres F, Freyland W (1998) Electrochemical scanning tunneling microscopy investigation of HOPG and silver electrodeposition on HOPG from the acid room-temperature molten salt aluminum chloride 1-methyl-3-butyl-imidazolium chloride. J Phys Chem B 102(50):10229–10233
6. Hultgren VM, Mariotti AWA, Bond AM, Wedd AG (2002) Reference potential calibration and voltammetry at macrodisk electrodes of metallocene derivatives in the ionic liquid [bmim][PF6]. Anal Chem 74(13):3151–3156
7. McFarlane DR, Sun J, Golding J, Meakin P, Forsyth M High conductivity molten salts based on the imide ion. In: 6th international symposium on polymer electrolytes (ISPE-6), Hayama, Japan, 01–06 Nov 1998. Pergamon-Elsevier Science Ltd., oxford, pp 1271–1278
8. Schroder U, Wadhawan JD, Compton RG, Marken F, Suarez PAZ, Consorti CS, de Souza RF, Dupont J (2000) Water-induced accelerated ion diffusion: voltammetric studies in 1-methyl3--[2,6-(S)-dimethylocten-2-yl]imidazolium tetrafluoroborate, 1-butyl-3-methylimidazolium tetrafluoroborate and hexafluorophosphate ionic liquids. New J Chem 24(12):1009–1015

9. Zell CA, Endres F, Freyland W (1999) Electrochemical in situ STM study of phase formation during Ag and Al electrodeposition on Au(111) from a room temperature molten salt. PCCP 1(4):697–704
10. Zhang J, Bond AM (2003) Conditions required to achieve the apparent equivalence of adhered solid- and solution-phase voltammetry for ferrocene and other redox-active solids in ionic liquids. Anal Chem 75(11):2694–2702. doi:10.1021/ac026329f
11. Zhang J, Bond AM (2005) Practical considerations associated with voltammetric studies in room temperature ionic liquids. Analyst 130(8):1132–1147. doi:10.1039/b504721h
12. Zhang J, Bond AM, Schumann H, Suhring K (2005) Voltammetric studies on decaphenylferrocene, substituted decaphenylferrocenes, and their oxidized forms in dichloromethane and ionic liquids. Organometallics 24(9):2188–2196. doi:10.1021/om049117c
13. Caban K, Donten M, Stojek Z (2004) Electroformation of microlayers of ionic liquids in undiluted nitromethane and its homologues. Unusual oscillations behind the range of limiting steady-state current. J Phys Chem B 108(3):1153–1159. doi:10.1021/jp036637x
14. Brown RJC, Dyson PJ, Ellis DJ, Welton T (2001) 1-Butyl-3-methylimidazolium cobalt tetracarbonyl [bmim][Co(CO)4]: a catalytically active organometallic ionic liquid. Chem Commun 18:1862–1863
15. Endres F, Schrodt C (2000) In situ STM studies on germanium tetraiodide electroreduction on Au(111) in the room temperature molten salt 1-butyl-3-methylimidazolium hexafluorophosphate. PCCP 2(24):5517–5520
16. Suarez PAZ, Selbach VM, Dullius JEL, Einloft S, Piatnicki CMS, Azambuja DS, deSouza RF, Dupont J (1997) Enlarged electrochemical window in dialkyl-imidazolium cation based room-temperature air and water-stable molten salts. Electrochim Acta 42(16):2533–2535
17. NuLi Y, Yang J, Wang JL, Xu JQ, Wang P (2005) Electrochemical magnesium deposition and dissolution with high efficiency in ionic liquid. Electrochem Solid-State Lett 8(11):C166–C169. doi:10.1149/1.2052048
18. Inzelt G, Lewenstam A, Scholz F (2013) Handbook of reference electrodes. Springer, Berlin
19. Snook GA, Best AS, Pandolfo AG, Hollenkamp AF (2006) Evaluation of a Ag|Ag+ reference electrode for use in room temperature ionic liquids. Electrochem Commun 8(9):1405–1411
20. Torriero AAJ, Sunarso J, Howlett PC (2012) Critical evaluation of reference systems for voltammetric measurements in ionic liquids. Electrochim Acta 82:60–68. doi:10.1016/j.electacta.2012.01.115
21. Huber B, Roling B (2011) Development of a Ag/Ag+ micro-reference electrode for electrochemical measurements in ionic liquids. Electrochim Acta 56(19):6569–6572. doi:10.1016/j.electacta.2011.02.055
22. Hapiot P, Lagrost C (2008) Electrochemical reactivity in room-temperature ionic liquids. Chem Rev 108(7):2238–2264. doi:10.1021/cr0680686
23. Silvester DS, Roger EI, Compton RG, McKenzie KJ, Ryder KS, Endres F, MacFarlane D, Abbott AJ (2008) Technical aspects. In: Endres F, MacFarlane D, Abbott A (eds) Electrodeposition from ionic liquids. Wiley-VCH, Weinheim
24. Weingarth D, Foelske-Schmitz A, Wokaun A, Kötz R (2012) PTFE bound activated carbon – a quasi-reference electrode for ionic liquids. Electrochem Commun 18(1):116–118
25. Torriero AAJ (2014) Characterization of decamethylferrocene and ferrocene in ionic liquids: argon and vacuum effect on their electrochemical properties. Electrochim Acta 137:235–244. doi:10.1016/j.electacta.2014.06.005
26. Lewandowski A, Waligora L, Galinski M (2009) Ferrocene as a reference redox couple for aprotic ionic liquids. Electroanalysis 21(20):2221–2227. doi:10.1002/elan.200904669
27. Saheb A, Janata J, Josowicz M (2006) Reference electrode for ionic liquids. Electroanalysis 18(4):405–409. doi:10.1002/elan.200503435
28. Katayama Y (2005) In: Ohno HE (ed) Electrochemical aspects of ionic liquids. Wiley, Hoboken, NJ
29. De Vreese P, Haerens K, Matthijs E, Binnemans K (2012) Redox reference systems in ionic liquids. Electrochim Acta 76:242–248. doi:10.1016/j.electacta.2012.04.108

30. Lewandowski A, Osinska M, Swiderska-Mocek A, Galinski M (2008) A cryptate reference electrode for ionic liquids. Electroanalysis 20(17):1903–1908. doi:10.1002/elan.200804267
31. Lewandowski A, Swiderska-Mocek A (2010) Lithium-metal potential in Li+ containing ionic liquids. J Appl Electrochem 40(3):515–524. doi:10.1007/s10800-009-0023-z
32. Rogers EI, Silvester DS, Jones SEW, Aldous L, Hardacre C, Russell AJ, Davies SG, Compton RG (2007) Electrochemical kinetics of Ag vertical bar Ag+ and TMPD vertical bar TMPD+center dot in the room-temperature ionic liquid [C(4)mpyrr][NTf2]; toward optimizing reference electrodes for voltammetry in RTILs. J Phys Chem C 111(37):13957–13966. doi:10.1021/jp0737754
33. Andriyko YO, Reischl W, Nauer GE (2009) Trialkyl-substituted imidazolium-based ionic liquids for electrochemical applications: basic physicochemical properties. J Chem Eng Data 54(3):855–860. doi:10.1021/je800636k
34. Ikeda Y, Hiroe K, Asanuma N, Shirai A (2009) Electrochemical studies on uranyl(VI) chloride complexes in ionic liquid, 1-butyl-3-methylimidazolium chloride. J Nucl Sci Technol 46(2):158–162
35. Kakiuchi T, Yoshimatsu T, Nishi N (2007) New class of Ag/AgCl electrodes based on hydrophobic ionic liquid saturated with AgCl. Anal Chem 79(18):7187–7191. doi:10.1021/ac070820v
36. Lodermeyer J, Multerer M, Zistler M, Jordan S, Gores HJ, Kipferl W, Diaconu E, Sperl M, Bayreuther G (2006) Electroplating of dysprosium, electrochemical investigations, and study of magnetic properties. J Electrochem Soc 153(4):C242–C248. doi:10.1149/1.2172548
37. Chu D, Qin GX, Yuan XM, Xu M, Zheng P, Lu J (2008) Fixation of CO2 by electrocatalytic reduction and electropolymerization in ionic liquid-H2O solution. ChemSusChem 1(3):205–209. doi:10.1002/cssc.200700052
38. Du YF, Wang H, Zhang AJ, Lu JX (2007) Electrosynthesis of poly(o-phenytenediamine) in ionic liquid and its properties. Chin Sci Bull 52(16):2174–2178. doi:10.1007/s11434-007-0331-9
39. Liu YZ, Lin MY, Xiao LP, Zhang K, Lu JX (2008) Electrochemical reduction of 2-nitroanisole in ionic liquid BMIm-BF4: synthesis of 2-anisidine. Chin J Chem 26(7):1168–1172
40. Shkurankov A, El Abedin SZ, Endres F (2007) AFM-assisted investigation of the corrosion behaviour of magnesium and AZ91 alloys in an ionic liquid with varying water content. Aust J Chem 60(1):35–42. doi:10.1071/ch06305
41. Bard AJ, Faulkner LR (2001) Electrochemical methods: fundamentals and applications, 2nd edn. Wiley, New York, NY
42. Izutzu K (2002) Electrochemistry in nonaqueous solutions. Wiley-VCH, New York
43. Smith TJ, Stevenson KJ (2007) Reference electrodes. In: Zoski CG (ed) Handbook of electrochemistry. Elsevier, Amsterdam, pp 73–110
44. Torriero AAJ, Bond AM (2009) Critical Evaluation of Electrochemistry in Ionic Liquids. In: Hayashi K (ed) Electroanalytical chemistry research trends. Nova Science, New York, NY
45. Torriero AAJ, Feldberg S, Zhang J, Simonov A, Bond A (2013) On choosing a reference redox system for electrochemical measurements: a cautionary tale. J Solid State Electrochem 17:3021–3026. doi:10.1007/s10008-013-2183-3
46. Rogers EI, Silvester DS, Poole DL, Aldous L, Hardacre C, Compton RG (2008) Voltammetric characterization of the ferrocene/ferrocenium and cobaltocenium/cobaltocene redox couples in RTILs. J Phys Chem C 112(7):2729–2735
47. Sukardi SK, Zhang J, Burgar I, Horne MD, Hollenkamp AF, MacFarlane DR, Bond AM (2008) Prospects for a widely applicable reference potential scale in ionic liquids based on ideal reversible reduction of the cobaltocenium cation. Electrochem Commun 10:250–254
48. Noviandri I, Brown KN, Fleming DS, Gulyas PT, Lay PA, Masters AF, Phillips L (1999) The decamethylferrocenium/decamethylferrocene redox couple: a superior redox standard to the ferrocenium/ferrocene redox couple for studying solvent effects on the thermodynamics of electron transfer. J Phys Chem B 103(32):6713–6722
49. Barriere F, Geiger WE (2006) Use of weakly coordinating anions to develop an integrated approach to the tuning of delta e1/2 values by medium effects. J Am Chem Soc 128(12): 3980–3989

50. Yang Y, Yu L (2013) Theoretical investigations of ferrocene/ferrocenium solvation in imidazolium-based room-temperature ionic liquids. PCCP 15(8):2669–2683
51. Ruiz-Diaz JJJ, Torriero AAJ, Salinas E, Marchevsky EJ, Sanz MI, Raba J (2006) Enzymatic rotating biosensor for cysteine and glutathione determination in a FIA system. Talanta 68(4):1343–1352
52. Aranzaes JR, Daniel M-C, Astruc D (2006) Metallocenes as references for the determination of redox potentials by cyclic voltammetry – permethylated iron and cobalt sandwich complexes, inhibition by polyamine dendrimers, and the role of hydroxy – containing ferrocenes. Can J Chem 84(2):288–299. doi:10.1139/v05-262
53. Ruiz J, Astruc D (1998) Permethylated electron-reservoir sandwich complexes as references for the determination of redox potentials. Suggestion of a new redox scale. Comptes Rendus de l'Academie des Sciences, Serie IIc: Chimie 1(1):21–27
54. Matsumoto M, Swaddle TW (2004) The decamethylferrocene(+/0) electrode reaction in organic solvents at variable pressure and temperature. Inorg Chem 43(8):2724–2735
55. Freyberg DP, Robbins JL, Raymond KN, Smart JC (1979) Crystal and molecular structures of decamethylmanganocene and decamethylferrocene. Static Jahn-Teller distortion in a metallocene. J Am Chem Soc 101(4):892–897. doi:10.1021/ja00498a017
56. Bennett MA, Bhargava SK, Bond AM, Burgar IM, Guo SX, Kar G, Priver SH, Wagler J, Willis AC, Torriero AAJ (2010) Synthesis, X-ray structure and electrochemical oxidation of palladium(II) complexes of ferrocenyldiphenylphosphine. Dalton Trans 39(38):9079–9090
57. Torriero AAJ, Howlett PC (2012) Ionic liquid effects on the redox potential of ferrocene. Electrochem Commun 16(1):84–87. doi:10.1016/j.elecom.2011.12.006
58. Torriero AAJ, Siriwardana AI, Bond AM, Burgar IM, Dunlop NF, Deacon GB, MacFarlane DR (2009) Physical and electrochemical properties of thioether-functionalized ionic liquids. J Phys Chem B 113:11222–11231
59. Lewandowski A, Waligora L, Galinski M (2013) Electrochemical behavior of cobaltocene in ionic liquids. J Solution Chem 42(2):251–262. doi:10.1007/s10953-013-9957-1
60. Waligora L, Lewandowski A, Gritzner G (2009) Electrochemical studies of four organometallic redox couples as possible reference redox systems in 1-ethyl-3-methylimidazolium tetrafluoroborate. Electrochim Acta 54(5):1414–1419
61. Koepp ZM, Wendt H, Strehlow H (1960) Der Vergleich der Spannungsreihen in verschiedenen Solventien. II. Elektrochemie 64:483
62. Stojanovic RS, Bond AM (1993) Examination of conditions under which the reduction of the cobaltocenium cation can be used as a standard voltammetric reference process in organic and aqueous solvents. Anal Chem 65(1):56–64. doi:10.1021/ac00049a012
63. Zhao C, Burrell G, Torriero AAJ, Separovic F, Dunlop NF, MacFarlane DR, Bond AM (2008) Electrochemistry of room temperature protic ionic liquids. J Phys Chem B 112(23):6923–6936
64. Duffy NW, Bond AM (2006) Macroelectrode voltammetry in toluene using a phosphonium–phosphate ionic liquid as the supporting electrolyte. Electrochem Commun 8:892–898
65. Torriero AAJ, Sunarso J, Forsyth M, Pozo-Gonzalo C (2013) Assessment of permethylated transition-metal sandwich complexes as internal reference redox systems in ionic liquids. PCCP 15(7):2547–2553
66. Torriero AAJ, Shiddiky MJA, Bullock JP, Boas JF, MacFarlane DR, Bond AM (2010) Electrooxidation of $[(\eta^5\text{-}C_5H_5)Fe(CO)_2]_2$ as a probe of the nucleophilic properties of ionic liquid anions. Inorg Chem 49(5):2502–2511
67. Tokuda H, Tsuzuki S, Susan MABH, Hayamizu K, Watanabe M (2006) How ionic are room-temperature ionic liquids? An indicator of the physicochemical properties. J Phys Chem B 110(39):19593–19600
68. Bhatt AI, Bond AM, MacFarlane DR, Zhang J, Scott JL, Strauss CR, Iotov PI, Kalcheva SV (2006) A critical assessment of electrochemistry in a distillable room temperature ionic liquid, DIMCARB. Green Chem 8:161–171
69. Fu C-P, Aldous L, Dickinson EJF, Manan NSA, Compton RG (2011) The kinetics of ferrocene volatilisation from an ionic liquid. ChemPhysChem 12(9):1708–1713

70. Fu C, Aldous L, Dickinson EJF, Manan NSA, Compton RG (2011) Volatilisation of ferrocene from ionic liquids: kinetics and mechanism. Chem Commun 47(25):7083–7085
71. Andrieux CP, Saveant JM (1970) Electro-oxidation of 1,2-ene-diamines. J Electroanal Chem 28:339–348
72. Strojek JW, Kuwana T, Feldberg SW (1968) Spectrochemical-electrochemical evaluation of kinetics using optically transparent electrodes. J Am Chem Soc 90:1353–1355
73. Amatore C, Klymenko O, Svir I (2010) A new strategy for simulation of electrochemical mechanisms involving acute reaction fronts in solution: principle. Electrochem Commun 12:1170–1173

Chapter 4
Ultramicroelectrode Voltammetry and Scanning Electrochemical Microscopy in Room Temperature Ionic Liquids

Darren A. Walsh

4.1 Introduction

This year (2014) marks the 25th anniversary of the invention of SECM in the Bard lab at the University of Texas [1, 2]. Such is the impact of its development that the scanning electrochemical microscope (SECM—the same abbreviation is used for the instrument and the method) is now a "textbook" electroanalytical tool used in a wide variety of applications, ranging from fundamental studies of mass and charge transfer dynamics to molecular transport across biological cell membranes, corrosion analysis and electrocatalyst screening [3, 4]. Given the increasing popularity of room-temperature ionic liquids (RTILs) in electrochemical studies and devices [5, 6], it was perhaps inevitable that SECM measurements would also be carried out in RTILs and a number of such studies have appeared in the recent literature [7–10]. However, the unique physicochemical properties of RTILs can lead to very unusual behaviour being observed during SECM. For example, drastic differences in the rates of diffusion of charged and uncharged species can arise in RTILs and such phenomena have been studied using SECM [11]. The high viscosities of RTILs also mean that unusual SECM responses are often obtained in RTILs, unless one carefully controls the experimental parameters [9].

In this chapter, those studies that have involved performing SECM in RTILs will be discussed. As we will see, a relatively small number of such studies have been carried out to date, but unique electrochemical insights have been obtained. In attempting to fully describe how the properties of RTILs affect SECM in these media, the voltammetric behaviour of UMEs must also be described as the operation of the SECM depends on the behaviour of UMEs during electrolytic reactions. The chapter begins with a brief description of the principles of UME voltammetry and

D.A. Walsh (✉)
School of Chemistry, University of Nottingham, Nottingham NG7 2RD, UK
e-mail: darren.walsh@nottingham.ac.uk

A.A.J. Torriero (ed.), *Electrochemistry in Ionic Liquids*,
DOI 10.1007/978-3-319-13485-7_4

SECM. The origin of the unusual voltammetric and SECM responses obtained in RTILs is discussed and the effects of a range of electrochemical parameters on the responses are then described. Following this discussion, some of the electrochemical insights that have come from UME voltammetry and SECM in RTILs are discussed.

4.2 Principles of Ultramicroelectrode Voltammetry

UMEs are often defined as electrodes with critical dimensions in the range 1–25 μm. Compared to conventional "macroelectrodes" with mm (or larger) dimensions, UMEs exhibit faster double-layer charging, lower ohmic losses and higher rates of mass transport, and these advantages have allowed electrochemists to use UMEs to explore a wide array of phenomena inaccessible to study using macroelectrodes [12–14]. One of the key advantages of UMEs is their ability to carry steady-state (time-independent) currents during the electrolysis of redox-active species, a phenomenon that has been exploited in the development of the SECM. In describing why steady-state currents are often observed at UMEs and not usually at macroelectrodes, we consider first the electrochemical reduction of species Ox to species Red at a spherical electrode of radius r_0 in a solution containing only dissolved Ox and an inert supporting electrolyte. A spherical electrode is considered here because its surface is uniformly accessible and a rigorous mathematical treatment leads to a closed-form solution of the mass transport equation. Reduction of Ox occurs by the addition of n electrons to each molecule:

$$\mathrm{Ox} + ne^- \rightarrow \mathrm{Red} \tag{4.1}$$

and, if the potential of the electrode is stepped from an initial potential where the reduction of Ox does not occur to one where Ox is reduced at the diffusion-controlled rate, the concentration gradient is given by integration of Fick's second law:

$$\frac{\partial c_{\mathrm{Ox}}(r,t)}{\partial t} = D_{\mathrm{Ox}}\left[\frac{\partial^2 c_{\mathrm{Ox}}(r,t)}{\partial r^2} + \frac{2}{r}\frac{\partial c_{\mathrm{Ox}}(r,t)}{\partial r}\right] \tag{4.2}$$

where $c_{\mathrm{Ox}}(r,t)$ is the concentration of Ox at a distance r from the centre of the sphere and time t after the potential step and D_{Ox} is the diffusion coefficient of Ox. The boundary conditions for the potential-step experiment are

$$\lim_{r\to\infty} c_{\mathrm{Ox}}(r,t) = c_{\mathrm{Ox}}^{\infty} \tag{4.3}$$

$$c_{\mathrm{Ox}}(r,0) = c_{\mathrm{Ox}}^{\infty} \text{ for } r > r_0 \tag{4.4}$$

$$c_{\mathrm{Ox}}(r_0,t) = 0 \text{ for } t > 0 \tag{4.5}$$

where c_{Ox}^{∞} is the bulk solution concentration of Ox. Equation (2) can be solved using Laplace transform methods to give the faradaic current, *i*, flowing through the electrode as a function of *t* [14]:

$$i(t) = \frac{nFAD_{\mathrm{Ox}} c_{\mathrm{Ox}}^{\infty}}{r_0} + \frac{nFAD_{\mathrm{Ox}}^{1/2} c_{\mathrm{Ox}}^{\infty}}{\pi^{1/2} t^{1/2}} \tag{4.6}$$

where *F* is the Faraday constant (96,485 C/mol) and *A* is the electrode area. Therefore, the current flowing through a spherical electrode during the reduction of Ox is the sum of time-independent (steady-state) and time-dependent (transient) contributions. At large spherical electrodes (i.e. when r_0 is large), the time-independent term (first term on the right of Eq. 4.6) is negligible, the electrode can be considered planar, linear diffusion of Ox to the electrode surface dominates and *i* decays with $t^{1/2}$. However, at spherical UMEs (i.e. when r_0 is small), the time-independent term can become significant depending on the experimental conditions. If, for example, the timescale of the experiment is very short (i.e. *t* is small), the diffusion field remains small relative to the electrode radius, the second term on the right side of Eq. (4.6) dominates and the current decays with $t^{1/2}$ as expected for linear diffusion to a planar electrode (Fig. 4.1a). However, at long timescales, the diffusion field becomes large relative to r_0, the second term on the right side of Eq. (4.6) decays such that its contribution to *i* becomes negligible and the spherical character of the diffusion field becomes important (Fig. 4.1b). In this case, a steady-state current, i_{ss}, flows through the UME, which is given by

$$i_{\mathrm{ss}} = 4\pi nFD_{\mathrm{Ox}} c_{\mathrm{Ox}}^{\infty} r_0 \tag{4.7}$$

where the area of the electrode, *A*, is given by $4\pi r_0^2$.

By far, the most common UME geometry is that of a microdisc inlaid in an insulating sheath but, unlike spherical electrodes, the surface of an inlaid microdisc UME is not uniformly accessible as electrolysis at the edges of the microdisc diminishes diffusion to the disc centre. A rigorous mathematical treatment of the steady-state current at a microdisc is not possible but similar considerations to those described above apply. The faradaic current flowing at a microdisc UME during the electrolysis of a redox-active species contains contributions from a transient component

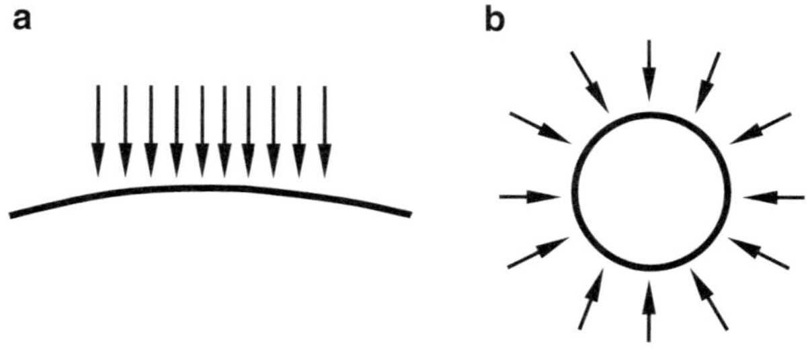

Fig. 4.1 (**a**) Linear diffusion to the surface of a spherical electrode at short timescales and (**b**) spherical diffusion to the electrode surface at long timescales

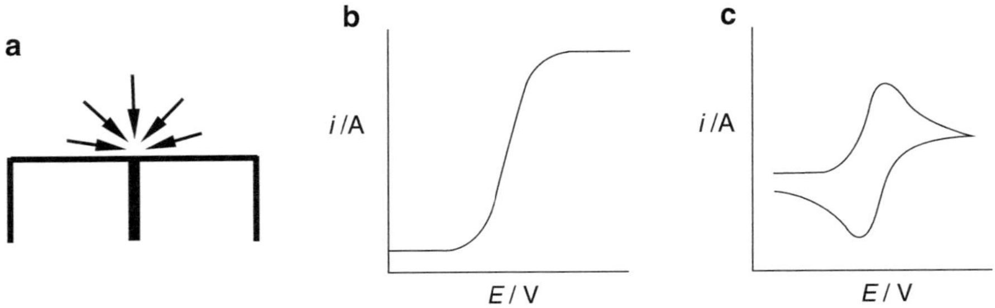

Fig. 4.2 (**a**) Hemispherical diffusion to a microdisk ultramicroelectrode. (**b**) A steady-state cyclic voltammogram recorded using an ultramicroelectrode. (**c**) A transient cyclic voltammogram recorded at a macroelectrode

and a steady-state component. The steady-state component arises due to hemispherical diffusion to the microdisc surface (Fig. 4.2a), which leads to a steady-state current given by

$$i_{ss} = 4nFD_{Ox}c^{\infty}_{Ox}a\beta \tag{4.8}$$

where a is the radius of the microdisc and β depends on the geometry of the UME. When the ratio of the radius of the non-conducting sheath to that of the microdisc ($R_g = r_g/a$, where r_g is the radius of the sheath) is large (>20), $\beta = 1.00$, but β increases until it equals 1.43 at $R_g = 1$ [15]. This means that i_{ss} at a microdisc UME with a very small R_g is approximately 40 % larger than that at a microdisc with large R_g due to the diffusion of redox species from behind the plane of the microdisc surface. A typical sigmoidal, steady-state cyclic voltammogram (CV) obtained at a microdisc UME is shown in Fig. 4.2b and differs from the peaked responses usually obtained at macroelectrodes (Fig. 4.2c), where linear diffusion dominates and the peak height, i_p, increases as the voltammetric scan rate, υ, increases ($i_p \propto \upsilon^{1/2}$ at macroelectrodes).

It is clear from the discussion above that diffusion of redox-active species to a disc UME can be dominated by a transient component or a steady-state component depending on the timescale of the experiment (which in a cyclic voltammetry experiment is controlled through υ). It can be shown that steady-state CVs should be recorded when [16]

$$\upsilon << \frac{RTD}{nFa^2} \tag{4.9}$$

where R is the universal gas constant, T is the absolute temperature and D is the diffusion coefficient of the redox species under study. Figure 4.3 shows a number of simulated CVs showing the effects of increasing υ during cyclic voltammetry at a 10-μm-radius UME (note that the diffusion coefficients D_{Ox} and D_{Red} used to generate the CVs in Fig. 4.3 are typical of those observed in “conventional” aqueous or organic electrolytes). At 0.01 V/s, a steady-state CV is obtained, demonstrating that

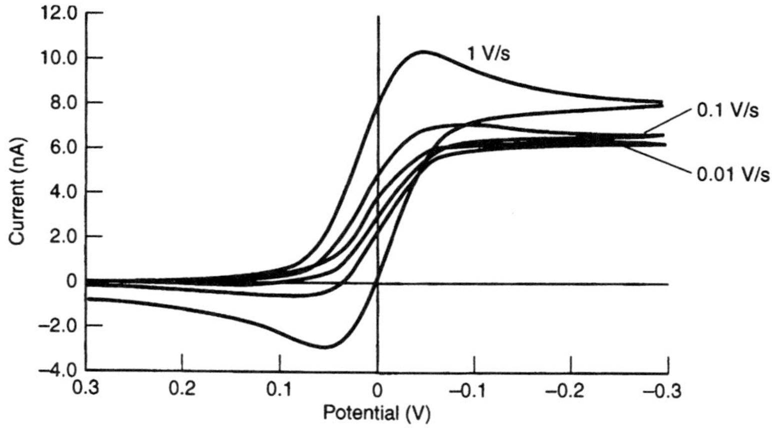

Fig. 4.3 Simulated CVs showing the effects of increasing the voltammetric scan rate during cyclic voltammetry using a 10-μm-radius UME. The responses were simulated for a reversible (Nernstian) reaction with $n = 1$, $D_{Ox} = D_{Red} = 10^{-5}$ cm^2/s, $c_{Ox} = 1.0$ mM, and $T = 25$ °C. Steady-state responses are observed at $\upsilon = 0.01$ V/s but at 0.1 V/s and above peaks begin to appear in the CVs. Reprinted with permission from Bard, A J; Faulkner, L R (2001) Electrochemical Methods: Fundamentals and Applications; 2nd ed. Copyright 2001 John Wiley and Sons

the steady-state component dominates at this υ. While the steady-state component is still dominant as υ increases to 1 V/s (note, for example, the high current at the switching potential), peaks emerge in the CV, illustrating that the transient component is significant at 1 V/s. In general, UME voltammetry is performed either at low or high scan rates, when the steady-state or transient component, respectively, dominates the observed behaviour [17, 18]. Equation (9) also illustrates why steady-state CVs cannot readily be recorded at macroelectrodes at reasonable timescales; if, for example, a is 1 mm and other experimental conditions are as in Fig. 4.3, a steady-state response should be observed at $\upsilon \ll 25$ μV/s, a scan rate that corresponds to an experimental "time window" longer than 10^3 s, in which convective effects will be significant and mass transport will no longer occur primarily by diffusion [16].

4.3 Principles of Scanning Electrochemical Microscopy

In SECM, the UME (which is usually called the SECM tip) can be moved in three dimensions, usually under piezo-control. The most common mode of SECM is the amperometric mode, in which a redox-active species is oxidised or reduced at the SECM tip and the effects of moving the tip close to the surface of a second electrode (which is usually called the substrate) are studied. To describe the operation of the amperometric mode of SECM, we again consider the reduction of species Ox to species Red at a disc-shaped SECM tip (Eq. 4.1), which results in the flow of a steady-state current through the tip (Eq. 4.8). In the SECM literature, the steady-state current at the tip (when the tip is far from any substrate) is usually identified as $i_{T,\infty}$ rather than i_{ss} and this convention will be used in the remainder of this chapter. If the SECM tip is brought close to the surface of the substrate, a number of effects can be observed depending on the composition of the substrate. If the substrate is made from an insulating material, diffusion of Ox to the SECM tip surface becomes hindered as the tip approaches the substrate and the tip current, i_T, decreases with decreasing tip–substrate distance, d (Fig. 4.4a). This effect is called negative feedback. Conversely, if the substrate is made from a conducting material, regeneration of Ox occurs at the substrate surface leading to an enhancement in the flux of Ox to the SECM tip as it approaches the substrate (Fig. 4.4b). This enhancement causes an increase in i_T as d decreases and this effect is called positive feedback. The curves shown in Fig. 4.4 are called SECM feedback approach curves and, when presenting such data, d is usually normalised to a and expressed as L and i_T is usually normalised to $i_{T,\infty}$. The theoretical basis for SECM is well developed [3, 4, 15, 19] and, while no exact analytical expressions have been found for positive or negative feedback, a number of approximate analytical equations describing the effect of L on $i_T/i_{T,\infty}$ have been developed. In 2006, an approximate equation describing the effect of positive feedback on the normalised current at a microdisc SECM tip, which included as variables R_g and L, was developed (earlier equations did not include R_g as a variable) [20]. The general expression, which is accurate to better than 2 % for any L and R_g, is as follows:

$$\frac{i_T}{i_{T,\infty}} = \alpha\left(R_g\right) + \frac{\pi}{4\beta\left(R_g\right)\arctan L} + \left(1 - \alpha\left(R_g\right) - \frac{1}{2\beta\left(R_g\right)}\right)\frac{2}{\pi}\arctan L \quad (4.10)$$

where α and β are given by Eqs. (4.11) and (4.12), respectively:

$$\alpha\left(R_g\right) = \ln 2 + \ln 2\left(1 - \frac{2}{\pi}\arccos\frac{1}{R_g}\right) - \ln 2\left[1 - \left(\frac{2}{\pi}\arccos\frac{1}{R_g}\right)^2\right] \quad (4.11)$$

$$\beta\left(R_g\right) = 1 + 0.639\left(1 - \frac{2}{\pi}\arccos\frac{1}{R_g}\right) - 0.186\left[1 - \left(\frac{2}{\pi}\arccos\frac{1}{R_g}\right)^2\right] \quad (4.12)$$

In the case of negative feedback at an insulating substrate, the following expression is accurate to better than 1 % [21]:

$$\frac{i_T}{i_{t,\infty}} = \frac{\frac{2.08}{R_g^{0.358}}\left(L - \frac{0.145}{Rg}\right) + 1.585}{\frac{2.08}{R_g^{0.358}}\left(L + 0.0023R_g\right) + 1.57 + \frac{\ln R_g}{L} + \frac{2}{\pi R_g}\ln\left(1 + \frac{\pi Rg}{2L}\right)} \quad (4.13)$$

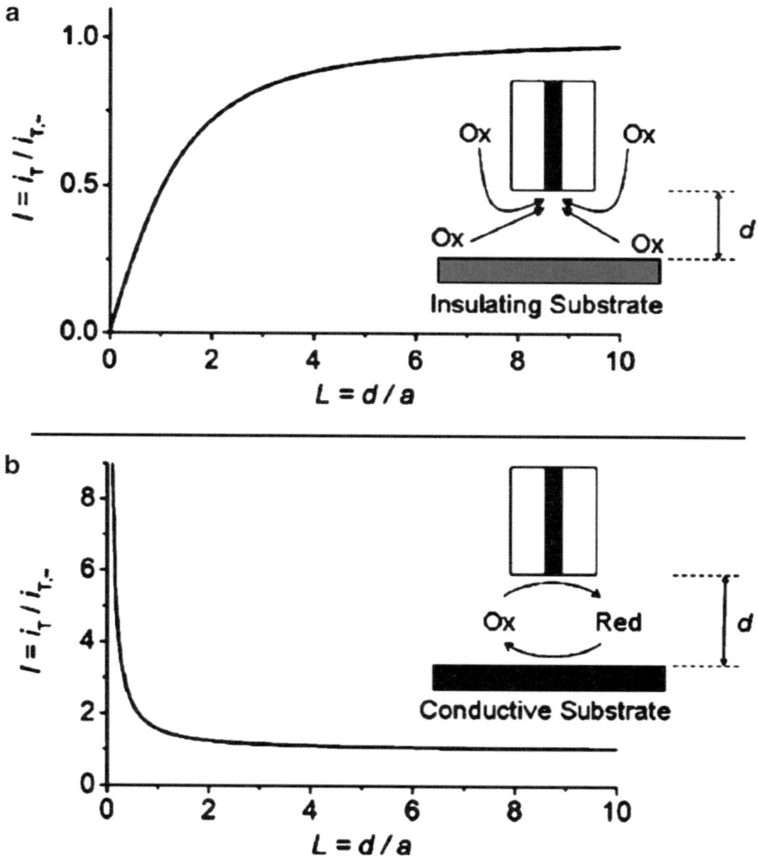

Fig. 4.4 SECM feedback approach for (**a**) negative feedback as an SECM tip approaches an insulating substrate and (**b**) positive feedback as an SECM tip approaches a conducting substrate. Reprinted with permission from Fan, F R; Fernández, J L; Liu, B; Mauzeroll, J (2007) Scanning Electrochemical Microscopy. In Zoski, C. G., Ed. Handbook of Electrochemistry. Copyright 2007 Elsevier BV

SECM feedback approach curves such as those shown in Fig. 4.4 are obtained when the rate of the substrate reaction is negligibly small (negative feedback) or very high (positive feedback). In the case of finite heterogeneous electron transfer (ET) kinetics at the substrate, intermediate SECM feedback approach curves are obtained, and theoretical kinetic SECM curves also exist that can be used to obtain kinetic information about the substrate reaction [4, 15]. The kinetics of the ET reaction occurring at an SECM tip can also be studied by positioning the SECM tip very close to the substrate and recording tip CVs and fitting the resulting data to theoretical voltammograms [22]. SECM is often also used in the amperometric imaging mode by scanning the tip across the surface and recording i_T as the tip passes over regions of varying conductivity or electrochemical activity [4].

Several alternative modes of SECM have also been developed that differ from the amperometric feedback modes described above. In generation-collection mode SECM, a redox-active species is electrochemically "generated" at either the tip or the substrate and is then "collected" at the opposing electrode by oxidising or reducing it. Generation-collection modes of SECM are especially useful for studying homogeneous reactions [23]. Chronoamperometric SECM, in which i_T is recorded as a function of time, is particularly useful for studying mass transport effects [24]. In alternating current (AC) mode SECM, an AC signal is superimposed onto the potentiostat input signal, which generates an alternating current signal that depends on L and the substrate properties [25]. A relatively recent advance is the development of a number of novel modes of SECM based on the use of hollow pipet tips filled with solutions. These new modes of SECM have allowed users to expand the applications of SECM to areas inaccessible to study using conventional SECM. For further information on these methods, the interested reader is directed to the cited works and references therein [26–29].

4.4 Instrumentation for Scanning Electrochemical Microscopy

A schematic of a typical SECM is shown in Fig. 4.5. The main components are the bipotentiostat, tip-positioning system, electrochemical cell and data acquisition/output system. While these basic components are common to almost all SECMs, the SECM tip can differ significantly depending on the particular system under investigation. Due to the ease of fabrication and polishing, microdisc SECM tips are most common. Microdisc tips are constructed by sealing microwires of metals such as Pt

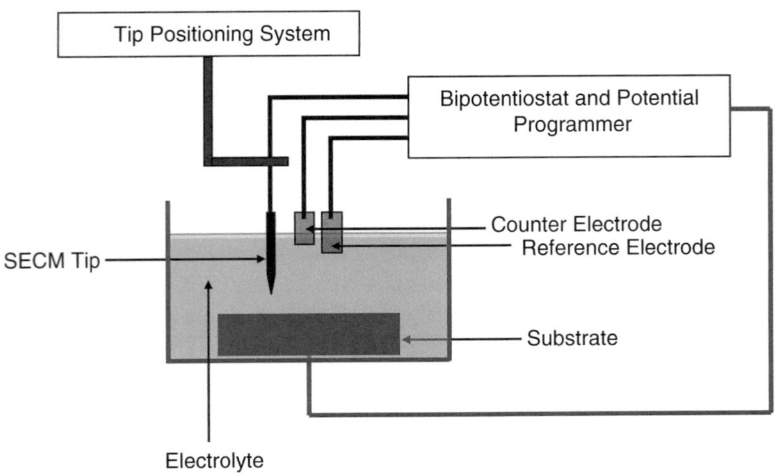

Fig. 4.5 A schematic diagram of a typical SECM

or Au into soft glass tubing and exposing and polishing the cross section of the wire. To ensure that the tips formed using this approach can be positioned close to substrate surfaces, they are usually sharpened to reduce R_g to below 10 (note that this sharpening affects $i_{T,\infty}$; see discussion above). For some applications, it is necessary to use SECM tips with $a \ll 10$ μm (e.g. for studying rapid charge transfer kinetics or for imaging sub-μm features on substrates) and fabricating, polishing and sharpening such very small SECM tips by hand are difficult. The most common method used for fabricating very small tips is to seal a metal microwire into quartz tubing and then to pull the quartz/microwire assembly to a fine point using a laser pipette puller [30, 31]. Other methods include insulating the sides of sharp tips with electrophoretic paint [32] and pushing sharpened wires through wax to leave just the tip exposed [33].

4.5 Steady-State Voltammetry at Ultramicroelectrodes in RTILs

If we consider again the parameters used to generate the CVs shown in Fig. 4.3, Eq. (4.9) suggests that steady-state CVs should be recorded in this system at $\upsilon \ll 0.25$ V/s and, indeed, Fig. 4.3 illustrates that linear diffusion to the UME affects the response at almost half that value. However, when performing electrochemical measurements in RTILs, one must consider that diffusion coefficients in RTILs are typically 2–3 orders of magnitude smaller than in conventional aqueous and organic electrolytes. If the diffusion coefficients used to generate the curves shown in Fig. 4.3 were typical of that expected in RTILs (e.g. 10^{-8} cm^2/s), Eq. (4.9) suggests that steady-state responses would only be recorded at $\upsilon \ll 0.3$ mV/s at a 10-μm-radius UME. This brief calculation suggests that transient responses may be obtained in RTILs at UMEs under typical UME voltammetry conditions (i.e. $\upsilon \sim 1$–10 mV/s and $a \sim 5$–12.5 μm). In an early investigation into such phenomena, Bond and co-workers showed that peak-shaped responses were obtained in the RTIL 1-butyl-3-methylimidazolium hexafluorophosphate, [C_4mim][PF_6], during ferrocene (Fc) oxidation at a 5-μm-radius Pt disc UME, even at 1 mV/s (Fig. 4.6a–c) [34]. The transient responses obtained in [C_4mim][PF_6] are due to the relatively high viscosity of the RTIL ($\eta = 381$ mPa s) [35], which leads to a diffusion coefficient of Fc, D_{Fc}, of 6×10^{-8} cm^2/s in this RTIL [36]. By moving to a smaller UME ($a = 1$ μm), the authors recorded CVs that resembled a steady-state response, but there was some hysteresis between the forward and reverse sweeps (Fig. 4.6d) [34]. As well as decreasing the UME dimensions, another strategy for moving between transient and steady-state responses in RTILs is to move to a less viscous RTIL. Figure 4.7 shows CVs recorded in [C_4mim][BF_4] ($\eta = 120$ mPa s and in which $D_{Fc} = 9 \times 10^{-8}$ cm^2/s [36]). Note that the CVs were simulated with a D_{Fc} of 5×10^{-7} cm^2/s, an earlier estimate of D_{Fc} in this RTIL [34]. In this case, an approximately steady-state CV (that showed some hysteresis) was recorded at the 5-μm-radius Pt disc UME at 1 mV/s but not at higher scan rates (Fig. 4.7a–c). However, at the 1-μm-radius UME, a steady-state CV was obtained at 10 mV/s.

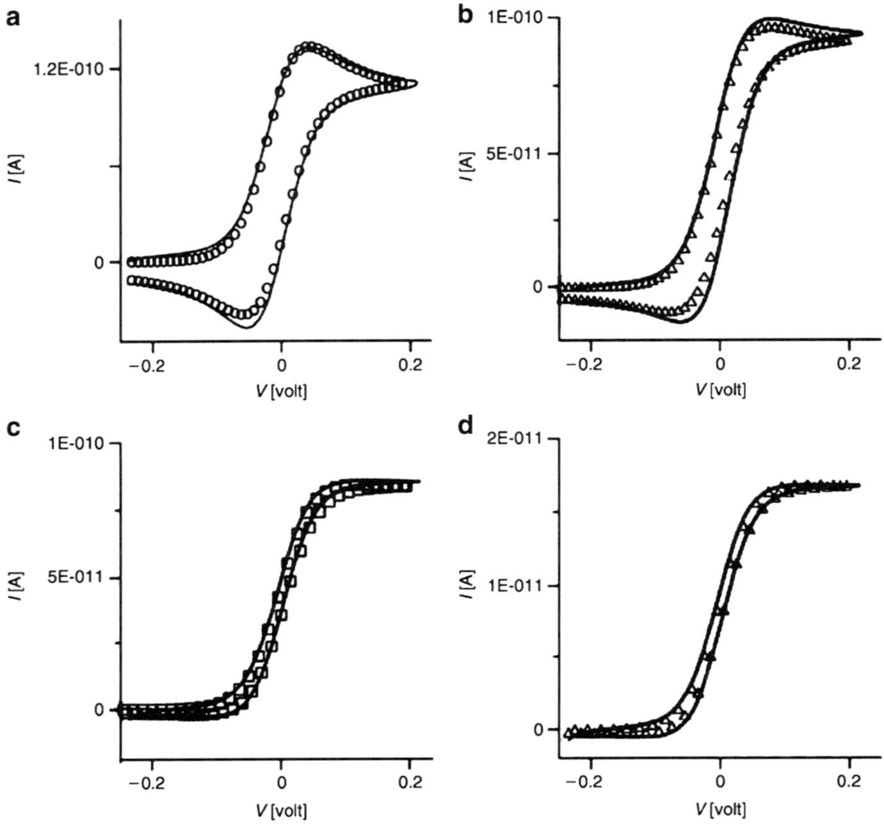

Fig. 4.6 Experimental (*solid line*) and simulated (*open circles, open triangles, open squares*) CVs recorded at a 5-μm-radius Pt microdisk UME at scan rates of (**a**) 50, (**b**) 10 and (**c**) 1 mV/s, and (**d**) at a 1-μm-radius microdisk at a scan rate of 10 mV/s during oxidation of 7 mM ferrocene (Fc) in [C_4mim][PF_6] at 25 °C. Potentials are vs. Fc/Fc^+ and the simulated voltammograms represent those calculated for a reversible process with a diffusion coefficient of 6×10^{-8} cm^2/s. Reprinted with permission from Carano, M; Bond, A M (2007) Aust. J. Chem. 60:29–34. Copyright 2007 CSIRO

Walsh and co-workers also analysed the effects of the tip dimensions and voltammetric scan rates during cyclic voltammetry in a range of imidazolium-based RTILs with widely varying viscosities. Fc was again used as the redox species and the data are shown in Fig. 4.8. In this study, a relatively large UME (a = 12.5 μm) and a small UME (a = 1.5 μm) were used. Figure 4.8a shows the CVs recorded at the 12.5-μm-radius UME at 1 mV/s ≤ υ ≤ 50 mV/s in [C_4mim][PF_6]. Transient responses were obtained at each υ due to linear diffusion of Fc to the UME surface. The application of Eq. (4.9) to this system indicates that transient responses should indeed have been observed at υ > 1 mV/s. Figure 4.8b shows the responses obtained at the 12.5-μm-radius UME in the less viscous RTIL 1-ethyl-3-methylimidazolium bis[trifluoromethanesulfonyl]imide, [C_2mim][N(Tf)$_2$] (which has η = 27 mPa s and in which $D_{Fc} = 4\times10^{-7}$ cm^2/s). According to Eq. (4.9), steady-state CVs should be

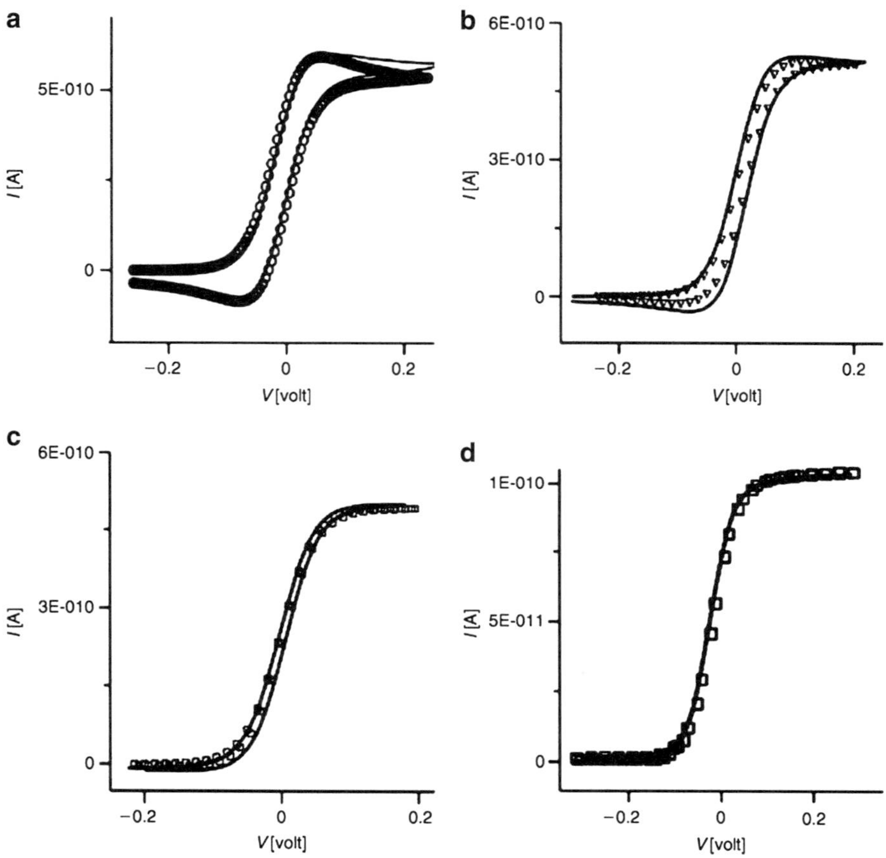

Fig. 4.7 Experimental (*solid line*) and simulated (*open circles*, *open triangles*, *open squares*) CVs recorded at a 5-μm-radius Pt microdisk UME at scan rates of (**a**) 50, (**b**) 10 and (**c**) 1 mV/s, and (**d**) at a 1-μm-radius microdisc at a scan rate of 10 mV/s during oxidation of 7 mM Fc in $[C_4mim][BF_4]$ at 25 °C. Potentials are vs. Fc/Fc^+ scale and the simulated voltammograms represent those calculated for a reversible process with a diffusion coefficient of 5×10^{-7} cm^2/s. Reprinted with permission from Carano, M; Bond, A M (2007) Aust. J. Chem. 60:29–34. Copyright 2007 CSIRO

recorded in this RTIL at $\upsilon \ll 6$ mV s^{-1} at a 12.5-μm UME and Fig. 4.8b shows that a steady-state CV (although with hysteresis between the forward and reverse sweeps) was recorded at $\upsilon = 1$ mV/s, but, as the scan rate increased, peak-shaped responses were clearly obtained, demonstrating that linear diffusion became significant at higher scan rates.

Figure 4.8c, d shows the CVs recorded using the 1.5-μm UME at 1, 10 and 50 mV/s in $[C_4mim][PF_6]$ and $[C_2mim][N(Tf)_2]$. Equation (9) predicts that steady-state CVs should be observed at $\upsilon \ll 70$ mV/s in $[C_4mim][PF_6]$ and Fig. 4.8c shows that such responses were recorded at $\upsilon < 10$ mV/s. However, when $\upsilon = 50$ mV/s, slightly peak-shaped responses were obtained, demonstrating that linear diffusion

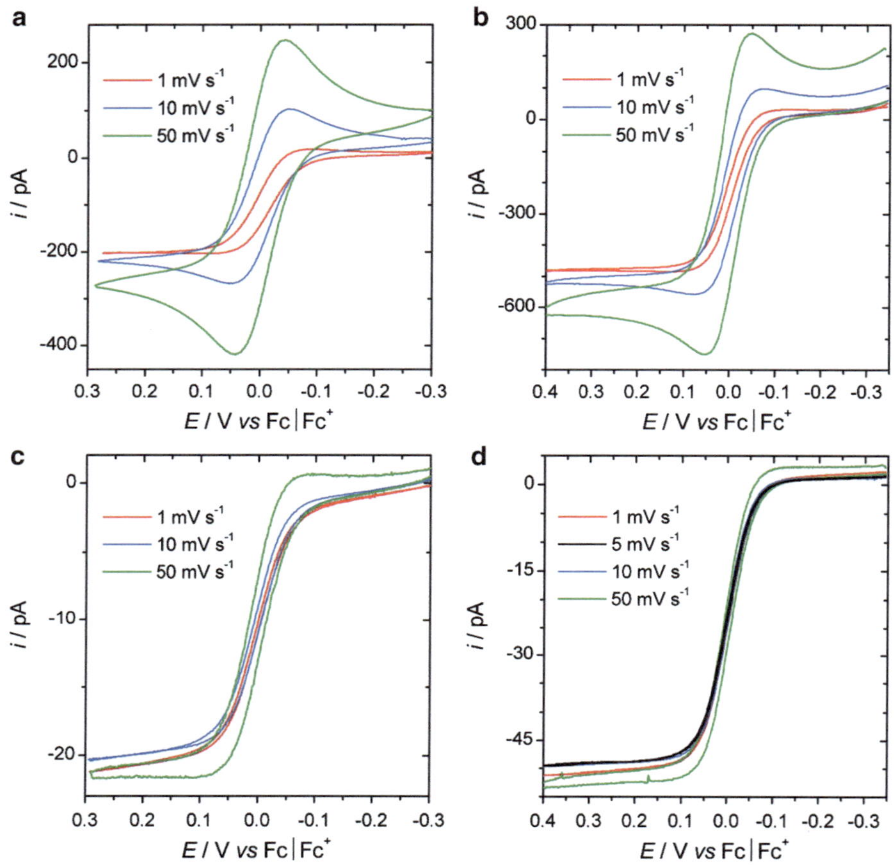

Fig. 4.8 Cyclic voltammograms obtained at a 12.5-μm-radius Pt disk UME in (**a**) 3.20 mM Fc in [C_4mim][PF_6] and (**b**) 1.56 mM Fc in [C_2mim][N$(Tf)_2$]. (**c**) and (**d**) Responses obtained at a 1.5-μm-radius Pt disk UME in 3.20 mM Fc in [C_4mim][PF_6] and 1.56 mM Fc in [C_2mim][N$(Tf)_2$], respectively. The voltammetric scan rates were 1, 10 and 50 mV/s and additionally 5 mV/s for (**d**). Reprinted with permission from Lovelock, K R J; Cowling, F N; Taylor, A W; Licence, P; Walsh, D A (2010) J. Phys. Chem. B 114:4442–4450. Copyright 2010 American Chemical Society

became significant as the voltammetric scan rate approached 70 mV/s, in agreement with the expected behaviour. In the case of the CVs recorded in [C_2mim][N$(Tf)_2$], Eq. (4.9) suggests that steady-state CVs should be obtained at $\upsilon \ll 400$ mV/s at a 1.5-μm UME and Fig. 4.8d shows that true steady-state CVs could be recorded at $\upsilon < 10$ mV/s. At 50 mV/s (green line in Fig. 4.8d), the shape of the CV changed slightly and the current at the switching potential increased above those observed at lower scan rates, suggesting that a purely steady-state response may not have been recorded at this scan rate.

In summary, the UME voltammetry that has been performed in RTILs has illustrated the crucial roles that the UME dimensions, the RTIL viscosity (and consequently the diffusion coefficient of the redox species) and the voltammetric scan rate play in such measurements. If one wants to record truly steady-state CVs in RTILs, it is important to consider each effect and, for electroanalysis in especially viscous RTILs, much smaller UMEs may be required to record steady-state currents.

4.6 Scanning Electrochemical Microscopy in RTILs

Prior to carrying out SECM experiments in "conventional" organic and aqueous electrolytes, one usually records steady-state CVs (with the tip positioned far from any substrate) to ensure that the tip is "behaving" as expected. Such measurements are usually not problematic, assuming that the SECM tip has been prepared carefully and does not allow, e.g. a leak of the electrolyte into the electrode interior. In conventional electrochemical solvents, such CVs are usually recorded at a scan rate of a few mV/s. However, the discussion in the previous section illustrated some of the challenges that one faces when attempting to record steady-state currents using UMEs in RTILs and the crucial roles that a, η and υ play during such measurements. During SECM measurements in RTILs, not only must a and η be considered, the fact that the SECM tip is usually moving within the RTIL must also be considered. The effects of moving an SECM tip within an RTIL were first explored by Bond and co-workers who recorded SECM feedback approach curves during Fc oxidation in $[C_4mim][BF_4]$ using a 5-μm-radius microdisc SECM tip at a tip approach speed, ν_0, of 0.1 μm/s (main frame of Fig. 4.9) [34]. Note that under these conditions the experimental curve agreed with that predicted by conventional SECM theory. However, in the more viscous $[C_4mim][PF_6]$, the experimental feedback approach curves deviated from the theoretical curves. During the positive feedback experiment, i_T was lower than that predicted by theory at all L values, and during the negative feedback experiment, i_T was higher than that predicted by theory at all L values (except at $L>8$).

Hapiot and co-workers also performed SECM measurements in a viscous RTIL and recorded unusual SECM feedback approach curves. In this case, SECM was performed in the pure ferrocenyl RTIL $[FcC_1mim][N(Tf)_2]$ at 50 °C (the high temperature was necessary due to the high melting point of this RTIL), and the results are shown in Fig. 4.10 [8]. The diffusion coefficient of $[FcC_1mim][N(Tf)_2]$ is 3.3×10^{-8} cm^2/s. In this RTIL, similar results to those observed by Bond and co-workers were observed. At $\nu_0=1$ μm/s, a negative feedback was observed as the tip approached an insulating substrate, but the curve did not fit the conventional SECM theory satisfactorily (i_T was higher than that predicted by theory at all L). During the positive feedback approach curve experiment, i_T was lower than that predicted by theory. Interestingly, at high ν_0 (10 μm/s), i_T increased as the tip approached an insulating substrate, an effect that has been attributed to convective mass transfer to the tip surface.

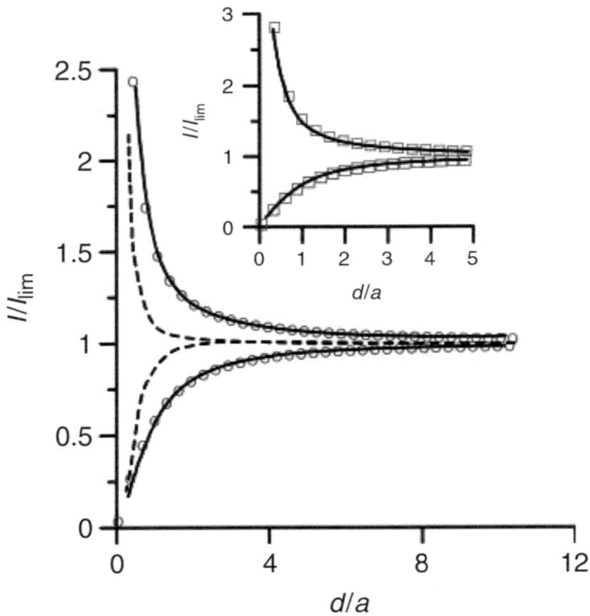

Fig. 4.9 Comparison of theoretical (*open circles*) and experimental SECM approach curves obtained using 5 mM Fc dissolved in [C_4mim][BF_4] (*solid line*) and [C_4mim][PF_6] (*dashed line*) at a 5-μm-radius Pt SECM tip with a tip speed of 0.1 μm/s. The equilibration time before commencing each approach was 50 s. The potential applied at the SECM tip was 0.6 V vs. Ag wire. The positive feedback curve was obtained by approaching a conducting Pt disk biased at 0.0 V vs. the Ag wire quasi-reference electrode and the negative feedback curve was obtained by approaching a glass slide. Inset: Theoretical (*open squares*) and experimental (*solid line*) data obtained in [C_4mim][PF_6] using a 1-μm-radius SECM tip and tip speed of 0.05 μm/s. Reprinted with permission from Carano, M; Bond, A M (2007) Aust. J. Chem. 60:29–34. Copyright 2007 CSIRO

LeSuer performed SECM in deep eutectic solvents (DES), which constitute a sub-class of RTILs, consisting of choline chloride and either trifluoroacetamide or malonic acids as the hydrogen-bond donor [37]. In LeSuer's DESs, the shape of the negative feedback approach curves differed as ν_0 was altered (Fig. 4.11) and the responses were reminiscent of those recorded by Bond in [C_4mim][PF_6] using a 5-μm-radius SECM tip (Fig. 4.9), i.e. in the positive feedback experiment, i_T was lower than that predicted by theory at all *L*, and during the negative feedback experiment, i_T was higher than that predicted by theory at all *L*.

In Bond's, Hapiot's, and LeSuer's negative feedback experiments, i_T was higher than that expected from theory at all *L* due to convective mass transfer of the redox species to the moving tip. In the case of the positive feedback approach curve experiments, sluggish diffusion of the redox species to the SECM tip meant that i_T was

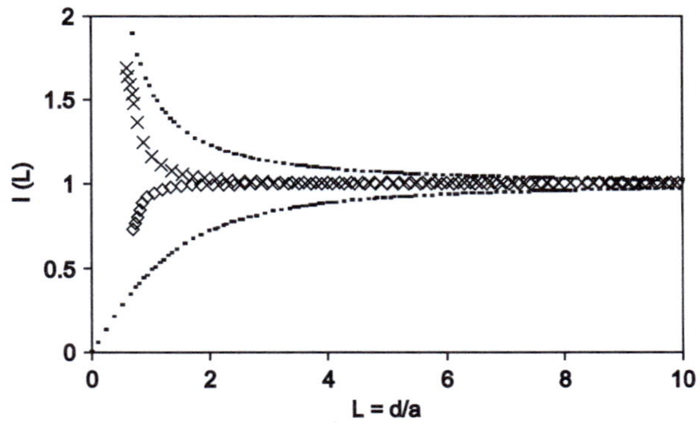

Fig. 4.10 SECM feedback approach curves at 1 μm/s at insulating (*open diamonds*) and conducting (*cross symbols*) substrates in [FcC$_1$mim][N(Tf)$_2$] at 50 °C. *Dashed lines* show the theoretical feedback approach curves. Reprinted with permission from Fontaine, O; Lagrost, C; Ghilane, J; Martin, P; Trippe, G; Fave, C; Lacroix, J C; Hapiot, P; Randriamahazaka, H N (2009) J. Electroanal. Chem. 632:88–96. Copyright 2009 Elsevier BV

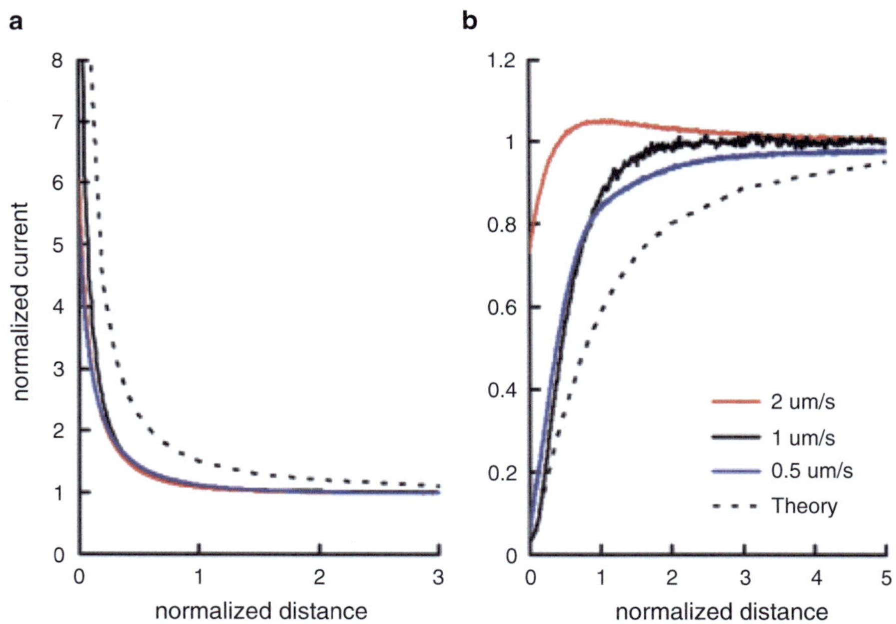

Fig. 4.11 (**a**) Positive feedback and (**b**) negative feedback approach curves in a 1:2 choline chloride/trifluoroacetamide deep eutectic solvent containing 15 mM Fc as the redox mediator. The *x*-axis represents normalised distance only for theoretical curves (*dashed lines*). Experimental curves were normalised by assuming that tip/substrate contact occurred at $L = 0$. The diffusion coefficient of Fc in the DES is 2.7×10^{-8} cm^2/s. Reprinted with permission from Nkuku, C A; LeSuer, R J (2007) J. Phys. Chem. B 111:13271–13277. Copyright 2007 American Chemical Society

lower at all L than that expected from theory. However, just as one can preferentially record steady-state CVs at UMEs in RTILs by moving to smaller electrodes, one can change the dimensions of the SECM tip in order to record steady-state SECM feedback approach curves. The inset of Fig. 4.9 shows that by decreasing the radius of their SECM tip from 5 to 1 μm and reducing ν_0 from 0.1 to 0.05 μm/s, Bond and co-workers could ensure that steady-state currents flowed during the feedback approach curve experiment in $[C_4mim][PF_6]$ and that the experimental curves agreed with theory [34].

The magnitudes of convective and diffusive mass transport to a moving SECM tip are related to the SECM tip dimensions by a dimensionless parameter known as the Péclet number, $Pe = \nu_0 a/D$ [38]. When $Pe \leq 0.01$ (i.e. low ν_0, small a and high D), convection does not contribute significantly to mass transport to the tip [37]. At $Pe \geq 1$ (i.e. high ν_0, large a and low D), convection contributes significantly to mass transfer to the tip surface, and feedback approach curves do not fit conventional SECM theory. Hapiot and co-workers calculated Pe values of 1.5 and 15 for their work on $[FcC_1mim][N(Tf)_2]$ at 50 °C at $\nu_0 = 1$ μm/s and the curves shown in Fig. 4.10 confirm that the feedback approach curves did not fit conventional steady-state SECM theory at this ν_0 [8]. Walsh and co-workers have explored the effects of experimental parameters during SECM in $[C_nmim][N(Tf)_2]$ RTILs and compared the observed behaviour with that expected from calculating Pe values. By increasing the length of the alkyl chain on the RTIL cation, they synthesised a series of RTILs of increasing viscosities. Figure 4.12 shows that steady-state negative feedback approach curves were recorded at small ν_0 values using a 1.5-μm-radius SECM tip in $[C_4mim][N(Tf)_2]$, $[C_6mim][N(Tf)_2]$ and $[C_8mim][N(Tf)_2]$ but not in $[C_4mim][PF_6]$ due to its especially high viscosity. In those experiments that did yield steady-state SECM feedback approach curves, Pe was less than 0.01, in agreement with the expected behaviour.

As well dealing with (presumably unwanted) convective effects during SECM in RTILs by changing the experimental parameters, one can incorporate convective effects into the model used to describe the SECM feedback approach curves. LeSuer has developed a method in which Pe can be incorporated as an adjustable parameter into the simulation of SECM feedback approach curves [37]. Figure 4.13 shows experimental feedback approach curves recorded in a DES at a range of ν_0 values and the accompanying simulations that describe the observed behaviour very well. Good agreement between simulated curves that incorporate Pe into the model and feedback approach curves recorded by Walsh and co-workers during the oxidation of ferrocenemethanol in $[C_8mim][PF_6]$ (for which $Pe = 2$) has also been demonstrated [10].

In summary, it is clear that SECM in RTILs can be complicated by the high viscosity of the liquids. Sluggish diffusion within RTILs, combined with convective effects that can become apparent at moving SECM tips in RTILs, can yield non-steady-state feedback approach curves. If one wants to combat such effects and record steady-state SECM feedback approach curves in RTILs, one can consider

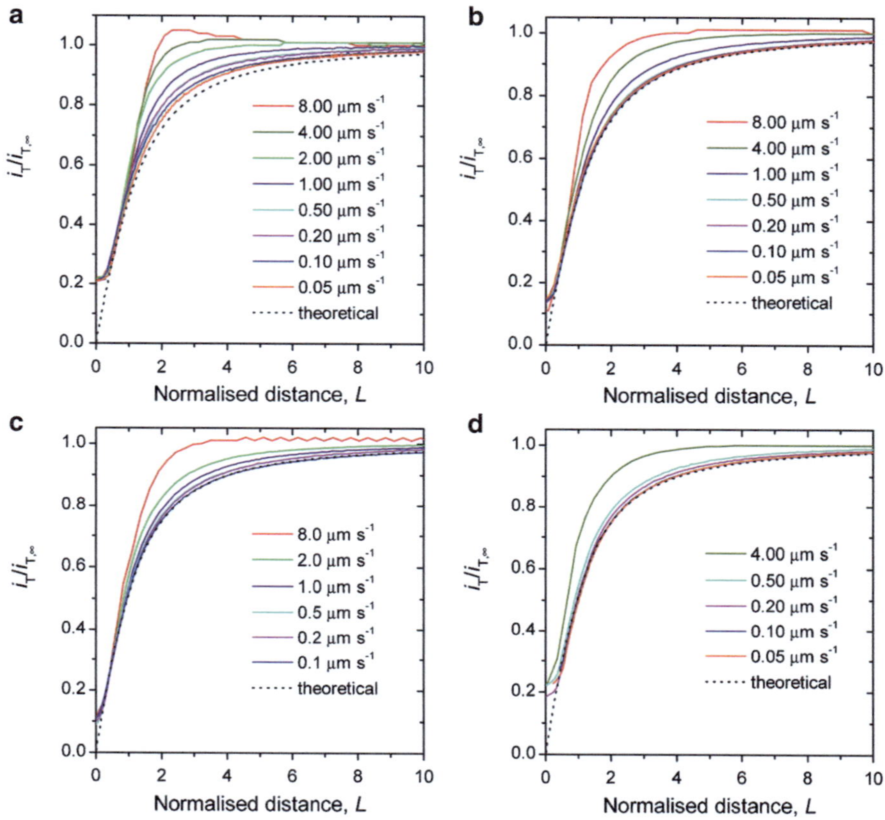

Fig. 4.12 Negative feedback SECM approach curves obtained in (**a**) 3.20 mM Fc in [C_4mim][PF_6] ($\eta=203$), (**b**) 3.50 mM Fc in [C_4mim][N(Tf)$_2$] ($\eta=43$), (**c**) 3.30 mM Fc in [C_6mim][N(Tf)$_2$] ($\eta=59$) and (**d**) 4.35 mM Fc in [C_8mim][N(Tf)$_2$] ($\eta=84$) as a 1.5-µm SECM tip approached a PTFE substrate. The tip was held at (**a**) +0.43 V, (**b**) +0.47 V, (**c**) +0.38 V and (**d**) +0.38 V vs. Fc/Fc^+. In each case, the viscosities were measured using a falling ball viscometer and the *dashed line* shows the theoretical negative feedback approach curve generated for an SECM tip with RG = 10 (RG = 8 in **d**). Reprinted with permission from Lovelock, K R J; Cowling, F N; Taylor, A W; Licence, P; Walsh, D A (2010) J. Phys. Chem. B 114:4442–4450. Copyright 2010 American Chemical Society

changing the tip dimensions, the RTIL viscosity (if possible) and the tip approach speed. However, it may well be that the fabrication and the use of especially small SECM tips will be necessary as users attempt to perform steady-state SECM in highly viscous RTILs. An alternative strategy, as proposed by LeSuer [37], is to incorporate convective effects into the model used to describe the shape of SECM feedback approach curves, and from the few studies in which such models have been used, this appears to be a very promising approach.

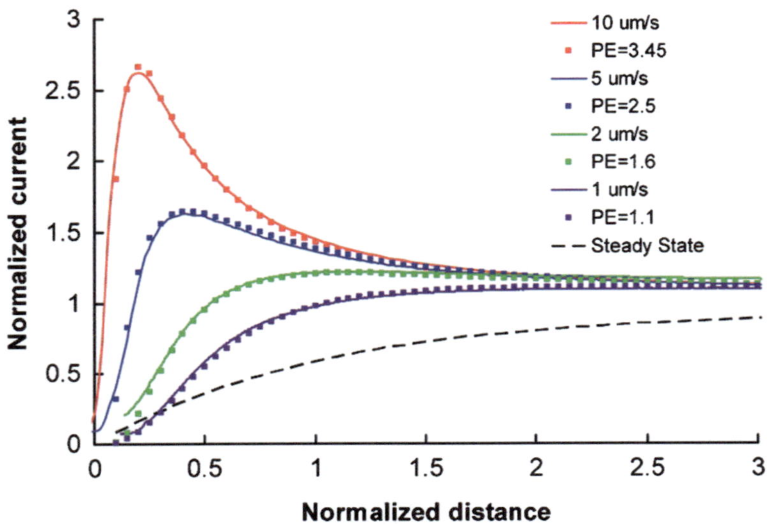

Fig. 4.13 Experimental (*solid lines*) and simulated (*squares*) SECM feedback approach curves recorded during oxidation of 15 mM Fc in a choline chloride/trifluoroacetamide DES. Reprinted with permission from Nkuku, C A; LeSuer, R J (2007) J. Phys. Chem. B 111:13271–13277. Copyright 2007 American Chemical Society

4.7 Effects of Non-equal Diffusion Coefficients on Scanning Electrochemical Microscopy

Due to the high density of charges within RTILs, the reduced and oxidised forms of a redox mediator can exhibit drastically different diffusion coefficients, a phenomenon that is less significant in organic solvents [11, 39]. An example of the effects of such an inequality was reported by Compton and co-workers who used cyclic voltammetry to study the $O_2/O_2^{\bullet-}$ system in the RTIL $[N_{2226}][N(Tf)_2]$, where $[N_{2226}]$ is the triethylhexylammonium ion [40]. In this RTIL, the diffusion coefficient of O_2 is more than an order of magnitude larger than that of $O_2^{\bullet-}$. Remarkably, the effects of this difference in diffusion coefficients can be observed in a single CV (Fig. 4.14) [41]. This CV shows that hemispherical diffusion of O_2 to the Au UME dominated during the reduction sweep (due to diffusion of O_2 from the RTIL from an expanding hemispherical diffusion field), while linear diffusion of $O_2^{\bullet-}$ dominated during the oxidation sweep (due to the formation of a relatively thin diffusion field).

Hapiot and co-workers studied the $O_2/O_2^{\bullet-}$ system using SECM and the feedback approach curves obtained using $O_2/O_2^{\bullet-}$ in $[N_{2224}][N(Tf)_2]$, where $[N_{2224}]$ is the triethylbutylammonium ion, are shown in Fig. 4.15 [11]. At $\nu_0 = 5$ µm/s, an unusual negative feedback approach curve was observed as the tip approached the conducting Pt substrate due to the low diffusion coefficient of $O_2^{\bullet-}$, which hindered the regeneration of O_2 at the substrate. However, at a slower ν_0 (0.2 µm/s), positive feedback was observed because tip movement was slow enough to allow sufficient time for

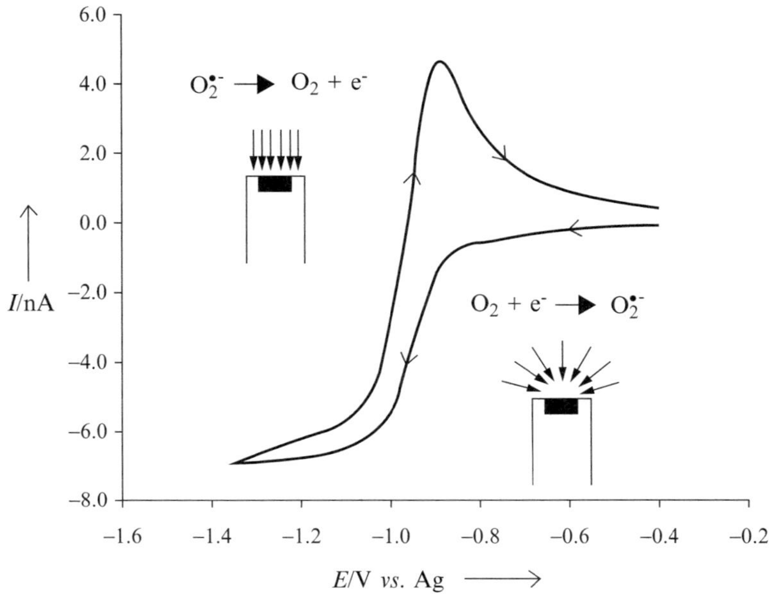

Fig. 4.14 A cyclic voltammogram showing the reduction of O_2 in $[N_{2226}][N(Tf)_2]$ at an Au microdisk UME with $a=5$ μm, $\upsilon=0.5$ V/s and $T=308$ K. Inset is a schematic indicating the dominant mode of diffusion to the electrode surface responsible for dictating the shape of the forward and reverse waves. Reprinted with permission from Buzzeo, M C; Evans, R G; Compton, R G (2004) ChemPhysChem 5:1106–1120. Copyright 2004 Wiley VCH

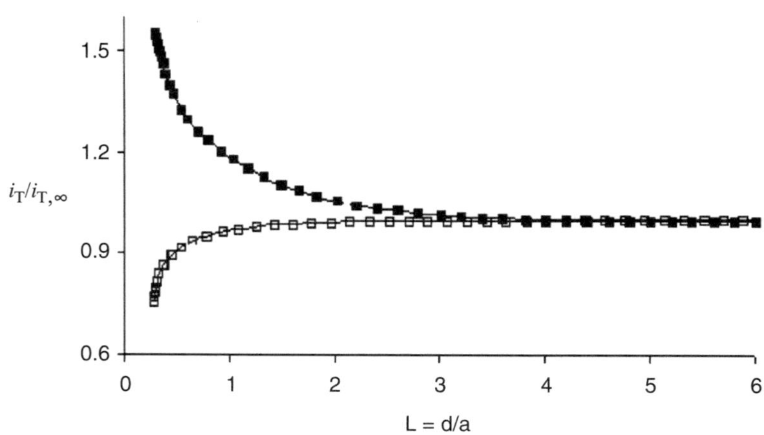

Fig. 4.15 SECM feedback approach curves obtained at an Au SECM tip as it approached a Pt substrate in $[N_{2224}][N(Tf)_2]$ in the presence of O_2. The substrate potential was −0.1 V vs. the saturated calomel electrode (SCE) and the tip potential was −1.1 V vs. SCE. $\nu_0=5$ μm/s (*open squares*). $\nu_0=0.2$ μm/s (*filled squares*). Reprinted with permission from Ghilane, J; Lagrost, C; Hapiot, P (2007) Anal. Chem. 79:7383–7391. Copyright 2007 American Chemical Society

the diffusion of $O_2^{\bullet-}$ to the substrate and regeneration of O_2. This example shows that SECM feedback approach curves can become complicated when the ratio of D_{red} to D_{ox} ($\gamma = D_{red}/D_{ox}$) is far from unity but are possible if ν_0 is sufficiently low. The effect of adding varying concentrations of dimethylformamide (DMF) as a co-solvent was also studied by SECM using the $O_2/O_2^{\bullet-}$ system in $[N_{2224}][N(Tf)_2]$ [42]. D_{ox} and D_{red} increased drastically with increasing DMF concentration, due to a decrease in the viscosity of the electrolyte. However, γ also increased exponentially with the addition of DMF, demonstrating that higher concentrations of DMF had a far greater effect on D_{red} than on D_{ox} and that the addition of the DMF reduced the strength of the interactions between the cation of the RTIL and $O_2^{\bullet-}$.

4.8 Studying Electron Transfer in RTILs Using SECM

When an SECM tip is positioned close (within about 1-tip radius) to a substrate, the mass transfer coefficient, *m*, is equal to *D*/*d*. By decreasing *d*, it is possible to achieve extremely high rates of mass transfer to the tip surface and measure the kinetics of very fast reactions [43]. For example, Mirkin and co-workers have fabricated SECM tips with radii as small as 10 nm and studied kinetics of fast ET reactions in aqueous and non-aqueous solutions [22]. Furthermore, using this approach, they estimate that rate constants as large as 200 cm/s may be measurable using nm-sized SECM tips.

For a simple quasi-reversible redox reaction at a disc-shaped SECM tip, coupled with the diffusion-controlled regeneration of the mediator at the substrate, the normalised tip current at potential *E* is given by [22]:

$$I_T(E,L) = \frac{k_2}{L(\theta + 1/\kappa)} + \frac{k_1 + k_3 \exp(k_4 / L)}{\theta\left[1 + \frac{\pi}{\kappa\theta}\frac{2\kappa\theta + 3\pi}{4\kappa\theta + 3\pi^2}\right]} \tag{4.14}$$

where k_1, k_2, k_3 and k_4 are constants. θ and κ for an oxidation reaction are given by

$$\theta = 1 + \exp\left[\frac{F}{RT}\left(E^{o'} - E\right)\right]\frac{D_{ox}}{D_{red}} \tag{4.15}$$

$$\kappa = \pi\lambda \exp\left[\frac{(1-\alpha)F}{RT}\left(E - E^{o'}\right)\right] / 4I_T^c \tag{4.16}$$

where $E^{0\prime}$ is the formal potential, α is the transfer coefficient and the kinetic parameter, $\lambda = k^0 a/D$, where k^0 is the standard heterogeneous electron transfer rate constant. I_T^c is the normalised diffusion-limiting tip current for the same *L* above a conducting substrate. By positioning SECM tips close to substrate surfaces, recording tip CVs, and fitting the resulting data to Eq. (4.14), Walsh and co-workers have

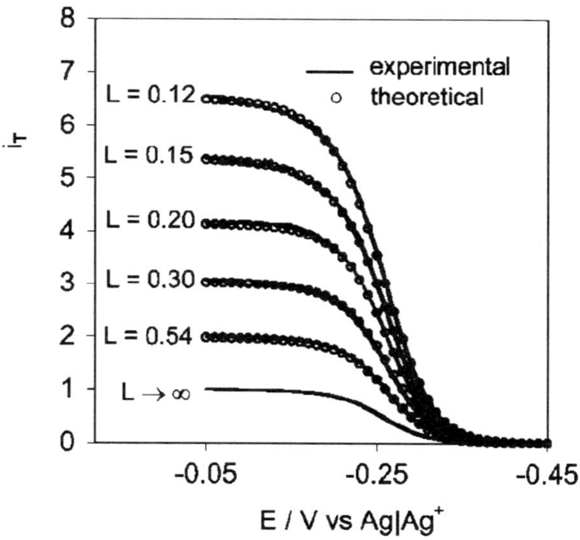

Fig. 4.16 Experimental voltammograms (*solid lines*) obtained at a 1.5-μm-radius Pt SECM tip at distances of $L=0.12$, 0.15, 0.20, 0.30, 0.54 and $L\rightarrow\infty$ from a 1-mm-radius Pt-disk substrate. The solution was 9 mM [FcC$_1$mim][N(Tf)$_2$] in [C$_2$mim][N(Tf)$_2$] and the substrate potential was −0.6 V vs. Ag/Ag$^+$. The theoretical voltammograms at each distance (*open circles*) were calculated using Eq. (4.14) in the text. Reprinted with permission from Taylor, A W; Qiu, F L; Hu, J P; Licence, P; Walsh, D A (2008) J. Phys. Chem. B 112:13292–13299. Copyright 2008 American Chemical Society

measured k^0 values for a number of electrochemical reactions. For example, Fig. 4.16 shows tip CVs recorded at a range of L values in a solution of [FcC$_1$mim][N(Tf)$_2$] when dissolved in [C$_2$mim][N(Tf)$_2$] [44]. In this case, the k^0 value obtained from fitting the experimental data to Eq. (4.14) (7.6×10^{-3} cm/s) was slightly higher than that obtained using cyclic voltammetry (4.25×10^{-3} cm/s) and was also slightly higher than that determined by Hapiot and co-workers for the same system using cyclic voltammetry and electrochemical impedance spectroscopy [8]. Walsh and co-workers have also used the same method to determine k^0 for Fc oxidation in a series of imidazolium-based RTILs with varying viscosities [9]. In general, the k^0 values determined using SECM were higher than those determined using cyclic voltammetry [45], suggesting that the use of cyclic voltammetry for kinetic measurements in RTILs can lead to significant underestimations of k^0. In this study, the k^0 values determined for Fc oxidation using SECM were also correlated with the RTIL viscosity, revealing an inverse relationship between k^0 and η (Fig. 4.17) [9].

Studies of ET, as well as ion and molecule transfer, across the interface between two immiscible electrolyte solutions (ITIES) have been of interest for a number of years. Traditionally, the phases involved are water and a water-immiscible organic solvent. However, in recent years a number of researchers have replaced the organic phase in such experiments with a water-immiscible RTIL. This approach has been

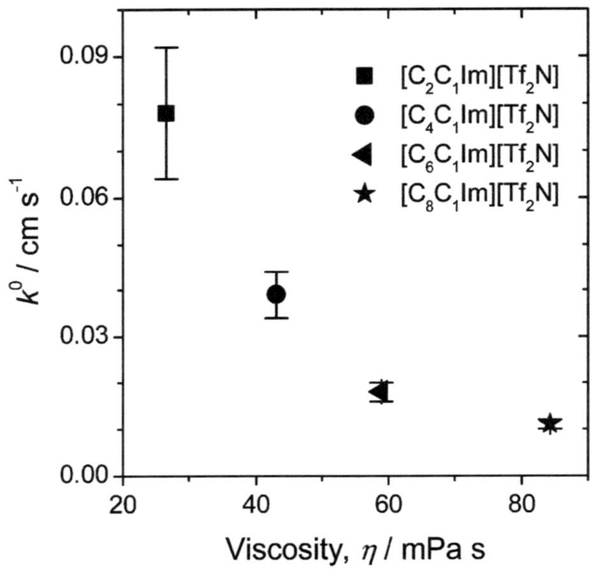

Fig. 4.17 Graph of the heterogeneous electron transfer rate constant, k^0, vs. viscosity for Fc oxidation in $[CnC_1Im][Tf_2N]$ RTILs, where $n = 2–8$. The *error bars* represent the standard deviation of four measurements of k^0. Reprinted with permission from Lovelock, K R J; Cowling, F N; Taylor, A W; Licence, P; Walsh, D A (2010) J. Phys. Chem. B 114:4442–4450. Copyright 2008 American Chemical Society

described by Mirkin and co-workers, who compared the rate of ET across the RTIL/water interface with that of ET across the organic/water interface using the $[Fe(CN)_6]^{4-}/[Fe(CN)_6]^{3-}$ system as the redox probe (Fig. 4.18) [46]. In this experiment, the upper aqueous phase contained $[Fe(CN)_6]^{4-}$, which was oxidised to $[Fe(CN)_6]^{3-}$ at a Pt SECM tip. The electrogenerated $[Fe(CN)_6]^{3-}$ diffused to the $[C_8mim][N(Tf)_2]$/water(w) interface where it was reduced by interfacial ET from Fc dissolved in the RTIL:

$$\mathrm{Fc(RTIL)} + \left[\mathrm{Fe(CN)_6}\right]^{3-}(\mathrm{w}) \rightarrow \mathrm{Fc^+(RTIL)} + \left[\mathrm{Fe(CN)_6}\right]^{4-}(\mathrm{w}) \quad (4.17)$$

The rate of ET across the RTIL/water interface was drastically higher than that across the 1,2-dichloroethane/water interface. In a later study using the same RTIL, the authors showed that Fc partitioning from the RTIL into the aqueous phase was significant and that there was a large contribution to the tip current from oxidation of Fc at the tip in the water phase [47]. This phenomenon is known as SECM-induced transfer (SECMIT) as Fc is transferred from the RTIL to the aqueous phase (Fig. 4.19). To study the kinetics of ET across the RTIL–water interface, the authors developed a method for separating the current due to partitioned Fc oxidation from the total current, allowing accurate measurement of the kinetics of interfacial

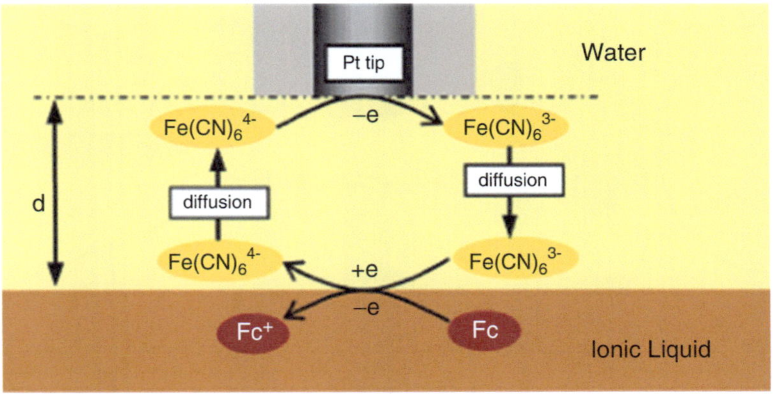

Fig. 4.18 Schematic diagram of the measurement of electron transfer kinetics between Fc dissolved in $[C_8mim][N(Tf)_2]$ and $Fe(CN)_6^{3-}$ dissolved in water. Reprinted with permission from Laforge, F O; Kakiuchi, T; Shigematsu, F; Mirkin, M V (2004) J. Am. Chem. Soc. 126:15380–15381. Copyright 2004 American Chemical Society

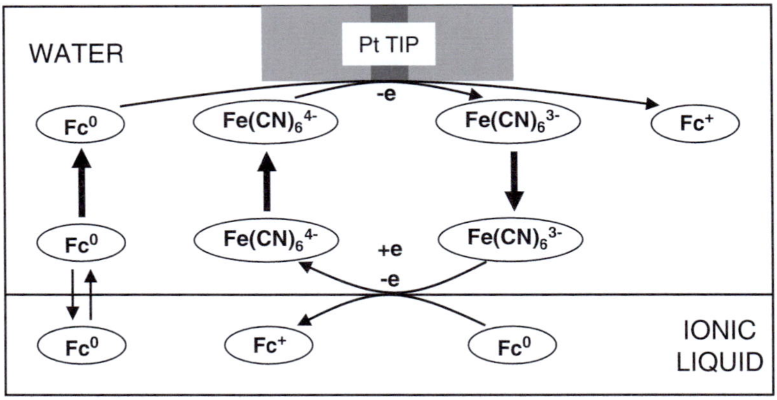

Fig. 4.19 Scheme of an SECMIT/feedback experiment at the water/$[C_8mim][N(Tf)_2]$ interface. Reprinted with permission from Laforge, F O; Kakiuchi, T; Shigematsu, F; Mirkin, M V (2006) Langmuir 22:10705–10710. Copyright 2006 American Chemical Society

ET. The same authors also studied ET reactions involving the $[IrCl_6]^{3-/2-}$ and $[Ru(CN)_6]^{4-/3-}$ systems, which were dissolved in the aqueous phase [47].

Wittstock and co-workers have used SECM to explore the kinetics of dye regeneration in dye-sensitised solar cells (DSSCs) using an RTIL electrolyte [48]. In this experiment, I^- was electrogenerated from I_3^- at a Pt SECM tip (Fig. 4.20). As the tip approached a TiO_2 substrate, which was coated with a DSSC dye (in this case *cis-di*(thiocyanato)bis(2,2′-bipyridyl-4,4″-dicarboxylate)ruthenium, which is usually

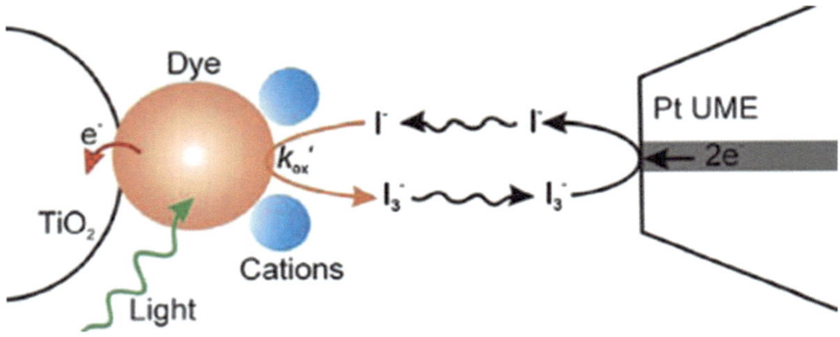

Fig. 4.20 Schematic diagram of the feedback-mode SECM experiment used to measure the rate of dye regeneration in dye-sensitised solar cells. Reprinted with permission from Tefashe, U M; Nonomura, K; Vlachopoulos, N; Hagfeldt, A; Wittstock, G (2012) J. Phys. Chem. C 116:4316–4323. Copyright 2012 American Chemical Society

labelled N719), I_3^- was regenerated by ET from I^- to the oxidised dye. Regeneration of I_3^- at the dye-coated substrate increased the flux of I_3^- to the tip which increased i_T above that expected for pure negative feedback. Using this method, the authors showed that dye regeneration kinetics are faster in acetonitrile-based electrolytes than in the exemplar RTIL [C_2mim][N(TF_2)]. By adding various ions to the RTIL, the group identified drastic decreases in the effective rate constant for dye regeneration, k_{eff}, with increasing cation concentration. For example, Fig. 4.21a shows SECM feedback approach curves recorded in [C_2mim][N$(Tf)_2$] containing varying amounts of Li[N$(Tf)_2$]. At the highest concentration of Li[N$(Tf)_2$], the feedback approach curve overlapped with the curve expected for the hindered diffusion of I_3^- to the tip (i.e. pure negative feedback). However, as the concentration of Li[N$(Tf)_2$] decreased, the rate of I_3^- regeneration increased resulting in an increase in i_T at all L values. By fitting the entire feedback approach curves to theoretical kinetic SECM feedback approach curves, the group extracted effective rate constants, k_{eff}, for dye regeneration and Fig. 4.21b shows the effect of the concentration of Li^+, as well as that of the 1,2,-dimethyl-3-propylimidazolium and tetrabutylammonium cations, on k_{eff}. In each case, increasing the concentration of the cations within the RTIL decreased k_{eff} and the effect was most prominent for the Li^+ cation. These effects were attributed to (1) cation adsorption onto the substrate, which can shift the energy of the conduction band of the TiO_2 (in the case of Li^+), (2) possible changes in the local I^- concentration near the substrate due to cation adsorption and (3) possible changes in the substrate structure due to interactions between the dye and the cations. For those interested in the development of new electrolytes for DSSCs, such insights are extremely valuable and could lead to the development of new RTIL electrolytes for DSSCs, which would benefit from the non-volatility of the RTILs and overcome the problem of solvent loss in DSSCs. The advantage of using the "photo-SECM" method for determining kinetic parameters is that the measurements are performed at a single sensitised electrode rather than within a complete DSSC, offering a potentially simpler route to kinetic data on such devices.

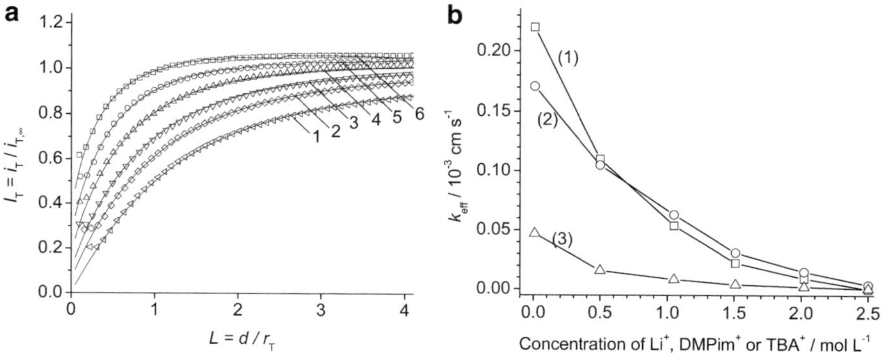

Fig. 4.21 (**a**) SECM feedback approach curves recorded using a Pt SECM tip ($a = 12.5$ μm) as it approached a TiO_2/N719 film illuminated with green LED with an intensity of 25.5×10^{-9} mol/cm^2/s and in solutions containing 0.064 mM $[I_3^-]$* and various concentration of Li[N(Tf)$_2$] in [C$_2$mim][N(Tf)$_2$]. The Li[N(Tf)$_2$] concentrations were (1) 2.5 M, (2) 2.02 M, (3) 1.51 M, (4) 1.05 M, (5) 0.5 M and (6) 0.01 M. (**b**) Plot of k_{eff} as a function of cation concentration for three different electrolytes compositions varying only in terms of cation counter ions (1) LiI/I$_2$/Li[N(Tf)$_2$], (2) [mmpim]I/I$_2$/[mmpim][N(Tf)$_2$] and (3) [TBA]I/I$_2$/[TBA][N(Tf)$_2$] in [C$_2$mim][N(Tf)$_2$], where [mmpim]$^+$ is the 1,2,-dimethyl-3-propylimidazolium cation and [TBA]$^+$ is the tetrabutylammonium cation. Reprinted with permission from Tefashe, U M; Nonomura, K; Vlachopoulos, N; Hagfeldt, A; Wittstock, G (2012) J. Phys. Chem. C 116:4316–4323. Copyright 2012 American Chemical Society

4.9 SECM Imaging in Ionic Liquids

The resolution that can be achieved during SECM imaging is predominantly governed by the magnitudes of *a* and *d* so surface features ~10 μm in diameter can be imaged by SECM routinely. To date, there are only a couple of reports on the use of SECM imaging in RTILs. In one example of SECM imaging in RTILs, Bond and co-workers imaged an Ag substrate using Fc dissolved in [C$_4$mim][BF$_4$] at a 5-μm-radius tip and a lateral tip speed of 0.2 μm/s [34]. Mirkin and co-workers imaged the surface of a compact disc using Fc dissolved in [C$_8$mim][N(Tf)$_2$] as the redox species with a 0.19-μm-radius tip and lateral tip speed of 0.5 μm/s (Fig. 4.22) [49].

4.10 Conclusions

While the number of papers describing the use of UME voltammetry and SECM in RTILs is relatively small, it is clear that these methods are very useful for studying electrochemical phenomena in RTILs. Given that this field is relatively young, it is

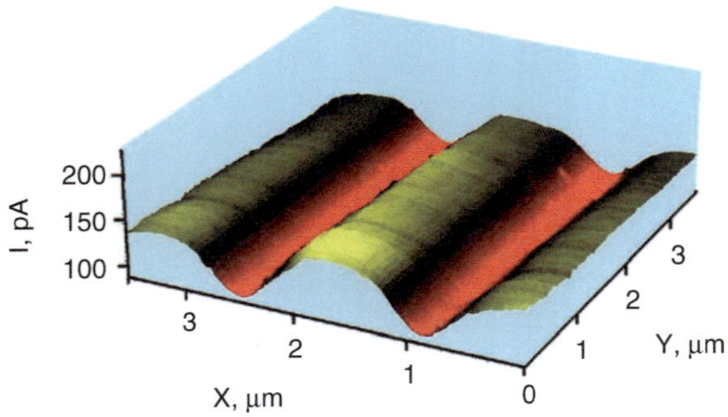

Fig. 4.22 Constant-height SECM image of a 3.5 μm × 3.5 μm portion of a gold CD surface obtained using a 190 nm Pt tip in [C_8mim][$N(Tf)_2$] containing 50 mM Fc. The lateral scan rate was 500 nm/s. Reprinted with permission from Laforge, F O; Velmurugan, J; Wang, Y X; Mirkin, M V (2009) Anal. Chem. 81:3143–3150. Copyright 2009 American Chemical Society

perhaps not surprising that much of the work done thus far is quite "fundamental" in nature, focusing on phenomena such as mass and charge transport in RTILs. The sluggishness of mass transport in RTILs, and how it affects voltammetry and SECM, is a particularly common theme and the effects of experimental parameters on mass transport to ultramicroelectrodes are now well understood. Even in modestly viscous RTILs, SECM feedback approach curves can be complicated by convective mass transport to the moving tip and the use of the Péclet number can guide users towards the appropriate tip dimensions and speeds to ensure that steady-state responses are obtained. However, as electrochemists move to especially viscous RTILs, it is likely that very small (nm-sized) SECM tips will be required to record steady-state responses, and it is fortunate that there are now several methods available for fabricating nanoscopic SECM tips. An interesting development is the recent availability of a model for producing SECM feedback approach curves, which includes the Péclet number as an adjustable parameter. In those RTILs that are too viscous for steady-state SECM to be performed, even using nm-sized SECM tips, this model can provide useful information about mass transport in RTILs.

There is little doubt that, as this field develops, new insights into the electrochemical behaviour of RTILs will be revealed by UME voltammetry and SECM. SECM has already been used to study the kinetics of electron transfer between RTILs and solid electrodes, and the use of SECM to probe the kinetics of dye regeneration in RTIL-based solar cells has demonstrated the power of these methods. As the electrochemical applications of RTILs now extend to sensors, batteries and fuel cells, it is likely that UME voltammetry and SECM will be used to study processes occurring within these devices and contribute to the development of the next-generation electrochemical technologies.

References

1. Bard AJ, Fan FRF, Kwak J, Lev O (1989) Scanning electrochemical microscopy. Introduction and principles. Anal Chem 61:132–138
2. Kwak J, Bard AJ (1989) Scanning electrochemical microscopy – theory of the feedback mode. Anal Chem 61:1221–1227
3. Bard AJ, Mirkin MV (eds) (2012) Scanning electrochemical microscopy, 2nd edn. CRC Press, Boca Raton, FL
4. Sun P, Laforge FO, Mirkin MV (2007) Scanning electrochemical microscopy in the 21st Century. Phys Chem Chem Phys 9:802–823
5. Armand M, Endres F, MacFarlane DR, Ohno H, Scrosati B (2009) Ionic-liquid materials for the electrochemical challenges of the future. Nat Mater 8:621–629
6. Fedorov MV, Kornyshev AA (2014) Ionic liquids at electrified interfaces. Chem Rev 114: 2978–3036
7. Walsh DA, Lovelock KRJ, Licence P (2010) Ultramicroelectrode voltammetry and scanning electrochemical microscopy in room-temperature ionic liquid electrolytes. Chem Soc Rev 39:4185–4194
8. Fontaine O, Lagrost C, Ghilane J, Martin P, Trippe G, Fave C, Lacroix JC, Hapiot P, Randriamahazaka HN (2009) Mass transport and heterogeneous electron transfer of a ferrocene derivative in a room-temperature ionic liquid. J Electroanal Chem 632:88–96
9. Lovelock KRJ, Cowling FN, Taylor AW, Licence P, Walsh DA (2010) Effect of viscosity on steady-state voltammetry and scanning electrochemical microscopy in room temperature ionic liquids. J Phys Chem B 114:4442–4450
10. Lovelock KRJ, Ejigu A, Loh SF, Men S, Licence P, Walsh DA (2011) On the diffusion of ferrocenemethanol in room-temperature ionic liquids: an electrochemical study. Phys Chem Chem Phys 13:10155–10164
11. Ghilane J, Lagrost C, Hapiot P (2007) Scanning electrochemical microscopy in nonusual solvents: inequality of diffusion coefficients problem. Anal Chem 79:7383–7391
12. Stulik K, Amatore C, Holub K, Marecek V, Kutner W (2000) Microelectrodes. definitions, characterization, and applications (technical report). Pure Appl Chem 72:1483–1492
13. Amatore C, Maisonhaute E (2005) When voltammetry reaches nanoseconds. Anal Chem 77:303A–311A
14. Forster RJ (1994) Microelectrodes – new dimensions in electrochemistry. Chem Soc Rev 23:289–297
15. Lefrou C, Cornut R (2010) Analytical expressions for quantitative scanning electrochemical microscopy (SECM). ChemPhysChem 11:547–556
16. Bard AJ, Faulkner LR (2001) Electrochemical methods: fundamentals and applications, 2nd edn. Wiley, London
17. Wipf DO, Kristensen EW, Deakin MR, Wightman RM (1988) Fast-scan cyclic voltammetry as a method to measure rapid, heterogeneous electron-transfer kinetics. Anal Chem 60:306–310
18. Mirkin MV, Bard AJ (1992) Simple analysis of quasi-reversible steady-state voltammograms. Anal Chem 64:2293–2302
19. Fan FR, Fernández JL, Liu B, Mauzeroll J (2007) Scanning electrochemical microscopy. In: Zoski CG (ed) Handbook of electrochemistry. Elsevier, Amsterdam
20. Lefrou C (2006) A unified new analytical-approximation for positive feedback currents with a microdisk SECM tip. J Electroanal Chem 592:103–112
21. Cornut R, Lefrou C (2007) A unified new analytical approximation for negative feedback currents with a microdisk SECM tip. J Electroanal Chem 608:59–66
22. Sun P, Mirkin MV (2006) Kinetics of electron-transfer reactions at nanoelectrodes. Anal Chem 78:6526–6534
23. Treichel DA, Mirkin MV, Bard AJ (1994) Scanning electrochemical microscopy. 27. Application of a simplified treatment of an irreversible homogeneous reaction following electron-transfer to the oxidative dimerization of 4-nitrophenolate in acetonitrile. J Phys Chem 98:5751–5757

24. Bard AJ, Denault G, Friesner RA, Dornblaser BC, Tuckerman LS (1991) Scanning electrochemical microscopy – theory and application of the transient (chronoamperometric) SECM response. Anal Chem 63:1282–1288
25. Etienne M, Schulte A, Schuhmann W (2004) High resolution constant-distance mode alternating current scanning electrochemical microscopy (AC-SECM). Electrochem Commun 6:288–293
26. Wang YX, Kececi K, Velmurugan J, Mirkin MV (2013) Electron transfer/ion transfer mode of scanning electrochemical microscopy (SECM): a new tool for imaging and kinetic studies. Chem Sci 4:3606–3616
27. Ebejer N, Gueell AG, Lai SCS, McKelvey K, Snowden ME, Unwin PR (2013) Scanning electrochemical cell microscopy: a versatile technique for nanoscale electrochemistry and functional imaging. Ann Rev Anal Chem 6:329–351
28. Shen M, Ishimatsu R, Kim J, Amemiya S (2012) Quantitative imaging of ion transport through single nanopores by high-resolution scanning electrochemical microscopy. J Am Chem Soc 134:9856–9859
29. Walsh DA, Fernandez JL, Mauzeroll J, Bard AJ (2005) Scanning electrochemical microscopy. 55. Fabrication and characterization of micropipet probes. Anal Chem 77:5182–5188
30. Mezour MA, Morin M, Mauzeroll J (2011) Fabrication and characterization of laser pulled platinum microelectrodes with controlled geometry. Anal Chem 83:2378–2382
31. Katemann BB, Schuhmann T (2002) Fabrication and characterization of needle-type Pt-disk nanoelectrodes. Electroanalysis 14:22–28
32. Tel-Vered R, Walsh DA, Mehrgardi MA, Bard AJ (2006) Carbon nanofiber electrodes and controlled nanogaps for scanning electrochemical microscopy experiments. Anal Chem 78:6959–6966
33. Mirkin MV, Ren FF, Bard AJ (1992) Scanning electrochemical microscopy. Part 13. Evaluation of the tip shapes of nanometer size microelectrodes. J Electroanal Chem 328:47–62
34. Carano M, Bond AM (2007) Prospects for the application of scanning electrochemical microscopy in ionic liquids. Aust J Chem 60:29–34
35. Laurenczy G, Dyson PJ (2008) Determination of the viscosity of the ionic liquids [Bmim][PF_6] and [Bmim][TF_2N] under high CO_2 gas pressure using sapphire NMR tubes. Zeitschrift Fur Naturforschung Section B-a. J Chem Sci 63:681–684
36. Shiddiky MJA, Torriero AAJ, Zhao C, Burgar I, Kennedy G, Bond AM (2009) Nonadditivity of faradaic currents and modification of capacitance currents in the voltammetry of mixtures of ferrocene and the cobaltocenium cation in protic and aprotic ionic liquids. J Am Chem Soc 131:7976–7989
37. Nkuku CA, LeSuer RJ (2007) Electrochemistry in deep eutectic solvents. J Phys Chem B 111:13271–13277
38. Combellas C, Fermigier M, Fuchs A, Kanoufi F (2005) Scanning electrochemical microscopy. Hydrodynamics generated by the motion of a scanning tip and its consequences on the tip current. Anal Chem 77:7966–7975
39. Wang YJ, Rogers EI, Compton RG (2010) The measurement of the diffusion coefficients of ferrocene and ferrocenium and their temperature dependence in acetonitrile using double potential step microdisk electrode chronoamperometry. J Electroanal Chem 648:15–19
40. Buzzeo MC, Klymenko OV, Wadhawan JD, Hardacre C, Seddon KR, Compton RG (2003) Voltammetry of oxygen in the room-temperature ionic liquids 1-ethyl-3-methylimidazolium bis((trifluoromethyl)sulfonyl)imide and hexyltriethylammonium bis((trifluoromethyl)sulfonyl)imide: one-electron reduction to form superoxide. Steady-state and transient behavior in the same cyclic voltammogram resulting from widely different diffusion coefficients of oxygen and superoxide. J Phys Chem A 107:8872–8878
41. Buzzeo MC, Evans RG, Compton RG (2004) Non-haloaluminate room-temperature ionic liquids in electrochemistry – a review. ChemPhysChem 5:1106–1120
42. Zigah D, Wang AF, Lagrost C, Hapiot P (2009) Diffusion of molecules in ionic liquids/organic solvent mixtures. Example of the reversible reduction of O_2 to superoxide. J Phys Chem B 113:2019–2023

43. Lu X, Wang Q, Liu X (2007) Review: recent applications of scanning electrochemical microscopy to the study of charge transfer kinetics. Anal Chim Acta 601:10–25
44. Taylor AW, Qiu FL, Hu JP, Licence P, Walsh DA (2008) Heterogeneous electron transfer kinetics at the ionic liquid/metal interface studied using cyclic voltammetry and scanning electrochemical microscopy. J Phys Chem B 112:13292–13299
45. Barnes AS, Rogers EI, Streeter I, Aldous L, Hardacre C, Compton RG (2008) Extraction of electrode kinetic parameters from microdisc voltammetric data measured under transport conditions intermediate between steady-state convergent and transient linear diffusion as typically applies to room temperature ionic liquids. J Phys Chem B 112:7560–7565
46. Laforge FO, Kakiuchi T, Shigematsu F, Mirkin MV (2004) Comparative study of electron transfer reactions at the ionic liquid/water and organic/water interfaces. J Am Chem Soc 126:15380–15381
47. Laforge FO, Kakiuchi T, Shigematsu F, Mirkin MV (2006) SECM study of solute partitioning and electron transfer at the ionic liquid/water interface. Langmuir 22:10705–10710
48. Tefashe UM, Nonomura K, Vlachopoulos N, Hagfeldt A, Wittstock G (2012) Effect of cation on dye regeneration kinetics of N719-sensitized TiO_2 films in acetonitrile-based and ionic-liquid-based electrolytes investigated by scanning electrochemical microscopy. J Phys Chem C 116:4316–4323
49. Laforge FO, Velmurugan J, Wang YX, Mirkin MV (2009) Nanoscale imaging of surface topography and reactivity with the scanning electrochemical microscope. Anal Chem 81:3143–3150

Chapter 5
Electroanalytical Applications of Semiintegral and Convolution Voltammetry in Room-Temperature Ionic Liquids

Cameron L. Bentley, Alan M. Bond, Anthony F. Hollenkamp, Peter J. Mahon, and Jie Zhang

5.1 Introduction

Room-temperature ionic liquids (RTILs) are intrinsic ionic conductors which have been successfully employed as nonflammable/nonreactive electrolytes in a range of electrochemical devices, including dye-sensitized solar cells [1, 2], lithium batteries [3], fuel cells [4], and supercapacitors [5]. The quantification of mass transport is of interest in any solvent, particularly those employed in electrochemical devices, as it affects the ultimate rate/speed at which the device can operate. The diffusivity or diffusion coefficient (D) of a redox active species, along with other thermodynamic parameters such as the bulk concentration (c) and the stoichiometric number of electrons (n) that are of fundamental significance in any study of an electrode reaction, can be determined experimentally using a range of electroanalytical techniques [6].

As with any analytical method, the ideal electroanalytical technique for parameter characterization should be accurate, reproducible, selective, and robust. In many respects voltammetric methods meet these requirements, since they can be

C.L. Bentley
School of Chemistry, Monash University, Clayton South, VIC 3800, Australia

Energy Flagship, Commonwealth Scientific and Industrial Research Organisation (CSIRO), Box 312, Clayton South, VIC 3169, Australia

A.M. Bond (✉) • J. Zhang
School of Chemistry, Monash University, Clayton South, VIC 3800, Australia
e-mail: Alan.Bond@monash.edu

A.F. Hollenkamp
Energy Flagship, Commonwealth Scientific and Industrial Research Organisation (CSIRO), Box 312, Clayton South, VIC 3169, Australia

P.J. Mahon
Faculty of Science, Engineering and Technology, Swinburne University of Technology, Hawthorn, VIC 3122, Australia

A.A.J. Torriero (ed.), *Electrochemistry in Ionic Liquids*,
DOI 10.1007/978-3-319-13485-7_5

successfully applied to a diverse range of electroactive species in a wide range of electrolyte media. In conventional low-viscosity molecular solvents such as water or acetonitrile containing added supporting electrolyte, steady-state techniques such as microelectrode voltammetry or rotating disk electrode (RDE) voltammetry are generally advantageous over transient methods for quantifying the mass transport parameters associated with an electrochemical system, since the magnitude of the steady-state mass transport limited current (I_L) is unaffected by sluggish heterogeneous kinetics and/or uncompensated resistance (R_u). In practice, this means D, c, or n can be determined straightforwardly from the voltammetric I_L plateau at a microdisk electrode or RDE using well-established analytical theories [6].

By contrast, RTILs possess viscosities that are typically 10 to 10,000 times higher than that of conventional molecular solvent (electrolyte) media. The Stokes–Einstein relationship [7] predicts that diffusivity is inversely proportional to the viscosity of the medium. As a result, mass transport is considerably slower in RTILs, with D often being in the range of 10^{-7} to 10^{-9} cm^2 s^{-1} compared to 10^{-5} to 10^{-6} cm^2 s^{-1} in molecular solvents. This means that in RTIL media, particularly those with high viscosity, it is much more difficult to achieve a true steady state at a microdisk electrode, preventing D, c, or n from being determined from the mass transport limited current [8–10]. It has also been shown that it is similarly difficult to achieve a true steady state at an RDE in RTILs, arising from the difficulty in rotating the electrode at sufficiently high rates under highly viscous conditions [9–11].

In theory, D, c, or n can be calculated from the peak current (I_p) of a dc linear-sweep or cyclic voltammogram by employing the relationships described by the Randles–Sevcik equation for a reversible electron-transfer process of the kind $\text{Ox} + ne^- \rightleftharpoons \text{Red}$, when the effect of R_u is negligible [6]. When working in a viscous RTIL, however, very few of the electrochemical systems of interest fulfill these criteria, which severely limits the direct application of dc voltammetry for this purpose. Assuming the Randles–Sevcik equation is not applicable, a best-fit approach with numerical simulation may be a viable alternative, although even this method has its limitations, as mechanistic complexities or other uncertainties can make the modeling process difficult [10].

Through data processing, transient peak-shaped voltammetric (I–E) curves can be transformed into a form which closely resembles that of a sigmoidal steady-state voltammogram; this outcome can be highly advantageous for further analysis [6]. Transforming the data in this manner is achieved using a convolution procedure [12], which, when applied under conditions where mass transport occurs solely by semi-infinite planar (1D) diffusion, is known as semiintegration [13]. This new form of data presentation provides the same advantages as outlined above for steady-state techniques (i.e., insensitivity to sluggish heterogeneous kinetics and R_u). Furthermore, an extension of this technique, convolution voltammetry, allows the method to be applied even when mass transport includes a contribution from nonplanar (2D) diffusion, permitting the use of a wide range of electrode geometries and voltammetric scan rates [14]. The convolutive transformation also allows the parameters D and nc to be determined simultaneously from two sets of I–E data obtained using two distinctly different voltammetric scan rates applied at the same electrode [15].

In this chapter, the difficulties associated with applying conventional electroanalytical techniques, such as steady-state microelectrode voltammetry, RDE voltammetry, and dc cyclic voltammetry with numerical simulation for the purpose of quantifying diffusivity in RTILs, are highlighted. In developing a strategy for overcoming these issues, we first outline the fundamentals of convolution voltammetry and then discuss a range of situations where electroanalytical applications of these techniques in RTILs have been employed successfully.

5.2 Conventional Voltammetric Techniques

5.2.1 Steady-State Microelectrode Voltammetry

Under true steady-state conditions, D can be calculated from the value of I_L measured at a microdisk electrode using the expression [6]:

$$I_L = 4nFcr_0D \tag{5.1}$$

where r_0 is the radius of the microdisk electrode and F is Faraday's constant. In order to approach a true steady state at a microdisk electrode during potential sweep voltammetry, the following inequality must be satisfied [16, 17]:

$$v \ll \frac{RTD}{nFr_0^2} \tag{5.2}$$

where ν is the voltammetric scan rate, T is the temperature, and R is the universal gas constant. Since D is inversely proportional to viscosity as per the Stokes–Einstein relationship [7], one consequence of this inequality is that relatively slow voltammetric scan rates and/or small electrode geometries are required to approach a true steady state in highly viscous RTILs [8, 9]. In practice, satisfying the inequality shown in Eq. (5.2) may not be feasible, as is demonstrated from the cyclic voltammograms shown in Fig. 5.1, obtained from the one-electron reduction of the cobaltocenium cation ($[Co(Cp)_2]^+$, where Cp=cyclopentadienyl) at a platinum microdisk electrode in the highly viscous RTIL 1-methyl-3-octylimidazolium hexafluorophosphate ($[C_8mim][PF_6]$, $\eta=866$ cP at 20 °C [18]). The shape of the voltammogram is intermediate between sigmoidal and peak-shaped, implying that radial diffusion contributes significantly to mass transport under these conditions. Despite the small electrode size ($r_0=5$ μm) and slow voltammetric scan rate ($\nu=5$ mV s^{-1}), near steady-state conditions have not been attained and therefore Eq. (5.1) cannot be applied, making it very difficult to quantify D without resorting to a "best-fit" approach with numerical simulation. It follows that under these conditions, an impractically small electrode (difficulty in both electrode fabrication and current measurement [19]) or slow scan rate would be required to achieve a good approximation to true steady-state measurements [10].

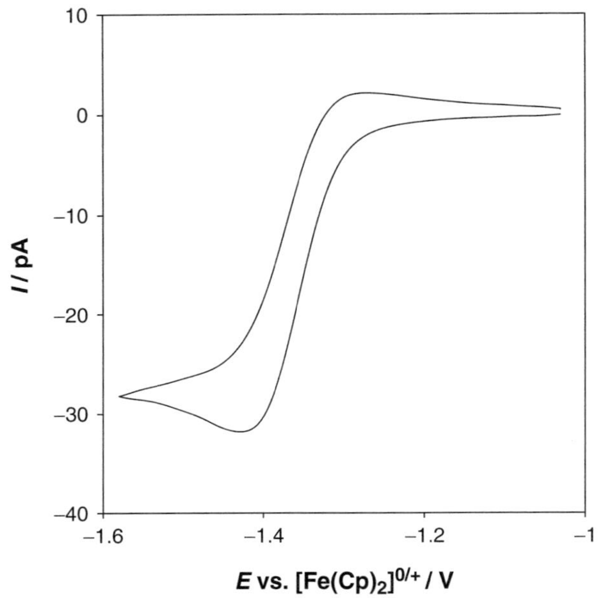

Fig. 5.1 Cyclic voltammogram obtained from the reduction of 10 mM $[Co(Cp)_2]^+$ in $[C_8mim][PF_6]$ using a 10 μm diameter Pt microdisk electrode with a scan rate of 5 mV s^{-1}. Reprinted with permission from Bentley et al., *Anal. Chem.* 2012, *85*, 2239–2245 [10]. Copyright 2013, American Chemical Society

5.2.2 Rotating Disk Electrode Voltammetry

With this technique, under true steady-state conditions, D can be calculated from the value of I_L measured at an RDE using the Levich equation [20]:

$$I_L = 0.620nFA\left(\frac{d}{\eta}\right)^{1/6} D^{2/3}\omega^{1/2}c \tag{5.3}$$

where A is the active electrode area, d is the density of the solvent (electrolyte), η is the hydrodynamic viscosity of the solvent (electrolyte), and ω is the angular rotation rate of the electrode. Much higher rotation speeds are required to approach steady-state conditions at an RDE in RTILs compared to conventional molecular solvents (electrolytes) due to the relative insensitivity of high-viscosity fluids to forced convection. This requirement is evident from the cyclic voltammograms shown in Fig. 5.2, obtained from the one-electron reduction of $[Co(Cp)_2]^+$ at a platinum RDE in $[C_8mim][PF_6]$. The shape of the voltammogram obtained with a rotation rate of 500 rpm is almost identical to that obtained at a stationary electrode (0 rpm), implying that mass transport is governed predominantly by planar diffusion. Once again, because steady-state conditions have not been attained, Eq. (5.3)

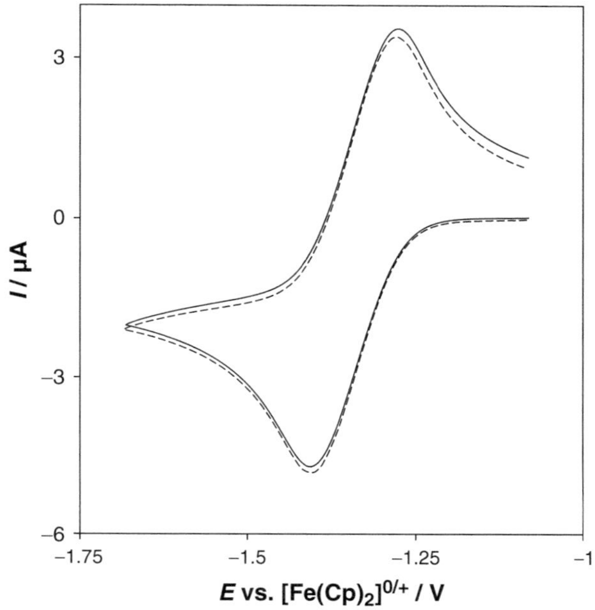

Fig. 5.2 Cyclic voltammograms obtained from the reduction of 10 mM $[Co(Cp)_2]^+$ in $[C_8mim]$ $[PF_6]$ using a 3.0 mm diameter Pt RDE with a scan rate of 50 mV s^{-1} and rotation speeds of 0 rpm (*solid line*) and 500 rpm (*dashed line*). Reprinted with permission from Bentley et al., *Anal. Chem.* 2012, *85*, 2239–2245 [10]. Copyright 2013, American Chemical Society

is not applicable, making it very difficult to quantify D without resorting to a "best-fit" approach with numerical simulation [10].

In principle, much higher rotation speeds could be employed in order to approach true steady-state conditions in highly viscous $[C_8mim][PF_6]$. Indeed, numerical simulations performed for the reduction of $[Co(Cp)_2]^+$ in the room-temperature protic ionic liquid (RTPIL) diethanolamine acetate ($\eta = 336$ cP at 25 °C) revealed that rotation speeds in excess of 5,000 rpm would be required to approach a true steady state at an RDE [11]. In practice, this approach is not feasible, as air bubbles become trapped at the electrode–electrolyte interface at high rotation speeds (>500 rpm), giving rise to noisy and non-reproducible RDE voltammetric data [10, 11].

5.2.3 DC Cyclic Voltammetry

The peak current (I_p) for a simple reversible process under purely planar diffusion conditions can be described by the Randles–Sevcik equation [6] when it is assumed that R_u is negligible:

$$I_p = 0.4463nFAc\left(\frac{nFDv}{RT}\right)^{1/2} \tag{5.4}$$

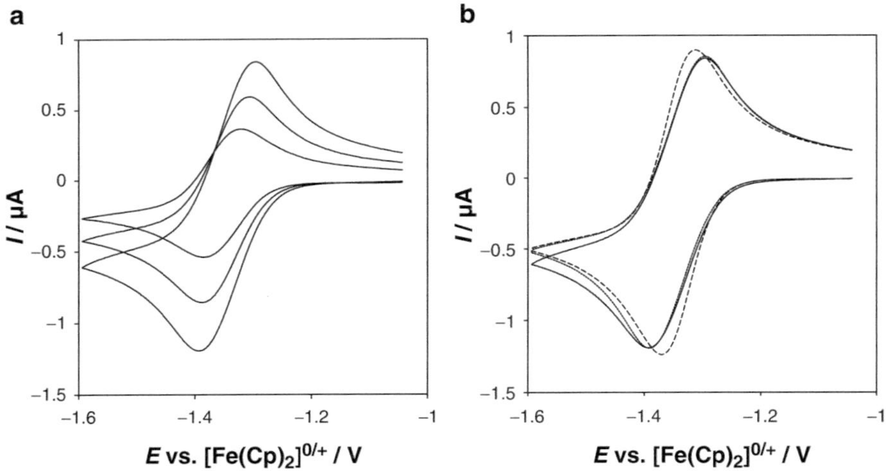

Fig. 5.3 (**a**) Cyclic voltammograms obtained from the reduction of 10 mM $[Co(Cp)_2]^+$ in $[C_8mim][PF_6]$ at a 1.6 mm diameter Pt macrodisk electrode with scan rates (from *top* to *bottom*) of 50, 25, and 10 mV s^{-1}. (**b**) Comparison of an experimental voltammogram (*solid line*) and simulated voltammograms for a reversible process with $R_u = 0\,\Omega$ (*dashed line*) and $R_u = 13\ k\Omega$ (*dotted line*). Other simulation parameters used were $\nu = 50$ mV s^{-1}, $D_{ox} = 9.1 \times 10^{-9}$ cm^2 s^{-1}, $D_{red} = 1.5 \times 10^{-8}$ cm^2 s^{-1} (where the subscripts 'ox' and 'red' refer to the oxidised and reduced forms of the electroactive species, respectively), $k_s = 1$ cm s^{-1}, $\alpha = 0.5$, $C_{dl} = 2 \times 10^{-7}$ F, $E^0 = -1.104$ V, and $T = 297$ K. Reprinted with permission from Bentley et al., *Anal. Chem.* 2012, *85*, 2239–2245 [10]. Copyright 2013, American Chemical Society

In reality, very few electrochemical systems fulfill the above criteria, as is demonstrated by the cyclic voltammograms shown in Fig. 5.3a, obtained from the one-electron reduction of $[Co(Cp)_2]^+$ at a platinum macrodisk electrode in $[C_8mim][PF_6]$. Despite the fact that $[Co(Cp)_2]^{+/0}$ is a well-known reversible process, peak-to-peak separations (ΔE_p) were larger than expected (56.8 mV at 24 °C) and increased significantly with scan rate, attributable to the high level of uncompensated resistance ($R_u \approx 13$ kΩ) in this highly viscous RTIL [10].

The value of D directly estimated from the slope of the I_p vs. $\nu^{1/2}$ plot (obtained by linear regression of the experimental data), 7.6 (±0.2) × 10^{-9} cm^2 s^{-1}, is expected to be somewhat smaller in magnitude than the true value due to the ohmic distortion in the voltammetric I–E data. A more accurate value for D can be obtained using a "best-fit" approach with numerical simulations that accommodate the contribution from R_u, as shown in Fig. 5.3b. As expected, the D value estimated from simulation/experiment comparison, 9.0 (±0.2) × 10^{-9} cm^2 s^{-1}, is significantly greater than the value obtained from the direct use of peak currents. Although this approach is straightforward in the present case (the $[Co(Cp)_2]^{+/0}$ process is reversible, simple, and well characterized), application to electroactive species undergoing more complex mechanisms may be problematic, due to uncertainties with the simulation, as will be demonstrated in Sect. 5.3.2.2.

5.3 Convolution Voltammetry

5.3.1 Theory

Post-measurement data processing techniques which can be applied to voltammetric I–t or I–E data can greatly enhance the ability to extract information from the raw data. One such data transformation approach makes use of the general convolution procedure initially described by Savéant and coworkers [12, 21–28], where semiintegral voltammetry [13] is a subset of a range of techniques collectively known as convolution voltammetry. As described by Oldham and coworkers [13, 29–38], semiintegration is specific to conditions under which mass transport occurs by semi-infinite planar diffusion and gives rise to a function, $M(t)$, which is the 1/2 order fractional integral of current, $I(t)$, where:

$$M(t) = \frac{d^{-1/2} I}{dt^{-1/2}}(t) = I(t) * \frac{1}{\sqrt{\pi t}} \tag{5.5}$$

and the asterisk represents the convolution operation. In Eq. (5.5), the current, $I(t)$ is said to have been semiintegrated with respect to time or convolved with the function $1/\sqrt{\pi t}$. The function, $M(t)$, known as the semiintegral of the current, is related to the concentration of electroactive species at the electrode surface:

$$c(0,t) = c(\infty,t) - \frac{M(t)}{nAF\sqrt{D}} \tag{5.6}$$

where $c(x,t)$ denotes the concentration of electroactive species at distance x from the electrode surface at time t. It follows that under purely diffusion-controlled conditions [i.e., $c(0,t)=0$], $M(t)$ reaches its maximum or limiting value, M_L:

$$M_L = nAFc\sqrt{D} \tag{5.7}$$

Therefore, semiintegrating a peak-shaped transient voltammogram (I–E) gives rise to a sigmoidal M–E curve which resembles a steady-state voltammogram, which in some circumstances is more amenable to data analysis [6]. Indeed, just like the I_L plateau of a steady-state voltammogram, the magnitude of the M_L plateau is unaffected by sluggish heterogeneous kinetics or R_u. The semiintegral at the time $t=J\Delta$ is calculable from current data evenly spaced in time, $I(0)$, $I(\Delta)$, $I(2\Delta)$, …, $I(j\Delta)$, …, $I(J\Delta)$, using the easily programmed formula [39–41]:

$$M(t) = M(J\Delta) = \frac{4}{3}\sqrt{\frac{\Delta}{\pi}}\left[I(J\Delta) + \sum_{J-1}^{j=1} \left\{ \begin{matrix} (J-j+1)^{3/2} - 2(J-j)^{3/2} \\ +(J-j-1)^{3/2} \end{matrix} \right\} I(j\Delta) \right] \tag{5.8}$$

or by equivalent methods [12, 20, 42, 43], where Δ is a small increment of the total time. A Microsoft Excel macro containing a semiintegration algorithm is available free online [44].

The fully planar diffusion restrictions associated with semiintegral voltammetry can be relaxed by replacing the function, $1/\sqrt{\pi t}$, in Eq. (5.5) with the generalized function, $g(t)$, whose form is dependent upon the details of the mechanism (e.g., electron transfer with coupled chemical reactions) and the geometry of the electrode [45]:

$$M(t) = I(t) * g(t) \tag{5.9}$$

Herein $M(t)$ will be referred to as the *semiintegrated current* when calculated based on Eq. (5.5) under planar diffusion conditions and as the *convolved current* for other electrode geometries when calculated more generally based on Eq. (5.9) taking into account the contribution of radial diffusion to the mass transport. Although the function library for $g(t)$ is quite extensive and can deal with first-order homogeneous mechanisms at a range of electrode geometries [45]; herein the discussion is restricted to diffusion limited conditions at an inlaid disk electrode [14, 46]. In order to accommodate radial or edge effects, the formalization applied to the inlaid disk geometry is the bipartite expression which at short times is given by [47]:

$$g(t) = \frac{\sqrt{D}}{r_0} \sum_{j=1}^{5} a_j \theta^{(j-2)/2} \tag{5.10}$$

and at long times is given by:

$$g(t) = \frac{\sqrt{D}}{r_0} \left(b_2 \theta^{-3/2} + \sum_{j=4}^{8} b_j \theta^{(3-2j)/2} \right) \tag{5.11}$$

where θ is the dimensionless time parameter equal to Dt/r_0^2 with the coefficients a_j and b_j listed in Table 5.1 and the transition from the short to the long time expression occurring when $\theta = 1.304$. The convolved current is readily calculable from current data evenly spaced in time using the programmable convolution algorithm [48]:

$$M(t) = \left[I(t) * g(t) \right]_{t=J\Delta} = \frac{1}{\Delta} \left(I_J G_1 + \sum_{j=1}^{J-1} I_{J-j} \left(G_{j-1} - 2G_j + G_{j+1} \right) \right) \tag{5.12}$$

where $G(t) = \int_0^t \int_0^t g(t)\,dt\,dt$ and the abbreviations I_j and G_j refer to $I(j\Delta)$ and $G(j\Delta)$. The *double* integral of $g(t)$ at short times for use in the convolution algorithm is obtained from:

$$G(t) = \frac{r_0^3}{D^{3/2}} \sum_{j=1}^{5} a_j \theta^{(j+2)/2} \tag{5.13}$$

Table 5.1 Values of important polynomial coefficients used in Eqs. (5.10), (5.11), (5.13), and (5.14)

	$g(t)$		$G(t)$	
j	a_j	b_j	a_j	b_j
1	$\frac{1}{\pi^{1/2}}$	0	$\frac{4}{3\pi^{1/2}}$	$\frac{\pi}{4}$
2	−1	$\frac{1}{4\pi^{1/2}}$	$\frac{-1}{2}$	$\frac{-1}{\pi^{1/2}}$
3	$\frac{3}{2\pi^{1/2}}$	0	$\frac{2}{5\pi^{1/2}}$	0.26180
4	−0.38584	-4.7016×10^{-2}	-6.4307×10^{-2}	-6.2688×10^{-2}
5	7.8535×10^{-2}	4.7019×10^{-3}	8.9754×10^{-3}	1.2538×10^{-3}
6		1.6079×10^{-2}		1.8376×10^{-3}
7		-2.2686×10^{-2}		-1.4404×10^{-3}
8		1.2948×10^{-2}		5.2315×10^{-4}
	$\theta=1.304$		$\theta=1.368$	

and for the long time series:

$$G(t)=\frac{r_0^3}{D^{3/2}}\left(\sum_{j=1}^{3}b_j\theta^{(3-j)/2}+\sum_{j=4}^{8}b_j\theta^{(7-2j)/2}\right) \tag{5.14}$$

with transition occurring when $\theta=1.368$. The coefficients b_1 and b_3 are integration constants obtained from Aoki et al. [49]. The double integral of the long-time expression contains eight terms due to the addition of the integration constant for each integration step. Again, the values for the a_j and b_j coefficients are listed in Table 5.1.

The complex time dependence of the diffusion process to an inlaid disk electrode is described by Eqs. (5.10) and (5.11), which are defined in terms of the dimensionless time parameter θ. It follows that under a specific set of experimental conditions, a unique ratio of $\sqrt{D}/r_0$ will apply and this ratio can be defined as a single parameter, ρ. In order to obtain the convolved current, $M(t)$, an initial estimate for ρ must be inputted into the convolution algorithm. If c is known and M_L can be obtained from the limiting convolved current plateau, then it is possible to iteratively refine ρ based on Eq. (5.7) for a known value of r_0:

$$\rho=\frac{M_L}{nF\pi r_0^3 c} \tag{5.15}$$

or D:

$$\rho=\frac{nFD^{3/2}c}{M_L} \tag{5.16}$$

Successive refinements of ρ oscillate and converge on an optimized value that can then be used to calculate either D or r_0 depending on which is known. Furthermore, if voltammograms are obtained using two distinctly different scan rates at the same electrode, then it is possible to calculate both D and nc simultaneously, as both voltammograms should result in identical M_L plateaus even though different time dependencies are involved. This approach assumes r_0 is known, as the optimized value of ρ provides D and this is then substituted into Eq. (5.7) in order to estimate the product of n and c. The discussed convolutive methods can effectively be thought of as "black box" data analysis methods, as the required transformations can be readily performed off-line on voltammetric I–t or I–E data using programmable digital algorithms [shown in Eqs. (5.8) and (5.12)].

In order to apply convolutive analysis to a particular set of experimental data, it is necessary to ensure that certain criteria apply, with uniformity of concentration across the electrode surface being the key requirement [14, 45, 46, 50] for the voltammetry to be describable by a temporal convolution integral and there are three circumstances where this might occur. The first case involves an electrode that is geometrically uniformly accessible. This is an electrode where symmetry dictates that the rate of transport be equal at all points on the surface (i.e., under conditions where mass transport to the electrode surface occurs solely by semi-infinite planar diffusion). When a diffusion limited current occurs at such an electrode, then a uniform concentration distribution will result at the electrode surface if the initial concentration distribution is uniform and this is the situation in most electrochemical experiments. In practice, the only such geometries are a suitably shrouded (or very large) plane, a capped cylinder, an isolated sphere, or a sessile hemisphere. Secondly, if the electrochemical reaction is reversible, then the uniformity of the potential and Nernst's law conspire to ensure that the ratio of activities for the substrate and product is also uniform across the electrode surface. The third situation is when the polarization is so intense that the concentration of the electroactive species at the electrode surface is virtually zero; this will lead to a uniform distribution of the product across the electrode surface even if the reaction is not reversible. In this case, conservation laws ensure that the product is generated uniformly.

A commonly encountered experimental scenario arises when an inlaid disk electrode is used and the electron-transfer kinetics are quasireversible. The implication in this circumstance is that the convolved current will no longer be directly proportional to the surface concentration in the vicinity of the standard potential as described by Eq. (5.6), because the current is kinetically limited in this potential region. However, once the diffusion limited plateau is attained, then the convolved current is once again directly proportional to the surface concentration, and Eq. (5.7), which is the limiting form of Eq. (5.6), is applicable [46, 50].

5.3.2 *Applications of Semiintegral Voltammetry*

5.3.2.1 $[Co(Cp)_2]^{+/0}$ Process in $[C_8mim][PF_6]$

The merits of semiintegration as an analytical technique can be tested by comparing the diffusivity of $[Co(Cp)_2]^+$ determined using this method to the value previously determined in Sect. 5.2.3 using numerical simulation. Linear-sweep semiintegral voltammograms obtained from the one-electron reduction of $[Co(Cp)_2]^+$ at a platinum macrodisk electrode in $[C_8mim][PF_6]$ are shown in Fig. 5.4. The *M–E* curves resemble true steady-state voltammograms, possessing a sigmoidal shape with a well-defined M_L plateau. However, there is a significant slope in the M_L plateau prior to background subtraction, attributable to the charging of the electrical double layer at the stationary electrode (i.e., nonfaradaic current) [10, 29]. Background subtraction can be easily performed by measuring the residual current in the absence of $[Co(Cp)_2]^+$ and subtracting this from the total current prior to semiintegration. The value for *D*, 9.4 (±0.3) × 10^{-9} $cm^2\ s^{-1}$, estimated from the background subtracted M_L plateau by substituting the appropriate values for known parameters into Eq. (5.7), is, as expected, significantly larger than the value previously estimated by direct use of the Randles–Sevcik method (underestimated due to R_u effects) and in good agreement with the value determined using numerical

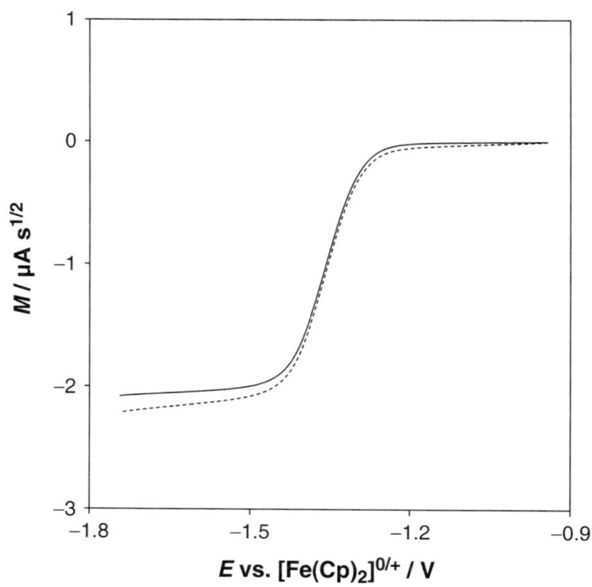

Fig. 5.4 Linear-sweep semiintegral voltammograms obtained from the reduction of 10 mM $[Co(Cp)_2]^+$ in $[C_8mim][PF_6]$ using a 1.6 mm diameter Pt macrodisk electrode with a scan rate of 50 mV s^{-1} before (*dashed line*) and after (*solid line*) background subtraction. Reprinted with permission from Bentley et al., *Anal. Chem.* 2012, *85*, 2239–2245 [10]. Copyright 2013, American Chemical Society

simulation. Despite the significant ohmic distortion present in the voltammetric *I–E* data (previously shown in Fig. 5.3), the diffusivity of $[Co(Cp)_2]^+$ has been successfully determined with high accuracy using the semiintegral method.

5.3.2.2 $I^-/I_3^-/I_2$ Processes in $[C_8mim][PF_6]$

In aprotic solvents such as acetonitrile [51, 52] and analogously in a range of RTILs [53–57], iodide has been shown to be oxidized to molecular iodine on platinum in two steps, giving rise to two resolved voltammetric processes:

$$3I^- \rightleftharpoons I_3^- + 2e^- \tag{5.17}$$

$$I_3^- \rightleftharpoons \frac{3}{2} I_2 + e^- \tag{5.18}$$

The reaction mechanism is complicated by the following homogeneous equilibrium process:

$$I^- + I_2 \rightleftharpoons I_3^- \tag{5.19}$$

Cyclic voltammograms obtained from the oxidation of iodide (from 1-ethyl-3-methylimidazolium iodide, ($[C_2mim]I$) at a platinum macrodisk electrode in $[C_8mim][PF_6]$ are shown in Fig. 5.5a. Evidently, the iodide oxidation process possesses much greater mechanistic complexity than the previously investigated $[Co(Cp)_2]^{+/0}$ process. In addition, the I^-/I_3^- process [reaction shown in Eq. (5.17)] is reportedly kinetically sluggish [54–56], evidenced by the large ΔE_p values associated with this process in the cyclic voltammograms.

The sluggish heterogeneous electron-transfer kinetics of the iodide oxidation process precludes the direct application of the relationships described by the Randles–Sevcik equation and mechanistic complexities make numerical simulation difficult. Fortunately, the M_L plateau of a linear-sweep semiintegral voltammogram is insensitive to both sluggish heterogeneous kinetics and R_u; *M–E* curves obtained from the oxidation of iodide at a platinum macrodisk electrode in $[C_8mim][PF_6]$ are shown in Fig. 5.5b. Once again, the *M–E* curves resemble steady-state voltammograms, possessing a sigmoidal shape with two M_L plateaus corresponding to the two oxidation processes shown in Eqs. (5.17) and (5.18), respectively. The diffusion coefficient of iodide can be estimated from the background subtracted M_L plateau by substituting the appropriate values for known parameters into Eq. (5.7); the determined value, $7.3\ (\pm 0.3) \times 10^{-9}$ cm^2 s^{-1}, is comparable to the diffusivity of $[Co(Cp)_2]^+$ previously determined in this highly viscous RTIL. Semiintegral voltammetry has successfully been employed to determine the diffusivity of an electroactive species under conditions where all of the "conventional voltammetric techniques" discussed in Sect. 5.2, including numerical simulation, are either not valid or unavailable, further highlighting the benefits of convolutive techniques for electroanalytical applications.

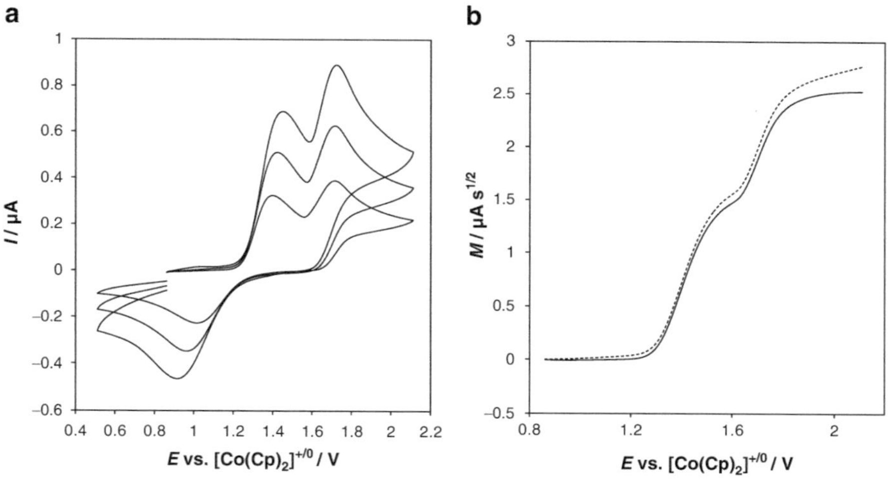

Fig. 5.5 (**a**) Cyclic voltammograms obtained from the oxidation of 14 mM I^- in $[C_8mim][PF_6]$ using a 1.6 mm diameter Pt macrodisk electrode with scan rates (from *top* to *bottom*) of 50, 25, and 10 mV s^{-1}. (**b**) *M–E* curves obtained by semiintegrating the forward sweep of the voltammogram obtained at a scan rate of 50 mV s^{-1} shown in (**a**) before (*dashed line*) and after (*solid line*) background subtraction. Reprinted with permission from Bentley et al., *Anal. Chem.* 2012, *85*, 2239–2245 [10]. Copyright 2013, American Chemical Society

5.3.2.3 Electro-oxidation of Iodide on a Glassy Carbon Electrode in $[C_2mim][N(Tf)_2]$

The discussions below pertain to the $I^-/I_3^-/I_2$ redox processes in RTILs, a subject which has led to significant research interest in recent years [1, 53, 57–64] due to the integral role that the I^-/I_3^- process plays in the operation of dye-sensitized solar cells (DSSCs). A DSSC, in essence, has three fundamental components: a semiconductor photoanode (TiO_2) with an adhered photoactive dye, a platinized cathode (counter electrode), and an electrolyte containing a suitable redox shuttle. As previously alluded to, the electrolyte is most commonly composed of the iodide/triiodide (I^-/I_3^-) couple dissolved in a volatile organic solvent, such as acetonitrile [2]. RTILs have shown promise as alternative electrolytes in DSSCs as they circumvent the long-term safety/stability problems associated with the traditional organic solvent-based electrolytes [1, 2, 65].

Platinum is the most commonly employed cathode material in DSSCs because it acts as an electrocatalyst for the I^-/I_3^- redox shuttle mechanism shown in Eq. (5.17). However, the high cost of noble metals is a significant driving force for the development of more cost-efficient cathode materials, notably carbon-based ones [2, 66, 67]. Cyclic voltammograms obtained for the oxidation of $[C_2mim]I$ on platinum and glassy carbon macrodisk electrodes in the relatively fluid RTIL 1-ethyl-3-methylimidazolium bis(trifluoromethanesulfonyl)imide ($[C_2mim][N(Tf)_2]$, $\eta = 34$ cP at 20 °C [9]) are shown in Fig. 5.6a. On platinum, two sets of peaks are observed,

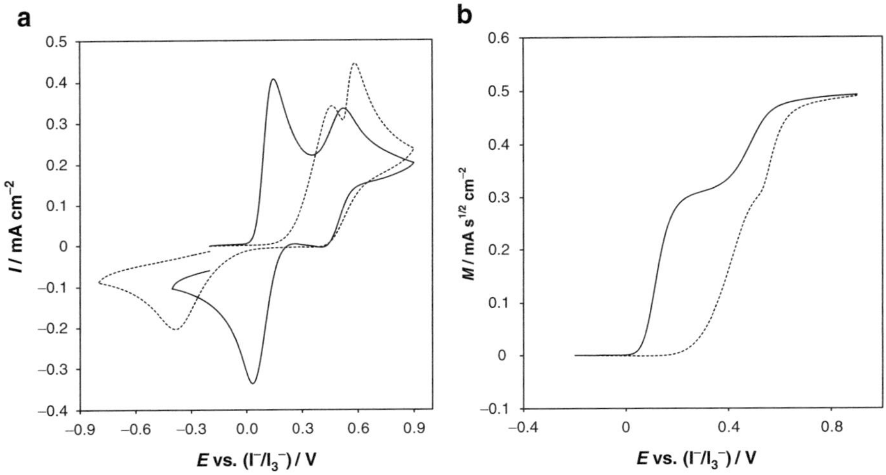

Fig. 5.6 (**a**) Cyclic voltammograms obtained from the oxidation of 10 mM I^- in $[C_2mim][N(Tf)_2]$ with a scan rate of 250 mV s^{-1} using a 1.6 mm diameter Pt macrodisk electrode (*solid line*) or a 3.0 mm diameter glassy carbon macrodisk electrode (*dashed line*). (**b**) *M–E* curves obtained by semiintegrating the forward sweep of the voltammograms shown in (**a**) after background subtraction. Reprinted with permission from Bentley et al., *Electrochim. Acta* 2013, *109*, 554–561 [53]. Copyright 2013, Elsevier

corresponding to the I^-/I_3^- process and I_3^-/I_2 process at less and more positive potentials, respectively, as shown in Eqs. (5.17) and (5.18). On a glassy carbon electrode, two sets of peaks also are observed, albeit with significant overpotentials when compared with the behavior at platinum. Unfortunately, it is not obvious from the transient voltammograms whether the number of electrons associated with each process (and hence, the overall redox mechanism) is the same on both electrode materials.

This information would be readily available from a sigmoidal steady-state voltammogram (I_L plateau magnitudes are proportional to n). However, performing these additional experiments is unnecessary, as the same information is available through semiintegration, as shown in Fig. 5.6b. As expected, on platinum, the relative magnitude of the background subtracted M_L plateaus is clearly proportional to the ratio of number of electrons involved in each process (i.e., 2:1). On glassy carbon, even though the I^-/I_3^- plateau is not as well defined, the two processes still convincingly exhibit a 2:1 ratio, strongly suggesting that the overall iodide oxidation processes are analogous on both electrode materials. The diffusion coefficient of iodide is estimated to be 2.68×10^{-7} cm^2 s^{-1} and 2.61×10^{-7} cm^2 s^{-1} from the background subtracted M_L plateaus on platinum and glassy carbon, respectively, in excellent agreement with each other and with the literature value of 2.68×10^{-7} cm^2 s^{-1} [53, 57]. This study clearly demonstrates that the information available from the semiintegrated *M–E* data is superior to that available from the unprocessed *I–E* data, further expanding upon the electroanalytical utility of cyclic voltammetry.

5.3.2.4 Electro-oxidation of Iodide on a Gold Electrode in $[C_2mim][N(Tf)_2]$

The limited number of studies available on the electro-oxidation of iodide on gold electrodes in conventional solvents such as water [68, 69] and acetonitrile [70] indicates that unique interactions between the I^-/I_2 couple and gold occur. Cyclic voltammograms obtained from the oxidation of $[C_2mim]I$ on platinum and gold macrodisk electrodes in $[C_2mim][N(Tf)_2]$ are shown in Fig. 5.7a. As expected, two sets of peaks are observed on platinum, corresponding to the I^-/I_3^- and I_3^-/I_2 processes as per Eqs. (5.17) and (5.18), respectively. Two sets of peaks are also observed on gold. However, unlike at a glassy carbon electrode, the positions of the peaks are comparable to those observed on platinum. Once again, it is difficult to ascertain whether the number of electrons associated with each process is identical on each material from the unprocessed transient cyclic voltammograms.

M–E curves obtained by semiintegrating the forward sweeps of the respective voltammograms (after background subtraction) are shown in Fig. 5.7b. Remarkably, the magnitude of the M_L plateaus relative to each other is different on platinum and gold; while platinum displays the expected 2:1 magnitude ratio, gold clearly exhibits a 1:1 magnitude ratio, indicative of a different oxidation mechanism for iodide. The magnitudes of the area-normalized M_L plateaus on gold and platinum are comparable, indicating the overall number of electrons involved in the electro-oxidation of each iodide ion (i.e., $n = 1$) is identical on both electrode materials. Therefore, the difference in the iodide oxidation mechanism

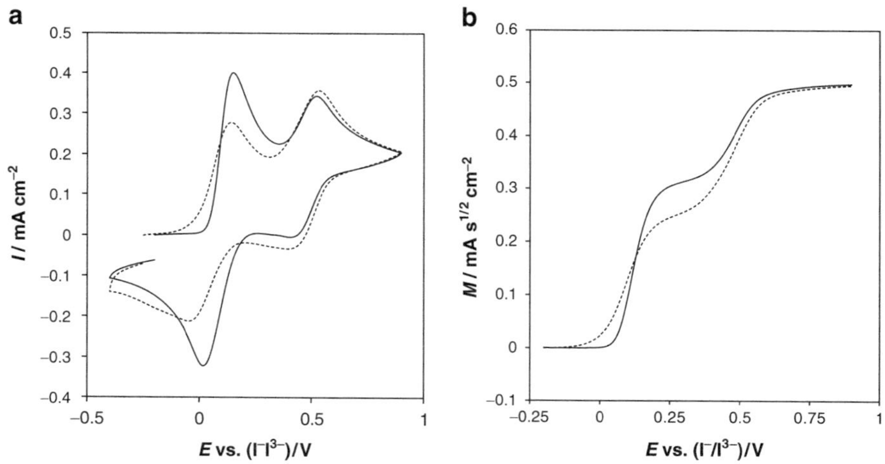

Fig. 5.7 (**a**) Cyclic voltammograms obtained from the oxidation of 10 mM I^- in $[C_2mim][N(Tf)_2]$ with a scan rate of 250 mV s^{-1} using a 1.6 mm diameter Pt macrodisk electrode (*solid line*) or a 3.0 mm diameter Au macrodisk electrode (*dashed line*). (**b**) *M–E* curves obtained by semiintegrating the forward sweep of the voltammograms shown in (**a**) after background subtraction. Reprinted with permission from Bentley et al., *Anal. Chem.* 2013, *85*, 11319–11325 [60]. Copyright 2013, American Chemical Society

on gold and platinum must arise from a disparity in the intermediate formed after the initial oxidation process, which is well known to be triiodide (I_3^-) on platinum [see the reaction shown in Eq. (5.17)]. Indeed, by performing further investigations with other electroanalytical techniques such as electrogravimetry and EQCM, it can be shown that iodide oxidation on gold in $[C_2mim][N(Tf)_2]$ proceeds predominantly via a diiodoaurate ($[AuI_2]^-$) intermediate [60]:

$$2I^- + Au \rightleftharpoons [AuI_2]^- + e^- \tag{5.20}$$

$$[AuI_2]^- \rightleftharpoons I_2 + Au + e^- \tag{5.21}$$

The 1:1 electron stoichiometry shown in Eqs. (5.20) and (5.21) is in agreement with the 1:1 M_L plateau magnitude ratio observed on gold, as demonstrated in Fig. 5.7b. This study and the study previously discussed in Sect. 5.3.2.3 clearly highlight the utility of the semiintegral technique to provide valuable quantitative (i.e., D and n) information about an electrochemical system, achievable simply through off-line data processing, transforming a peak-shaped transient I–E curve into a sigmoidal M–E curve.

5.3.3 *Applications of Convolution Voltammetry*

5.3.3.1 Electro-oxidation of Iodide at a Platinum Microdisk Electrode

The requirement of semiintegration that mass transport due to nonplanar (radial) diffusion must be negligible, limits the use of this technique to relatively large planar electrodes and/or high scan rates, as demonstrated in Sect. 5.3.2. Unfortunately, when working with a macrodisk electrode in RTILs, nonfaradaic current makes a significant contribution to the total measured current, manifesting as a severely sloped M_L plateau, making background subtraction essential for quantitative semiintegral work. As addressed in Sect. 5.3.1, the strictly planar diffusion requirement of semiintegration can be relaxed through the extension available via. Eqs. (5.10) and (11). This allows the convolution algorithm to be applied when working with microdisk electrodes, while retaining all of the advantages of the original semiintegration approach. Changing from a macrodisk to a microdisk electrode has obvious benefits, as it minimizes difficulties with potential control (i.e., IR_u distortion) and reduces the capacitive background current contribution [71–74].

The advantages of convolution voltammetry are illustrated in Fig. 5.8, obtained from the oxidation of $[C_2mim]I$ on a platinum microdisk electrode in $[C_2mim][N(Tf)_2]$. In Fig. 5.8a, the shape of the voltammogram is intermediate between sigmoidal and peak-shaped, implying mixed planar/radial diffusion conditions,

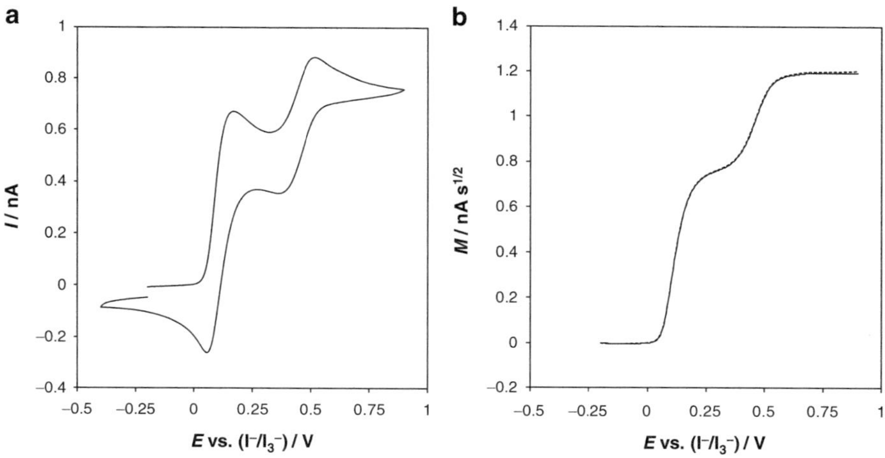

Fig. 5.8 (**a**) Cyclic voltammogram obtained from the oxidation of 4.3 mM I^- in $[C_2mim][N(Tf)_2]$ using a 25 μm diameter Pt microdisk electrode with a scan rate of 50 mV s^{-1}. (**b**) The convolved current of the voltammogram (forward sweep) shown in (**a**) before (*dashed line*) and after (*solid line*) background subtraction. Reprinted with permission from Bentley et al., *Anal. Chem.* 2014, *86*, 2073–2081 [15]. Copyright 2014, American Chemical Society

and as a result, steady-state microdisk electrode theory is not valid and semiintegration is not applicable. Nonetheless, convolution with Eq. (5.9) taking into account the contribution of radial diffusion to the mass transport is applicable, as illustrated in Fig. 5.8b. The convolved current voltammogram is sigmoidal in shape and features two well-defined, scan rate independent, diffusion-controlled M_L plateaus with the expected size ratio (i.e., 2:1). Under these conditions, the charging current is negligible and therefore background subtraction is not required. The diffusion coefficient of iodide is estimated to be 2.70 $(\pm 0.03) \times 10^{-7}$ cm^2 s^{-1} from the M_L plateau, in excellent agreement with the values previously determined on platinum/glassy carbon macrodisk electrodes using the semiintegration technique (after background subtraction).

Use of the technique known generally as convolution voltammetry is obviously advantageous over the previously discussed semiintegration method, as avoiding the need to background subtract greatly simplifies the analytical procedure. Although $[C_2mim][N(Tf)_2]$ is sufficiently fluid to allow a true steady state to be approached at a microdisk electrode over a practical experimental timescale (i.e., substituting $T = 298$ K, $D = 3 \times 10^{-7}$ cm^2 s^{-1}, $n = 1$, and $r_0 = 0.00025$ cm into Eq. (5.2) results in $\nu \ll 123$ mV s^{-1}), convolution voltammetry remains an attractive alternative. This is because, unlike steady-state microelectrode voltammetry, there are no r_0 or ν restrictions associated with the convolution technique, therefore avoiding the practical difficulties associated with measuring small currents over extended experimental timescales.

5.3.3.2 $[Fe(Cp)_2]^{0/+}$ Process in $[C_2mim][N(Tf)_2]$: Determination of *D* and *nc*

In Sects. 5.3.2 and 5.3.3.1, it was shown how the value of *D* may be determined from the M_L plateau after semiintegration/convolution of the experimental *I–E* data when the parameters r_0, n, and c are known. As highlighted in Sect. 5.3.1, convolution can also be employed to determine the constants *D* and *nc* simultaneously from two *I–E* datasets obtained under conditions with differing sphericity factors (σ):

$$\sigma = \sqrt{\frac{RTD}{nFvr_0^2}} \tag{5.22}$$

where the dimensionless sphericity factor describes the relative contribution of planar and radial diffusion to an inlaid disk electrode under a given set of voltammetric conditions. σ is varied in practice simply by changing either the electrode dimensions (i.e., r_0) or the experimental timescale (i.e., ν) [46, 75]. This convolutive technique is most sensitive in the "mixed diffusion region" ($0.012<\sigma<40$) where the contributions from both planar and radial diffusion are significant.

Ferrocene ($[Fe(Cp)_2]$) is a suitable redox species to test the merits of this convolutive technique because while it is reasonably soluble in a range of RTILs [76], it dissolves at a very slow rate, making quantitative preparation of solutions (i.e., where *c* is known) problematic and time-consuming [8]. The need to prepare solutions of known ferrocene concentration can be avoided entirely if the constants *D* and *c* are determined simultaneously, and since in the present case *n* is known, the convolution method is ideal. Cyclic voltammograms (*I–E*) and the corresponding convolved current voltammograms (*M–E*) obtained from the one-electron oxidation of an unknown concentration of $[Fe(Cp)_2]$ in $[C_2mim][N(Tf)_2]$ on a platinum microdisk electrode are shown in Fig. 5.9. Once again, the shape of the cyclic voltammograms indicates mixed planar/radial diffusion conditions, ideal circumstances to apply the convolutive transformation. Due to the reversible nature of the $[Fe(Cp)_2]^{0/+}$ process and the negligible effects of C_{dl} and R_u, the resulting *M–E* curves at the two investigated scan rates almost perfectly coincide, possessing identical M_L plateaus. Using the convolution method described in Sect. 5.3.1, the values for *D* and *c* are estimated to be 4.49×10^{-7} cm^2 s^{-1} and 10.5 mM, respectively.

Since the $[Fe(Cp)]^{0/+}$ electrode mechanism parameters are known, their accuracy can be confirmed by checking the "goodness of fit" with numerical simulation of the cyclic voltammograms, as shown in Fig. 5.10. The excellent agreement between the experimental and simulated data indicates that the values determined for *D* and *c* are valid. This method has also been applied successfully to determine the parameters *D* and *n* simultaneously (i.e., when *c* is known) for more complex

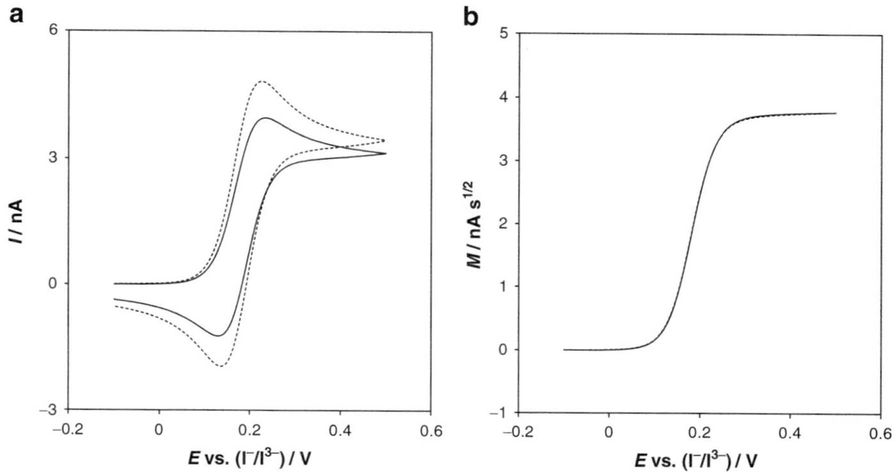

Fig. 5.9 (**a**) Cyclic voltammograms and (**b**) linear-sweep convolved current voltammograms obtained from the oxidation of an unknown concentration of $[Fe(Cp)_2]$ in $[C_2mim][N(Tf)_2]$ using a 25 μm diameter Pt microdisk electrode with scan rates of 50 mV s^{-1} (*solid line*) and 100 mV s^{-1} (*dashed line*). Reprinted with permission from Bentley et al., *Anal. Chem.* 2014, *86*, 2073–2081 [15]. Copyright 2014, American Chemical Society

redox mechanisms in conventional molecular solvents [15]. Although it is possible to obtain the same information using chronoamperometry [55, 76, 77], the voltammetric convolution approach described here avoids the instrumental problems and ohmic (IR_u) effects arising from the initially large currents encountered with potential-step methods.

5.4 Conclusions

The utility of several voltammetric methods has been evaluated for the purpose of quantifying the parameters (i.e., diffusivity, stoichiometric number of electrons, and/or bulk concentration), associated with a range of electroactive species in room-temperature ionic liquids. The application of conventional steady-state methods in this kind of medium is problematic under the highly viscous conditions that commonly are encountered, due to the difficulty in approaching a true steady-state under reasonable experimental conditions. The peak currents of dc cyclic voltammograms are sensitive to heterogeneous kinetics/uncompensated resistance, making analysis on the basis of the relationships outlined by the Randles–Sevcik equation problematic and while numerical simulation is relatively straightforward and represents a

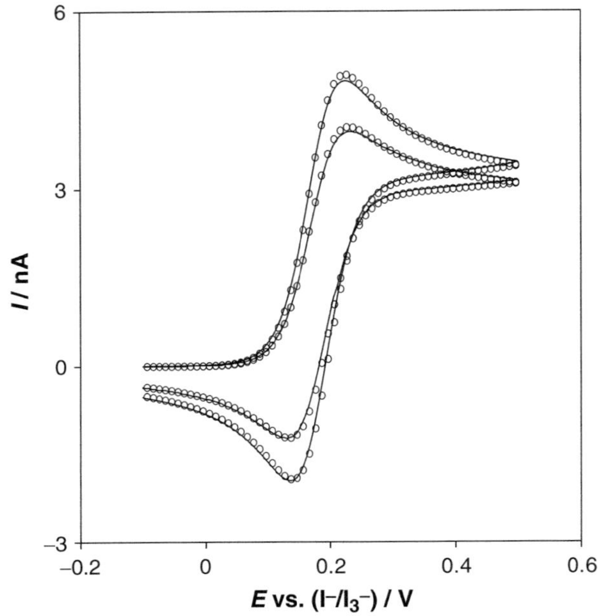

Fig. 5.10 Comparison of simulated (*open circles*) and experimental (*solid line*) cyclic voltammograms obtained from the oxidation of $[Fe(Cp)_2]$ (concentration determined by convolution voltammetry) in $[C_2mim][N(Tf)_2]$ using a 25 μm diameter Pt microdisk electrode with scan rates of (from *top* to *bottom*) 100 and 50 mV s^{-1}. The parameters used in the simulations were $D_{red}=4.49\times10^{-7}$ cm^2 s^{-1}, $D_{ox}=4.20\times10^{-7}$ cm^2 s^{-1}, $c=0.0105$ M, $k_s=10$ cm/s, $r_0=0.00133$ cm, $\alpha=0.5$, $T=296$ K, and $E^0=0.179$ V. Reprinted with permission from Bentley et al., *Anal. Chem.* 2014, *86*, 2073–2081 [15]. Copyright 2014, American Chemical Society

robust method of analysis; on some occasions it is severely hampered by mechanistic complexities or other uncertainties.

Semiintegral voltammetry, a convolution method specific to conditions under which mass transport occurs solely by semi-infinite planar diffusion, is a powerful and also robust analytical method which can be applied under highly viscous conditions to quantify the diffusivity. However, the fully planar diffusion requirement of semiintegration limits the use of this technique to conditions where background charging current is significant, making its subtraction essential for quantitative work. Fortunately, an extension of this method, known generally as convolution voltammetry, accommodates radial diffusion, allowing the experimental conditions to be tailored to minimize the charging current, thus avoiding the need to subtract the background contribution. Overall, the convolutive techniques discussed in this chapter provide powerful analytical tools which have great versatility, being applicable under experimental conditions where the direct use of conventional steady-state and transient voltammetry is severely limited.

References

1. Papageorgiou N, Athanassov Y, Armand M, Bonhote P, Pettersson H, Azam A, Grätzel M (1996) The performance and stability of ambient temperature molten salts for solar cell applications. J Electrochem Soc 143(10):3099–3108. doi:10.1149/1.1837171
2. Hagfeldt A, Boschloo G, Sun LC, Kloo L, Pettersson H (2010) Dye-sensitized solar cells. Chem Rev 110(11):6595–6663. doi:10.1021/cr900356p
3. Howlett PC, MacFarlane DR, Hollenkamp AF (2004) High lithium metal cycling efficiency in a room-temperature ionic liquid. Electrochem Solid State Lett 7(5):A97–A101. doi:10.1149/1.1664051
4. de Souza RF, Padilha JC, Goncalves RS, Dupont J (2003) Room temperature dialkylimidazolium ionic liquid-based fuel cells. Electrochem Commun 5(8):728–731. doi:10.1016/s1388-2481(03)00173-5
5. Balducci A, Dugas R, Taberna PL, Simon P, Plee D, Mastragostino M, Passerini S (2007) High temperature carbon-carbon supercapacitor using ionic liquid as electrolyte. J Power Sources 165(2):922–927. doi:10.1016/j.jpowsour.2006.12.048
6. Bard AJ, Faulkner LR (2001) Electrochemical methods: fundamentals and applications, 2nd edn. Wiley, New York, NY
7. Atkins PW, De Paula J (2010) Atkins' physical chemistry, 9th edn. Oxford University Press, Oxford
8. Zhang J, Bond AM (2005) Practical considerations associated with voltammetric studies in room temperature ionic liquids. Analyst 130(8):1132–1147. doi:10.1039/b504721h
9. Barrosse-Antle LE, Bond AM, Compton RG, O'Mahony AM, Rogers EI, Silvester DS (2010) Voltammetry in room temperature ionic liquids: comparisons and contrasts with conventional electrochemical solvents. Chem-Asian J 5(2):202–230. doi:10.1002/asia.200900191
10. Bentley CL, Bond AM, Hollenkamp AF, Mahon PJ, Zhang J (2012) Advantages available in the application of the semi-integral electroanalysis technique for the determination of diffusion coefficients in the highly viscous ionic liquid 1-methyl-3-octylimidazolium hexafluorophosphate. Anal Chem 85(4):2239–2245. doi:10.1021/ac303042r
11. Zhao C, Burrell G, Torriero AAJ, Separovic F, Dunlop NF, MacFarlane DR, Bond AM (2008) Electrochemistry of room temperature protic ionic liquids. J Phys Chem B 112(23):6923–6936. doi:10.1021/jp711804j
12. Imbeaux JC, Savéant JM (1973) Convolution potential sweep voltammetry: I. Introduction. J Electroanal Chem 44(2):169–187. doi:10.1016/S0022-0728(73)80244-X
13. Grenness M, Oldham KB (1972) Semiintegral electroanalysis: theory and verification. Anal Chem 44(7):1121–1129. doi:10.1021/ac60315a037
14. Mahon PJ (2009) Convolutive reshaping with applications for voltammetry. J Solid State Electrochem 13(4):573–582. doi:10.1007/s10008-008-0664-6
15. Bentley CL, Bond AM, Hollenkamp AF, Mahon PJ, Zhang J (2014) Applications of convolution voltammetry in electroanalytical chemistry. Anal Chem 86(4):2073–2081. doi:10.1021/ac4036422
16. Zoski CG, Bond AM, Colyer CL, Myland JC, Oldham KB (1989) Near-steady-state cyclic voltammetry at microelectrodes. J Electroanal Chem 263(1):1–21. doi:10.1016/0022-0728(89)80120-2
17. Zoski CG, Bond AM, Allinson ET, Oldham KB (1990) How long does it take a microelectrode to reach a voltammetric steady-state. Anal Chem 62(1):37–45. doi:10.1021/ac00200a008
18. Seddon KR, Stark A, Torres MJ (2002) Viscosity and density of 1-alkyl-3-methylimidazolium ionic liquids. In: Abraham MA, Moens L (eds) Clean solvents: alternative media for chemical reactions and processing, vol 819, ACS symposium series. American Chemical Society, Washington, DC, pp 34–49

19. Mauzeroll J, LeSuer RJ (2007) 6.3.3-Laser-pulled ultramicroelectrodes. In: Zoski CG (ed) Handbook of electrochemistry. Elsevier, Amsterdam, pp 199–211
20. Oldham KB, Myland JC (1994) Fundamentals of electrochemical science. Academic Press, San Diego, CA
21. Savéant JM, Vianello E (1967) Potential-sweep voltammetry: theoretical analysis of monomerization and dimerization mechanisms. Electrochim Acta 12(12):1545–1561. doi:10.1016/0013-4686(67)80070-7
22. Savéant JM, Vianello E (1967) Potential-sweep voltammetry: general theory of chemical polarization. Electrochim Acta 12(6):629–646. doi:10.1016/0013-4686(67)85031-x
23. Mastragostino M, Nadjo L, Savéant JM (1968) Disproportionation and ECE mechanisms—I. Theoretical analysis. Relationships for linear sweep voltammetry. Electrochim Acta 13(4):721–749. doi:10.1016/0013-4686(68)85007-8
24. Ammar F, Savéant JM (1973) Convolution potential sweep voltammetry: II. Multistep nernstian waves. J Electroanal Chem 47(2):215–221. doi:10.1016/S0022-0728(73)80448-6
25. Nadjo L, Savéant JM, Tessier D (1974) Convolution potential sweep voltammetry: III. Effect of sweep rate cyclic voltammetry. J Electroanal Chem 52(3):403–412. doi:10.1016/S0022-0728(74)80450-X
26. Savéant JM, Tessier D (1975) Convolution potential sweep voltammetry: V. Determination of charge transfer kinetics deviating from the Butler-Volmer behaviour. J Electroanal Chem 65(1):57–66. doi:10.1016/0368-1874(75)85105-7
27. Savéant JM, Tessier D (1975) Convolution potential sweep voltammetry: Part IV. Homogeneous follow-up chemical reactions. J Electroanal Chem 61(3):251–263. doi:10.1016/S0022-0728(75)80226-9
28. Amatore C, Nadjo L, Savéant JM (1978) Convolution and finite difference approach: application to cyclic voltammetry and spectroelectrochemistry. J Electroanal Chem 90(3):321–331. doi:10.1016/S0022-0728(78)80069-2
29. Oldham KB (1972) Signal-independent electroanalytical method. Anal Chem 44(1):196–198. doi:10.1021/ac60309a028
30. Goto M, Oldham KB (1973) Semiintegral electroanalysis: shapes of neopolarograms. Anal Chem 45(12):2043–2050. doi:10.1021/ac60334a027
31. Goto M, Oldham KB (1974) Semiintegral electroanalysis: studies on the neopolarographic plateau. Anal Chem 46(11):1522–1530. doi:10.1021/ac60347a058
32. Oldham KB (1976) Semi-integration of cyclic voltammograms. J Electroanal Chem 72(3):371–378. doi:10.1016/s0022-0728(76)80323-3
33. Goto M, Oldham KB (1976) Semiintegral electroanalysis: the shape of irreversible neopolarograms. Anal Chem 48(12):1671–1676. doi:10.1021/ac50006a012
34. Dalrymple-Alford P, Goto M, Oldham KB (1977) Shapes of derivative neo-polarograms. J Electroanal Chem 85(1):1–15. doi:10.1016/s0022-0728(77)80148-4
35. Dalrymple-Alford P, Goto M, Oldham KB (1977) Peak shapes in semi-differential electroanalysis. Anal Chem 49(9):1390–1394. doi:10.1021/ac50017a025
36. Oldham KB (1973) Semiintegral electroanalysis. Analog implementation. Anal Chem 45(1):39–47. doi:10.1021/ac60323a005
37. Oldham KB, Zoski CG (1983) Semiintegral electroanalysis with discrimination against charging current. J Electroanal Chem 145(2):265–278. doi:10.1016/S0022-0728(83)80086-2
38. Oldham KB (1997) Tables of semiintegrals. J Electroanal Chem 430(1–2):1–14. doi:10.1016/S0022-0728(96)04827-9
39. Oldham KB, Spanier J (1970) The replacement of Fick's laws by a formulation involving semidifferentiation. J Electroanal Chem 26(2–3):331–341. doi:10.1016/S0022-0728(70)80316-3
40. Oldham KB, Spanier J (1974) The fractional calculus: theory and applications of differentiation and integration to arbitrary order. Academic Press, San Diego, CA

41. Bond AM, Oldham KB, Snook GA (2000) Use of the ferrocene oxidation process to provide both reference electrode potential calibration and a simple measurement (via semiintegration) of the uncompensated resistance in cyclic voltammetric studies in high resistance organic solvents. Anal Chem 72(15):3492–3496. doi:10.1021/ac000020j
42. Whitson PE, VandenBorn HW, Evans DH (1973) Acquisition and analysis of cyclic voltammetric data. Anal Chem 45(8):1298–1306. doi:10.1021/ac60330a016
43. VandenBorn HW, Evans DH (1974) Determination of electrochemical kinetic parameters in high resistance solutions using semiintegration. Anal Chem 46(6):643–646. doi:10.1021/ac60342a005
44. de Levie R (2012) Excellaneous. http://www.bowdoin.edu/~rdelevie/excellaneous/. Accessed 19 Sept 2013
45. Mahon PJ, Oldham KB (1999) Convolutive modelling of electrochemical processes based on the relationship between the current and the surface concentration. J Electroanal Chem 464(1):1–13. doi:10.1016/s0022-0728(98)00450-1
46. Mahon PJ (2010) Application of global analysis for the measurement of electrochemical parameters with disk electrodes. Electrochim Acta 55(3):673–680. doi:10.1016/j.electacta.2009.09.035
47. Mahon PJ (2011) Convolutive reshaping as a way to simulate voltammetry at disk electrodes. Electrochim Acta 56(5):2190–2200. doi:10.1016/j.electacta.2010.11.096
48. Oldham KB (1986) Convolution: a general electrochemical procedure implemented by a universal algorithm. Anal Chem 58(11):2296–2300. doi:10.1021/ac00124a040
49. Aoki K, Osteryoung J (1984) Formulation of the diffusion-controlled current at very small stationary disk electrodes. J Electroanal Chem 160(1–2):335–339. doi:10.1016/0368-1874(84)83278-5
50. Mahon PJ, Oldham KB (2006) An examination of the conditions resulting in a temporal convolution related to electrochemical processes. J Solid State Electrochem 10(10):785–791. doi:10.1007/s10008-006-0169-0
51. Hanson KJ, Tobias CW (1987) Electrochemistry of iodide in propylene carbonate. I. Cyclic voltammetry monitored by optical spectroscopy. J Electrochem Soc 134(9):2204–2210. doi:10.1149/1.2100852
52. Macagno VA, Giordano MC, Arvia AJ (1969) Kinetics and mechanisms of electrochemical reactions on platinum with solutions of iodine-sodium iodide in acetonitrile. Electrochim Acta 14(4):335–357. doi:10.1016/0013-4686(69)85005-x
53. Bentley CL, Bond AM, Hollenkamp AF, Mahon PJ, Zhang J (2013) Concentration and electrode material dependence of the voltammetric response of iodide on platinum, glassy carbon and boron-doped diamond in the room temperature ionic liquid 1-ethyl-3-methylimidazolium bis(trifluoromethanesulfonyl)imide. Electrochim Acta 109:554–561. doi:10.1016/j.electacta.2013.07.101
54. Zhang Y, Zheng JB (2007) Investigation on the electro-oxidation of iodide in the room temperature ionic liquid, 1-butyl-3-methylimidazolium tetrafluoroborate at platinum electrode. Electrochim Acta 52(12):4082–4086. doi:10.1016/j.electacta.2006.11.025
55. Silvester DS, Aldous L, Hardacre C, Compton RG (2008) Electrooxidation of the iodides [C4mim]I, LiI, NaI, KI, RbI, and CsI in the room temperature ionic liquid [C_4mim][NTf_2]. J Phys Chem C 112(16):6551–6557. doi:10.1021/jp800145k
56. Ejigu A, Lovelock KRJ, Licence P, Walsh DA (2011) Iodide/triiodide electrochemistry in ionic liquids: effect of viscosity on mass transport, voltammetry and scanning electrochemical microscopy. Electrochim Acta 56(28):10313–10320. doi:10.1016/j.electacta.2011.03.108
57. Kawano R, Watanabe M (2003) Equilibrium potentials and charge transport of an I^-/I_3^- redox couple in an ionic liquid. Chem Commun 3:330–331. doi:10.1039/b208388b

58. Kawano R, Matsui H, Matsuyama C, Sato A, Susan M, Tanabe N, Watanabe M (2004) High performance dye-sensitized solar cells using ionic liquids as their electrolytes. J Photochem Photobiol A-Chem 164(1–3):87–92. doi:10.1016/j.jphotochem.2003.12.019
59. Kawano R, Watanabe M (2005) Anomaly of charge transport of an iodide/tri-iodide redox couple in an ionic liquid and its importance in dye-sensitized solar cells. Chem Commun 16:2107–2109. doi:10.1039/b418031c
60. Bentley CL, Bond AM, Hollenkamp AF, Mahon PJ, Zhang J (2013) Unexpected complexity in the electro-oxidation of iodide on gold in the ionic liquid 1-ethyl-3-methylimidazolium bis(trifluoromethanesulfonyl)imide. Anal Chem 85(23):11319–11325. doi:10.1021/ac402150y
61. Wachter P, Schreiner C, Zistler M, Gerhard D, Wasserscheid P, Gores HJ (2008) A microelectrode study of triiodide diffusion coefficients in mixtures of room temperature ionic liquids, useful for dye-sensitised solar cells. Microchim Acta 160(1–2):125–133. doi:10.1007/s00604-007-0803-2
62. Wachter P, Zistler M, Schreiner C, Fleischmann M, Gerhard D, Wasserscheid P, Barthel J, Gores HJ (2009) Temperature dependence of the non-stokesian charge transport in binary blends of ionic liquids. J Chem Eng Data 54(2):491–497. doi:10.1021/je800480d
63. Zistler M, Wachter P, Schreiner C, Gores HJ (2010) Electrochemical measurement of triiodide diffusion coefficients in blends of ionic liquids Results for improving a critical parameter of dye-sensitized solar cells. J Mol Liq 156(1):52–57. doi:10.1016/j.molliq.2010.04.021
64. Thorsmolle VK, Rothenberger G, Topgaard D, Brauer JC, Kuang DB, Zakeeruddin SM, Lindman B, Grätzel M, Moser JE (2011) Extraordinarily efficient conduction in a redox-active ionic liquid. ChemPhysChem 12(1):145–149. doi:10.1002/cphc.201000819
65. Wang P, Zakeeruddin SM, Moser JE, Humphry-Baker R, Grätzel M (2004) A solvent-free, $SeCN^-/(SeCN)_3^-$ based ionic liquid electrolyte for high-efficiency dye-sensitized nanocrystalline solar cells. J Am Chem Soc 126(23):7164–7165. doi:10.1021/ja048472r
66. Kay A, Grätzel M (1996) Low cost photovoltaic modules based on dye sensitized nanocrystalline titanium dioxide and carbon powder. Sol Energy Mater Sol Cells 44(1):99–117. doi:10.1016/0927-0248(96)00063-3
67. Murakami TN, Ito S, Wang Q, Nazeeruddin MK, Bessho T (2006) Highly efficient dye-sensitized solar cells based on carbon black counter electrodes. J Electrochem Soc 153(12):A2255–A2261. doi:10.1149/1.2358087
68. Loo BH (1982) In situ identification of halide-complexes on gold electrode by surface-enhanced raman-spectroscopy. J Phys Chem 86(4):433–437. doi:10.1021/j100393a002
69. Qi PH, Hiskey JB (1993) Electrochemical-behavior of gold in iodide solutions. Hydrometallurgy 32(2):161–179. doi:10.1016/0304-386x(93)90021-5
70. Zhao SF, Lu JX, Bond AM, Zhang J (2012) Remarkable sensitivity of the electrochemical reduction of benzophenone to proton availability in ionic liquids. Chem Eur J 18(17):5290–5301. doi:10.1002/chem.201103365
71. Forster RJ (2006) Ultrafast electrochemical techniques. Encyclopedia of analytical chemistry. Wiley, New York, NY. doi:10.1002/9780470027318.a5319
72. Heinze J (1993) Ultramicroelectrodes in electrochemistry. Angew Chem Int Ed Engl 32(9):1268–1288. doi:10.1002/anie.199312681
73. Bond AM (1994) Past, present and future contributions of microelectrodes to analytical studies employing voltammetric detection. A review. Analyst 119(11):1R–21R. doi:10.1039/an994190001r
74. Amatore C, Maisonhaute E, Simonneau G (2000) Ultrafast cyclic voltammetry: performing in the few megavolts per second range without ohmic drop. Electrochem Commun 2(2):81–84. doi:10.1016/s1388-2481(99)00150-2
75. Mahon PJ, Phillips WRC (2012) Global analysis for the measurement of electrochemical parameters with cylindrical electrodes. Electrochim Acta 74:16–22. doi:10.1016/j.electacta.2012.03.117

76. Rogers EI, Silvester DS, Poole DL, Aldous L, Hardacre C, Compton RG (2008) Voltammetric characterization of the ferrocene|ferrocenium and cobaltocenium|cobaltocene redox couples in RTILs. J Phys Chem C 112(7):2729–2735. doi:10.1021/jp710134e
77. Xiong L, Aldous L, Henstridge MC, Compton RG (2012) Investigation of the optimal transient times for chronoamperometric analysis of diffusion coefficients and concentrations in non-aqueous solvents and ionic liquids. Anal Methods 4(2):371–376. doi:10.1039/c1ay05667k

Chapter 6
Small-Angle X-Ray Scattering of Ionic Liquids

Joshua A. Hammons, Jan Ilavsky, and Fan Zhang

6.1 Introduction

As in most experimental techniques, an experimenter observes how a sample reacts to some kind of external stimulus. For example, in a simple electrochemical experiment, one applies a potential and observes a current response. Similarly, in SAXS, one exposes a sample to a beam of X-rays and measures the intensity, I, of the scattered X-rays as a function of the scattering angle, 2θ, shown in Fig. 6.1.

This is the basic concept of a small-angle scattering (SAS) experiment, whether X-rays or neutrons are being used. The fundamental difference between the two is that neutrons scatter from nuclei whilst X-rays scatter from electrons. With this in mind, the theory behind SAXS will be presented here, keeping in mind that the general theory can be applied to both, accounting only for the difference in the definition of scattering contrast.

With the increasingly high X-ray brilliance provided by synchrotrons around the world and the accessibility of bench-top SAXS instruments, we will only concern ourselves with X-rays. Regardless of how they are produced, the X-rays that interact with the sample are collimated (Fig. 6.1). Here, collimation means that the X-rays must propagate with a minimal amount of angular divergence so that when they interact with the sample, they do so at the same relative angle. The incident X-rays should be monochromatic, as polychromatic X-rays complicate analysis of the SAXS data, as we will show in the General Theory section below. However,

J.A. Hammons (✉) • J. Ilavsky
X-Ray Science Division, Advanced Photon Source, Argonne National Laboratory,
9700 S. Cass Ave., Argonne, IL 60439, USA
e-mail: hammons@aps.anl.gov; josh.hammons@gmail.com

F. Zhang
Materials Measurement Science Division, Material Measurement Laboratory, National Institute of Standards and Technology, 100 Bureau Drive, Gaithersburg, MD 20899, USA

A.A.J. Torriero (ed.), *Electrochemistry in Ionic Liquids*,
DOI 10.1007/978-3-319-13485-7_6

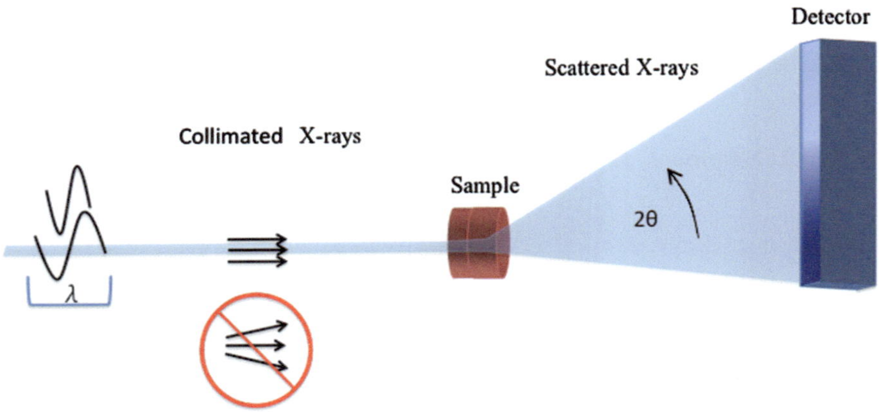

Fig. 6.1 A simplified schematic showing a typical SAXS experiment showing the incoming and scattered X-rays from a sample

so-called pink beams can also be used to increase the incident flux of the X-rays, which have a wider spread of wavelengths (typically up to 10 %). This results in wavelength smearing of the data and can be treated with a priori knowledge of the beam profile. In most cases, the X-ray wavelengths are sufficiently monochromatic that wavelength smearing is negligible.

In this chapter, we aim to provide the reader with an understanding of how SAXS has been used to study a range of ionic liquid (IL) systems as well as the full potential that X-ray scattering experiments have in the study of ILs. For brevity, we will only present a brief introduction to SAS so that the reader can imagine a physical picture when looking at SAXS data, as well as understand both its potential and limits. A more comprehensive account of SAS analysis can be found in books by Guinier [1, 2], Glatter and Kratky [3], Giles [4], and (for biological macromolecules) Svergun [5], as well as review papers [6, 7]. Readers may also be interested in data reduction [8], corrections [9, 10], and analysis [11–17] work that is also very useful.

6.2 General Theory

6.2.1 Elastic X-Ray Scattering

Elastic X-ray scattering occurs when an incoming X-ray causes an electron to (briefly) accelerate, emitting a radially propagating X-ray of the same wavelength [18]. The resulting X-ray intensity scattered by a single electron depends on the polarization of the incoming X-rays and is given by the equations:

$$I_e = r_e^2 I_0 \cos 2\theta \quad (parallel) \tag{6.1}$$

$$I_e = r_e^2 I_0 \quad (perpendicular) \tag{6.2}$$

where r_e is the classical electron radius, I_0 is the incoming intensity, and 2θ is the angle of intensity measurement [2]. In the small-angle regime, $2\theta < 5°$, the $\cos(2\theta)$ term does not deviate from one by more than a percent. Therefore, Eqs. (6.1) and (6.2) show that the polarization state of the X-rays does not affect the outcome of a SAXS experiment. However, the measured intensity should be normalized by I_e to obtain SAXS data on an absolute scale, if this is needed (recommended).

In general, SAXS data can be divided into two different types: diffraction in the small-angle regime and true small-angle scattering. The discrete diffraction peaks in the small-angle regime originate from ordered nanostructures, which are often found in ILs. The scattering angle where the diffraction peak reaches a maximum is related to a repeat distance, *d*, of an ordered phase. The SAXS data, on the other hand, usually vary smoothly with scattering angle and reflect the morphology and/or microstructure of the scattering objects. In other words, diffraction peaks indicate the presence of crystalline or semi-crystalline phases, whereas SAS does not. The analyses of the two cases are different. Both will be discussed here, with emphasis on SAS analysis.

6.2.2 Data Reduction

Before one can analyze the raw scattering intensity, the data must be properly reduced and corrected. This can be a rather involved discussion and will only be briefly discussed here. More detail on data reduction and correction can be found in Glatter and Kratky [3] or an excellent review by Pauw [9]. Briefly, one has to calibrate the angle of measurement, 2θ, and the scaled intensity; large angles of measurement require additional geometrical and possibly polarization corrections (Eqs. 6.1 and 6.2). The angular calibration allows the experimenter to obtain the intensity as a function of 2θ. Common small-angle diffraction standards include silver behenate and rat-tail collagen for SAS measurements. In most cases, this is sufficient and the scaled intensity is not calibrated. However, when done properly [19], an absolute calibration of the intensity can provide the experimenter with information about the volume and contrast of the scattering objects, which may be important. especially when trying to identify unknown phases. In this case, one obtains the differential scattering cross section, $\frac{\partial \Sigma}{\partial \Omega}(q)$, which contains all of the relevant SAS information. The magnitude of the scattering vector q is defined as $q = 4\pi/\lambda \times \sin(\theta)$, where λ is the X-ray wavelength and θ is one-half of the scattering angle 2θ. In this chapter, we will follow the convention of using $I(q)$, instead of $\frac{\partial \Sigma}{\partial \Omega}(q)$.

6.2.3 Diffraction

From a SAXS experiment, one can resolve ordered inhomogeneities within the sample. Specific to ILs, such ordering may be inherent to the IL itself, whether it be a consequence of the cation [20, 21], anion [22], or the aggregation of amphiphilic molecules in ILs [23, 24]. In these cases, the structural order is analyzed by resolving the repeat distance, *d*, of the phases. Such structural order is often a key feature of ILs, making SAXS an important technique for their study. This order is observed directly from the SAXS pattern as "peaks" in the $I(2\theta)$ curve. These peaks correspond to the distance, *d*, of the ordered structure through Bragg's Law:

$$d = \frac{n\lambda}{2\sin\theta} \tag{6.3}$$

where λ is the wavelength of X-rays and n is an integer. Equation (6.3) has been used fruitfully in the study of ionic liquids using SAXS and wide-angle X-ray scattering (WAXS), which uses scattering profiles at larger scattering angles to resolve smaller distances. For $n=1$, Eq. (6.3) can also be written as $d=2\pi/q$, where q is the magnitude of the scattering vector at the center of the peak. We should note at this point that while the ordered distances, *d*, are directly obtained through Eq. (6.3), Bragg's law says nothing about the source of the ordering. However, because of the simplicity and ubiquitous application of Eq. (6.3) to SAXS in ILs, Bragg's law is often employed where better models should be used. A good example of this concerns ordered micelles, for which Eq. (6.3) is used to analyze the inter-micelle distance [25–27]. However, in these cases an analytical solution is readily available to model the SAXS data [28]. For this reason, we will focus mainly on SAXS analysis.

6.2.4 Fundamental SAXS Theory

As mentioned previously, SAXS provides information about the physical geometry of the system. for example, feature size, shape, and internal and/or external structure. With the assumption that λ is much smaller than the dimensions of the inhomogeneity in the sample, each scattered wave will constructively interfere with other scattered waves to some degree. The degree of constructive interference is determined by their relative phase, φ.

Figure 6.2 shows the scattering of an X-ray wave-front by two points in an inhomogeneity, separated by a vector, **r**, where the scattered intensity is measured at a scattering angle, 2θ. When a monochromatic X-ray wave is scattered from different points in the inhomogeneity, the interference of the scattered waves will depend on the phase shift, φ, between the scattered and incident waves, which is given by equation:

$$\varphi = \frac{2\pi\left(\left(\mathbf{s}-\mathbf{s}_0\right)\cdot\mathbf{r}\right)}{\lambda} = \mathbf{q}\cdot\mathbf{r}, \tag{6.4}$$

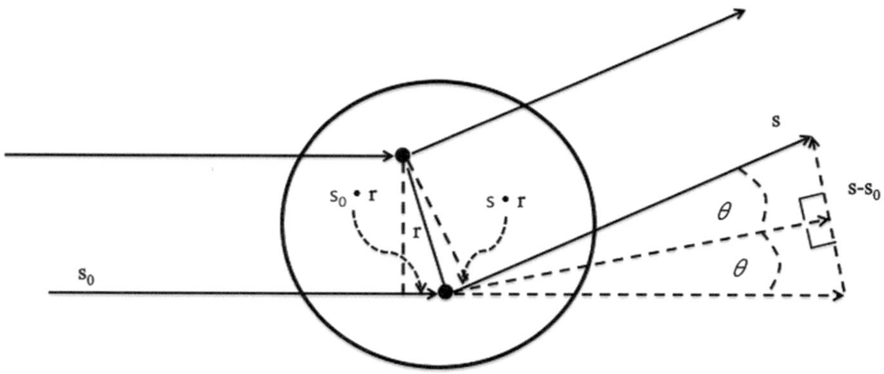

Fig. 6.2 A simplified diagram showing how two electromagnetic waves can either constructively or destructively interfere, depending on the angle of measurement (for a specific wavelength)

s and $\mathbf{s_0}$ are unit vectors in the directions of the scattered and incoming waves, respectively, and the vector **q** is defined accordingly [1–3, 18]. Figure 6.2 also shows that the magnitude of $\mathbf{s}-\mathbf{s_0}$ is given by $2\sin\theta$, resulting in the conventional equation for $|\mathbf{q}|(=q)$:

$$q = 2\pi \frac{2\sin(\theta)}{\lambda} \tag{6.5}$$

where the factors 2π and λ are included from Eq. (6.4). If the X-ray phase difference along the path between any two points is known, the resulting scattering amplitude from the two points in Fig. 6.2 can be calculated using complex notation:

$$A(\mathbf{q}) = \sum f_j e^{-i(\mathbf{q}\cdot\mathbf{r}_j)} \tag{6.6}$$

where f_j is the scattering power of a single electron j. The variable f is 1 for a free electron, but will not be the case near an adsorption edge [1–3, 18]. Thus, between Fig. 6.2 and Eq. (6.5), we can now relate real space dimensions to the scattering amplitude measured at a certain **q** value. At this point, **q** is a vector and not a scalar. However, it is important to note that the phase difference along the path of a vector, **r**, which is orthogonal to **q**, is 0 (Eq. 6.4). Thus, only the component of **r** parallel to **q** is relevant for scattering [3]. This is a particularly important concept for anisotropic systems where dimensions perpendicular to **q** do not affect the scattering intensity.

While Eq. (6.6) does relate the physical dimensions to the X-ray amplitude, it cannot be used directly because an X-ray scattering experiment measures scattering intensity (where scattering amplitudes are summed and multiplied by their complex conjugates), instead of scattering amplitude. The continuous distribution of electrons in the scattering object allows us to replace the summation in Eq. (6.6) with a scattering length density function, $\rho(\mathbf{r})$, and rewrite Eq. (6.6) as $A(\mathbf{q}) = \int \rho(\mathbf{r}) e^{-i\mathbf{q}\cdot\mathbf{r}}\, dV$, where the integral is taken over the scattering volume. The measured scattering intensity, I, will therefore be:

$$I(\mathbf{q}) = A(\mathbf{q})A(\mathbf{q})^* = \iiint \iiint \rho(\mathbf{r}_1)\rho(\mathbf{r}_2)e^{-i\mathbf{q}\cdot(\mathbf{r}_1-\mathbf{r}_2)}dV_1 dV_2 \tag{6.7}$$

where A^* is the complex conjugate of A, and $\mathbf{r}_1$ and $\mathbf{r}_2$ are two separate points in the sample. Equation (6.7) relates the measured scattering intensity, $I(\mathbf{q})$, to real space dimensions, $\mathbf{r}$. We can further simplify Eq. (6.7) by redefining $\mathbf{r}=\mathbf{r}_1-\mathbf{r}_2$ and introducing the auto-correlation function, $\bar{\rho}^2(\mathbf{r})$ by the equation:

$$\bar{\rho}^2(\mathbf{r}) = \iiint \rho(\mathbf{r}_1)\rho(\mathbf{r}_1-\mathbf{r})dV_1 \tag{6.8}$$

where the function is integrated over the sample volume once. Combining Eq. (6.8) with Eq. (6.7), the scattering intensity can be written:

$$I(\mathbf{q}) = \iiint \bar{\rho}^2(\mathbf{r})e^{-i\mathbf{q}\cdot\mathbf{r}}dV, \tag{6.9}$$

where the function is again integrated over the sample volume just once. Equation (6.9) now provides the experimenter with a mathematical link between the scattering intensity and autocorrelation function of the scattering length density [3]. We derived Eq. (6.9) without making assumptions. However, this equation is not very useful because it contains vectors and imaginary numbers that are difficult to deal with analytically for many systems. By applying appropriate assumptions concerning the material system being studied, more useful forms of Eq. (6.9) can be derived and applied to extract physically meaningful parameters.

6.2.5 *Scattering Inhomogeneities*

At this point, we should point out an important aspect of SAXS—the scattering object, as exemplified in Fig. 6.2, needs to have a different electron density than that of its immediate surroundings in order to give a finite scattering intensity. The difference in electron density, however, can be either positive or negative, and the sign cannot be distinguished by SAXS alone. For example, the object in Fig. 6.2 could just as easily be a pore inside a matrix instead of a solid particle within a suspension. This ambiguity is a consequence of the well-known Babinet's principal of reciprocity [1]. For this reason, we will refer to the scattering inhomogeneities as such, instead of particles or pores.

6.3 SAXS Analysis

SAXS analysis is a very broad topic and contains within it a wide variety of ways to analyze the $I(q)$ curve. In many cases, the type of applicable analysis depends on the system, and on the specific assumptions that can be made about the system. In this chapter, we do not intend to cover all of the available SAXS analysis methods. Instead, we will provide an overview of the analysis methods that are relevant to ILs and offer some case examples. The goal is to familiarize the reader with fundamental aspects of SAXS. For analysis, we encourage the readers to use any one of the free SAXS software packages available [11, 12, 16, 29–31].

6.3.1 Common SAXS Assumptions

The first and most common assumption is that of *statistical isotropy*. It is assumed that there is no preferred orientation of the scattering inhomogeneities relative to the scattering vector, **q**. As a general rule, this means that if we collect the scattering intensity from different sample orientations, we will get the same $I(\mathbf{q})$ curve. This concept is illustrated in Fig. 6.3. For spherical shapes (far left), the inhomogeneities themselves are isotropic so they satisfy the assumption of statistical isotropy. If, however, these inhomogeneities are anisotropic (middle and far right), then they must be randomly oriented to be considered as statistically isotropic (far right). In the following sections, we will present some useful analytical techniques one can use when this assumption is valid.

The *dilute limit* assumption is also very common in SAXS analysis. This assumption requires that interaction or ordering between scattering inhomogeneities is negligible. We would strongly caution against making this assumption liberally. Instead,

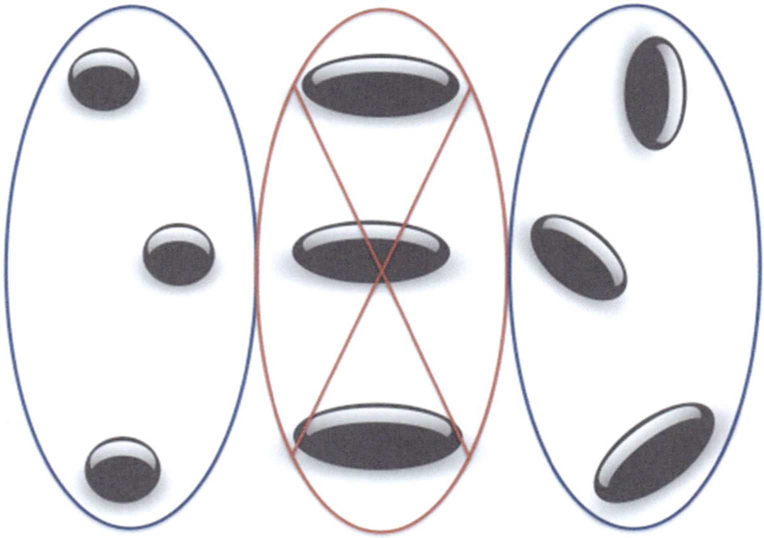

Fig. 6.3 Illustrates the concept of statistical isotropy (*left* and *right*), while the *center* is anisotropic

we encourage using a priori knowledge of the system to determine the validity of this assumption. As a rule of thumb, if the inhomogeneity is evenly dispersed, its volume fraction needs to be less than a few percent to be considered "dilute."

Centro-symmetry assumes that the scattering inhomogeneities, themselves, are *centro-symmetric*. One can think of this as a shape-dependent assumption whereby a radial vector (from the center of mass of the inhomogeneity) has the same chord length within the inhomogeneity as a radial vector in the opposite direction. This is shown in Fig. 6.4 with a parallelepiped. Another *centro-symmetric* shape would be an ellipsoid of revolution.

6.3.2 Indirect Fourier Transformation Analysis

Indirect Fourier Transformation (IFT) Analysis is a method that has been applied in studies of protein stability in a hydrated choline di-hydrogen phosphate ionic liquid [32]. The IFT analysis can also be useful in identifying unknown phases, as well as understanding the link between real space and reciprocal space. This analysis assumes *statistical isotropy*, as well as the *dilute limit* [33].

If these assumptions are valid for a given system, then the phase factor in Eq. (6.9) can be averaged over all orientations and vectors replaced by scalars. The scattering intensity can be represented by the equation:

$$I(q) = V\overline{\eta^2}\int \mathrm{PDDF}(r)\frac{\sin qr}{qr}dr = V\overline{\eta^2}\int \gamma_o(r)4\pi r^2\frac{\sin qr}{qr}dr \tag{6.10}$$

$$\gamma_o(r) = \frac{\overline{\rho}^2(r)}{V\overline{\eta^2}} \tag{6.11}$$

$$\overline{\eta^2} = \left\langle (\rho - \overline{\rho})^2 \right\rangle \tag{6.12}$$

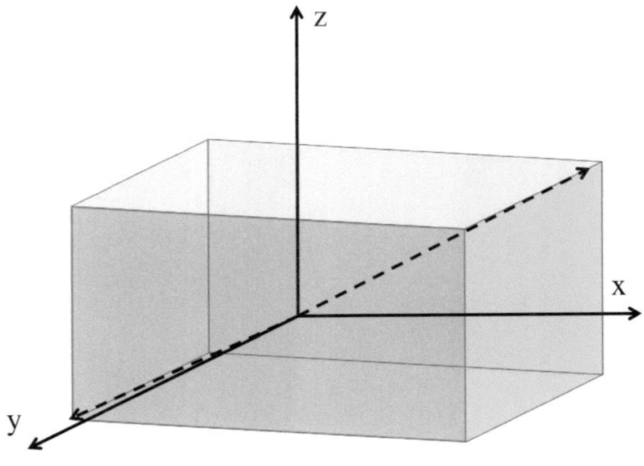

Fig. 6.4 Diagram depicting the concept of centro-symmetry

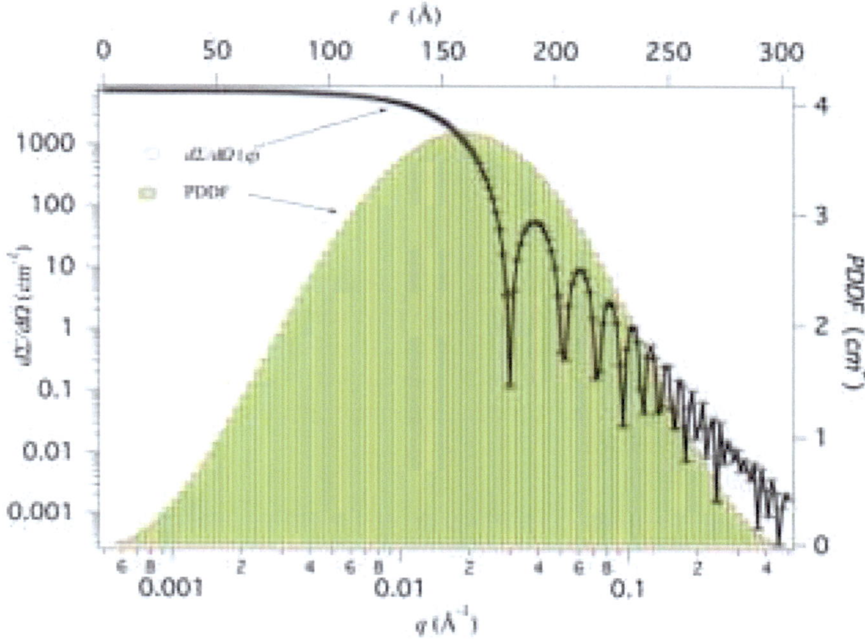

Fig. 6.5 Log–log plot of the theoretical scattering intensity from a sphere 300 Å in diameter together with its PDDF function

where V is the total volume of the scattering inhomogeneities, PDDF(r) is the pair distance distribution function, $\gamma_o(r)$ is the normalized correlation function, ρ is the scattering length density of the inhomogeneity, $\bar{\rho}$ is the volume average scattering length density of the sample, η is the difference in scattering length densities, and $\overline{\eta^2}$ represents the scattering contrast [3]. Though one might propose that the PDDF be calculated by taking the inverse Fourier transform of $I(q)$, this is not advisable [3]. Instead, one can generate the PDDF function with a series of cubic spline functions with unknown coefficients. These unknown coefficients can then be varied to fit the raw $I(q)$ data through Eq. (6.10). Of course, if a system cannot be considered *dilute*, the Generalized Indirect Fourier Transformation (GIFT) should be used instead of IFT [33].

Figure 6.5 shows both $I(q)$ and the PDDF for a system of monodisperse spheres, 300 Å in diameter. Quantitatively, this is a good moment to discuss the units of $\frac{\partial \Sigma}{\partial \Omega}(q)$ and PDDF. Such a system of monodisperse spheres would be represented by the equation:

$$\frac{\partial \Sigma}{\partial \Omega}(q) = N I_{\text{sphere}}(q) \tag{6.13}$$

where $I_{sphere}(q)$ is the theoretical scattered intensity of a single sphere and N is the number of scatterers per unit volume. The resulting units are *always* cm^{-1}. It follows then from Eq. (6.10) that the units for the PDDF are cm^2. Equation (6.13) therefore shows, albeit for a simple case, the relationship between $\frac{\partial\Sigma}{\partial\Omega}(q)$ and $I(q)$. Of course, more complex systems require slightly different relationships. Regardless, the units for $\frac{\partial\Sigma}{\partial\Omega}(q)$ are always the same.

Qualitatively, one can see that the PDDF function approaches zero at 300 Å because of the finite size of the sphere. Other physically meaningful parameters can also be obtained from the PDDF, such as the radius of gyration, R_g, of the scattering population [3]. The radius of gyration, R_g, can be thought of as the root mean square distance from the center of the scattering inhomogeneity and will be discussed further in the following section.

6.3.3 *Useful Approximations and Relationships*

In the previous section, it was shown that the IFT could be used to analyze the entire (or a portion of the) SAXS curve ($I(q)$ vs. q) to obtain physical information about a system. In this section, we will present quantitative approximations that can also be used to evaluate different portions of the $I(q)$ curve. These approximations are extremely useful to an experimenter, particularly in the initial analysis where one needs to identify the source of the scattering. Here we will discuss portions of the $I(q)$ curve representing different aspects of a structural inhomogeneity. Briefly, the Guinier approximation provides the experimenter with information about the size and overall dimensions of the system. Power-Law decays in the $I(q)$ curve can provide information about the surface of the inhomogeneity, as well as the shape of the inhomogeneity. Finally, we will discuss the scattering invariant, Q, which contains information about the total volume and contrast of the inhomogeneity.

6.3.3.1 Guinier Approximation

The Guinier approximation provides information about volume parameters such as the particle size *via* the radius of gyration, R_g. The Guinier approximation assumes that the inhomogeneity is *dilute* and *centro-symmetric*, such that the single particle scattering intensity can be represented by the equation:

$$\text{single particle } I(\mathbf{q}) = \left[\eta \int \cos(\mathbf{q} \cdot \mathbf{r}) dV\right]^2 \quad (6.14)$$

where again η is the contrast. The Guinier approximation is valid only at very small $(\mathbf{q} \cdot \mathbf{r})$, where the $\cos(\mathbf{q} \cdot \mathbf{r})$ can be approximated. If one considers only the magnitude of $\mathbf{q}$ (whether this be a consequence of *statistical isotropy* or known orientation of the inhomogeneity), the resulting approximation of the intensity at low q is:

$$I(q) \cong G e^{\frac{-q^2 R_g^2}{3}} \tag{6.15}$$

$$G = V^2 \overline{\eta^2} \tag{6.16}$$

where G is a constant containing $\overline{\eta^2}$, the electron density contrast between the particle and surrounding matrix and the inhomogeneity volume, V, and R_g is the radius of gyration [1, 3]. Equation (6.15) predicts a constant intensity at very low q, followed by a rapid decay proportional to $e^{-q^2 R_g^2}$, resulting in a "knee" in the scattering curve that is related to the size of the inhomogeneity. This "knee" is shown as the black dotted curve in Fig. 6.6. Note that Eq. (6.15) only describes the central portion of the $I(q)$ curve that contains information about the volume parameters of the inhomogeneity. Also, from Eq. (6.15), the scattering intensity falls off quickly as qR_g term increases. This implies a reciprocal relationship between q and R_g. Therefore, for large values of R_g, one must obtain the scattering intensity $I(q)$ at lower q values. Conversely, if one does not measure to low enough q values, little to no information about the volume parameters of an inhomogeneity can be resolved.

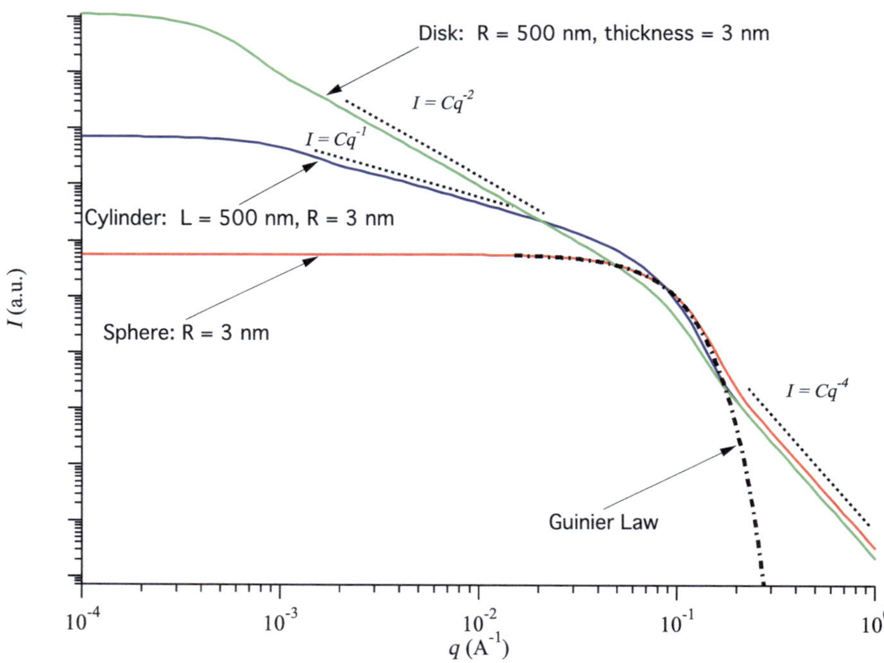

Fig. 6.6 Log–log plot of the scattering intensity from simple geometric shapes to illustrate the concept of multiple dimensions

As a general rule, to resolve the largest inhomogeneity size, L, one must measure the scattering intensity at q values of at least π/L [3] regardless of the model used to analyze the data. This is important to remember, if one is reporting sizes from SAXS data that were not measured. For example, micelles formed in ILs on the order of 1,500 Å cannot be resolved in a q-range that only goes down to 0.01 Å^{-1} [34].

A least squares fit can be used to determine the R_g using Eq. (6.15); however, it is difficult for one to select the applicable q-range to fit. Alternatively, one can use a Guinier plot of $\ln(I(q))$ vs. q^2 to determine the R_g value. This is a simple and quick way to determine particle size. For example, in the photo-reduction of $AgNO_3$ in water-ionic liquid solutions, a Guinier plot was used to determine the size of the Ag nanoparticles as a function of composition and reduction time [35]. In this case, the Ag particles were approximated as spheres; thus the R_g is directly related to the radius of the particles following $R_g = \sqrt{\frac{3}{5}} \times R$, where R is the radius of the spheres.

In the case of a sphere, there is only one volume parameter: the radius, R. But what about shapes that have more dimensions, such as a rod, disk, or parallelepiped? In these cases, simple analysis can show that R_g is related to all the physical dimensions of the scattering object. For example, for a cylindrical rod of length L and radius R, we have $R_g = \sqrt{\frac{R^2}{2} + \frac{L^2}{12}}$. In practice, for particles with extreme shapes, the scattering curve will show characteristics of two or more separated dimensions [36, 37]. An example can be found in Fig. 6.6. Here, in the case of rod and disk, the Guinier knees at $q \sim 0.0009$ Å^{-1} and $q \sim 0.0005$ Å^{-1} represent the cylindrical or discotic radius of gyration, respectively. Following these knees are power-law decays, followed by knees in the scattering curves that are associated with their smaller dimensions.

For simple geometric shapes, one can obtain each dimension separately by fitting Eq. (6.15) to the $I(q)$ curve where each knee is observed. In these cases, one obtains a value of R_g and must use the appropriate relationship to obtain the actual dimension of the inhomogeneity [1, 3]. Therefore, the Guinier approximation can be used to determine the volume parameters of a dilute scattering inhomogeneity.

6.3.3.2 Power-Law Decays

In Fig. 6.6, one will notice that for rods and disks, a power-law decay separates the two Guinier knees. Put simply, these power-law exponents in q can be associated with the presence of one-dimensional rods or two-dimensional disks at q values less than those associated with the smallest Guinier knee at high q. It is shown elsewhere that the intensity decays according to q^{-1} and q^{-2} for a rod and a disk, respectively [3]. These relationships are extremely useful since it would otherwise be difficult to determine whether a scattering inhomogeneity is rod-shaped or disk-shaped. It is, however, important to note that the presence of a finite distribution in size or significant shape distortion (bending) causes departures from these power-law slopes and that it is relatively rare to observe exact q^{-1} or q^{-2} exponents; higher exponents are more common.

In contrast to the Guinier approximation, the Porod approximation is derived for large $(\mathbf{q}\cdot\mathbf{r})$. Unlike the Guinier approximation, the Porod approximation is derived for small r-values near the surface. For a certain scattering object, Porod approximation gives rise to a general approximation for scattering at high q, whereas its corresponding Guinier approximation describes the knee at low q. Also, Porod approximation is valid only for a *statistically isotropic* system. The principal assumptions behind this approximation are that the surface is *smooth* and *well defined*; this means that the electron density function across the inhomogeneity surface must be a Heaviside (step) function and the surface itself is smooth [3]. The resulting approximation, at high q, is:

$$I(q) \cong Bq^{-4} \tag{6.17}$$

$$B = 2\pi\overline{\eta^2}S \tag{6.18}$$

where B is a constant proportional to the contrast and surface area [1, 3]. Equation (6.17) predicts a power-law decay in q of q^{-4} and shown in Fig. 6.7 as a straight line on a log–log plot of Iq^4 vs. q. It should be pointed out that if the particle surface is not smooth or isotropic or the surface electron density function is not continuous, Eq. (6.17) would not be valid. These cases are discussed below.

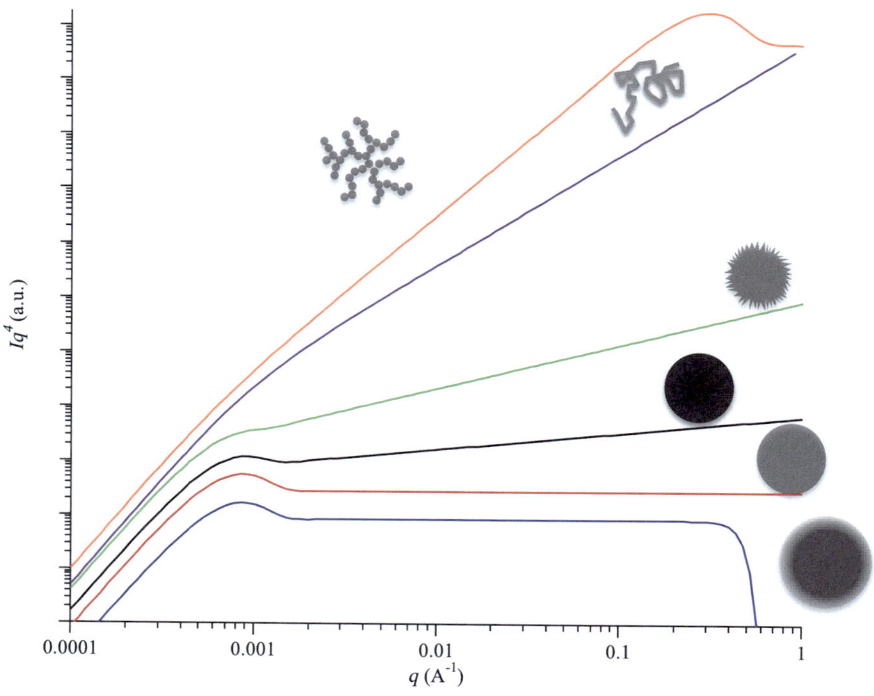

Fig. 6.7 Porod plot showing different power-law decays that are common in SAXS: interface with a well-defined electron density gradient (*blue curve*), Porod surface scattering (*red*), internal electron density fluctuations (*black*), surface fractal (*green*), Gaussian coil (*purple*), and mass fractal agglomerate (*orange*)

Though the Porod approximation has proven to be robust for many systems, deviations from this behavior are common and are thus briefly discussed here. Such deviations from Porod's law can be found in systems where the electron density function is continuous at the inhomogeneity surface (i.e., not a sharp boundary), the surface is rough (on the angstrom scale), or convoluted anisotropic inhomogeneities or mass fractals are present. These deviations can be described as either negative or positive deviations. In the former case, the scattering intensity at high q decreases more steeply than q^{-4} (e.g., $q^{-4.5}$), while positive deviations result in decays less than q^{-4} (e.g., q^{-3}).

Negative deviations from Porod decays (Fig. 6.7) can occur when the inhomogeneity has a diffuse interface, as opposed to a sharp, well-defined one. Such a system is commonly found in low-crystalline polymers where there is a gradual transition of $\rho(r)$ at the phase boundary [3]. This phenomenon is also found in nanoparticle electrodeposition in deep eutectic solvents [38]. In this study, the observed negative deviations were associated with charge-neutralizing layers above particle ensembles on the surface.

These boundary gradients can be accounted for with only a slight modification to Eq. (6.17) [39]. Using the convolution property of Fourier Transforms, it can be shown that the scattering intensity (at high q) for an isotropic system can then be approximated as:

$$I(q) \cong \frac{B}{q^4}\left(1-\sigma^2 q^2\right) \tag{6.19}$$

where B is the same as in Eq. (6.18) and σ is the thickness of the boundary over which the electron density gradient exists. Equation (6.19) is applicable for *isotropic systems* and assumes that the gradient can be described as the convolution of a step function and a Gaussian function, where the Gaussian function is approximated by a Taylor series expansion, though other functions can be used [40]. Similar relationships also exist for anisotropic systems [39].

Positive deviations from Porod's law are much more common in SAXS. We have already seen that scattering from anisotropic inhomogeneities can lead to power-law decays in the scattering intensity with q (Fig. 6.6). Others include but are not limited to electron density fluctuations within the inhomogeneity, polymers, surface fractals, and mass fractal agglomerates.

Spatial electron density fluctuations, within a phase, can cause positive deviations from Porod's law [41]. This effect is caused by electron density fluctuations that are small enough to interfere with the Porod scattering. This phenomenon has been reported for certain $Al(OH)_3$ gels, whereby the composition of the gel is the Al_{13} polymer, which itself contains large fluctuations, resulting in a power-law decay of $q^{-3.7}$ [41, 42]. While this phenomenon is rare, it is certainly possible in ILs that have a tendency to aggregate into complex structures [43]. This power-law decay is similar to surface fractal scattering, which is physically very different.

Large molecules (i.e., polymers) may also lead to power-law decays, for example, a polymer with Gaussian statistics [44]. An example would be aggregation in ILs. The scattering intensity is given by the equation:

$$I(q) = 2\frac{\left(e^{-R_g q} + R_g q - 1\right)}{\left(R_g q\right)^2} \tag{6.20}$$

where R_g is the average radius of gyration of the polymer. Strictly speaking, a polymer can be thought of as a special case of a one-dimensional rod that is randomly coiled. However, because the polymer is coiled, the scattering intensity, at high q, decreases according to q^{-2}, instead of q^{-1}, as shown in Fig. 6.7. Equation (6.21) is only correct for molecules with no self-avoidance. Polymers with self-avoidance will cause the intensity to decay closer to $q^{-5/3}$ [45].

Another important class of morphology is fractal, which includes mass fractal and surface fractal. Although mass and surface fractals appear remarkably different, they share a common trace essential to fractal objects: self-affinity. Self-affinity means that the morphology of interest looks the same under different length scales, i.e., within certain limits, the essential geometric characteristics of a fractal object are unchanged by magnification.

For surface fractals, mathematically, this self-affinity is described by $S \sim R^{ds}$ for surface fractals, where R is the scale of probing yardstick, S is the total surface area within the dimension specified by the yardstick, and d_s is the surface fractal exponent [46]. For an ordinary smooth object, $d_s = 2$, consistent with a smooth surface being two-dimensional. For surface fractal dimensions, however, d_s can be non-integer and vary between 2 and 3, which then provides a quantitative measure of surface roughness.

Surface fractal analysis has been applied to the study of the breakdown of cellulose fibers using ILs, where it is shown that the IL pretreatment time influenced the surface morphology of the resulting biomass [47]. Such a surface is also common in glasses [48] and rocks [49] where the pore wall is rough as shown in Fig. 6.7. For surface fractals, it can be shown that the scattering intensity decays at high q by the equation:

$$I(q) = Bq^{d_s - 6} \tag{6.21}$$

where B carries the same definition as in Eq. (6.18). Given that d_s varies between 2 and 3, surface fractal scattering is confined to power-law decays between q^{-3} and q^{-4}.

Mass fractal scattering, on the other hand, is associated with the structure of the inhomogeneities and not their surface. Physically, mass fractal scattering is observed when small particles agglomerate together to form a non-crystalline superstructure similar to that shown in Fig. 6.7. These structures are known to exist in aerogels, whereby particles are very close to each other but do not possess any order. The structure of these particles can be described as a fractal-like object [50]. The degree to which the particles are packed can be expressed in terms of the mass density, ρ_m, of the agglomerate:

$$\rho_{\mathrm{m}} \propto R_{\mathrm{g}}^{d_{\mathrm{m}}-3} \tag{6.22}$$

where R_g is the radius of gyration of the agglomerate and d_m is the mass fractal dimension, which ranges from 1 to 3. Therefore, as the value of d_m increases toward 3, the density, ρ_m, will also increase. Just as for regular inhomogeneities, the scattering intensity would show a typical Guinier knee at low q that corresponds to the R_g of the entire agglomerate. However, as q increases, it can be shown that the scattering intensity will decay according to:

$$I(q) \cong Bq^{-d_{\mathrm{m}}} \tag{6.23}$$

where the intensity decays according to the value of d_m, which can only be between 1 and 3 [50]. Scattering from agglomerates is well established in many systems that are prone to agglomeration.

6.3.4 *Unified Equations*

The Guinier and Porod approximations have proven to be robust for scattering inhomogeneities, but neither one can model the entire scattering curve. These approximations have been combined to describe the full $I(q)$ curve for spheres, rods, disks, micelles, and intermediate objects [51]. A slightly more flexible unified equation has also been developed and will be briefly discussed here. Using both the Guinier approximation at low q and a power-law decay at high q, the scattering intensity can be represented by the equation:

$$I(q) = \sum_{i=1}^{N} \left(\mathrm{LQ}_i + \mathrm{HQ}_i\right) \tag{6.24}$$

$$\mathrm{LQ}_i = G_i e^{\frac{-q^2 R_{g_i}^2}{3}} \tag{6.25}$$

$$\mathrm{HQ}_i = B_i e^{\frac{-q^2 R_{g\mathrm{CO}}^2}{3}} \left(\frac{\left(erf\left(\frac{qR_{g_i}}{\sqrt{6}} \right) \right)^3}{q} \right)^{P_i} \tag{6.26}$$

where LQ_i is the low q scattering contribution, HQ_i is the high q scattering contribution, i is the level number, and P_i is the exponential power-law decay that dominates only the high q region, using the error function in Eq. (6.26). Using Eq. (6.24), one can model the scattering from a variety of systems or combination thereof by using

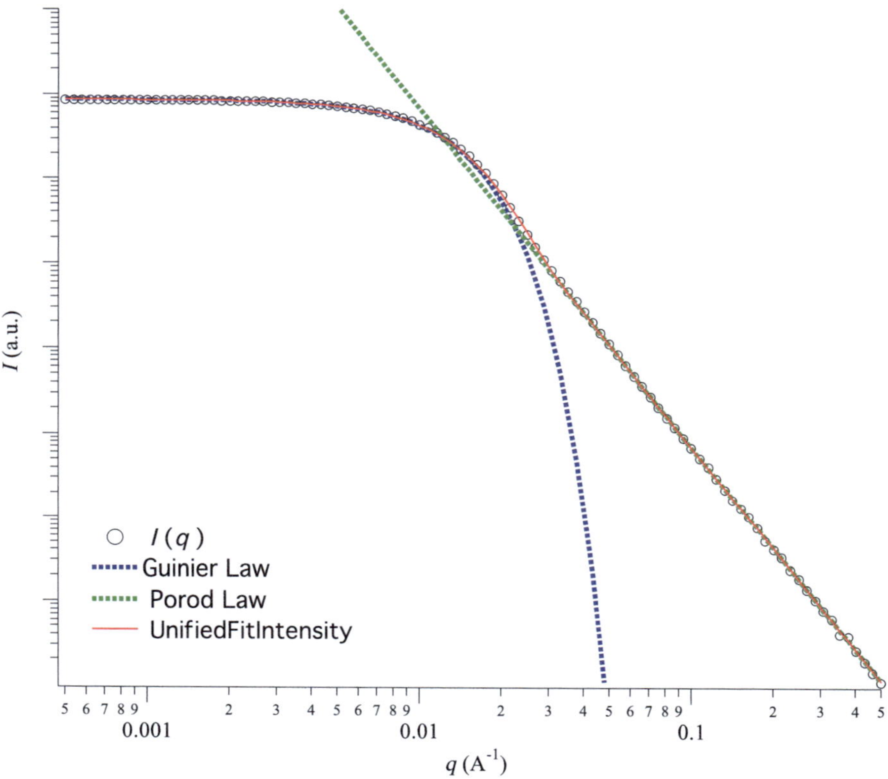

Fig. 6.8 Log–log plot of the scattering intensity from a Gaussian distribution of spheres (*black circles*) and the associated Guinier approximation (*blue*), Porod approximation (*green*), and the unified equation (*red*)

multiple levels, *i*. An example of the unified equation is shown in Fig. 6.8 where Eq. (6.24) is used to model a size distribution of spheres. For more complicated objects, such as fractal agglomerates, different levels can be linked together with the R_{gCO}, which is used to terminate the high-q scattering in the case that the scattering intensity from another level dominates that portion of the $I(q)$ curve; fractal agglomerates and anisotropic inhomogeneities are good examples of this [52]. The value of P_i in Eq. (6.26) would be 4, if Porod's law is valid or any of the previous mentioned decays [52]. However, one will note that the definition of B depends on the value of P_i chosen. For example, if the P_i equals 4, Eq. (6.18) can be used; different expressions must be used for different decays.

In practice, one can use Eq. (6.24) to model the scattering intensity to obtain information about the inhomogeneity size (R_g), total volume or contrast (G), and surface or structure (B, P_i). In certain cases, parameters obtained from Eq. (6.24) can be used to provide information about agglomerates [53], size distribution [54], or a two-phase system [55].

6.3.5 Invariant

In previous analyses, we have seen that the shape of the $I(q)$ curve provides information about the physical dimensions of an inhomogeneity. When the relative scaling of the scattering intensity is important, one can use the invariant, Q_{inv}, to analyze changes in the total scattering volume or contrast of a system. Irrespective of the size, shape, or structure of the inhomogeneity, Q_{inv} can be calculated by the equation:

$$Q_{inv} = \frac{1}{2\pi^2}\int_0^\infty q^2 I(q)\,dq = CV\overline{\eta^2} \tag{6.27}$$

where V is the total volume of the scattering objects, and C would be a calibration constant that is normally acquired using an intensity calibration standard [3]. The only requirement of Eq. (6.27) is that the system be *statistically isotropic*. For anisotropic systems, a pseudo-invariant, Q_φ, can also be used [56].

6.3.6 Composite Inhomogeneities, Form Factors, and Structure Factors

In Sect. 6.2.4,we saw that the total scattering amplitude can be calculated by simply summing all other amplitudes, with the weighted phase factor (Eq. 6.6). For a system of N particulate inhomogeneities, each at a position vector, $\mathbf{P}_\mathrm{n}$, Eq. (6.6) can be modified to calculate the total scattering amplitude of these inhomogeneities by the equation:

$$A(\mathbf{q}) = \sum_j f_j(\mathbf{q})\,e^{i(\mathbf{q}\cdot\mathbf{P}_\mathbf{n})}\sum_n e^{i(\mathbf{q}\cdot\mathbf{r}_\mathbf{j})} \tag{6.28}$$

where the second term is a summation over each particle, and the first term accounts for inter-particle displacements, $\mathbf{P}_\mathrm{n}$. For known shapes, the second summation has an analytical solution. Therefore, if the second term can be replaced by the scattering amplitude of a single inhomogeneity, then the amplitude is equal to:

$$A(\mathbf{q}) = \sum_n V_n(R\ldots)F_n(q,R\ldots)\,e^{i(\mathbf{q}\cdot\mathbf{P}_\mathbf{n})} \tag{6.29}$$

where V_n is the volume, $R\ldots$ is the volume parameters of the inhomogeneity, and $F_n(q,R\ldots)$ is the so-called form factor of the inhomogeneity. Assuming *statistical isotropy* leads to the well-known Debye Equation for composite inhomogeneities:

$$I(q,R\ldots) = \sum I_j(q,R\ldots) + 2\sum_j\sum_{n\neq j} A_j(q,R\ldots)A_n(q,R\ldots)\frac{\sin qP_{jn}}{qP_{jn}} \tag{6.30}$$

$$A(q,R\ldots)=V(R\ldots)F(q,R\ldots) \tag{6.31}$$

where P_{jn} is the magnitude of the vector that separates particle j from particle n. Equation (6.30) is a useful equation when one wants to calculate the scattering intensity of a collection of inhomogeneities. Note that if large distances, P_{jn}, separate the inhomogeneities, then the second term is zero and the total scattering intensity can be represented as the sum of the intensity of each inhomogeneity; in this case, the system is considered *dilute*.

In reality, small inhomogeneities measured by SAXS are not always in their dilute limit, making the second term in Eq. (6.30) necessary. We note that this second term is difficult to calculate directly. However, Eq. (6.30) can be rewritten, using a correlation function, $g(P)$:

$$I(q,R\ldots)=NI(q,R\ldots)\left[1+\frac{N}{V}\int(g(P)-1)\frac{\sin qP}{qP}\right] \tag{6.32}$$

where N is the total number of inhomogeneities. One can further simplify Eq. (6.32) to a term for the inhomogeneities, F, and a term for their structure, S:

$$I(q,R\ldots)=V^2(R\ldots)F^2(q,R\ldots)S(q,X\ldots) \tag{6.33}$$

where V is the volume, F is the form factor, S is the structure factor, and $X\ldots$ represent the parameters that describe the structures of the inhomogeneities. Nowadays there is a wide variety of form factors and structure factors available [12, 57]. These analytical expressions are very simple, yet powerful, tools that the reader is strongly encouraged to review. While these analytical expressions provide a valuable link between the physical aspects of the system and the scattering intensity, it is important that they be used correctly. We will discuss a few notable examples below. However, it is impossible to discuss every case and the reader is encouraged to be aware of the assumptions that go into the form factor and structure factor they wish to use.

6.3.6.1 Size Distribution and Anisotropic Inhomogeneities

For a dilute isotropic system, the contribution of each particle to the total scattering intensity is additive. When the distribution of inhomogeneities is known, the scattering intensity can be modeled as:

$$I(q)=CV\overline{\eta^2}\int D_n(R)V_p(R)^2\left|F(q,R)\right|^2 dR \tag{6.34}$$

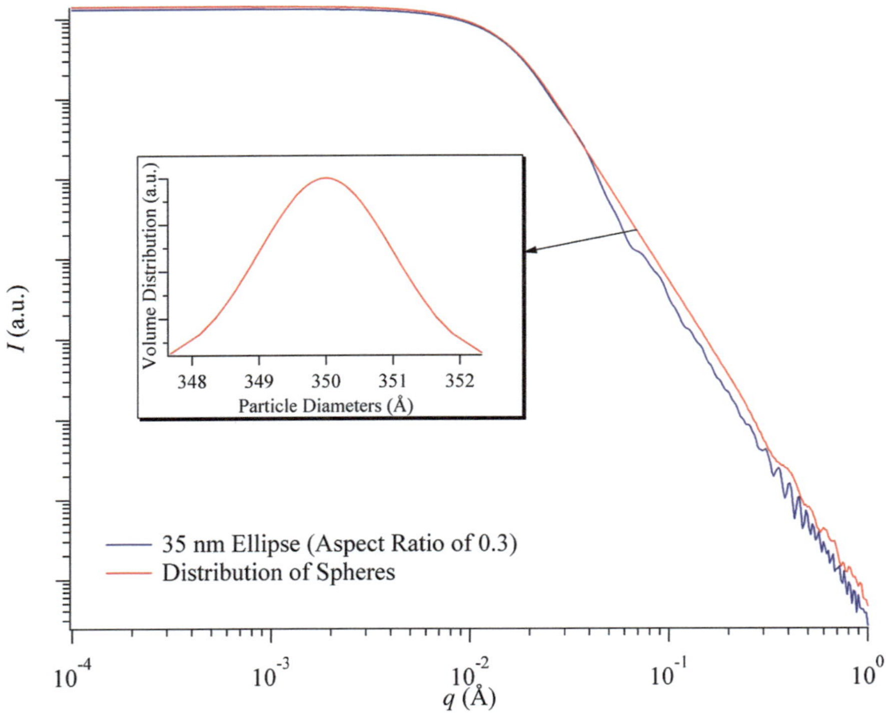

Fig. 6.9 Log–log plots of the scattering intensity calculated from (1) a single 35 nm-diametered ellipsoid particle (*blue*) and (2) spheres with a Gaussian size distribution (*red*). The spherical mean diameter is 20 nm

where D_n is the number distribution of inhomogeneities of size, R, V_p is the particle volume, $|F(q,R)|^2$ is the form factor, and k is a general constant that depends on the contrast and volume [3]. While Eq. (6.34) can be easily fit to an $I(q)$ curve, how certain can one be of the resulting parameters? The answer to this question is that it depends on the data and model. For example, Fig. 6.9 is a plot of the intensity *vs.* q for a broad distribution of spheres and a single 350 Å ellipsoid with an aspect ratio of 0.3. The effects of polydispersity and particle shape on the scattering intensity are very similar, as both broaden the Guinier knee as well as dampen the oscillations at high q. Though particle shape and polydispersity may become distinguishable at extreme aspect ratios (i.e., disks and rods), both effects are indistinguishable if the system has inhomogeneities of moderate aspect ratios and an unknown distribution, as both traces in Fig. 6.9 are similar. One can imagine then that if both the size distribution and aspect ratio were fit simultaneously, there would be large uncertainties in the fit parameters. For this reason, an uncertainty analysis of the fit parameters should be performed. This can be accomplished by holding one of the fit parameters (i.e., parameter A) constant and fitting all other parameters and then looking at how much the chi-square changes with different values of parameter A. This method is incorporated in the Irena package [11] for Igor Pro but can easily be accomplished in any least squares fitting routine.

Solving both the distribution and shape from a single 1–D $I(q)$ curve is not possible without a priori knowledge of one or the other. In the case that the distribution is known through some other technique, the scattering curve can be fit to a system of inhomogeneities with either the same aspect ratio or with aspect ratio a function of size. Alternatively, if an aspect ratio is either assumed or known, the distribution can be solved for, directly, from the discretized form of Eq. (6.34). In this case, other methods such as the method of maximum entropy [58] can be used to determine the size distribution.

For a polydisperse system that contains inter-particle (inhomogeneity) interferences, Eq. (6.34) cannot be used by simple multiplication by the structured factor, as in Eq. (6.33). Although this has been used for significantly narrow distributions [37, 38, 59], there are better ways to properly account for structure in a polydisperse system. One such method, for relatively narrow size distributions, is by using the so-called *local monodisperse* approximation [60]. This approximation ignores the interference between different sized inhomogeneities and assumes that only like-sized inhomogeneities interact.

$$I(q) = CV\overline{\eta^2}\sum D_n(R_j)V_p(R_j)^2\left|F(q,R_j)\right|^2 S(q,,R_{\mathrm{HS}_j},,\nu)\Delta R \tag{6.35}$$

where $S(q,R_{\mathrm{HS}},\nu)$ is the hard sphere structure factor [61]. This particular structure factor assumes that the inhomogeneities interact with a hard sphere radius, R_{HS}, and have a local volume fraction, ν. For these "hard spheres," the potential energy of each particle (in normalized units) is everywhere one, except for within the hard-sphere radius, R_{HS}, where it is infinite (i.e., the hard spheres are impenetrable) [61]. Equation (6.35) is just one example of how a structured-polydisperse system can be treated [62]. In addition, other structure factors have been derived for different cases, each with their own definition of the potential energy function acting between the inhomogeneities [12, 57].

6.3.7 Aggregation

Aggregation in ILs is a phenomenon by which either or both of the cations and anions aggregate together in solution to form local, highly charged phases. In these cases, aggregation is often described by the order in the system, which is observable by SAXS. On the other hand, when a third component is added (i.e., an amphiphilic molecule), other aggregated phases can form, which can also be observed with SAXS. Both of these cases will be discussed below.

IL aggregation leads to a diffraction peak, arising from the ordered distances between phases. The observed peaks can then be used to understand the effect of different cations and anions. For example, it was shown that the ordering of 1-alkyl-3-methylimidazolium ILs depends on both the anion and the length of the alkyl chain [63]. While Bragg's law has been used extensively in this area to analyze the peak position(s), it is not adequate to interpret the entire SAXS curve, specifically, the shape of the inhomogeneities making up the individual phases. For this, structure factors must be used in Eq. (6.33) with the appropriate form factor. While no new analytical structure factors exist specifically for neat ILs, molecular dynamic simulations have been used to compare the simulated and observed structure factors in pyrrolidinium-based ILs [64]. These structure factors are not in readily useable analytical forms. However, they can be calculated numerically using Eq. (6.32). Finally, the structure factor of any system, on its own, can also be obtained through Eq. (6.33) by simply dividing the scattering intensity, $I(q)$, by the form factor, if this is known a priori.

Aggregated phases, formed by the addition of third components, have also been studied in ILs. Often, these phases form micelles, which can easily be modeled by Eq. (6.35), using the appropriate form factor [28] and structure factor [65]. Using this approach, one can obtain the micelle size, hard sphere radius, chain aggregation number, and chain size [66]. However, this is not always possible because of the limited q-range available—a problem that is sometimes unavoidable depending on the instrument used.

In some cases Bragg's Law has been used in place of Eq. (6.35), to analyze the peak in the SAXS curve [67]. While this does provide information about the local order of the micelles, information, such as micelle size, is not provided. In other cases, the form factor for an ellipsoid has been used instead to resolve micelle size [68, 69]. In these cases, not all the micelle information was resolved from the SAXS data. However, for large micelles a significantly wide q-range is required to resolve the micelle size and individual chain dimensions using the appropriate form factor [28]. With a limited available q-range, this is just one example of how sometimes the simplest model is best. As stated previously, an uncertainty analysis on the fit parameters can also be useful.

6.3.8 Microemulsions

Models have been developed to describe the SAXS from microemulsions, which contain polar, nonpolar, and amphiphilic constituents [70]. There are a few different ways that one can analyze the SAXS from these systems. These approaches are reviewed here. Qualitatively, the Teubner-Strey phenomenological model accounts for a bi-continuous repeat distance, d, as well as the average length scale of the polar/nonpolar domains, ξ, using only three fit parameters: a_2, c_1, and c_2 [70]. This model has found application in the study of ILs, which can either be polar or nonpolar. To this end, the Teubner-Strey equation has been used to characterize and compare microemulsions of ILs [71] and even magnetic ILs [72] as well as microemulsions formed with water. Scattering from these systems typically contains a peak in the intensity, characteristic of concentrated systems, as well as a decay in the scattering intensity of q^{-4}, which corresponds to surface scattering of the separate phases. These features are accounted for by the equations:

$$I(q) = \frac{8\pi / \xi c_2 V \overline{\eta^2}}{a_2 + c_1 q^2 + c_2 q^4} \tag{6.36}$$

$$\xi = \left[\frac{1}{2} \left(\frac{a_2}{c_2} \right)^{1/2} + \frac{1}{4} \frac{c_1}{c_2} \right]^{-1/2} \tag{6.37}$$

$$d = 2\pi \left[\frac{1}{2} \left(\frac{a_2}{c_2} \right)^{1/2} - \frac{1}{4} \frac{c_1}{c_2} \right]^{-1/2} \tag{6.38}$$

This equation has been applied successfully to many microemulsions, regardless of the concentration of the polar/nonpolar phase. However, in the case of dilute systems, the Teubner-Strey model does not reveal the size distribution of the inhomogeneities making up the dilute phase.

As pointed out by Klee et al., such a system can also be modeled using Eq. (6.33). This was accomplished by approximating the dilute phase as a polydisperse system of core-shell particles [37] and assuming a hard sphere structure factor; however, the effect of polydispersity on the structure factor was not accounted for [62]. As well, the GIFT analysis has also been used to verify the presence of a core-shell structure [73]. Ultimately, there is more than one single model that can accurately describe the SAXS in terms of the peak intensity, average size, and Porod scattering law. For example, a core-shell structure can correctly account for the shell of amphiphilic molecules at the surface of the microemulsion, which Eq. (6.36) cannot.

Yet another case for microemulsions is when colloidal particles are used to stabilize the emulsion, instead of amphiphilic molecules. To our knowledge, such a study has yet to be carried out for ILs. For such a system, an analytical scattering model that can account for the self-correlation of a spherical core and surface-adsorbed particles has been developed [74]. Again, this model assumes that the system is in its dilute limit.

6.3.9 Monte-Carlo Simulations

A Monte-Carlo analysis may be required when analytical models are not available to analyze the scattering profile. In most cases, this happens because one or more assumptions in Sect. 6.3.1 cannot be met, which would otherwise allow for an analytical solution. In Small-angle scattering (SAS), a Monte-Carlo analysis includes the following steps: (1) random number generation within a sample volume, (2) assignment of the appropriate contrast to each point, and (3) numerical calculation of the scattering intensity [75]. In this section, a Monte-Carlo approach will be introduced, followed by one simple example.

The small-angle scattering can be calculated using the numerical forms of Eq. (6.7) or (6.14), depending on whether or not the system can be considered *centro-symmetric*. In all cases, the scattering intensity can be calculated using the numerical form of Eq. (6.7). If the system can be considered *isotropic*, the complex phase factor in Eq. (6.7) will be replaced by the "average" phase factor: $\frac{\sin qr}{qr}$. However, this involves two summations, which for a large number of points can be computationally expensive. The calculation time can be reduced by effectively enlarging each point such that it occupies a larger volume. In this case, the numerical form of Eq. (6.7) would then be the Debye equation for composite particles (Eq. 6.30), where instead of points, objects of specific shape can be used [76].

In the case that the scattering inhomogeneity is a single *isotropic* object, it is possible to generate the PDDF, which can then be used to calculate the scattered intensity through Eq. (6.10). This is accomplished by first filling the object with randomly positioned points, followed by generating a histogram of distances, r, between each point. The resulting histogram is proportional to the PDDF and can then be used to calculate the scattered intensity or compared to the PDDF obtained from the raw data using IFT [76].

If one can assume *centro-symmetry*, the numerical form of Eq. (6.14) can be used to calculate the scattering intensity from systems that are anisotropic. Significantly lower calculation times can also be achieved because only one summation over the real space is needed at each $\mathbf{q}$ value. To illustrate how a realistic system can be modeled using this method, we will consider a system of spheres that have a Gaussian size distribution with a mean of 150 A and standard deviation of 15 A. Similar to Eq. 6.34, the scattered intensity of the system can be calculated by the equations:

$$I(\mathbf{q}) = C\sum I_i(q, R_i) D_{n_i}(R_i)\Delta R_i \tag{6.39}$$

$$I_i(\mathbf{q}) = \left[\sum \eta_j v_p \cos \mathbf{r}\cdot\mathbf{q}\right]^2 \tag{6.40}$$

where η is the scattering length contrast of point j, ν is the volume simulated by each point (equally, the inverse of the point density), and C is again a scaling factor.

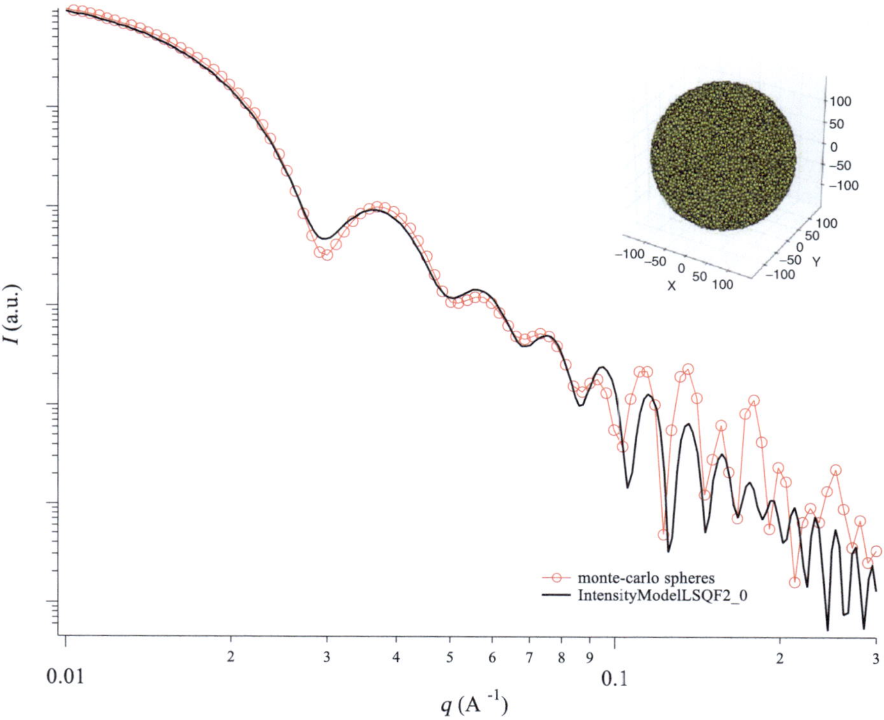

Fig. 6.10 Log–log plot showing the simulated SAS from a Gaussian distribution of spheres, having a mean radius of 150 A and standard deviation of 15 A. The scattered intensity is calculated from Eq. (6.39) (*circles*) and compared with the theoretical scattering (*black*), which was generated using the Irena package for Igor pro. In both cases, 50 spheres ranging in sizes between 4 Å and 15,000 Å (log binned) are used to simulate the distribution

Figure 6.10 is a log–log plot that compares the simulated scattered intensity with the analytical solution. One will note that large deviations between the Monte-Carlo curve and theoretical curve increase with increasing q. This can be improved by using more than the 10^5 points used in the simulation; this corresponded to a point density of 7 points per nm^3. While there is no need to use Eq. (6.39) for a system of spheres, the real advantage of this approach is for anisotropic systems [77].

In section, the general Monte-Carlo approach to small-angle scattering has been presented. A Monte-Carlo simulation, however, involves careful consideration of more than just the fundamental equations behind it. In addition to the computational algorithm, the uniqueness of a feature shape [78] can be difficult to determine. Fortunately, there are software programs that address these issues [17]. There is also software in this field devoted to biological systems as well as other systems that cannot easily be analytically modeled [16].

6.4 Experimental

A typical SAXS experiment involves a sample and necessary optics that measure the scattered X-rays. Nowadays, with the accessibility of desktop SAXS machines and synchrotron radiation, the latter is taken care of by the manufacturer, technician, or beamline scientist. In this section, we will discuss practical aspects that the experimenter (as a user) should be concerned with. This entails planning and executing a SAXS experiment, followed by the subsequent data reduction and analysis.

6.4.1 Experimental Preparation

In a typical SAXS experiment, one has to first determine what sample geometry is necessary to achieve what one is looking for. For example, in most cases, one wants to look at phases in solution. These phases are typically *statistically isotropic* so it makes no difference what the sample orientation is relative to the incoming X-rays. In this case, a simple transmission geometry is sufficient to obtain the volume average dimensions of the scattering inhomogeneities. On the other hand, if there is reason to believe that scattering inhomogeneities are preferentially oriented near an interface, then a grazing transmission geometry should be considered [79]. A case example would be an electrochemical experiment whereby the system is exposed to an external electrochemical signal at the working electrode. These geometries are shown schematically in Fig. 6.11.

6.4.1.1 Energy Selection

In some cases, for example, at a synchrotron, a user can perform their SAXS experiments at an energy that is optimal for their system. This is particularly useful when one has a weakly scattering system, due to either a small volume fraction or contrast. The optimal X-ray energy is one where no X-ray absorption edges are present and the measured scattering intensity is maximized. This requires some prior knowledge of the system, namely, the elemental composition and thickness of the sample. Once the composition is known, the scattering power, $\rho^2(\lambda)$, can be obtained.

In the case of ILs, the X-rays will need to pass through a sample holder that contains the liquid. In most cases, researchers use polyamide film windows for the X-rays to pass through. However, one should be careful with such windows because they scatter strongly at high q (~0.4 Å^{-1}). Also, the longer the liquid path that the X-rays pass through, the greater the attenuation due to sample absorption, as given by the equation:

$$\mathrm{Tr} = e^{-\left(\left(\mu/\rho\right)\rho_s t_s\right)} \tag{6.41}$$

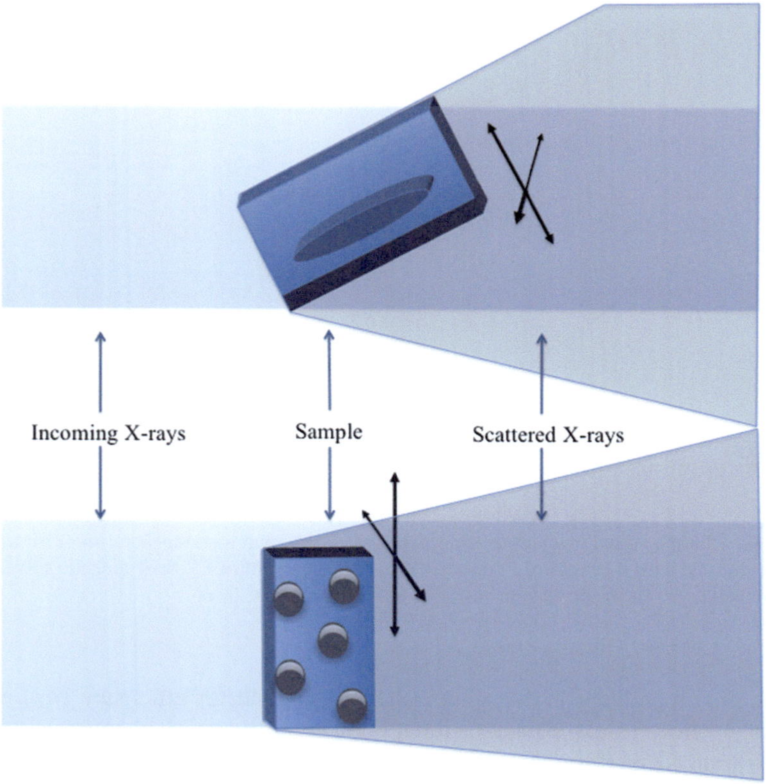

Fig. 6.11 A schematic illustrating the difference in a Grazing Transmission (*top*) and Transmission (*bottom*) SAXS. The resolvable dimensions in each case are shown in *black*

where μ/ρ is the mass absorption coefficient for the material, ρ_s is its density, and t_s is the thickness through which the X-rays pass. To calculate Tr from Eq. (6.41), the experimenter should be able to provide ρ_s and t_s with reasonable accuracy. The mass absorption coefficient, however, depends on the composition and energy by the equation:

$$\mu/\rho(\lambda) = \sum \mu/\rho(\lambda)_i \, w_i \tag{6.42}$$

where w_i is the mass fraction of element i and $\mu/\rho(\lambda)_i$ is its mass absorption coefficient. Using tabulated values of each element's mass absorption coefficient, one can calculate the transmitted intensity, I, from the sample.

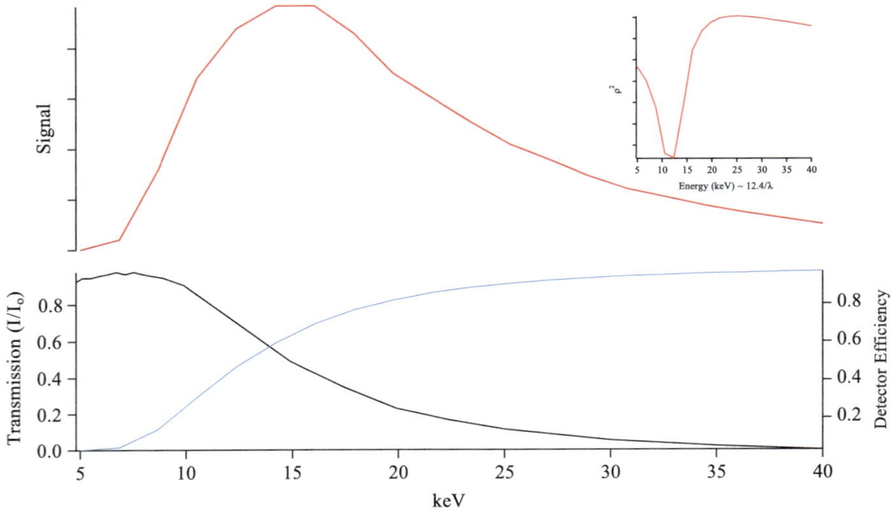

Fig. 6.12 Plot the signal (*top*), detector efficiency (*black*), and transmission (*blue*) for a simple liquid cell filled with a 10 mM of Pt dissolved in a 2:1 urea:choline Cl^- deep eutectic solvent with a 1 mm thickness. The scattering power, ρ^2 of Pt is shown in the inset

The final ingredient in the optimization is the detector efficiency $D(\lambda)$, which must be obtained from the manufacturer. For a given scattering population, the final optimization is therefore:

$$\text{Signal} = \left[\rho^2(\lambda)D(\lambda)\text{Tr}(\lambda)\right] \tag{6.43}$$

The optimal energy is then taken to be the one that maximizes the left side of Eq. (6.29). Of course, because we are analyzing the elastically scattered X-rays, absorption edges should be avoided in a normal SAXS experiment; this can be observed in the function $\rho^2(\lambda)$. Figure 6.12 shows a case example of a simple cell that contains a deep eutectic solvent (DES) and is 1 mm thick. If, for example, we wanted to observe Pt nanoparticles, then the optimal energy would be approximately 16 keV, which will provide a high signal and is far enough away from any Pt absorption edge.

6.5 Future of SAXS and XPCS

In the earlier parts of this chapter, we have introduced the general principles of SAXS. It is worth noting, however, that SAXS is a *static* probe that can resolve the structural inhomogeneities within the specimen of interest. SAXS cannot capture the underlying cause for the structural development or alteration of a physical system over time, i.e., dynamics. In this section, we will briefly introduce X-ray photon correlation spectroscopy (XPCS), a recently emerged technique that is both closely related to SAXS and capable of measuring nano-scale and micro-scale material dynamics.

XPCS probes the *dynamic* properties of matter by analyzing the temporal correlations between X-rays scattered by the sample. To achieve this, XPCS relies on the availability of coherent X-rays, which is not required for *static* SAXS measurements. Here, coherence refers to properties of X-rays that preserve the X-ray *phase* information commonly lost in conventional X-ray measurements. Preservation of the phase information enables temporally and spatially coherent interference.

Due to their stringent requirements on coherent X-rays, XPCS measurements cannot be performed using lab X-ray sources or even first- or second-generation synchrotron sources, where X-rays produced by these sources possess very low degrees of coherence. XPCS measurements were not possible until the operation of third-generation synchrotron sources, which have lower emittance and long straight sections of undulators that can achieve high X-ray brightness and a considerable degree of spatial coherence. The resulting partially coherent X-rays are several orders of magnitude more intense than previously available. Despite this, synchrotron undulator sources remain chaotic sources because of the spontaneous nature of the X-ray emission.

Overall, the characteristics of third-generation synchrotron sources have led to major progress in structural determinations of objects with low X-ray scattering or imaging contrast. Many groundbreaking discoveries in recent years [80–82] were facilitated by applications of coherent X-ray scattering, diffraction, and imaging techniques such as X-ray phase contrast imaging [83], X-ray holography [84], and coherent X-ray diffraction imaging [85]. More relevant to the topic of this chapter is the revolutionary development of the temporal probe of XPCS, which has made possible direct measurements of slow equilibrium and near-equilibrium dynamics in disordered systems.

From a technique point of view, in many aspects, XPCS is a natural translation of long-wavelength laser photon correlation spectroscopy (PCS), also known as dynamic light scattering (DLS), into the X-ray domain. Both XPCS and DLS share the same set of physical principles and exploit the coherence of their respective radiations by restricting the dimensions of the scattering sample volume to be similar to the coherent volume of the beam. The dynamic timescales of the monitored processes are acquired by analyzing the temporal correlations of photons scattered by the specimen [86].

There also exist key differences between XPCS and DLS. Visible light DLS, as a laser technique, suffers from two major limitations: resolvable size and adsorption. The wavelength of visible light ranges from 400 nm to 700 nm, which (from Eq. 6.5) necessarily restricts the collection of the scattering intensity to a very small scattering vector q. At the same time, the visible light scattering cross section is far

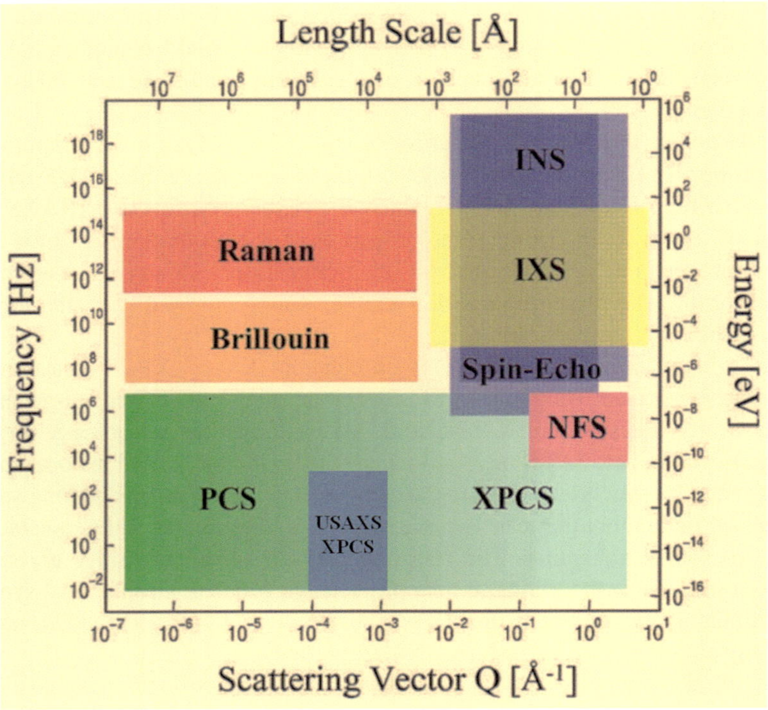

Fig. 6.13 Frequency-scattering vector domains of available techniques for dynamic studies. The techniques in the map are Raman spectroscopy, Brillouin spectroscopy, inelastic neutron scattering (INS), inelastic X-ray scattering (IXS), neutron spin-echo spectroscopy, nuclear forward scattering (NFS), XPCS, ultrasmall-angle X-ray scattering-based XPCS (USAXS XPCS), and DLS [also known as photon correlation spectroscopy (PCS)]

greater than that for X-rays. Therefore, it is very difficult, if not impossible, for visible light DLS to probe the dynamics of optically opaque or highly absorbing media due to strong absorption and/or multiple scattering effects. XPCS, on the other hand, being an X-ray technique, is not subject to these limitations and offers an unprecedented capability to measure low-frequency (10^{-3} to 10^6 Hz) dynamics of structures in the range of 1–10,000 Å (as things stands currently). Figure 6.13 shows the frequency-wavevector range accessible to XPCS and DLS as well as other scattering-based techniques that are frequently used to study dynamics in disordered systems [87, 88].

Since its emergence less than two decades ago [89, 90], XPCS has made great strides in advancing many aspects of statistical physics and providing access to the slow dynamics of a wide range of physical events. Research areas that have benefitted considerably from XPCS include but are not limited to equilibrium dynamics of concentrated colloidal suspensions [90–95], glassy dynamics [96, 97], atomic diffusion [98, 99], dynamical heterogeneities in soft matter [100, 101], critical fluctuations of liquid crystals [102, 103], near-equilibrium coarsening fluctuation in alloys [104] and weakly disturbed soft condensed matter systems [96, 105–107], and surface dynamics of fluidic and polymeric systems [108–114]. Most of these systems share some common traits, namely, the dynamical fluctuations happen on a length scale between nanometers and micrometers, and the dynamic timescale is greater than a microsecond.

The capabilities of XPCS have not been fully exploited by the ionic liquid research community, which may be attributed to two main reasons. Firstly, despite its rapid development, XPCS remains a highly specialized technique and its availability is limited to a handful of third-generation synchrotron facilities and X-ray free electron laser facilities around the world. Secondly, and perhaps more importantly, the current version of XPCS imposes a rather strict limitation on its detectable time resolution (1 μs), and thus cannot detect dynamics faster than this limit. However, a significant effort is under way in the global synchrotron community to generate X-rays with much higher brightness and degree of coherence. This effort would allow sub-nanosecond time resolution for XPCS measurements, which will enable quantification of particle mobility in ionic liquids, as well as determination of the effects that various factors such as particle solvent interaction have on the equilibrium fluctuations of ionic liquids.

While it will take another few years for XPCS development to become fully capable of addressing problems central to the understanding of dynamic behavior in ILs, XPCS is already capable of revealing aspects of the fundamental properties of ILs. In the rest of this section, we will briefly introduce the fundamental theories behind XPCS. We will also present a couple of examples to illustrate the *current* capabilities of XPCS that are relevant to the IL research.

6.5.1 Theory

When a thin, disordered sample system is illuminated by coherent radiation, the coherent scattering gives rise to a random coherent diffraction pattern, also known as a "speckle pattern". Speckle patterns, which have been long known from laser scattering [115], were first exhibited in the X-ray regime by Sutton et al. in 1991 [116]. Since then, speckle patterns have been observed in a wide range of materials under coherent X-ray illumination [117–119]. The far-field coherent scattering intensity is given by the coherent superposition of scattering probability amplitudes:

$$I(\mathbf{q},t) = \left| \sum_j f_j(\mathbf{q}) e^{i(\mathbf{q}\cdot\mathbf{r}(t))} \right|^2 \qquad (6.44)$$

where $\mathbf{q}$ is the scattering vector, and $f_j(\mathbf{q})$ is the scattering amplitude of the ith scatterer located at position $\mathbf{r}$ at time t. This summation takes place within the coherent volume, which is defined by the longitudinal coherent length and spatial coherent lengths of the given radiation. From a detection point of view, the measured intensity is an average of $I(\mathbf{q},t)$ during the entire duration of data acquisition T. Distinct and sharp variations in intensities can only be captured when the data acquisition time is comparable to the inherent fluctuation timescale of the underlying material system. Otherwise, the measured intensity becomes an ensemble average, similar to intensity profiles acquired in a conventional SAXS experiment with incoherent illumination.

It is also worth noting that Eq. (6.44) assumes that the X-ray beam is fully coherent. As mentioned previously, the X-rays from a synchrotron undulator source are partially coherent in nature. It is often the case that the illuminated sample volume is greater than the coherent volume. The detected intensity therefore becomes the incoherent sum of a number of coherent scattering intensities from small volume elements, with each volume element having the same size as the coherent volume. Obviously, when the illuminated sample volume becomes too large compared with the coherent volume, the speckle pattern will be smeared out and we once again arrive at an intensity profile that resembles that from a conventional SAXS experiment discussed previously. To preserve the speckle pattern while allowing the maximum coherent flux, common practice at synchrotron beamlines is to use an aperture in front of the sample as a second coherent source [120–122]. The size of the aperture ideally needs to match the spatial coherent lengths. In reality, this requirement is normally relaxed and the aperture often has a size greater than that defined by the spatial coherent lengths, because XPCS is a photon-starved technique.

We illustrate a static speckle pattern from a colloidal suspension of gold nanoparticles in Fig. 6.14 [123]. The left panel shows a single coherent scattering speckle pattern, and the right panel shows the incoherent summation from 50 consecutive measurements, which, as discussed above, is analogous to data from a conventional incoherent SAXS measurement. Figure 6.14 clearly displays the distinctive differences between coherent and incoherent SAXS results.

If the spatial arrangement of the scattering features in the illuminated sample volume changes with time, the corresponding coherent speckle pattern will also change accordingly, following Eq. (6.44). The measurement of intensity fluctuations of the speckles can reveal the underlying dynamics of the sample, and forms the basis of XPCS. An example of the XPCS intensity fluctuation at one reciprocal vector is shown in Fig. 6.15.

The data reduction and analysis of a typical XPCS data of an equilibrium dynamic system is well documented [87, 124, 125]. The quantification of the equilibrium dynamic timescale is assisted by the normalized second-order intensity correlation function $g_2(q,\tau)$, which is defined by

$$g_2(q,\tau) = \frac{\langle I(q,t+\tau)I(q,\tau)\rangle_t}{\langle I(q,t)\rangle^2} \tag{6.45}$$

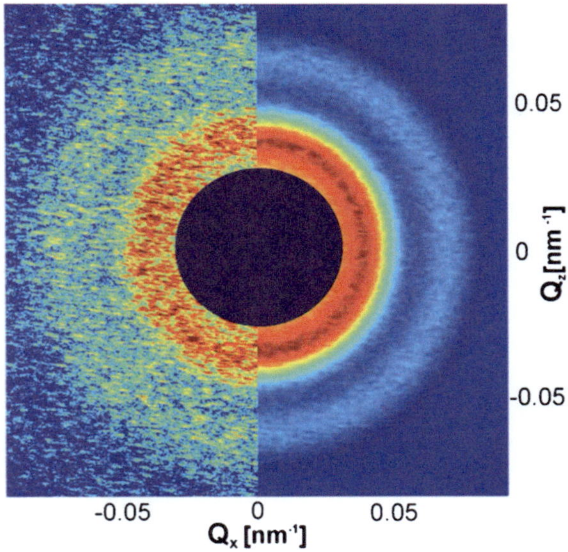

Fig. 6.14 Coherent X-ray scattering pattern from a colloidal suspension of gold nanoparticles. *Left half* shows a single-shot coherent pattern, where distinct speckles are clearly visible. *Right half* shows a scattering ring obtained from a sum of 50 successive shots, which is smooth and largely resembles data from conventional SAXS experiments

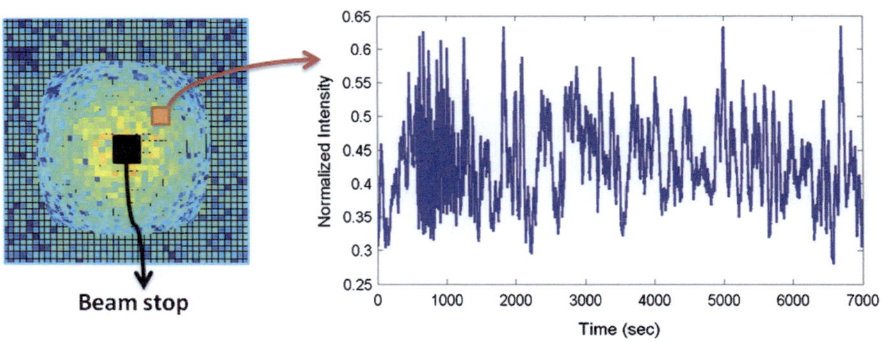

Fig. 6.15 Illustration of typical XPCS data. *Left panel* shows a simulated coherent speckle pattern. *Right panel* shows the fluctuation of the coherent scattering intensity from the highlighted pixel as a function time

where τ is the time delay between two measurements, and $\langle\rangle t$ denotes an ensemble average over all times.

It is also known that the normalized second-order intensity correlation function is related to the intermediate scattering function following

$$g_2(q,\tau) = 1 + \beta(q) \times |f(q,\tau)|^2 \tag{6.46}$$

where $\beta(q)$ is commonly referred to as the optical transfer function, and is related to the setup of the scattering instrument. For a fully coherent beam with no disturbance to the coherence from the optical elements, $\beta(q)=1$.

The function, $f(q,\tau)$, is the time-dependent intermediate scattering function. It contains time-dependent spatial structural information, and can be attributed to scattering that arises from the non-periodic pair correlation function within the electron density function and it is further defined by

$$f(q,\tau)=\frac{S(q,\tau)}{S(q,0)} \tag{6.47}$$

Here $S(q, 0)$ is the static structure factor, and $S(q, \tau)$ is the time-dependent structure factor. In other words, $f(q, \tau)$ can be seen as equivalent to the normalized dynamic structure factor in the time domain, which is fundamental in the description of dynamic behavior of condensed matter and is often the key parameter through which detailed comparisons between experiments and theories can be made. Access to this key parameter is the most important role of XPCS.

Depending on the underlying physical process that is responsible for the measured dynamics, the intensity correlation function $g_2(q, \tau)$ [or intermediate scattering function $f(q, \tau)$] can be analyzed using different models. For example, in the simplest case, when the scatterers undergo free diffusion (Brownian motion), the intermediate scattering function follows $f(q,\tau)=\exp\left(-D_0 q^2 \tau\right)$, where D_0 is the diffusion coefficient. For simple monodisperse spherical objects, $D_0 = k_B T / 6\pi\eta_s r$. Here k_B is the Boltzmann constant, T is temperature, r is particle radius, and η_s is the viscosity of the solvent. Therefore, XPCS measurements of dilute particle dynamics provide a direct probe of solvent viscosity, and when coupled with rheology experiments can help uncover the dynamic viscoelastic modulus of the solvent in the frequency domain.

We note that free diffusion is an exceptionally simple case. Most dynamic processes are more complicated and their intermediate scattering functions have to be treated accordingly and differently. Here we present another empirical and analytic form of the intermediate function that can be relevant to ILs, namely, the Kohlrausch–Williams–Watts (KWW) function $g_2(t)=\beta\exp\left(-2(t/\tau)^{\gamma}\right)+1$. Here τ is the characteristic relaxation time, and γ as an exponent indicates whether the decay is faster ($\gamma > 1$) or slower ($\gamma < 1$) than the exponential decay expected from a free diffusion.

6.5.2 *Examples*

6.5.2.1 Example 1: Particle Dynamics

Recently, novel colloidal systems with ionic liquids as solvents have gained attraction because the colloidal nanoparticles allow fabrication of novel nanohybrid materials [126, 127], which can find applications such as in IL-based solid electrolytes

[128, 129]. However, because ILs can act to modify the interparticle interactions between colloidal particles, the effects that different ILs have on the colloidal stability are not well understood [130]. To have a proper comprehension of the colloidal stability in ILs, it is critical to understand the dynamic behaviors of such colloids. XPCS, due to its suitable time and length scale, is unquestionably a technique that can offer a unique perspective to unveil the hidden mechanism governing the colloidal stabilities. While such studies do not yet exist, we will use the following example to illustrate the capabilities of XPCS in elucidating particle dynamics.

In colloidal suspensions, as colloidal particles move, they excite long-ranged flow in the fluid. Similarly, colloidal particles move in response to fluid motion. By generating and reacting to the local velocity of the fluid, colloidal particles experience hydrodynamic interaction, which, despite its long-recognized ubiquity, is not completely understood. Banchio et al. [93] studied the hydrodynamic interaction of concentrated ($\nu \approx 0.14$), monodisperse fluorinated latex spheres of radius 82.5 nm in solvents of different NaCl concentrations. Due to the apparent high volume concentrations, the effective diffusion constant $D(q)$ was obtained using a limiting form related to the intermediate scattering function, following $D(q) = -1/q^2 \times \lim_{t\to 0} \partial\left(\log\left|f(q,t)\right|\right)/\partial t$. The hydrodynamic interaction follows $H(q) = D(q)S(q)/D_0$; here $S(q)$ is the static structure factor and D_0 is the single particle diffusion constant in the dilute limit.

Figure 6.16 shows the q dependence of the static structure factor $S(q)$, effective diffusion constant $D(q)$, and the hydrodynamic interaction $H(q)$ in a solvent with 0.5 mM NaCl. The structure factor was found to follow the Rogers-Young model, which describes standard macroion fluids of charge-stabilized spheres. More importantly, by comparing XPCS data with simulation results using a novel accelerated Stokesian Dynamics method, the authors found that there is no hydrodynamic screening present in low-salinity systems. Through this effort, the influence of many-body hydrodynamic interactions on the diffusion properties of charge-stabilized colloidal spheres was explained and quantified. This result may carry potential impact on the transport properties of complex colloidal particles suspended in various solvents of interest to industry, including ILs.

6.5.2.2 Example 2: Surface Dynamics of Capillary Wave

ILs often possess peculiar molecular structure and composition, which can lead to unusual molecular packing. This also extends to the liquid/air interface of ILs, where a surface layer with a higher mass density than that of the bulk has been identified due to the low dimensionality at the free surface [131]. A subsequent question is on how this unusual packing affects the dynamics at the interface. To address this question, Sloutskin et al. conducted the first and the only XPCS investigation of ILs to date to probe the dynamics and critical damping of the capillary waves in ILs [132]. Due to the surface geometry, a grazing-incidence version of XPCS was employed, where the incidence angle of the X-rays was below the critical angle for total X-ray reflection so

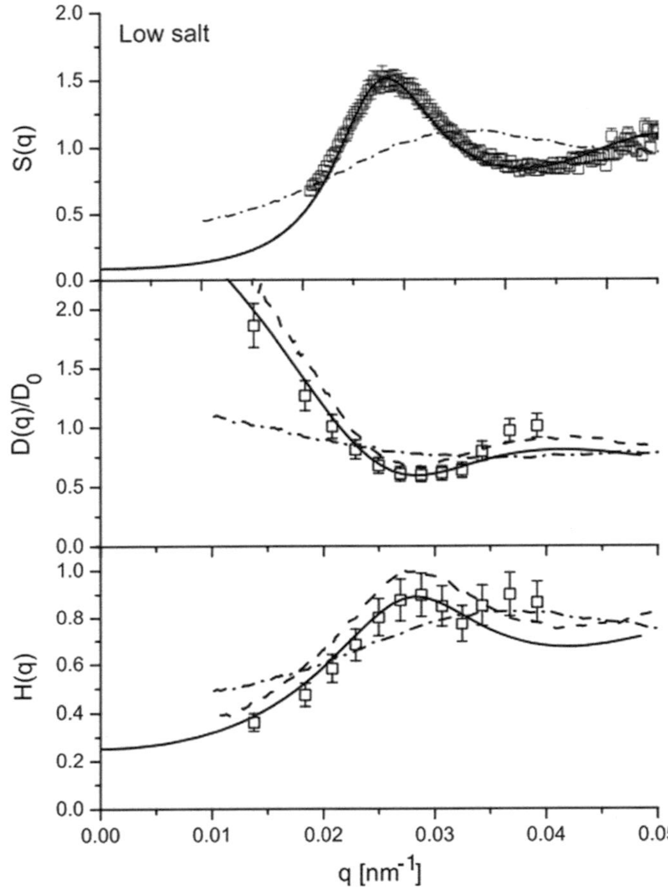

Fig. 6.16 Static structure factor $S(q)$, short-time collective diffusion coefficient $D(q)$, and hydrodynamic function $H(q)$ for a suspension of charged latex spheres in a 0.5 mM NaCl solvent

that the surface dynamics on the top layer of the ionic liquid butylmethylimidazolium tetrafluoroborate ([C_4mim][BF_4]) could be measured.

Similar to the transmission geometry, grazing incident intensity correlation function yields the dynamic structure factor. The dynamic structure factors of [C_4mim][BF_4] at different temperatures and reciprocal vectors are shown in Fig. 6.17. The experimental data reveal that capillary wave excitations at shorter wavelengths have higher frequencies, and capillary waves are over-damped at high q and low temperature. Moreover, it is found that in order to fully account for the line shape of the dynamic structure factor, both the homodyne and heterodyne components of the scattering have to be taken into account. Such findings are important to the understanding of the fundamental mechanic behaviors of ionic liquids, and can only be accessed through XPCS.

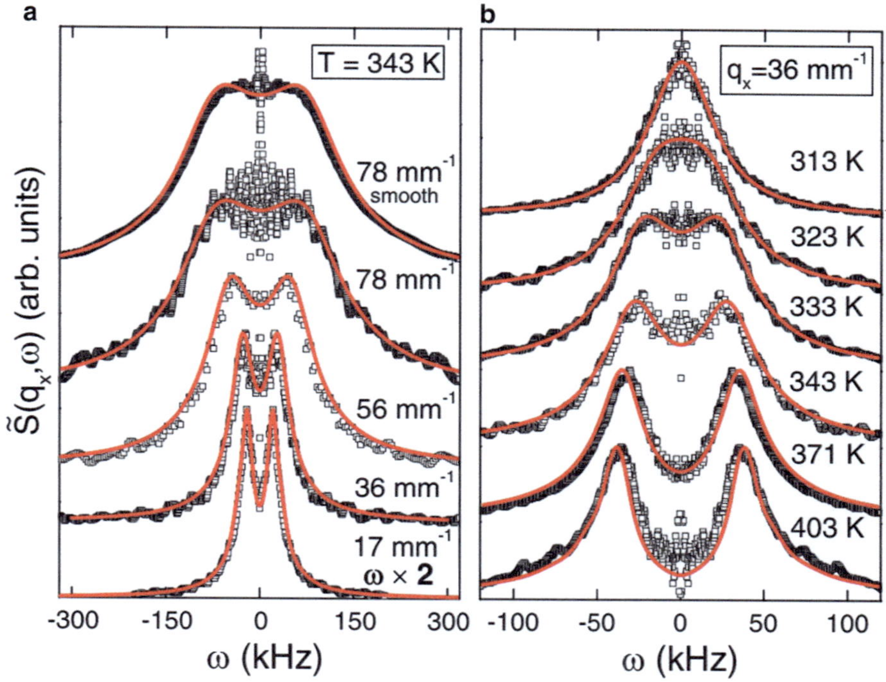

Fig. 6.17 The grazing incidence dynamic structure factors for [C_4mim][BF_4] for (**a**) listed *q*s and (**b**) listed temperatures

6.6 Conclusion and Outlook

XPCS is a promising young technique that can monitor the dynamics of condensed matter on length scales shorter than that accessible to light scattering techniques. In spite of the tremendous progress made since its discovery, XPCS, being a signal-to-noise limited technique, has limited spatiotemporal resolution because of the relatively low degree of coherence in the X-ray beam generated by undulator sources at synchrotrons. This fact constrains its applications in many fields, including that of ILs. However, as we identified in this section, even at this stage, XPCS can contribute to the understanding of dynamics in concentrated colloidal particles in various ILs and surface dynamics of ionic liquids.

An important feature of XPCS is that the available time resolution scales as the square of the strength of scattering intensity, which is proportional to source brightness. For example, a 100-fold increase in brightness will provide a 10^4 increase in dynamic range, which in turn will enable nanosecond time and nanometer space resolution. This has already been achieved using X-ray free electron laser (XFEL) facilities in the United States and Europe. However, access to XFEL is very limited. Currently, worldwide efforts are under way to upgrade the existing third-generation synchrotron facilities with multi-bend achromat lattices, which will provide orders of magnitude improvement in X-ray brightness and coherent flux. This upgrade, once completed, will allow XPCS measurements to a full spectrum of dynamic investigations of ionic liquids.

Acknowledgment We thank Dr. Andrew J. Allen for his careful reading of our chapter. ChemMatCARS Sector 15 is principally supported by the National Science Foundation/Department of Energy under grant number NSF/CHE-0822838. Use of the Advanced Photon Source was supported by the U. S. Department of Energy, Office of Science, Office of Basic Energy Sciences, under Contract No. DE-AC02-06CH11357.

References

1. Guinier A, Fournet G (1955) Small-angle scattering of X-rays. Structure of matter series. Wiley, New York, NY
2. Guinier A (1963) X-ray diffraction in crystals, imperfect crystals, and amorphous bodies, A series of books in physics. W.H Freeman, San Francisco, CA
3. Glatter O, Kratky O (1982) Small angle X-ray scattering. Academic Press, London
4. Gille W (2014) Particle and particle systems characterization: small-angle scattering (SAS) applications. CRC Press, Boca Raton, FL
5. Svergun DI, Koch MHJ, Timmins PA, May RP (2013) Small angle X-ray and neutron scattering from solutions of biological macromolecules, 1st edn. Oxford University Press, Oxford
6. Svergun DI, Koch MHJ (2003) Small-angle scattering studies of biological macromolecules in solution. Rep Progr Phys 66:1735–1782
7. Pedersen JS (1997) Analysis of small-angle scattering data from colloids and polymer solutions: modeling and least-squares fitting. Adv Colloid Interface Sci 70:171–210. doi:10.1016/S0001-8686(97)00312-6
8. Ilavsky J (2012) Nika: software for two-dimensional data reduction. J Appl Crystallogr 45(2):324–328. doi:10.1107/S0021889812004037
9. Pauw BR (2013) Everything SAXS: small-angle scattering pattern collection and correction. J Phys Condens Matter 25(38):383201. doi:10.1088/0953-8984/25/38/383201
10. Fan L, Degen M, Bendle S, Grupido N, Ilavsky J (2010) The absolute calibration of a small-angle scattering instrument with a laboratory X-ray source. J Phys Conf Ser 247:012005. doi:10.1088/1742-6596/247/1/012005
11. Ilavsky J, Jemian PR (2009) Irena: tool suite for modeling and analysis of small-angle scattering. J Appl Crystallogr 42:347–353
12. Kline SR (2006) Reduction and analysis of SANS and USANS data using IGOR Pro. J Appl Crystallogr 39:895–900
13. Glatter O (1980) Computation of distance distribution-functions and scattering functions of models for small-angle scattering experiments. Acta Phys Austriaca 52(3–4):243–256
14. Petoukhov MV, Svergun DI (2005) Global rigid body modeling of macromolecular complexes against small-angle scattering data. Biophys J 89(2):1237–1250. doi:10.1529/biophysj.105.064154

15. Svergun DI, Stuhrmann HB (1991) New developments in direct shape determination from small-angle scattering. Theory and model-calculations. Acta Crystallogr A 47:736–744. doi:10.1107/s0108767391006414
16. Konarev PV, Petoukhov MV, Volkov VV, Svergun DI (2006) ATSAS 2.1, a program package for small-angle scattering data analysis. J Appl Crystallogr 39:277–286. doi:10.1107/s0021889806004699
17. Franke D, Svergun DI (2009) DAMMIF, a program for rapid ab-initio shape determination in small-angle scattering. J Appl Crystallogr 42:342–346. doi:10.1107/s0021889809000338
18. Als-Nielsen J, McMorrow D (2001) Elements of modern X-ray physics. Wiley, New York, NY
19. Dreiss CA, Jack KS, Parker AP (2006) On the absolute calibration of bench-top small-angle X-ray scattering instruments: a comparison of different standard methods. J Appl Crystallogr 39(1):32–38. doi:10.1107/s0021889805033091
20. Greaves TL, Kennedy DF, Weerawardena A, Tse NMK, Kirby N, Drummond CJ (2011) Nanostructured protic ionic liquids retain nanoscale features in aqueous solution while precursor bronsted acids and bases exhibit different behavior. J Phys Chem B 115(9):2055–2066. doi:10.1021/jp1112203
21. Yan S, Kennedy DF, Greaves TL, Weerawardena A, Mulder RJ, Kirby N, Gonghua S, Drummond CJ (2012) Protic ionic liquids with fluorous anions: physicochemical properties and self-assembly nanostructure. Phys Chem Chem Phys 14(22):7981–7992. doi:10.1039/c2cp40463j
22. Hettige JJ, Kashyap HK, Annapureddy HVR, Margulis CJ (2013) Anions, the reporters of structure in ionic liquids. J Phys Chem Lett 4(1):105–110. doi:10.1021/jz301866f
23. Yue X, Chen X, Li QT (2012) Comparison of aggregation behaviors of a phytosterol ethoxylate surfactant in protic and aprotic ionic liquids. J Phys Chem B 116(31):9439–9444. doi:10.1021/jp305230r
24. Greaves TL, Weerawardena A, Fong C, Drummond CJ (2007) Many protic ionic liquids mediate hydrocarbon-solvent interactions and promote amphiphile self-assembly. Langmuir 23(2):402–404. doi:10.1021/la062895k
25. Wu FG, Yu JS, Sun SF, Sun HY, Luo JJ, Yu ZW (2012) Stepwise ordering of imidazolium-based cationic surfactants during cooling-induced crystallization. Langmuir 28(19):7350–7359. doi:10.1021/la300739x
26. Shi LJ, Zhao MW, Zheng LQ (2012) Lyotropic liquid crystalline phases formed in ternary mixtures of N-alkyl-N-methylpyrrolidinium bromide/1-decanol/water. RSC Adv 2(31): 11922–11929. doi:10.1039/c2ra21467a
27. Zhao XY, Cao YR, Cao GR, Xiao RJ (2012) Lamellar liquid crystalline phase formed by triblock copolymer L64 in 1-butyl-3-methylimidazolium tetrafluoroborate ionic liquid. Acta Physico-Chimica Sinica 28(6):1411–1417. doi:10.3866/pku.whxb201203262
28. Pedersen JS, Gerstenberg MC (1996) Scattering form factor of block copolymer micelles. Macromolecules 29(4):1363–1365
29. Foerster S, Apostol L, Bras W (2010) Scatter: software for the analysis of nano- and meso-scale small-angle scattering. J Appl Crystallogr 43:639–646. doi:10.1107/s0021889810008289
30. Liu H, Hexemer A, Zwart PH (2012) The small angle scattering toolbox (SASTBX): an open-source software for biomolecular small-angle scattering. J Appl Crystallogr 45:587–593. doi:10.1107/s0021889812015786
31. Tjioe E, Heller WT (2007) ORNL_SAS: software for calculation of small-angle scattering intensities of proteins and protein complexes. J Appl Crystallogr 40:782–785. doi:10.1107/s002188980702420x
32. Weaver KD, Vrikkis RM, Van Vorst MP, Trullinger J, Vijayaraghavan R, Foureau DM, McKillop IH, MacFarlane DR, Krueger JK, Elliott GD (2012) Structure and function of proteins in hydrated choline dihydrogen phosphate ionic liquid. Phys Chem Chem Phys 14(2):790–801. doi:10.1039/c1cp22965f

33. Brunner-Popela J, Glatter O (1997) Small-angle scattering of interacting particles. I. Basic principles of a global evaluation technique. J Appl Crystallogr 30:431–442
34. Soni SS, Vekariya RL, Aswal VK (2013) Ionic liquid induced sphere-to-ribbon transition in the block copolymer mediated synthesis of silver nanoparticles. RSC Adv 3(22):8398–8406. doi:10.1039/c3ra41138a
35. Harada M, Yamada M, Kimura Y, Saijo K (2013) Influence of the organization of water-in-ionic liquid microemulsions on the size of silver particles during photoreduction. J Colloid Interface Sci 406:94–104. doi:10.1016/j.jcis.2013.05.068
36. Teixeira J (1988) Small-angle scattering by fractal systems. J Appl Crystallogr 21:781–785
37. Wagner J (2004) Small-angle scattering from spherical core-shell particles: an analytical scattering function for particles with Schulz-Flory size distribution. J Appl Crystallogr 37(5):750–756
38. Hammons JA, Muselle T, Ustarroz J, Tzedaki M, Rats M, Hubin A, Terryn H (2013) Stability, assembly, and particle/solvent interactions of Pd nanoparticles electrodeposited from a deep eutectic solvent. J Phys Chem C 117(27):14381–14389. doi:10.1021/jp403739y
39. Ruland W (1971) Small-angle scattering of 2-phase systems – determination and significance of systematic deviations from Porod's law. J Appl Crystallogr 4:70–73
40. Kim MH (2004) Modified Porod's law estimate of the transition-layer thickness between two phases: test of triangular smoothing function. J Appl Crystallogr 37:643–651. doi:10.1107/s0021889804013196
41. Bottero JY, Tchoubar D, Cases JM, Flessinger F (1982) Investigation of the hydrolysis of aqueous-solutions of aluminum-chloride. 2. Nature and structure by small-angle X-ray scattering. J Phys Chem 86(18):3667–3673
42. Hammons JA, Rayment T, Vandendael I, Blajiev O, Hubin A, Davenport AJ, Raes M, Terryn H (2010) A method to detect retained gas during AC electrograining using in-situ small angle X-ray scattering. Electrochem Commun 12(6):717–719
43. Atkin R, Warr GG (2008) The smallest amphiphiles: nanostructure in protic room-temperature ionic liquids with short alkyl groups. J Phys Chem B 112(14):4164–4166. doi:10.1021/jp801190u
44. Debye P (1947) Molecular-weight determination by light scattering. J Phys Colloid Chem 51(1):18–32. doi:10.1021/j150451a002
45. Pedersen JS, Schurtenberger P (1996) Scattering functions of semiflexible polymers with and without excluded volume effects. Macromolecules 29:7602–7612
46. Pfeifer P (1984) Fractal dimension as working tool for surface-roughness problems. Appl Surf Sci 18(1–2):146–164
47. Cheng G, Varanasi P, Li CL, Liu HB, Menichenko YB, Simmons BA, Kent MS, Singh S (2011) Transition of cellulose crystalline structure and surface morphology of biomass as a function of ionic liquid pretreatment and its relation to enzymatic hydrolysis. Biomacromolecules 12(4):933–941. doi:10.1021/bm101240z
48. Hohr A, Neumann HB, Schmidt PW, Pfeifer P, Avnir D (1988) Fractal surface and cluster structure of controlled-pore glasses and vycor porous-glass as revealed by small-angle X-ray and neutron scattering. Phys Rev B 38(2):1462–1467. doi:10.1103/PhysRevB.38.1462
49. Wong PZ, Howard J (1986) Surface roughening and the fractal nature of rocks. Phys Rev Lett 57(5):637–640. doi:10.1103/PhysRevLett.57.637
50. Martin JE, Hurd AJ (1987) Scattering from fractals. J Appl Crystallogr 20:61–78. doi:10.1107/s0021889887087107
51. Hammouda B (2010) A new Guinier-Porod model. J Appl Crystallogr 43:716–719. doi:10.1107/s0021889810015773
52. Beaucage G (1995) Approximations leading to a unified exponential power-law approach to small-angle scattering. J Appl Crystallogr 28:717–728
53. Beaucage G (2004) Determination of branch fraction and minimum dimension of mass-fractal aggregates. Phys Rev E 70(3):031401–031411. doi:10.1103/PhysRevE.70.031401

54. Beaucage G, Kammler HK, Pratsinis SE (2004) Particle size distributions from small-angle scattering using global scattering functions. J Appl Crystallogr 37:523–535. doi:10.1107/s0021889804008969
55. Hu N, Borkar N, Kohls D, Schaefer DW (2011) Characterization of porous materials using combined small-angle X-ray and neutron scattering techniques. J Membr Sci 379(1–2):138–145. doi:10.1016/j.memsci.2011.05.053
56. Effler LJ, Fellers JF (1992) Structural orientation functions for anisotropic small-angle scattering. J Phys D Appl Phys 25(1):74–78. doi:10.1088/0022-3727/25/1/010
57. Pedersen JS (1997) Analysis of small-angle scattering data from colloids and polymer solutions: modeling and least-squares fitting. Adv Colloid Interface Sci 70:171–210
58. Potton JA, Daniell GJ, Rainford BD (1988) Particle-size distributions from SANS data using the maximum-entropy method. J Appl Crystallogr 21:663–668. doi:10.1107/s0021889888004819
59. Takamuku T, Shimomura T, Sadakane K, Koga M, Seto H (2012) Aggregation of 1-dodecyl-3-methylimidazolium nitrate in water and benzene studied by SANS and H-1 NMR. Phys Chem Chem Phys 14(31):11070–11080. doi:10.1039/c2cp40891k
60. Pedersen JS (1993) Small-angle scattering from precipitates-analysis by use of a polydisperse hard-sphere model. Phys Rev B 47(2):657–665. doi:10.1103/PhysRevB.47.657
61. Ashcroft NW, Lekner J (1966) Structure and resistivity of liquid metals. Phys Rev 145(1):83–90. doi:10.1103/PhysRev.145.83
62. Weyerich B, Brunner-Popela J, Glatter O (1999) Small-angle scattering of interacting particles. II. Generalized indirect Fourier transformation under consideration of the effective structure factor for polydisperse systems. J Appl Crystallogr 32:197–209
63. Bradley AE, Hardacre C, Holbrey JD, Johnston S, McMath SEJ, Nieuwenhuyzen M (2002) Small-angle X-ray scattering studies of liquid crystalline 1-alkyl-3-methylimidazolium salts. Chem Mater 14(2):629–635. doi:10.1021/cm010542v
64. Li S, Banuelos JL, Guo J, Anovitz L, Rother G, Shaw RW, Hillesheim PC, Dai S, Baker GA, Cummings PT (2012) Alkyl chain length and temperature effects on structural properties of pyrrolidinium-based ionic liquids: a combined atomistic simulation and small-angle X-ray scattering study. J Phys Chem Lett 3(1):125–130. doi:10.1021/jz2013209
65. Pedersen JS (2001) Structure factors effects in small-angle scattering from block copolymer micelles and star polymers. J Chem Phys 114(6):2839. doi:10.1063/1.1339221
66. Parmar A, Aswal VK, Bahadur P (2012) Interaction between the ionic liquids 1-alkyl-3-methylimidazolium tetrafluoroborate and Pluronic (R) P103 in aqueous solution: a DLS, SANS and NMR study. Spectrochim Acta A Mol Biomol Spectrosc 97:137–143. doi:10.1016/j.saa.2012.05.075
67. Sastry NV, Vaghela NM, Macwan PM, Soni SS, Aswal VK, Gibaud A (2012) Aggregation behavior of pyridinium based ionic liquids in water – surface tension, H-1 NMR chemical shifts, SANS and SAXS measurements. J Colloid Interface Sci 371:52–61. doi:10.1016/j.jcis.2011.12.077
68. Lopez-Barron CR, Wagner NJ (2012) Structural transitions of CTAB micelles in a protic ionic liquid. Langmuir 28(35):12722–12730. doi:10.1021/la302231w
69. Sastry NV, Vaghela NM, Aswal VK (2012) Effect of alkyl chain length and head group on surface active and aggregation behavior of ionic liquids in water. Fluid Phase Equilib 327:22–29. doi:10.1016/j.fluid.2012.04.013
70. Teubner M, Strey R (1987) Origin of the scattering peak in microemulsions. J Chem Phys 87(5):3195–3200. doi:10.1063/1.453006
71. Atkin R, Warr GG (2007) Phase behavior and microstructure of microemulsions with a room-temperature ionic liquid as the polar phase. J Phys Chem B 111(31):9309–9316. doi:10.1021/jp065020n
72. Klee A, Prevost S, Kunz W, Schweins R, Kiefer K, Gradzielski M (2012) Magnetic microemulsions based on magnetic ionic liquids. Phys Chem Chem Phys 14(44):15355–15360. doi:10.1039/c2cp43048g

73. Rojas O, Tiersch B, Rabe C, Stehle R, Hoell A, Arlt B, Koetz J (2013) Nonaqueous microemulsions based on N, N′-alkylimidazolium alkylsulfate ionic liquids. Langmuir 29(23):6833–6839. doi:10.1021/la401080q
74. Larson-Smith K, Jackson A, Pozzo DC (2010) Small angle scattering model for pickering emulsions and raspberry particles. J Colloid Interface Sci 343(1):36–41. doi:10.1016/j.jcis.2009.11.033
75. McAlister BC, Grady BP (2002) The use of Monte-Carlo simulations to calculate small-angle scattering patterns. Macromol Symp 190:117–129
76. Henderson SJ (1996) Monte Carlo modeling of small-angle scattering data from non-interacting homogeneous and heterogeneous particles in solution. Biophys J 70(4):1618–1627
77. McAlister BC, Grady BP (1998) Simulation of small-angle X-ray scattering from single-particle systems. J Appl Cryst 31:594–599
78. Volkov VV, Svergun DI (2003) Uniqueness of ab initio shape determination in small-angle scattering. J Appl Crystallogr 36:860–864
79. Lu X, Yager KG, Johnston D, Black CT, Ocko BM (2013) Grazing-incidence transmission X-ray scattering: surface scattering in the Born approximation. J Appl Crystallogr 46(1):165–172. doi:10.1107/s0021889812047887
80. Pfeifer MA, Williams GJ, Vartanyants IA, Harder R, Robinson IK (2006) Three-dimensional mapping of a deformation field inside a nanocrystal. Nature 442(7098):63–66. doi:10.1038/nature04867
81. Willmott PR, Pauli SA, Herger R, Schleputz CM, Martoccia D, Patterson BD, Delley B, Clarke R, Kumah D, Cionca C, Yacoby Y (2007) Structural basis for the conducting interface between LaAlO3 and SrTiO3. Phys Rev Lett 99(15). doi:10.1103/PhysRevLett.99.155502
82. Westneat MW, Betz O, Blob RW, Fezzaa K, Cooper WJ, Lee WK (2003) Tracheal respiration in insects visualized with synchrotron X-ray imaging. Science 299(5606):558–560. doi:10.1126/science.1078008
83. Davis TJ, Gao D, Gureyev TE, Stevenson AW, Wilkins SW (1995) Phase-contrast imaging of weakly absorbing materials using hard X-rays. Nature 373(6515):595–598
84. Faigel G, Tegze M (1999) X-ray holography. Rep Prog Phys 62(3):355–393
85. Miao JW, Charalambous P, Kirz J, Sayre D (1999) Extending the methodology of X-ray crystallography to allow imaging of micrometre-sized non-crystalline specimens. Nature 400(6742):342–344
86. Prasad V, Semwogerere D, Weeks ER (2007) Confocal microscopy of colloids. J Phys Condens Matter 19(11):113102. doi:10.1088/0953-8984/19/11/113102
87. Grubel G, Zontone F (2004) Correlation spectroscopy with coherent X-rays. J Alloys Compd 362(1–2):3–11. doi:10.1016/s0925-8388(03)00555-3
88. Zhang F, Allen AJ, Levine LE, Ilavsky J, Long GG (2012) Ultra-small-angle X-ray scattering-X-ray photon correlation spectroscopy: a new measurement technique for in-situ studies of equilibrium and nonequilibrium dynamics. Metall Mater Trans A Phys Metall Mater Sci 43A(5):1445–1453. doi:10.1007/s11661-011-0790-0
89. Brauer S, Stephenson GB, Sutton M, Bruning R, Dufresne E, Mochrie SGJ, Grubel G, Alsnielsen J, Abernathy DL (1995) X-ray-intensity fluctuation spectroscopy observations of critical-dynamics in Fe3al. Phys Rev Lett 74(11):2010–2013
90. Dierker SB, Pindak R, Fleming RM, Robinson IK, Berman L (1995) X-ray photon-correlation spectroscopy study of Brownian-motion of gold colloids in glycerol. Phys Rev Lett 75(3):449–452. doi:10.1103/PhysRevLett.75.449
91. ThurnAlbrecht T, Steffen W, Patkowski A, Meier G, Fischer EW, Grubel G, Abernathy DL (1996) Photon correlation spectroscopy of colloidal palladium using a coherent x-ray beam. Phys Rev Lett 77(27):5437–5440. doi:10.1103/PhysRevLett.77.5437
92. Lurio LB, Lumma D, Sandy AR, Borthwick MA, Falus P, Mochrie SGJ, Pelletier JF, Sutton M, Regan L, Malik A, Stephenson GB (2000) Absence of scaling for the intermediate scattering function of a hard-sphere suspension: static and dynamic X-ray scattering from concentrated polystyrene latex spheres. Phys Rev Lett 84(4):785–788. doi:10.1103/PhysRevLett.84.785

93. Banchio AJ, Gapinski J, Patkowski A, Haussler W, Fluerasu A, Sacanna S, Holmqvist P, Meier G, Lettinga MP, Nagele G (2006) Many-body hydrodynamic interactions in charge-stabilized suspensions. Phys Rev Lett 96(13):138303. doi:10.1103/PhysRevLett.96.138303
94. Orsi D, Fluerasu A, Moussaid A, Zontone F, Cristofolini L, Madsen A (2012) Dynamics in dense hard-sphere colloidal suspensions. Phys Rev E 85(1):011402. doi:10.1103/PhysRevE.85.011402
95. Zhang F, Allen AJ, Levine LE, Ilavsky J, Long GG (2013) Structure and dynamics studies of concentrated micrometer-sized colloidal suspensions. Langmuir 29(5):1379–1387. doi:10.1021/la3044768
96. Robert A, Wandersman E, Dubois E, Dupuis V, Perzynski R (2006) Glassy dynamics and aging in a dense ferrofluid. Europhys Lett 75(5):764–770. doi:10.1209/epl/i2006-10179-4
97. Guo HY, Bourret G, Corbierre MK, Rucareanu S, Lennox RB, Laaziri K, Piche L, Sutton M, Harden JL, Leheny RL (2009) Nanoparticle motion within glassy polymer melts. Phys Rev Lett 102(7):075702. doi:10.1103/PhysRevLett.102.075702
98. Leitner M, Sepiol B, Stadler LM, Pfau B, Vogl G (2009) Atomic diffusion studied with coherent X-rays. Nat Mater 8(9):717–720. doi:10.1038/nmat2506
99. Ruta B, Chushkin Y, Monaco G, Cipelletti L, Pineda E, Bruna P, Giordano VM, Gonzalez-Silveira M (2012) Atomic-scale relaxation dynamics and aging in a metallic glass probed by X-ray photon correlation spectroscopy. Phys Rev Lett 109(16). doi:10.1103/PhysRevLett.109.165701
100. Trappe V, Pitard E, Ramos L, Robert A, Bissig H, Cipelletti L (2007) Investigation of q-dependent dynamical heterogeneity in a colloidal gel by X-ray photon correlation spectroscopy. Phys Rev E 76(5). doi:10.1103/PhysRevE.76.051404
101. Guo HY, Ramakrishnan S, Harden JL, Leheny RL (2011) Gel formation and aging in weakly attractive nanocolloid suspensions at intermediate concentrations. J Chem Phys 135(15). doi:10.1063/1.3653380
102. Poniewierski A, Holyst R, Price AC, Sorensen LB, Kevan SD, Toner J (1998) Dynamic correlation functions for finite and infinite smectic-A systems: theory and experiment. Phys Rev E 58(2):2027–2040
103. Madsen A, Als-Nielsen J, Grubel G (2003) Viscosity of a liquid crystal near the nematic-smectic A phase transition. Phys Rev Lett 90(8):085701. doi:10.1103/PhysRevLett.90.085701
104. Livet F, Bley F, Caudron R, Geissler E, Abernathy D, Detlefs C, Grubel G, Sutton M (2001) Kinetic evolution of unmixing in an AlLi alloy using x-ray intensity fluctuation spectroscopy. Phys Rev E 63(3):036108
105. Chung B, Ramakrishnan S, Bandyopadhyay R, Liang D, Zukoski CF, Harden JL, Leheny RL (2006) Microscopic dynamics of recovery in sheared depletion gels. Phys Rev Lett 96(22):228301. doi:10.1103/PhysRevLett.96.228301
106. Bandyopadhyay R, Liang D, Yardimci H, Sessoms DA, Borthwick MA, Mochrie SGJ, Harden JL, Leheny RL (2004) Evolution of particle-scale dynamics in an aging clay suspension. Phys Rev Lett 93(22):228302. doi:10.1103/PhysRevLett.93.228302
107. Lu XH, Mochrie SGJ, Narayanan S, Sandy AR, Sprung M (2008) How a liquid becomes a glass both on cooling and on heating. Phys Rev Lett 100(4):045701. doi:10.1103/PhysRevLett.100.045701
108. Seydel T, Madsen A, Tolan M, Grubel G, Press W (2001) Capillary waves in slow motion. Phys Rev B 63(7):073409
109. Madsen A, Seydel T, Sprung M, Gutt C, Tolan M, Grubel G (2004) Capillary waves at the transition from propagating to overdamped behavior. Phys Rev Lett 92(9):096104. doi:10.1103/PhysRevLett.92.096104
110. Jiang Z, Kim H, Jiao X, Lee H, Lee YJ, Byun Y, Song S, Eom D, Li C, Rafailovich MH, Lurio LB, Sinha SK (2007) Evidence for viscoelastic effects in surface capillary waves of molten polymer films. Phys Rev Lett 98(22):227801. doi:10.1103/PhysRevLett.98.227801
111. Mukhopadhyay MK, Jiao X, Lurio LB, Jiang Z, Stark J, Sprung M, Narayanan S, Sandy AR, Sinha SK (2008) Thickness induced structural changes in polystyrene films. Phys Rev Lett 101(11):115501. doi:10.1103/PhysRevLett.101.115501

112. Narayanan S, Lee DR, Hagman A, Li XF, Wang J (2007) Particle dynamics in polymer-metal nanocomposite thin films on nanometer-length scales. Phys Rev Lett 98(18):185506. doi:10.1103/PhysRevLett.98.185506
113. Streit S, Gutt C, Chamard V, Robert A, Sprung M, Sternemann H, Tolan M (2007) Two-dimensional dynamics of metal nanoparticles on the surface of thin polymer films studied with coherent X rays. Phys Rev Lett 98(4):047801. doi:10.1103/PhysRevLett.98.047801
114. Duri A, Autenrieth T, Stadler LM, Leupold O, Chushkin Y, Grubel G, Gutt C (2009) Two-dimensional heterogeneous dynamics at the surface of a colloidal suspension. Phys Rev Lett 102(14):145701. doi:10.1103/PhysRevLett.102.145701
115. Goodman JW (1975) Statistical properties of laser speckle patterns. In: Laser speckle and related phenomena, vol 9. Topics in applied physics. Springer, Berlin, pp 9–75. doi:10.1007/BFb0111436
116. Sutton M, Mochrie SGJ, Greytak T, Nagler SE, Berman LE, Held GA, Stephenson GB (1991) Observation of speckle by diffraction with coherent X-rays. Nature 352(6336):608–610
117. Williams GJ, Pfeifer MA, Vartanyants IA, Robinson IK (2003) Three-dimensional imaging of microstructure in Au nanocrystals. Phys Rev Lett 90(17). doi:10.1103/PhysRevLett.90.175501
118. Malik A, Sandy AR, Lurio LB, Stephenson GB, Mochrie SGJ, McNulty I, Sutton M (1998) Coherent X-ray study of fluctuations during domain coarsening. Phys Rev Lett 81(26):5832–5835
119. Robinson I, Harder R (2009) Coherent X-ray diffraction imaging of strain at the nanoscale. Nat Mater 8(4):291–298. doi:10.1038/nmat2400
120. Abernathy DL, Grubel G, Brauer S, McNulty I, Stephenson GA, Mochrie SGJ, Sandy AR, Mulders N, Sutton M (1998) Small-angle X-ray scattering using coherent undulator radiation at the ESRF. J Synchrotron Radiat 5:37–47. doi:10.1107/s0909049597015835
121. Sandy AR, Lurio LB, Mochrie SGJ, Malik A, Stephenson GB, Pelletier JF, Sutton M (1999) Design and characterization of an undulator beamline optimized for small-angle coherent X-ray scattering at the Advanced Photon Source. J Synchrotron Radiat 6:1174–1184. doi:10.1107/s0909049599009590
122. Zhang F, Allen AJ, Levine LE, Ilavsky J, Long GG, Sandy AR (2011) Development of ultra-small-angle X-ray scattering-X-ray photon correlation spectroscopy. J Appl Crystallogr 44:200–212. doi:10.1107/s0021889810053446
123. Lee S, Roseker W, Gutt C, Fischer B, Conrad H, Lehmkuhler F, Steinke I, Zhu D, Lemke H, Cammarata M, Fritz DM, Wochner P, Castro-Colin M, Hruszkewycz SO, Fuoss PH, Stephenson GB, Grubel G, Robert A (2013) Single shot speckle and coherence analysis of the hard X-ray free electron laser LCLS. Opt Express 21(21):24647–24664. doi:10.1364/oe.21.024647
124. Sutton M (2008) A review of X-ray intensity fluctuation spectroscopy. Comptes Rendus Physique 9(5–6):657–667. doi:10.1016/j.crhy.2007.04.008
125. Livet F (2007) Diffraction with a coherent X-ray beam: dynamics and imaging. Acta Crystallogr A 63:87–107. doi:10.1107/s010876730605570x
126. Nakashima T, Kawai T (2005) Quantum dots-ionic liquid hybrids: efficient extraction of cationic CdTe nanocrystals into an ionic liquid. Chem Commun 12:1643–1645. doi:10.1039/b418001a
127. Guerrero-Sanchez C, Lara-Ceniceros T, Jimenez-Regalado E, Rasa M, Schubert US (2007) Magnetorheological fluids based on ionic liquids. Adv Mater 19(13):1740. doi:10.1002/adma.200700302
128. Wang P, Zakeeruddin SM, Comte P, Exnar I, Gratzel M (2003) Gelation of ionic liquid-based electrolytes with silica nanoparticles for quasi-solid-state dye-sensitized solar cells. J Am Chem Soc 125(5):1166–1167. doi:10.1021/ja029294+
129. Shimano S, Zhou H, Honma I (2007) Preparation of nanohybrid solid-state electrolytes with liquidlike mobilities by solidifying ionic liquids with silica particles. Chem Mater 19(22):5216–5221. doi:10.1021/cm0707814

130. Ueno K, Watanabe M (2011) From colloidal stability in ionic liquids to advanced soft materials using unique media. Langmuir 27(15):9105–9115. doi:10.1021/la103942f
131. Sloutskin E, Ocko BM, Tamam L, Kuzmenko I, Gog T, Deutsch M (2005) Surface layering in ionic liquids: an X-ray reflectivity study. J Am Chem Soc 127(21):7796–7804. doi:10.1021/ja0509679
132. Sloutskin E, Huber P, Wolff M, Ocko BM, Madsen A, Sprung M, Schon V, Baumert J, Deutsch M (2008) Dynamics and critical damping of capillary waves in an ionic liquid. Phys Rev E 77(6). doi:10.1103/PhysRevE.77.060601

Part II
Structure and Properties of Ionic Liquids Relevant to Electrochemistry

Chapter 7
Room-Temperature Molten Salts: Protic Ionic Liquids and Deep Eutectic Solvents as Media for Electrochemical Application

Mérièm Anouti

7.1 Introduction

Ionic liquids (ILs) are room-temperature molten salts (RTMS) composed mostly of organic ions that may undergo almost unlimited structural variations. ILs and deep eutectic solvents (DESs) have been applied in various fields, such as electrolytes for lithium ion batteries, redox flow batteries (RFBs), electrodeposition, and electropolishing, and even in fuel cells. This chapter covers the newest aspects of H-donor ILs in applications where their transport properties (ion conductivity, diffusion, and viscosity) are exploited, for example, as electrochemical solvents for energy storage where conventional media, organic solvents (in batteries), or water (in supercapacitors) fail. Metal nanoparticles' (NPs) formulation and stabilization in H-donor RTMS with some unique size, shape, and properties are also discussed. In these decidedly different materials, a mixture with a molecular solvent shows electroactivity that is not exhibited in more traditional systems, creating huge potential for energy storage and electrocatalysis. The remarkable recent increase in the level of interest in RTMS (protic ionic liquids, PILs, and DESs) in academics and industry is illustrated by the rapid growth in the number of publications on the topic (Fig. 7.1), with more than a 100-fold increase observed over the period 2004 to 2014.

The first main objective of this chapter is to summarize the synthesis and physical properties of PILs and DESs and discuss their suitability as green solvents for different applications. The second main objective is to review the electrochemical applications of RTMS and discuss their advantages and disadvantages as electrolytes in comparison with an aprotic homologue for energy storage applications. I hope to generate interest in the wider community and encourage others to make use of proton-donor RTMS—PILs and DESs—in tackling scientific challenges.

M. Anouti (✉)
Laboratoire PCM2E (EA 6296), Université François Rabelais de Tours,
Parc de Grandmont, 37200 Tours, France
e-mail: meriem.anouti@univ-tours.fr

A.A.J. Torriero (ed.), *Electrochemistry in Ionic Liquids*,
DOI 10.1007/978-3-319-13485-7_7

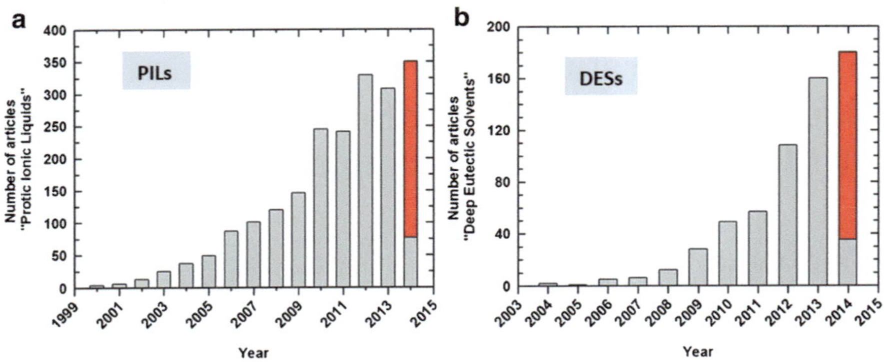

Fig. 7.1 Rapid growth of PIL and DES publications over the period 2000 to 2014; data obtained from a literature search using SciFinder scholar

7.2 Advances in Proton-Donor Room-Temperature Molten salts

7.2.1 Protic Ionic Liquids

ILs are organic salts with unusually low melting points, well below 100 °C [1, 2]. These substances have been suggested as potentially "green" replacements for traditional molecular solvents since some ILs are nonvolatile, nonflammable, thermally stable, and recyclable. Additionally, they open the path to novel methods that are not possible using classical solvents. Those formed by the transfer of a proton between a Brønsted acid and a Brønsted base form the protic subgroup now referred to as "Protic Ionic Liquids (PILs)" [3, 4].

$$HA + B = A^{-} + BH^{+} \tag{7.1}$$

While one of the first IL described in 1914 by Walden was of "the protic type" $[H_3NEt][NO_3]$ ethylammonium nitrate, conventionally called (NEA) (with a mp = 12.5 °C), described in nearly all reviews on ILs [5], aprotic ILs largely dominate the open literature. Recently, there has been a resurgence of interest for these Protic ILs essentially because of their great potential for proton-transfer applications in fuel-cell technologies as an electrolyte in nonhumidified fuel-cell systems [6, 7]. In recent years, real progress has been made in the electrochemical application of PILs as media thanks to their proton activity [8, 9]. The physicochemical properties of pure PILs and their mixtures with protic and aprotic molecular solvents are now systematically studied as proton-conducting electrolytes. PILs are starting to emerge as useful materials owing to various potential applications [10] including electrolytes in aqueous batteries, fuel cells, double-layer capacitors, and biosensors. There is, therefore, much promise of new and large application fields for PILs [11].

PILs differ from other types of ILs in that the ionic moiety is formed from the transfer of a proton [12]. However, neutral aggregates of PILs were hypothesized and later directly observed, both in the gas phase and at the PIL–solid interface [13]. Strong charge ordering can lead to a long-range order even in the presence of a solute and may arise from the competition among electrostatic, dispersion, and hydrogen-bonding interactions [14].

They present the advantage of being cost-effective and are easily prepared as their formation does not involve the formation of residual products. Different series of liquid crystalline have been isolated from the reaction of nitric, hydrochloric formic, and tetrafluoroboric acids with the corresponding alkylimidazoles, pyrrolidines, pyridines, or alkylamines. Those based on the imidazolium cation induce a specific stereoselectivity for the catalyzed Diels-Alder condensation reaction of cyclopentadiene and diethyl maleate [2, 15]. [The protonated Hünig base, $[NHEt(i\text{-}Pr)_2][CF_3CO_2]$, is reported to be an efficient medium for the nitration of arenes. Ohno and Yoshiwaza demonstrated for PILs prepared by neutralization of 1-alkylimidazoles the synergic effect between protonated alkylimidazoles and palladium, rhodium, or ruthenium catalysts observed in reactions where the presence of a proton is essential for good performance [16].

Many of these PILs involve very strong acids, such as $H[N(Tf)_2]$, H[NTf], etc., and hence the equilibrium (Eq. 7.1) is heavily shifted to the right, thus producing ILs as completely ionic salts.

The discussion concerning the labile proton and its acidity is still under debate [3, 17, 18]. It is possible to relate the acidity of PILs with their catalytic activities in different acid–base reactions. The efficiency of PILs as catalyzing media can further be improved by adding Brønsted acids. The strength of the added acid in a medium depends on the dissociation of acid and the stabilization of the conjugate base. Since different solvents solvate protons (H^+) to different extents, the strength of an acid varies with the medium.

Because of the various specific and nonspecific interactions, ILs have a unique solvation behavior. However, it is difficult to correlate the solvation behavior of ILs with its solvent properties, such as polarity (normalized charge transfer E_T^N) and relative permittivity (ε_r) [19–21]. Determining pKa PILs values aims to understand the proton properties [3]. Since proton acidity is mainly defined by their solvation, an environment other than water results in different pKa IL values. Adams et al. showed recently that the role of the solvent in the PIL/solvent mixture turns out to be preponderant in the chemical equilibrium. They demonstrated that pKa PIL values depend on the medium and give information about the relative feasibility of the proton to interact with a given solute. Their results concluded that it is possible to predict hydrogen bond strength between the cation and the anion and, ultimately, the type of ionic network they form. This property is the key in the activity of PILs as shown by the Kumar group [3], who noted that the ability of PILs to promote

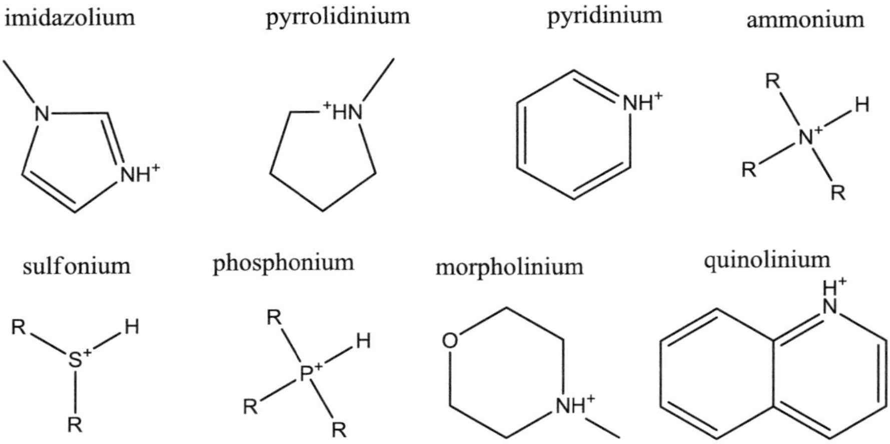

Fig. 7.2 Common cations in PILs

acidity is higher in [Hmim][HCO_2] than other PILs: $[Hmim]^+ > [Hpy]^+ > [HPyr]^+$. The observations drawn above are expected to be useful in synthesizing an efficient IL for acid catalysis. The most common cations for PILs are presented in Fig. 7.2.

7.2.2 Deep Eutectic Solvents

DESs are liquid media obtained by mixing, in certain ratios, a metal salt with a hydrogen bond donor (HBD), such as alcohol, amide, and carboxylic acid [22–27], as a complexing agent. This results in the formation of a eutectic mixture with a melting point temperature that is considerably lower than its original precursors, mainly due to the generation of intermolecular hydrogen bonds. The best glass-forming composition in a simple binary eutectic system emphasizes the composition range between the eutectic and the medium points as shown in Fig. 7.3. DESs, thus, have gained great attention to replace current organic solvents and have been applied to many chemical processes.

At room temperature, a eutectic mixture can fall under ionic liquid; it has the general properties of ILs, such as nonvolatility, high fluidity, low melting temperature, high boiling temperature, and nonflammability [26, 28–32]. These eutectic mixtures are made up of organic salts, organic acids, organic neutral molecules, or inorganic salts, which is the only difference from the traditional ILs comprising cations and anions. In order to differentiate these eutectic mixtures from traditional ILs, they have been defined as DESs. However, they have their own special properties compared with aprotic ionic liquids (AILs). For instance, first, the synthesis of the eutectic mixture is simple. The pure state can be obtained only through mixing the two components together mechanically. Second, in a eutectic mixture, many

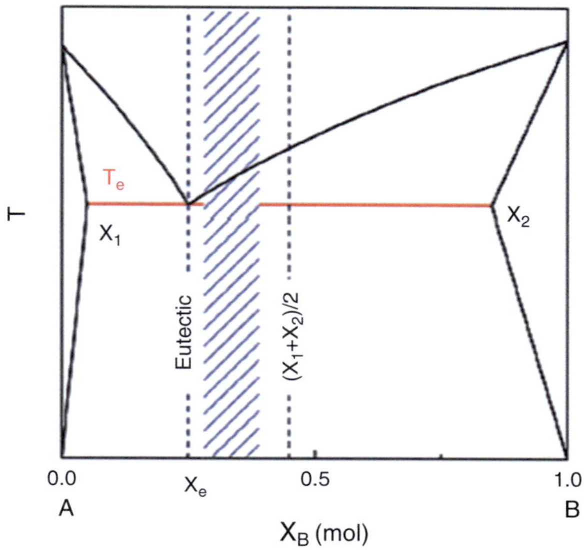

Fig. 7.3 Schematic plot of the best glass-forming composition in a simple binary eutectic system

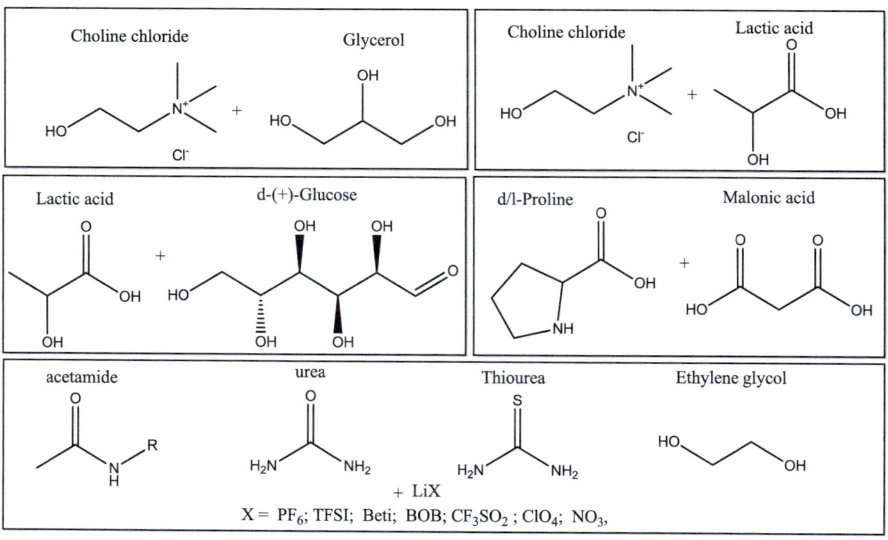

Fig. 7.4 Common DESs and NADESs

inorganic and organic chemicals are more soluble than in water. Even some "insoluble" metal oxides have been found to have significant solubility in a eutectic mixture. Third, it does not react with water and it is biodegradable, like the choline chloride/urea eutectic mixture that has very low toxicity [31]. Figure 7.4 shows the most common DESs and natural DES (NADESs).

Similar combinations are also observed in plant secretions, and plants in drought or cold conditions. In fact, the ingredients for natural ILs and DESs are abundant in organisms [33]. Young et al. demonstrated in an exhaustive description of the NADESs that these compounds play important physiological roles as a third medium polar liquid in living cells and organisms [34]. For example, flower nectar liquid is composed mainly of sugars that are individual in their solid form at room temperature; however, in liquid form, the composition of this sugar mixture is similar to the proposed composition, a fructose–glucose–sucrose NADES [34], or to the components found in honey, composed of glucose and fructose. NADES components were also observed in desert plants of the *Selaginella* species and also in microorganisms, lichen, and various other organisms that can survive longer periods without water.

During the last decade, a series of novel and cheap complex systems based on lithium salt (LiX; X = [$N(Tf)_2$], [ClO_4], [CF_3SO_2]) and organic compounds with an acylamino group (Urea, acetamide, glycol) [23, 33, 35–38] have been studied. A timeline of reported developments, both on applications and fundamental studies of deep eutectic solvents, is summarized in Fig. 7.5. The data indicate that complex systems are believed to be promising electrolytes for electrochemical devices due to their excellent thermal stability, low viscosity at room temperature, high ionic conductivity, and broad electrochemical stability. The Li[$N(Tf)_2$]–acetamide composite was formerly chosen as the electrolyte applied in the energy storage application by Chen [39]. Thereafter, Anouti et al. demonstrated that *N*-methylacetamide (MAc) was able to decrease the melting point of a complex system containing a molten salt due to its "water-like" physical properties, e.g., a very high-dielectric constant and dipolar moment [40] can be considered as a good H-bond donor for the formulation

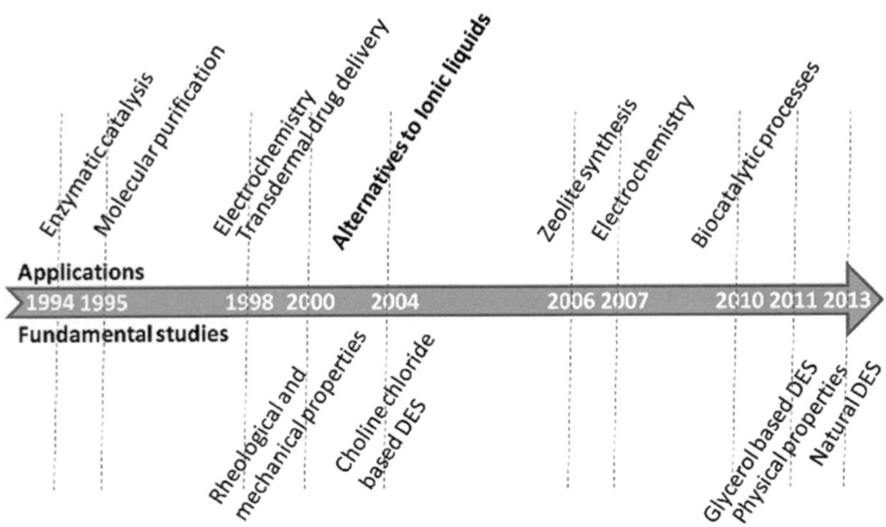

Fig. 7.5 Timeline of reported developments, both on applications and fundamental studies of deep eutectic solvents

of new DES-based electrolytes [40, 41]. We also demonstrated that each formulated DES-based electrolyte containing the MAc and a LiX salt (X=[NO_3], [$N(Tf)_2$], or [PF_6]) allows the realization of energy storage systems as LIBs or EDLC devices that show promising performances even at 80 °C. Interestingly, a discussion for the mechanism action of such electrolytes was centered on the mechanism of pore accessibility for non-solvated lithium ions in these media.

7.2.3 Natural Ionic Liquids

As our mineral resources continue to be consumed at a prodigious pace, it is necessary now that the chemical industry adopts renewable natural sources as starting materials. Natural amino acids [42, 43] and their derivatives provide the most abundant renewable natural chiral pool and can form an efficient, practical, and facile precursor for the preparation of chiral compounds [44–47]. Nevertheless, current studies have also raised doubts over whether traditional ILs are genuine green reagents. For example, some ILs are volatile, combustible, and toxic and could exert negative effects on the environment [48]. Rogers and Voth reported that the decomposition of some ILs could form corrosive HF [49]. As far as we are concerned, the development of biological (Fig. 7.6), environmentally compatible, or biodegradable ILs or Natural Ionic Liquids (NILs) is of paramount importance [50].

At the same time, it is hard to prepare high-purity RTILs by conventional anion exchange reactions because the by-products, such as metal salts, were previously difficult to eliminate until pure RTILs were prepared by neutralization reactions of tertiary amines and organic acids. Furthermore, task-specific [51] and biodegradable [52–56] ionic liquids have an attractive future in green chemistry because their utilization will greatly reduce the risk to the environment.

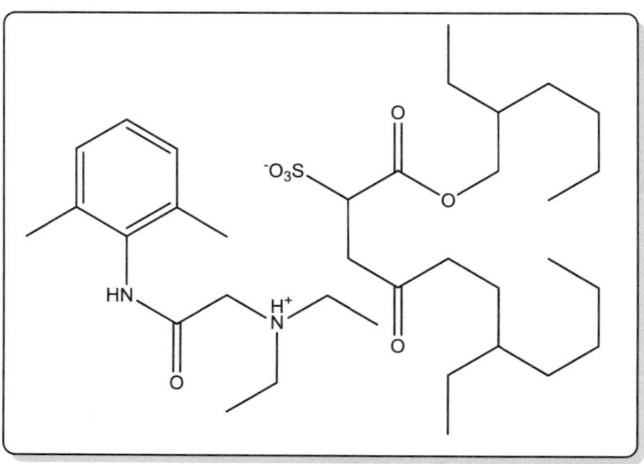

Fig. 7.6 Lidocaine docusate: biological ILs, the third generation of ILs

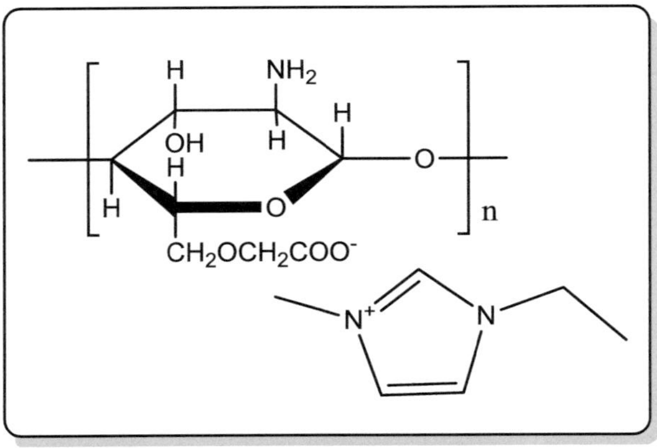

Fig. 7.7 Biological ILs based on chitosan for biosensors and biocatalysis application

NILs chiral and biodegradable ionic liquids can be synthesized from natural amino acids [45]. Some characteristics of these RTILs, such as good miscibility with organic solvents, high ionic conductivities, and low glass transition temperatures (Tg), have been reported. For example, Chitosan is obtained by full or partial deacetylation of chitin (Fig. 7.7).

Carboxymethylated chitosan (CM-chitosan) has a similar molecular structure with amino acids and can be obtained by carboxymethylated reactions between chitosan and monochloroacetic acid [57]. Chitosan and CM-chitosan are attractive nontoxic, biocompatible, biodegradable, and film-forming biomaterials that can be used as biosensors and biocatalysis [58]. Despite a composite system based on chitosan and RTILs for electrochemistry applications presented by Lu et al. [59], to date, there are no other reports on the synthesis of ionic liquids composed of natural oligosaccharides. Furthermore, it will be helpful to improve the biocompatibility of ionic liquid by introducing CM-chitosan as an anion compound. Recently, RTIL based on 1-ethyl-3-methylimidazolium hydroxide, [C_2mim][OH], and CM-chitosan was prepared by a neutralization reaction (Fig. 7.7) [57].

7.3 Physicochemical Properties of H-donor RTMS

To design any process involving ILs on an industrial scale, especially in electrochemical process, some thermophysical properties must be known, such as viscosity, conductivity, and diffusivity. Furthermore, these experimental data provide useful fundamental information on molecular interactions and can also be applied to evaluate thermodynamic models [60–62]. These models provide better understanding of the nature of the molecular aggregation that exists in binary mixtures. However, the relatively high viscosities of pure ILs and to a lesser degree PILs compared to traditional solvents have resulted in limited commercial use to date. This problem is partially resolved by mixing ILs with water or organic solvents [63–66]. In fact, the main advantage of using binary mixtures (PIL + molecular solvent) is to decrease the viscosity and increase the conductivity of the solution, which leads to most ionic liquids used as electrolytes having conductivities less than those of water with added electrolyte, but comparable to those of organic solvents with added supporting electrolyte (e.g., acetonitrile containing 0.1 M Bu_4NClO_4). Since ILs are composed of ions, it is frequently assumed that they should have high conductivities. However, conductivity depends upon both the number of charge carriers and their motilities. The high viscosity and large size of the constituent ions of ILs result in reduced ion mobility and moderate conductivities.

7.3.1 Transport Properties and Ionicity Classification

7.3.1.1 Conductivity

Conductivity determines the level of ohmic drop in voltammetric experiments. If the conductivity is too low, the voltammograms will be highly distorted. Among the most important properties of ILs is their conductivity for potential applications as electrolytes in electrochemical devices [67–69]. In the binary system where ILs are mixed with a molecular solvent (PIL + MS), solvent molecules separate the ions, and the mean distance between them depends on the IL concentration. The thermodynamic properties of such solutions are described by ion–solvent, ion–ion, and solvent–solvent interactions. This leads to various well-defined species present in the solution, such as free ions, complex ions, neutral ion pairs (ions of opposite sign are separated by the solvent molecule), or nondissociated neutral salt molecules (ions of the opposite sign are not separated by the solvent molecules).

As PILs can be mixed with molecular solvents, this provides new opportunities to study the conductivity of electrolyte solutions in salt-rich regions up to 15 mol. L^{-1}. Numerous studies concern the specific conductivity and viscosities of pure PILs [70–73], since it is desirable and of great importance to understand and predict the transport properties of PIL + molecular solvent solutions using classical and theoretical models.

The observed shape of the conductivity curve as a function of the solvent mole fraction follows the same trend for protic, aprotic, or deep eutectic solvents, as shown in Fig. 7.8 [74], which shows the change in conductivity following the addition of the solvents to $[S_{111}][N(Tf)_2]$, $[HN_{111}][N(Tf)_2]$, and $Li[N(Tf)_2]$ at 25 °C, respectively.

In this figure, the influence of the addition of the organic solvent, acetonitrile (ACN), gamma butyrolactone (γ-BL), and propylene carbonate (PC) is shown. The conductivity of each binary mixture increases quickly to reach a maximum conductivity in the solvent-rich region. Furthermore, it appears that all mixtures containing ACN have higher conductivity than those based on γ-BL; at the same time, all investigated PC solutions showed three times lower conductivities compared to ACN.

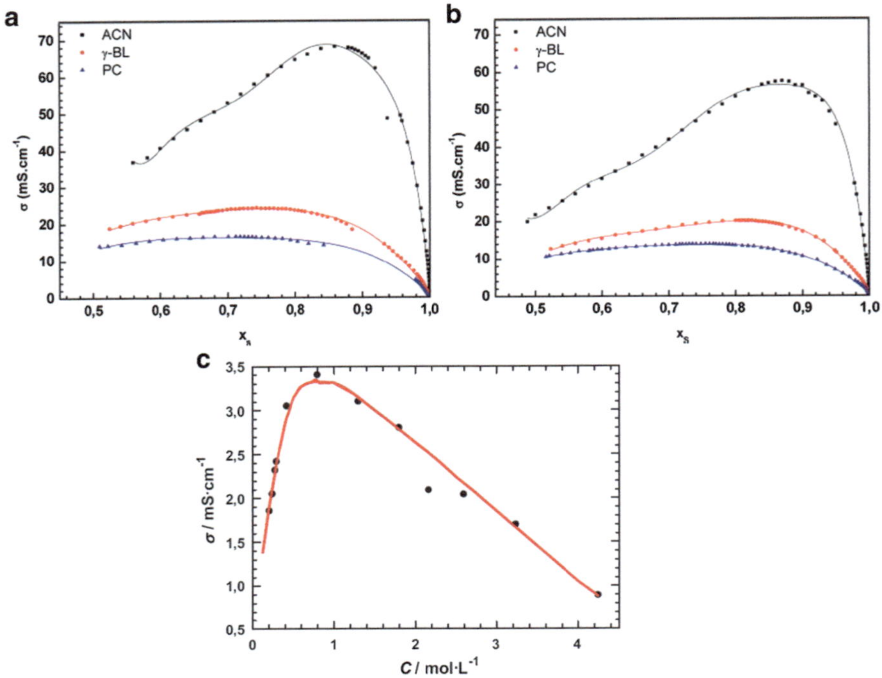

Fig. 7.8 Change in conductivity following the addition of solvents to $[S_{111}][N(Tf)_2]$ (**a**), $[HN_{111}][N(Tf)_2]$ (**b**), and $Li[N(Tf)_2]$ (**c**) at 25 °C

Generally, PILs have higher conductivities compared with those based on the homologue AIL. This may be due to the contribution of labile hydrogen driven by the Grotthuss mechanism [75]. In the case of DESs (Fig. 7.8c) [40], the strong interaction between N-methylacetamide (NMA) and the ion pair explains ion mobility and conductivity. The similarity between the graphs $\sigma = f(xs)$ reported for each RTMS indicates that the mechanisms of the charge transport must be similar for all investigated IL + solvent binary mixtures. When the solvent is added to an IL, the charge density, as well as the viscosity of the solution, is reduced. At solvent concentrations close to the maximum conductivity, the effect of dilution by the solvent is dominant.

7.3.1.2 Viscosity

Viscosity is influenced by several parameters. With respect to the anion, higher basicity, size, and capacity to form hydrogen bonds increase the IL viscosity. Moreover, due to the acid/base character of PILs, their viscosity is highly dependent on hydrogen bonds between ions [76–78]. In general, AIL viscosity is higher than their homologue PILs. The viscosity at 25 °C of the $[HN_{222}][N(Tf)_2]$ has been

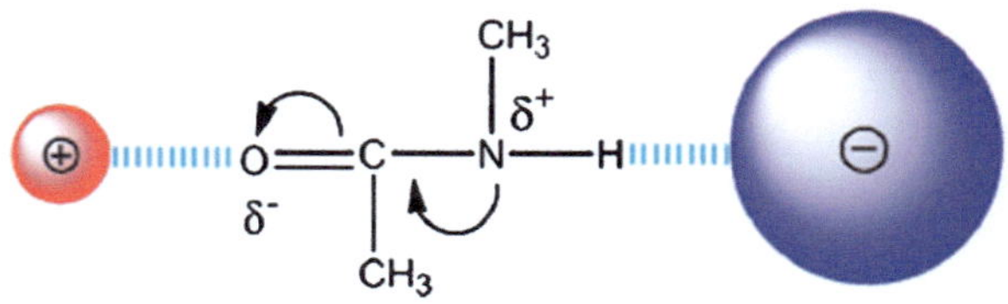

Fig. 7.9 Schematized intermolecular interactions between the NMA and ions in DESs. Reproduced from [41] with permission from the PCCP Owner Societies

reported as ~ 39 mPa·s [74]. This value is similar to that value measured for $[S_{222}][N(Tf)_2]$ [74, 79] (33 mPa·s at 25 °C). Both sulfonium AIL and ammonium PIL with homologous structures have lower viscosities than those reported for pyrrolidinium or imidazolium $[N(Tf)_2]^-$-based ILs, which have viscosities generally between 50 and 100 mPa·s at 25 °C [80].

At 25 °C, the viscosity values of different DESs reported in the literature range from 50 mPa s to 5,000 mPa, for example, viscosity up to 500 mPa·s is reported in the case of the equimolar oxalic acid/choline chloride DES system [23, 81, 82]. In general, at 25 °C the DES viscosity values are similar to those observed in the case of classical ILs. In other words, DES viscosity is high and is strongly affected by the selection of the solvent and salt driven by their structures that determine the nature of intermolecular interactions between the molecules and ions in solution as shown in Fig. 7.9 [41].

For example, the high viscosity of the acetamide-based DESs is mainly due to the Coulombic interactions between the Li^+ cation and the carbonyl group, as well as between the amide group and the $[N(Tf)_2]$ anion (Fig. 7.9). The relatively high viscosity, compared to traditional solvents, can also be interpreted as a decrease in conformational degrees of freedom in solution due to the stable trans-structure of the N-methylacetamide. Takano and Nakamura showed that the trans-state is more stable than the cis-state by 2.8 kcal mol^{-1}. Thus, van der Walls interactions are more stable, reinforcing the friction between species [83]. The influence of the addition of the organic solvent, acetonitrile (ACN), gamma butyrolactone (γ-BL), and propylene carbonate (PC), on the rheological properties of ILs can be analyzed by the rheological approach. In general, the viscosity of electrolytes varies with concentration C according to the Jones–Dole equation [84, 85]

$$\frac{\eta}{\eta_0} = 1 + A\sqrt{C} + BC \tag{7.2}$$

where η and C are, respectively, the viscosity and concentration of the solution and η_0 is the viscosity of the solvent. *A* and *B* are Jones–Dole coefficient values: the slope (*B*) and ordinate (*A*) are deduced from a plot of the reduced viscosity $\frac{\left(\frac{\eta}{\eta_0} - 1\right)}{\sqrt{C}} = f\left(\sqrt{C}\right)$. *A* is generally independent of concentration, *C*, and *B* is related to the effect of the ions on the structure of the solvent. *B* is generally interpreted as a measure of the structure-forming and structure-breaking capacity of an electrolyte in solution [86–88]. A positive value of the *B* coefficient indicates a structure-forming effect, and a negative value shows a structure-breaking effect. In a comparative study concerning sulfonium AILs and ammonium PILs, negative B values were obtained for both AILs and PILs in mixture with γ-BL and ACN, while positive B values was obtained in the case of ILs/PC mixtures [74]. This result can be interpreted as a structure-breaking effect on the addition of these ILs to γ-BL and ACN and as a structure-forming effect on the addition of these ILs to PC. In the same solvent, this effect is more pronounced with the aprotic IL $[S_{111}][N(Tf)_2]$ than with the protic $[HN_{111}][N(Tf)_2]$. Therefore, the structure-forming capability of the sulfonium cation is higher than that observed for the equivalent ammonium IL. This is probably due to the size of the sulfonium cation and to the fact that the sulfur d-orbitals may be able to interact with the solvent through a coordinative bond. Finally, for each salt, the *B* values vary according to the nature of the solvent in the following order: PC > γ-BL > ACN.

The addition of water in PIL solutions decreases drastically the viscosity of the media compared to the viscosity of pure PILs. For example, the viscosity of pure $[HPyr][HSO_4]$ is close to 156.3 cP, whereas when mixed with water, the equimolar composition has a viscosity close to 33 cP. Additionally, for high concentrations of PILs ($x_{IL} > 0.5$), anion choice also affects dramatically the solution viscosity [89]. This effect is driven by the fact that the HSO_4-based PIL is around nine times more viscous than PIL containing the $[CF_3CO_2]$ anion. For a lower PIL mole fraction, such a difference tends to decrease, since the viscosity of each aqueous solution tends to converge toward the viscosity of pure water. The dependence of the viscosity deviations from ideality, η^E, ($\eta^E = \eta - (x_1\,\eta_1 + x_2\,\eta_2)$) as a function of the PIL mole fraction composition, x_{PIL}, expresses the modification of ionic association by the addition of water (7.2) in PILs (7.1).

For example, Fig. 7.10 shows that η^E is positive over the entire composition range in the case of the ($[HPyr][CF_3CO_2] + H_2O$) binary mixture, with a small amplitude, whereas, for the ($[HPyr][HSO_4] + H_2O$) binary mixture, the deviation is negative with a high amplitude 14 times higher than the maximum viscosity deviation value reported for the fluorinated anion [89]. In fact, the strong ionic association and the hydrophobicity of the fluorinated anion could explain the low impact of the addition of water on the viscosity of the trifluoroacetate-based PIL solution. In contrast, the acidic equilibrium of $[HSO_4]$ anion and water promotes a drastic change in viscosity.

7.3.1.3 Ionicity and Diffusion

One way of assessing the ionicity of ILs, RTMSs, or DESs is to use the classification diagram based on the Walden rule [90, 91]. The Walden rule relates to the ionic mobility represented by the equivalent conductivity Λ ($\Lambda = \sigma/C$) to the fluidity η^{-1} of the medium through which the ions move [71, 92]. Figure 7.13 shows the variation of log (Λ) versus log(1/η) at various temperatures from 20 to 80 °C in the case of the $Li[N(Tf)_2]$/MAc system at x $Li[N(Tf)_2] = 1/4$ compared to those reported by our group in the case of the pyrrolidinium nitrate PIL $[HPyr][NO_3]$ and its mixture with

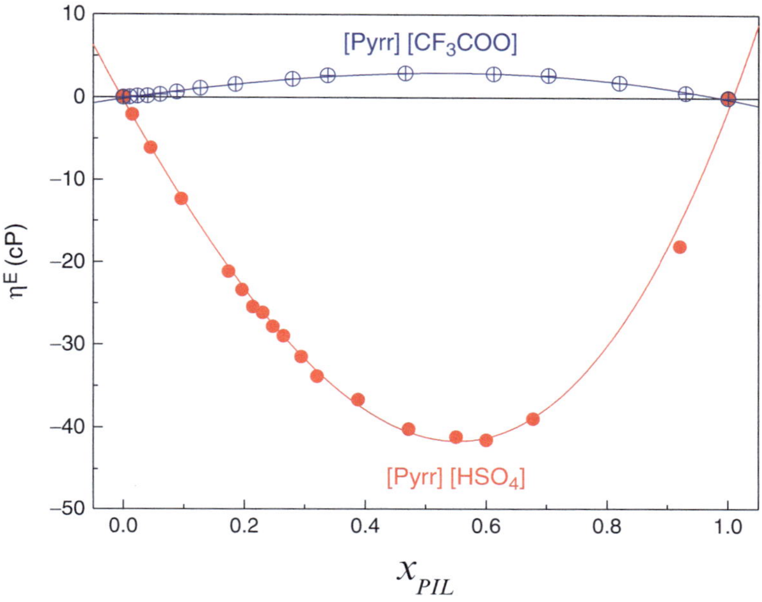

Fig. 7.10 Viscosity deviations from ideality, η^E, as a function of the PIL mole fraction composition, x_{PIL} for the aqueous mixture [HPyr][HSO_4] and [HPyr][CF_3CO_2] ([HPyr] = [Pyrr]). Reproduced from [89] with permission from Elsevier

propylene carbonate [40]. The ideal line is obtained on the basis that ions have mobility that is determined only by the viscosity of the medium and that the number of ions present in the equivalent volume is that indicated by the salt composition, i.e., all ions contribute equally. The position of the ideal line is established using aqueous KCl solutions at a high dilution. The results obtained in Fig. 7.11 indicate that the LiX-based DESs are a superionic liquid and these solutions become more "ionic" according to the Walden classification with temperature [40]. This observation can be linked, as discussed above, to the low pseudo energy of the viscous flow value associated with this solution, which further suggests the predisposition of this mixture for use as an electrolyte for energy storage application.

Several authors have then used self-diffusion coefficients determined by Pulsed-gradient spin-echo NMR (PGSE-NMR) to calculate the molar conductivity (Λ_{NMR}) according to the Nernst–Einstein equation [63, 93, 94]

$$\Lambda_{NMR} = \frac{F^2\left[D_{C^+} + D_{A^-}\right]}{RT} \tag{7.3}$$

where F is the Faraday constant, R is the universal gas constant, and D_{cation} and D_{anion} are the self-diffusion coefficients obtained by PGSE-NMR from the translational motion (self-diffusion) of the NMR-sensitive nuclei observed. Thus, the Λ_{NMR} values were derived from the assumption that all diffusing species detected during the

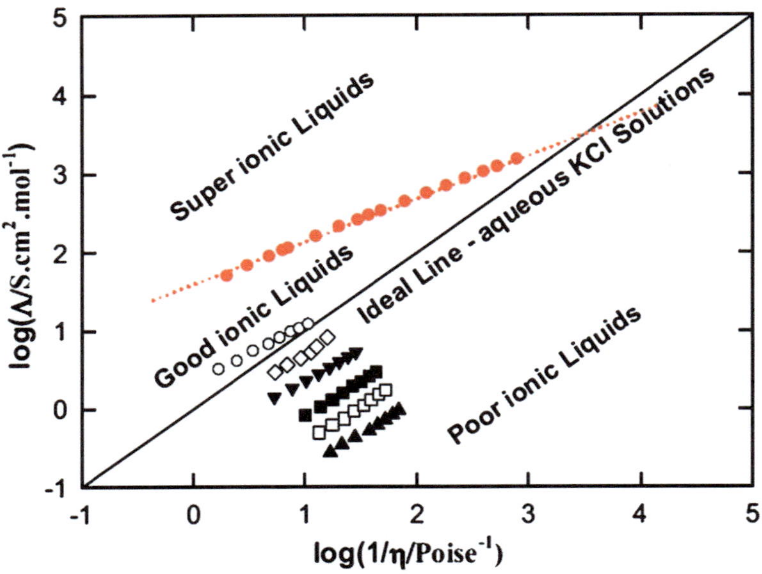

Fig. 7.11 The Walden rule for the ionicity of Walden plot log(Λ) versus log(η^{-1}) of the Li[N(Tf)$_2$]/MAc DES at $x_{LiTFSI}=1/4$: *filled circle* and Pyrrolidinium nitrate protic ionic liquid [HPyr][NO_3] and its mixture with propylene carbonate ([HPyr][NO_3]+PC) binary mixtures as a function of the temperature at $x_{PIL}=0.25$: *filled triangle*, 0.10; *open square*, 0.22; *filled square*, 0.30; *Reverse filled triangle*, 0.50; *open diamond*, 0.80; *open circle*, 1. Reproduced from [40] with permission from Elsevier

PGSE-NMR measurements contribute to the molar conductivity. On the other hand, the Λ_{cond} values can be evaluated by the migration of charged species under an electric field. In the presence of neutral ion pairs, the Λ_{NMR} values were expected to differ from those determined by conductivity (Λ_{cond}) [75]. In order to calculate the molar conductivity, it was necessary to take into account all mobile species: the mobile proton (H^+), the cation (C^+), and the anion (A^-) as in the following equation:

$$\Lambda_{NMR} = \frac{F^2}{RT}\left[(1-\alpha)D_{C^+} + \alpha\left(D_{H^+}\right) + D_{A^-}\right] \tag{7.4}$$

where T is the absolute temperature, F is Faraday's constant, and α a parameter ranging from 0 to 1 according the dissociation of pyrrolidinium cation.

For the sake of comparison, the experimental conductivities Λ_{cond}, as well as Λ_{RMN} for anion, cation, and proton for the ([HPyr][CF_3CO_2]+H_2O) binary system, have been also plotted in the same graph, illustrated in Fig. 7.12 [75]. We can be shown that, at a low composition range (i.e., $ww<50$ % w/w), irrespective of the choice of the α value (e.g., $\alpha=1$ or $\alpha=0$), Λ_{cond} was found to be similar to the calculated Λ_{NMR} values. In fact, in this case, ions provide the major contribution to diffusion in the solution. However, for a higher composition range (i.e., $ww>50$ % w/w), it is not the case since certain discrepancies have been found between the calculated

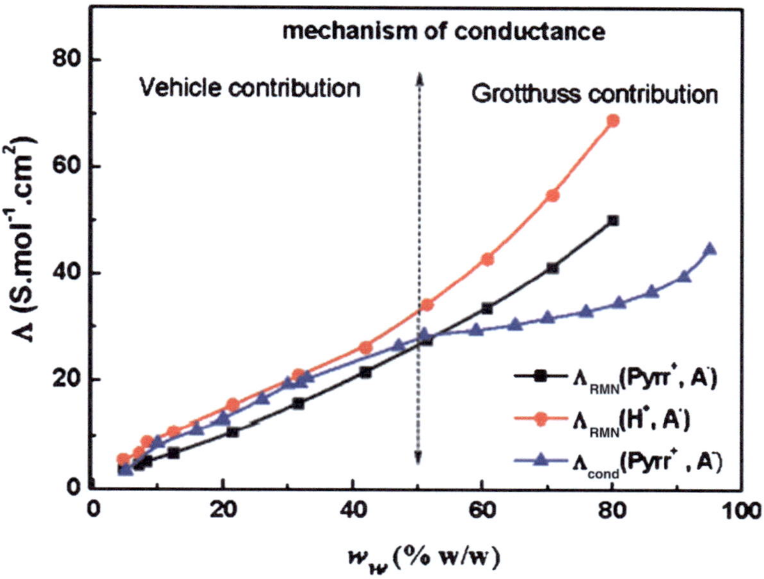

Fig. 7.12 Experimental conductivities Λ_{cond}, as well as calculated Λ_{RMN} for anion, cation, and proton for the ([HPyr][CF_3CO_2]+H_2O) binary system as function of mass fraction. Reproduced from [75] with permission from Elsevier

and experimental diffusion values. In other words, for such concentrations, diffusion in solution is certainly driven by the ion pairs, as well as by neutral species, which renders the Nernst equation invalid for the high composition range (i.e., *ww*>50 % w/w). The cation self-diffusivity was then decomposed into two contributions, which corresponded to the cation self-diffusivity contributions of the single ion (D_{Pyrr+}) and the ion pair (D_{PyrrA}), respectively. Apparently, at a low composition range, both corresponding curves are almost similar because the contribution of the Grotthuss is low [95].

7.3.2 Structural Properties and Application to Materials Chemistry

7.3.2.1 Self-Assembled Structures in PILs

In recent years, it has been established that many ILs are nanostructured, possessing populations of cations and anions that are ordered on specific length scales [96]. This nanostructure typically occurs due to segregation of the ionic liquid into polar and nonpolar domains. Ionic liquid cations usually consist of an ionic and an alkyl component which, similar to amphiphile self-assembly, drives the segregation.

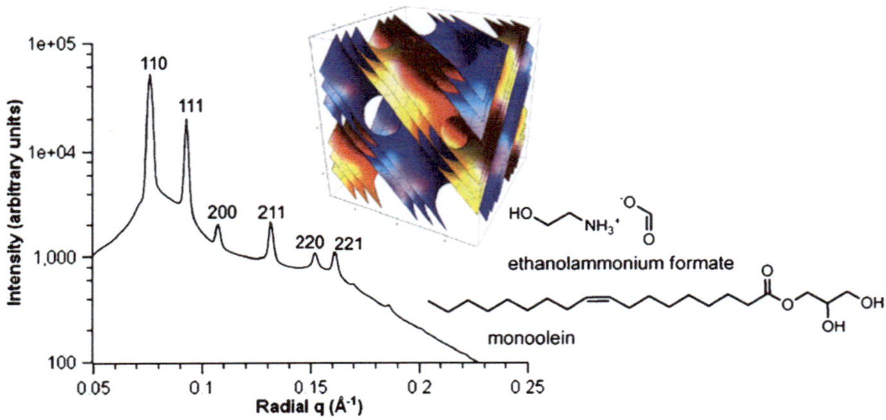

Fig. 7.13 DRX profiles for the double diamond (Pn3m) symmetry system formed by the monoolein (50.8 wt%) / ethanolammonium formate $[H_3N(CH_2CH_2OH)][HCO_2]$ system at 25 °C

Meanwhile, self-assembly in ILs and their application to materials chemistry have been emerging as a promising area of research. Although ILs exhibit smaller cohesive energy compared to water [97, 98], the formation of ordered molecular assemblies has been demonstrated for bilayer membranes [60, 99–101] and low-molecular weight ionogels [102]. PILs and water mixtures have been shown to display unique physicochemical properties, which are different from the pure counterparts [103–105]. They relate to the formation of solvent clusters and continuous superstructures [106] in binary aqueous IL mixtures, which vary depending on the component chemical structures of IL, composition of the binary mixtures, and physical conditions, such as temperature.

Several PILs, including ethyl ammonium nitrate, have been found to have intermediate range ordering, consisting of polar and nonpolar domains analogous to those observed in amphiphilic self-assembly processes. Greaves et al. reported an absence of such structure in ethanolammonium formate, $[H_3N(CH_2CH_2OH)][HCO_2]$, and correlated this lack of ordering of the ions as an influential factor in small molecule amphiphile self-assembly [107, 108]. In contrast, highly structured PILs have an increased propensity toward amphiphile solubilization as show in Fig. 7.13. By consequence higher concentrations of amphiphile are required to form self-assembled structures.

Observed changes in the physicochemical and micellar properties of the pyrrolidinium imidazolium or alkylammonium alkylcarboxylates protic ionic liquid surfactants in mixtures with water can be linked to the nature of the cation. Figure 7.14 shows, for example, lamellar patterns bilayer system of diisopropylammonium octanoate in water (50 % w/w). By comparison with classical anionic surfactants having inorganic counterions (Li^+, Na^+...), these monomers exhibit a higher ability to aggregate in aqueous solution, demonstrating their potential applicability as a surfactant.

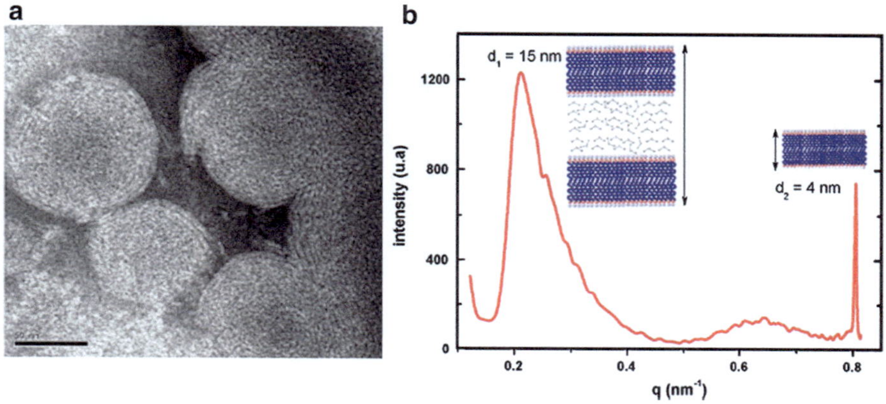

Fig. 7.14 (**a**) TEM of $[NHEt(i\text{-}Pr)_2][C_9H_{19}CO_2]$ in water (50 % w/w); bar = 50 nm. (**b**) The Bragg peaks by SAXS of lamellar patterns bilayer system

7.3.2.2 Application of PILs Structuring Reductive Media to Nanoparticle Synthesis

Understanding of ordered self-assembling phenomena in binary PIL–water systems has been largely lagging behind. The formation of clusters and continuous structures of PIL molecules in aqueous systems and their influence on molecular self-assembly are of pivotal importance not only for fundamental science but also for their wide range of technological applications.

For example, thanks to their "organized" and tunable structure, ILs-based media have already been reported in the literature as predisposed to the synthesis of different nanomaterials [109] including silver, gold, ruthenium, iron, and platinum nanoparticles [110].

Recently, a facile method for the formation of mono-dispersed metal nanoparticles (NPs) at room temperature from $M^{(III)}Cl_3$ (with M = Au, Ru, Mn, Fe, or V) in different media based on N,N-dimethylformamide (DMF) or water solutions containing a PIL, namely the octylammonium formate and the bis (2-ethyl-hexyl)ammonium formate, was presented [111, 112].

These PILs present redox-active structuring properties that influence their interactions with selected molecular compounds (DMF or water), as well as the shape and the size of formed metal NPs in these solutions. Figure 7.15 shows the different shape and size of gold and Pt NPs obtained in a water–octylammonium formate mixture.

By the same approach employing low-energy sputtering, several groups generated novel electrolytes comprising nanoparticle dispersions within ionic liquids [113]. These studies also afforded the unique opportunity to investigate nanoscale growth mechanisms occurring within the IL. Cyclic voltammetry and electrochemical impedance spectroscopy analyses revealed that when the IL contained a substantial fraction of sub-nanometer-sized particles, the double-layer capacitance was increased

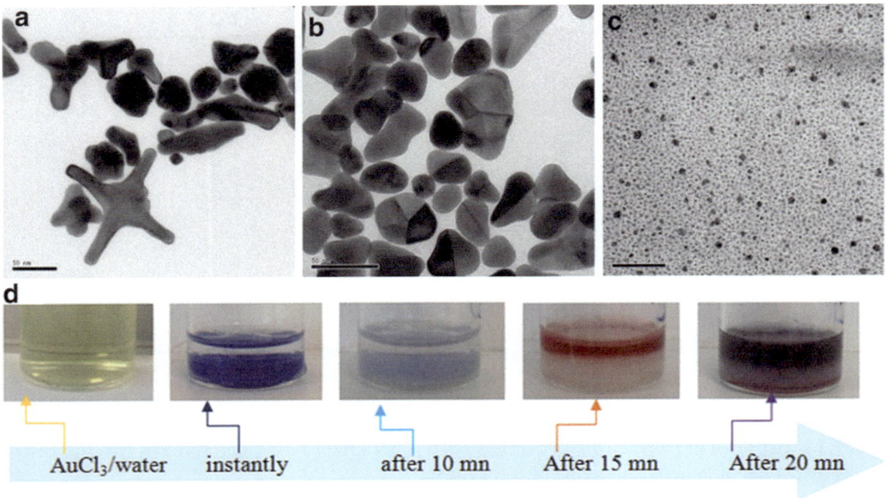

Fig. 7.15 Mono-dispersed colorless nanoparticles. (**a**) Gold NPs in aqueous bis(2-ethyl-hexyl) ammonium formate. (**b**) Gold NPs in octylammonium formate (OAF). (**c**) Vanadium NPs in bis(2-ethyl-hexyl)ammonium formate + dimethylformamide. (**d**) Pictures showing the various stages of the gold NPs' formation by adding fresh solution of $AuCl_3$ dissolved in DMF and bis(2-ethyl-hexyl)ammonium formate + water solution over the time

by close 200 %, concomitant with a bulk electrolyte resistance decrease of 70 % with respect to a gold-free control. These promising results reveal an anomalous capacitance increase and low internal resistance for nanoparticle-in-IL dispersions, suggesting intriguing potential as electrolytes for next-generation Electrical Double-Layer capacitors (EDLCs), fuel cells, and sensors. A very large potential development in the nanomaterial domain is expected in the PIL structured domain.

7.3.3 Electrochemical Stability and Metal Protection Against Corrosion in RTMS

Electrolytes for electrochemical devices should resist reduction and oxidation and exhibit a larger electrochemical window. The electrochemical stability of RTMS is generally measured by cyclic voltammetry at room temperature recorded in a three-electrode cell configuration using a platinum, gold, or vitreous carbon electrode as the working electrode, and Ag/AgCl (saturated in solvent) for triethylammonium formate PILs (Fig. 7.16a) or Li for DESs based on lithium salt (Fig. 7.16b) as the reference electrode.

The stability window of PILs ranges from 1.5 to 5.0 V according to the acidity of ammonium cation and the stability of the anion. In general, reversible and small peaks observed in the case of the PILs-based system, as shown in Fig. 7.18, is attributed to the presence of water (>1,000 ppm).

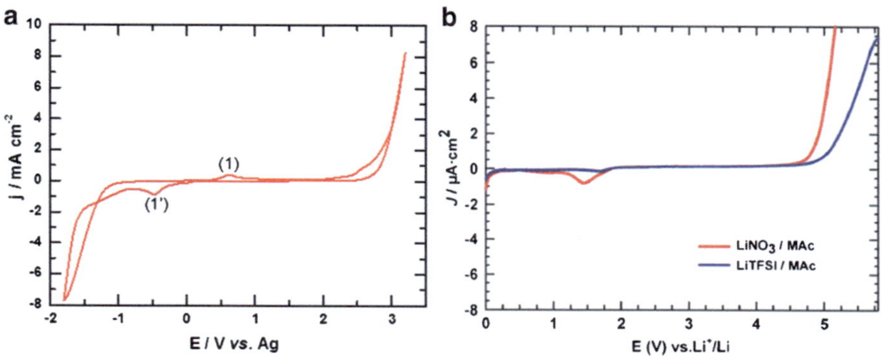

Fig. 7.16 Electrochemical stability of RTMS measured by cyclic voltammetry. Triethylammonium formate PILs (**a**) and DESs based on lithium salt (**b**)

The cathodic limit observed in an ionic liquid is generally due to the reduction of the ammonium cations [114, 115]. Such a reduction proceeds by the CE mechanism: first, the deprotonation of ammonium cation occurred (Eq. 7.5), followed by the reduction of the proton (Eq. 7.6). For the anodic limit, the ammonium could, in this case, be oxidized in the presence of oxygen.

$$R_3NH^+ \leftrightharpoons Et_3N + H^+ \tag{7.5}$$

$$H^+ + le^- \leftrightharpoons 1/2H_2 \tag{7.6}$$

In oxidation, the anion associated with the PILs is oxidized. In the case of carboxylates, the relatively low potential (1.5 V (vs. Ag/AgCl) mechanism proceeds according to the Kolbe reaction followed by the formation of dimers, whereas the oxidation of the alkyl radical at more anodic potential can be leading to products deriving from the carbocation, named the non-Kolbe reaction. The mechanism is described as follows:

$$RCOO^- - le^- \leftrightharpoons R^{\cdot} + CO_2 \tag{7.7}$$

$$2R^{\cdot} \leftrightharpoons R - R \tag{7.8}$$

$$R^{\cdot} - le^- \leftrightharpoons R^+ \leftrightharpoons \text{products} \tag{7.9}$$

The electrochemical stability window of the Lithium-based DESs on Pt is generally from 4.5 to 5.5 V [41]. The cathodic limit shown by these electrolytes was significantly higher, comparable to that observed for aprotic ionic liquid (see Fig. 7.16b). Furthermore, the anodic limit was comparable to electrolytes containing the X$^-$anion for lithium salt LiX as [NO$_3$] or [N(Tf)$_2$] anions. Considering these results, each DES appears to display an electrochemical stability window (ESW) large enough to be used as an electrolyte for the energy storage systems as described below.

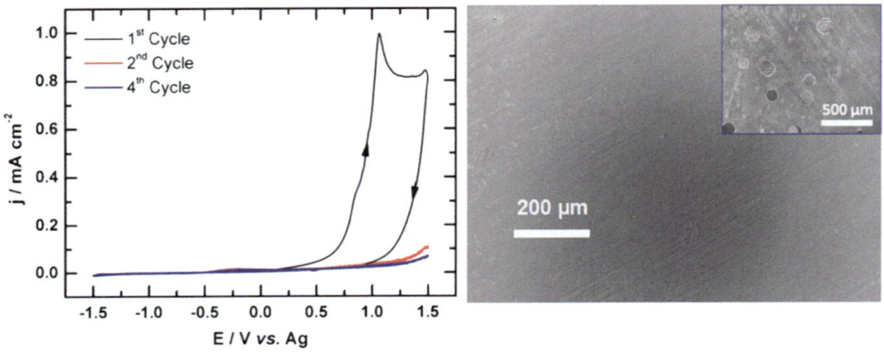

Fig. 7.17 Cyclic voltammograms of the Al electrode in $[HN_{222}][N(Tf)_2]$ at 20 °C (**a**) and scan rate: 5 mV s^{-1}. SEM images of aluminum current collectors after cycling in $Li[N(Tf)_2]$–MAc at $x_{Li+}=0.20$, inset: after cycling in 0.3 M $Li[N(Tf)_2]$ in PC. Reproduced from [41] with permission from the PCCP Owner Societies

The native passive Al_2O_3 layer existing on the metal surface provides protection against corrosion. It is well known that, depending on the nature on the electrolyte anions, this passive layer can be broken down leading to aluminum corrosion at a high potential. The $[N(Tf)_2]$ anion shows a corrosive effect on the aluminum collector, corroded around 3.8 V vs. Li/Li^+ when the $Li[N(Tf)_2]$ electrolyte was contained in lithium-ion batteries (LIBs) or electrochemical double-layer capacitor (EDLC) systems [116–119] with an aluminum-coated positive electrode. Therefore, it is very important before any use of RTMS as an electrolyte to control its effect on aluminum corrosion.

Aluminum's corrosive behavior was evaluated in several PILs, for example, $[N(Tf)_2]^-$-based PILs [120]. Figure 7.17 shows cyclic voltammograms of an aluminum electrode in $[HN_{222}][N(Tf)_2]$ from the first to the fourth cycle [120]. The first cycle illustrates a hysteresis loop initiated at about 0.25 V vs. Ag with a large irreversible current peak (1 mA cm^{-2}) at about 1.1 V vs. Ag. For the second cycle, and the following, the initiation potential of the hysteresis loop was significantly shifted to the more anodic side (1.25 V vs. Ag), while the current sharply decreased (0.1 mA cm^{-2}). Al passivation could be achieved in this PIL electrolyte from the second cyclic voltammogram.

Figure 7.17b shows SEM images of an aluminum current collector of the LFP electrode cycled in a cell containing DESs based on $Li[N(Tf)_2]$ taken at the end of the cycling [41]. As shown, the current collector displayed a quite uniform morphology, and only some residues were visible. In comparison, the inset presents the collector cycled in 0.3 M $Li[N(Tf)_2]$ in PC, displaying holes of different sizes and depth distributed all over the surface of the sample.

These holes were generated from the strong corrosion process occurring during the cycling in electrolytes containing the $[N(Tf)_2]$ anion, which is able to solubilize $Al[N(Tf)_2]_3$ formed at the surface of the collector [117], as illustrated in Fig. 7.18. From these observations, it appears evident that the $Li[N(Tf)_2]$ DESs-based electrolyte displays the ability to prevent aluminum corrosion and can be considered as a suitable electrolyte.

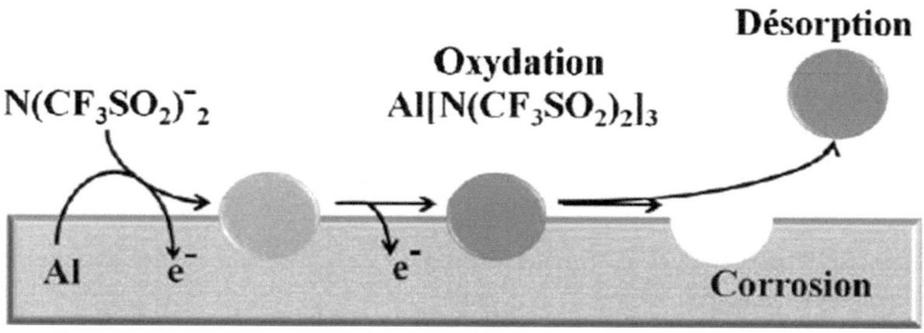

Fig. 7.18 Corrosion mechanism in organic solvent based on $Li[N(Tf)_2]$ as electrolytes. The dissolution and desorption of Al-complex formed not occurring in RTMS

7.4 Application of H-Donor RTMS as Electrolytes in Energy Storage

Accordingly, investments for the exploitation of renewable energy resources are increasing worldwide, with particular attention to wind and solar power energy, which are the most mature technologies. The intermittence of these resources requires high-efficiency energy storage systems. Electrochemical systems, such as batteries and supercapacitors that can efficiently store and deliver energy on demand in stand-alone power plants, as well as provide power quality and load leveling of the electrical grid in integrated systems, play a crucial role in this field. Indeed, the advantage of using electrochemical storage systems has been demonstrated for both wind and photovoltaic technologies [121, 122]. The efficacy of batteries in integrated systems is directly related to their content in energy efficiency and their lifetime. Indeed, in virtue of their high value of energy efficiency, lithium batteries are expected to provide an energy return factor higher than that assured by conventional batteries, e.g., lead-acid batteries [123–127]. Supercapacitors have attracted increasing interest because of their high-power storage capability, which is highly desirable for applications in electric vehicles (EVs) and hybrid electric vehicles (HEVs) [128–133]. Supercapacitors can be coupled with fuel cells or batteries to deliver the high power needed during acceleration and to recover energy during braking. In this section, we describe the latest developments regarding the use of H-Bond donor RTMS application in supercapacitors and batteries focusing on the use of lithium-based DESs and PILs.

7.4.1 *Use of H-Bond Donor for Supercapacitors: Advantage and Limitations*

Electrical double-layer capacitors are well known for their good power performance and extreme stability in terms of life cycle. Activated carbons (ACs) are the most commonly used electrode materials for EDLC applications because of their relatively low cost and large surface area compared to other carbon-based materials. Another factor that influences supercapacitor properties is the selected electrolyte.

Currently, aqueous solutions have been mostly utilized [134]. However, the drawback of the aqueous electrolyte-based supercapacitor is also obvious mainly due to the narrow cell voltage ($\Delta E \approx 1$ V) and low energy $E = \frac{1}{2} C_{sp} \Delta V^2$ (C_{sp} is the specific capacitance and ΔV is the operating voltage). Organic electrolytes and AILs have a higher breakdown voltage, but have greater resistance. The advantages of a higher cell voltage are therefore usually countered by the greater resistance. Another factor limiting the use of aqueous media is temperature. Today, manufacturers typically specify an operational temperature range from −40 to 70 °C for capacitors based on ACN or PC. In this context, PILs have naturally attracted a great deal of attention due to their suitable properties.

7.4.1.1 Application of PILs as Electrolytes for Supercapacitors

PIL-based electrolytes exhibit intermediate conductivity compared to conventional electrolytes (especially in ACN) and AILs, but they might have interesting conductivity at low temperatures [135]. Additionally, some works have indicated that in PIL-based electrolytes, ruthenium oxide [136], as well as activated carbon [137], might exhibit pseudo-capacitance due to the labile proton in PILs, which participates on the fast redox reactions at the electrode/electrolyte interface. Taking into account these properties, PILs could be interesting electrolyte candidates in order to use them as electrolytes for supercapacitors.

Cyclic voltammograms and galvanostatic curves obtained using activated carbon in neat PILs as well as in their mixture (Fig. 7.19) demonstrate that PIL-based

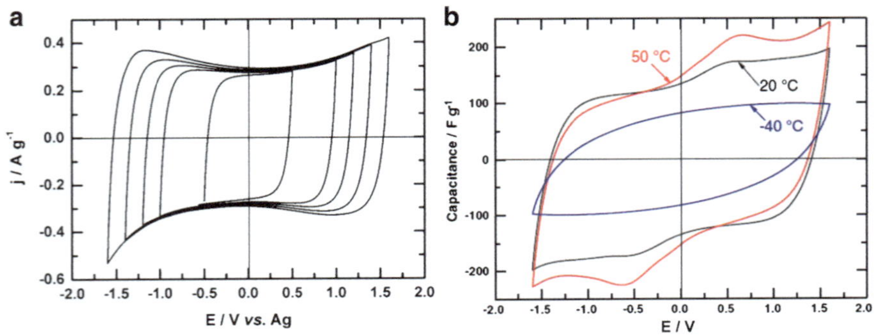

Fig. 7.19 (**a**) Cyclic voltammograms for three-electrode cell with activated carbon in neat $[HN_{222}][N(Tf)_2]$ with Ag wire as a reference electrode at a scan rate of $v = 5$ mV s^{-1}. Reproduced from [120] with permission from the PCCP Owner Societies; and (**b**) In $[HP_{444}][BF_4]$ with ($x_{CH3CN} = 0.86$) at a scan rate of $v = 100$ mV s^{-1}. Reproduced from [139] with permission from Elsevier

supercapacitors might exhibit good cycling stability (over 30,000 cycles) [138] and an interesting performance in a rather broad range of temperatures.

The mixture solvent-$[HN_{111}][N(Tf)_2]$ (solvent=PC, γ-BL, or ACN) can be successfully used at temperatures as low as −30 °C and −40 °C, while the solvent-free supercapacitor containing the PIL $[HN_{222}][N(Tf)_2]$ is able to feature stable performance in the temperature range from 10 to 60 °C. At the same time, when PILs with low-water content are used in combination with AC containing few surface groups, such supercapacitors do not exhibit any evident pseudo-capacitive behavior in comparison with phosphonium-based PILs, as shown in Fig. 7.19b. Taking into account these results, PIL and activated carbon-based supercapacitors appear as stable devices, but, due to their lower operative voltages in comparison to conventional organic solvents, their energy appears lower than that possible using organic electrolytes or aprotic ILs.

7.4.1.2 Parameters Influencing the Storage Mechanism According to PILs

Several parameters can govern the storage mechanism in the EDLC for PILs according to the electrolytes/material couple. First, a reduction of the ammonium cation in the presence of an activated carbon electrode is presumably linked to electronic transfer kinetics. Reactions in which electron and proton transfer is performed have been described either by two distinct steps, Electron Proton Transfer or Proton Electron Transfer (EPT or PET), or in the same concerted step, Concerted Proton Electron Transfer (CPET) [140–142]. Contrary to simple proton or electron transfer, CPET is more complicated and the coupling at the activated carbon /molecule (cation) interface influences the process both thermodynamically and kinetically.

In a concerted mechanism (Fig. 7.20), the electrons and protons are transferred in the same step; this mechanism is defined by only one transition state,

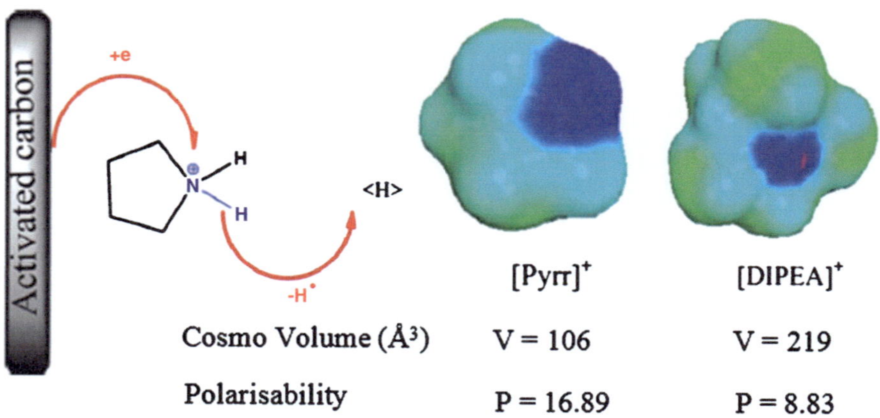

Fig. 7.20 Concerted mechanism of electrons and protons transferred to the activated carbon linked to the Cosmo volume and (Pyrrolidinium (HPyr) or diisopropylethylammonium $[NHEt(i\text{-}Pr)_2]$ cation polarizability in PILs

and consequently, it avoids high-energy reaction intermediates. However, microscopically, the kinetic description is not the same for both processes. The electron transfer is governed by both intramolecular cation and also solvent reorganization. Consequently, potential energies of reactants (here, cations) and products (here, hydrogen and amine) can be described by the internal deformations (variation of both bond lengths and angles) and external structuration of systems (solvent reorganization) as illustrated in Fig. 7.20.

Two other parameters influence the electrolyte behavior in the EDLC system: the van der Waals volume of ions, which must adapt itself to the material porosity, and their polarizability (Fig. 7.22). These two parameters are correlated, e.g., the volume of the [NHEt(i-Pr)$_2$] cation (207.37 Å^3) is twice that of the pyrrolidinium cation (106.01 Å^3), but at the same time, [NHEt(i-Pr)$_2$]$^+$ displays higher polarizability (Fig. 7.22), which allows better distortion in order to adapt its size to the carbon pore size [143]. Thus, [NHEt(i-Pr)$_2$]$^+$ counterbalances its larger volume by its higher polarizability. This believable hypothesis linking reduction stability limit to the geometry and the volume of cations could explain the different behavior observed in Fig. 7.21.

7.4.1.3 Specific Behavior of DESs as Electrolytes for Supercapacitors

This "superionicity" behavior of DESs based on N-methylacetamide and lithium ion used as electrolytes for EDLC at an activated carbon explains their good cycling ability (Fig. 7.22a), observed by cyclic voltammetry and galvanostic charge–discharge methods, with high up to 380 F g^{-1} at elevated voltage and temperature, i.e., $\Delta E = 2.8$ V and 80 °C for the Li[N(Tf)$_2$]–MAc mixture at, for example, $x_{Li} = 0.25$ (Fig. 7.22b) [144].

The electrochemical resistances ESR (equivalent series resistance) and EDR (equivalent diffusion resistance) evaluated using electrochemical impedance spectroscopy (EIS) measurements clearly demonstrate that according to the nature of the anion, the mechanism of ion adsorption can be described by pure double-layer adsorption at the specific surface, or by the insertion of desolvated ions into the

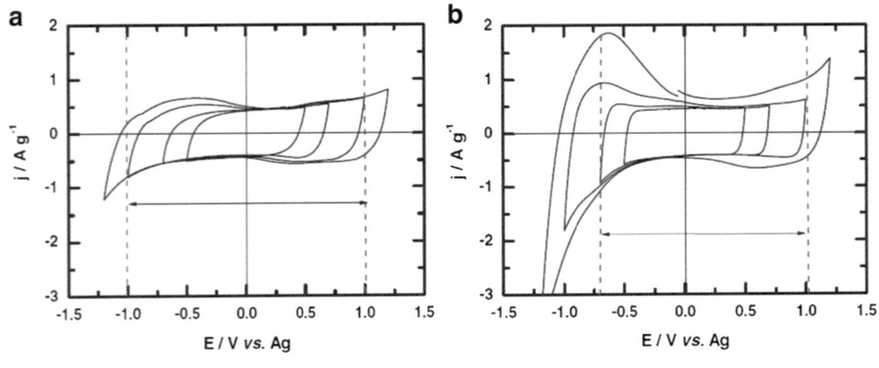

Fig. 7.21 Three-electrode cyclic voltammograms of a PVDF-AC electrode in [NHEt(i-Pr)$_2$][N(Tf)$_2$] (**a**) and [HPyr][N(Tf)$_2$] (**b**), with Ag wire as pseudo-reference, at 5 mV s^{-1}, at 30 °C

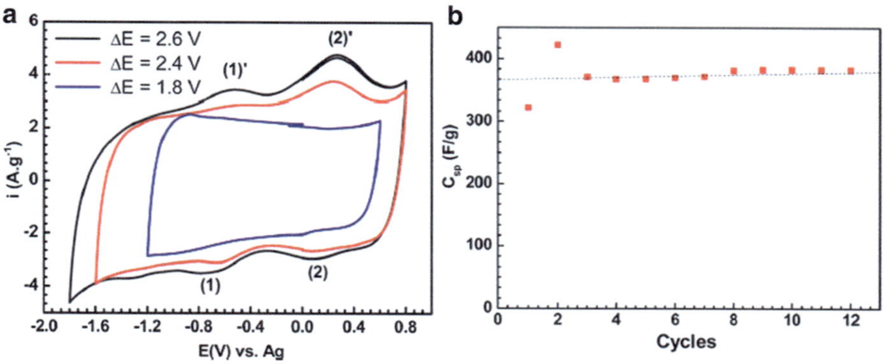

Fig. 7.22 (**a**) Three-electrode cyclic voltammograms of activated carbon in $Li[N(Tf)_2]-MAc$ ($x_{Li+}=0.25$) at scan rate 5 mV s^{-1}. (**b**) Specific capacitance calculated according to

$$\left(C_{sp} = \frac{2\int idt}{\Delta E.m}; i \text{ is applied current}, \Delta E \text{ is the tension voltage and } m \text{ is the active mass of } Ac \right)$$

for $\Delta E = 2.6$ V. Reprinted with permission from [142] with permission from the American Chemical Society

ultramicropores of the activated carbon material. Insertion of lithium ions is observed by the presence of two reversible peaks in the CVs when the operating voltage exceeds 2 V (Fig. 7.22) [144, 145]. The decreasing capacitance of symmetric AC/AC systems despite good efficiency can be interpreted by the imbalanced mass of carbon electrodes caused by important lithium insertion at the negative electrode and by saturation of the positive electrode by anions; both mechanisms prevent, in fact, the system being operational.

Considering the promising properties, especially their cost, hazard, and the risks of these DESs series, their introduction as safer electrolytes could represent an important challenge for the realization of environmentally friendly EDLCs operating at high temperatures. The study of the original ions' adsorption mechanism of desolvated lithium ions provides an interesting opportunity for fundamental studies even beyond the applicative interest of DESs as electrolytes for supercapacitors.

7.4.2 Application of the PILs for Lithium Ion Batteries

"Due to the high value of the energy content, lithium ion batteries have triggered the growth of the market of popular devices, such as mobile phones, laptop computers, MP3s, and others. Indeed, lithium ion batteries are today produced in billions of units per year [146, 147].

At first sight, the electrochemical process which drives the lithium ion battery appears quite simple, apparently consisting of the reversible exchange of lithium ions between two electrodes. However, in practice, the operation of this battery requires the mastery of key side processes. Accordingly, worldwide research and development efforts are directed toward the replacement of the present battery components with materials having higher performance in terms of energy, power, cost, reliability, lifetime, and safety.

Approaches to reach this goal are focused along two main lines:

- Replacement of graphite and of lithium cobalt oxide with alternative, higher capacity, lower cost anode and cathode materials;
- Replacement of the organic carbonate liquid electrolyte solutions with safer and more reliable electrolyte systems.

Among the electrolytes proposed so far, those based on aprotic ILs appear to be very promising since their introduction can improve both the safety as well as the operative temperature range of the use of LIBs [148–150]. Nevertheless, the performance of IL-based LIBs still needs to be improved in order to be fully competitive with that of the conventional electrolytes. Moreover, the cost of ILs might also represent an obstacle for the introduction of these electrolytes in LIBs.

7.4.2.1 Use of PILs as Electrolytes in Li-Ion Batteries

Recently, the Anouti and Balducci groups reported for the first time the use of PILs as electrolytes for LIBs [151]. The investigated electrolytes consisted of a solution containing PIL $[HN_{222}][N(Tf)_2)]$ and the lithium salt $Li[N(Tf)_2]$ (1 M). Considering all beneficial properties of PILs, easy to synthesize and cheaper than aprotic liquids, conductivity comparable to that of aprotic ILs, and electrochemical stability window large enough to allow the realization of LIBs containing LFP as cathode and Lithium Titanium Oxide ($Li_4Ti_5O_{12}$ (LTO)) as anode, the PIL-based electrolyte allows the realization of devices able to deliver capacities comparable with those possible with the aprotic IL-based electrolyte as shown in Fig. 7.23 by constant-current charge–discharge behavior [151].

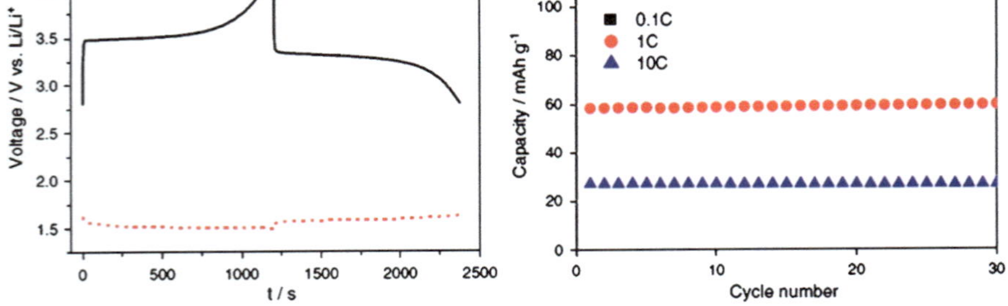

Fig. 7.23 (**a**) Charge–discharge voltage profiles of an LFP cathode and LTO anode and (**b**) discharge capacity of LIBs containing 1 M $Li[N(Tf)_2]$ in $[HN_{222}][N(Tf)_2]$ at 0.1 C, 1 C, and 10 °C

The performance of PILs for LIBs certainly needs to be improved. Nevertheless, the results of this work clearly show that PILs can be successfully introduced in LIBs. Considering the beneficial properties (and the cost) of these ILs, such introduction could represent an important contribution for the realization of safer and environmentally friendly LIBs.

7.4.2.2 Use of DESs as Electrolytes in Li-Ion Batteries

Considering that the presence of labile proton does not prevent the use of PIL-based electrolytes in lithium-ion batteries, it was therefore possible to imagine the possibility of using more broadly the "H-bond Donors" as electrolytes replacing those based on alkylcarbonates. The successful introduction of DESs-based lithium salts is an illustration of this new way of research into electrolytes replacing conventional organic solvents.

Recently, Anouti et al. demonstrated that these electrolytes have good compatibility with both electrodes ($Li_4Ti_5O_{12}$ and $LiFePO_4$) [40, 41]. Tested cells show good electrochemical performances in term of specific capacitance, rate, and cycling ability, indicating that the Li[N(Tf)$_2$]/MAc DES can be a promising electrolyte for EDLCs and LiBs applications, especially for those requiring high safety and stability.

As shown in Fig. 7.24, the $LiFePO_4$ electrode displays the typical voltage profile, the plateau corresponding to the lithium insertion–extraction at around 3.45 V vs. Li/Li^+; the difference between the charge and the discharge voltage plateau is exceptionally small. Obviously, the presence of water affects the specific capacity of the electrode: the lower the water content, the higher the capacity. Nevertheless, the presence of different amounts of water does not seem to have a significant effect on the voltage plateau. This is surprising because in general electrolytes for batteries do not work in

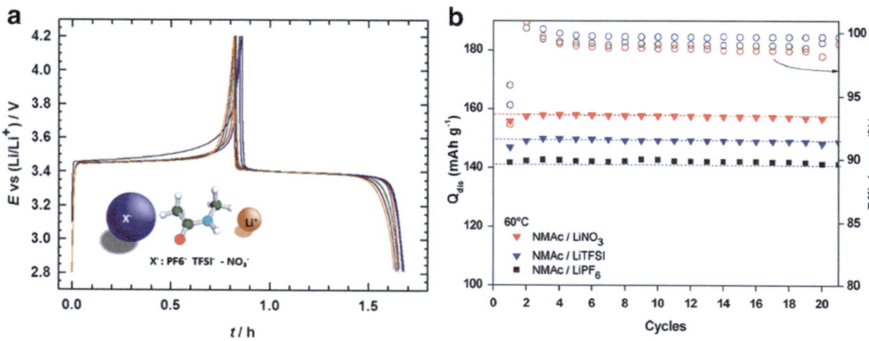

Fig. 7.24 (**a**) Constant-current charge–discharge profiles of LFP electrode in LiX–NMAc. X=NO_3, [N(Tf)$_2$] (TFSI), PF_6; NMAc is the N-methylacetamide. (**b**) Comparative discharge capacity and efficiency as obtained during tests carried out at a 1 C rate and 60 °C. Reproduced from [41] with permission from the PCCP Owner Societies

the presence of traces of water exceeding 20 ppm. Here, the amount of water is evaluated at 500 ppm, which does not prevent the function of electrode LFP. These observations are considered in light of free residual water or water attached by H-bonding. In DESs and PILs, the presence of a labile proton does not necessarily imply a hydrogen potential with the standard reaction of free water molecules.

7.5 Conclusion

This chapter presents an overview of PILs and DESs and discusses the possibility of employing them as electrolytes for renewable energy storage as lithium ion batteries and supercapacitors and for utility-scale load leveling applications. Use of ILs as electrolytes for energy storage application is provided, building on the feasibility of replacing ILs with DESs in the near future, based on recent publications on the topic. H-Bond donor RTMS may offer significant advantages over their generally more volatile organic molecular solvent counterparts. This kind of ionic liquid possesses several inherent advantages, such as cheap raw materials, simple preparation procedures, and better biocompatibility. For example, PIL properties, prepared by the proton-transfer reaction between the base and proton-donor acid, are easy to tune for target electrochemical application because of the diversity of bases and proton donors that are available. PILs often have high intrinsic conductivities—similar to aqueous solvents with added supporting electrolytes—resulting in simplified experimental arrangements. Many RTMS with selected ions have wide electrochemical windows (dictated by the high stability of the cation and anion) without requiring the presence of a labile proton.

PILs and DESs also have reductive structuring properties contributing to their ability to solubilize a wide range of polar and nonpolar species as precursors of metallic nanoparticles. All the above advantages, together with the competitive performance, make H-bond RTMS a strong candidate for future electrolyte development for energy storage systems. The RTMS-based electrolyte allows the realization of devices, lithium ion batteries, redox flow, and supercapacitors to deliver capacities comparable with those possible with organic solvent- or aprotic IL-based electrolytes. Especially, DESs show unusual solvent properties for metal oxide, inorganic salts, and solid organic compounds, which could make these binary eutectic mixtures the new medium for electrochemical application in the future. However, drawbacks include the high viscosity of DESs, resulting in small diffusion coefficients at ambient temperatures, and the hydrophilic properties of PILs, which limit their generalization.

The performance of RTMS and their purification certainly needs to be improved and the investigation of their potentialities should be continued. Nevertheless, the newest aspects addressed by the research community clearly conclude that RTMS can be successfully introduced in the energy storage application. Considering the beneficial properties (and the cost) of these media, such introduction could represent an important contribution for the realization of safer and environmentally friendly systems.

References

1. Ho TD, Zhang C, Hantao LW, Anderson JL (2013) Ionic liquids in analytical chemistry: fundamentals, advances, and perspectives. Anal Chem 86(1):262–285. doi:10.1021/ac4035554
2. Hallett JP, Welton T (2011) Room-temperature ionic liquids: solvents for synthesis and catalysis. 2. Chem Rev 111(5):3508–3576. doi:10.1021/cr1003248
3. Shukla SK, Kumar A (2013) Probing the acidity of carboxylic acids in protic ionic liquids, water, and their binary mixtures: activation energy of proton transfer. J Phys Chem B 117(8):2456–2465. doi:10.1021/jp310927w
4. Greaves TL, Drummond CJ (2007) Protic ionic liquids: properties and applications. Chem Rev 108(1):206–237. doi:10.1021/cr068040u
5. Walden, Bull. Acad. Sci. St. Petersbourg 1914, 405–422
6. Yasuda T, Watanabe M (2013) Protic ionic liquids: fuel cell applications. MRS Bull 38(07):560–566. doi:10.1557/mrs.2013.153
7. Lee S-Y, Ogawa A, Kanno M, Nakamoto H, Yasuda T, Watanabe M (2010) Nonhumidified intermediate temperature fuel cells using protic ionic liquids. J Am Chem Soc 132(28):9764–9773. doi:10.1021/ja102367x
8. Zhao C, Burrell G, Torriero AAJ, Separovic F, Dunlop NF, MacFarlane DR, Bond AM (2008) Electrochemistry of room temperature protic ionic liquids. J Phys Chem B 112(23):6923–6936. doi:10.1021/jp711804j
9. Johnson L, Ejigu A, Licence P, Walsh DA (2012) Hydrogen oxidation and oxygen reduction at platinum in protic ionic liquids. J Phys Chem C 116(34):18048–18056. doi:10.1021/jp303749k
10. Lu X, Burrell G, Separovic F, Zhao C (2012) Electrochemistry of room temperature protic ionic liquids: a critical assessment for use as electrolytes in electrochemical applications. J Phys Chem B 116(30):9160–9170. doi:10.1021/jp304735p
11. Zhang S, Miran MS, Ikoma A, Dokko K, Watanabe M (2014) Protic ionic liquids and salts as versatile carbon precursors. J Am Chem Soc 136(5):1690–1693. doi:10.1021/ja411981c
12. Angell CA, Byrne N, Belieres J-P (2007) Parallel developments in aprotic and protic ionic liquids: physical chemistry and applications. Acc Chem Res 40(11):1228–1236. doi:10.1021/ar7001842
13. Yue X, Chen X, Li Q (2012) Comparison of aggregation behaviors of a phytosterol ethoxylate surfactant in protic and aprotic ionic liquids. J Phys Chem B 116(31):9439–9444. doi:10.1021/jp305230r
14. Marin TW, Shkrob IA, Dietz ML (2011) Hydrogen-bonding interactions and protic equilibria in room-temperature ionic liquids containing crown ethers. J Phys Chem B 115(14):3912–3918. doi:10.1021/jp201193f
15. Gelbard G (2005) Organic synthesis by catalysis with ion-exchange resins. Ind Eng Chem Res 44(23):8468–8498. doi:10.1021/ie0580405
16. Ohno H, Yoshizawa M (2002) Ion conductive characteristics of ionic liquids prepared by neutralization of alkylimidazoles. Solid State Ionics 154–155:303–309. doi:10.1016/S0167-2738(02)00526-X
17. Adam C, Bravo MV, Mancini PME (2014) Molecular solvent effect on the acidity constant of protic ionic liquids. Tetrahedron Lett 55(1):148–150. doi:10.1016/j.tetlet.2013.10.137
18. Shukla SK, Kumar A (2013) Do protic ionic liquids and water display analogous behavior in terms of Hammett acidity function? Chem Phys Lett 566:12–16. doi:10.1016/j.cplett.2013.02.029
19. Jin H, Baker GA, Arzhantsev S, Dong J, Maroncelli M (2007) Solvation and rotational dynamics of coumarin 153 in ionic liquids: comparisons to conventional solvents. J Phys Chem B 111(25):7291–7302. doi:10.1021/jp070923h
20. Singh A, Kumar A (2012) Probing the mechanism of Baylis-Hillman reaction in ionic liquids. J Org Chem 77(19):8775–8779. doi:10.1021/jo301348k

21. Zhang X-X, Liang M, Ernsting NP, Maroncelli M (2013) Conductivity and solvation dynamics in ionic liquids. J Phys Chem Lett 4(7):1205–1210. doi:10.1021/jz400359r
22. Abbott AP, Capper G, Davies DL, Rasheed RK, Tambyrajah V (2003) Novel solvent properties of choline chloride/urea mixtures. Chem Commun 1:70–71. doi:10.1039/b210714g
23. Abbott AP, Boothby D, Capper G, Davies DL, Rasheed RK (2004) Deep eutectic solvents formed between choline chloride and carboxylic acids: versatile alternatives to ionic liquids. J Am Chem Soc 126(29):9142–9147. doi:10.1021/ja048266j
24. Hayyan A, Hashim MA, Hayyan M, Mjalli FS, AlNashef IM (2013) A novel ammonium based eutectic solvent for the treatment of free fatty acid and synthesis of biodiesel fuel. Ind Crop Prod 46:392–398. doi:10.1016/j.indcrop.2013.01.033
25. Salomé S, Pereira NM, Ferreira ES, Pereira CM, Silva AF (2013) Tin electrodeposition from choline chloride based solvent: Influence of the hydrogen bond donors. J Electroanal Chem 703:80–87. doi:10.1016/j.jelechem.2013.05.007
26. Chakrabarti MH, Mjalli FS, AlNashef IM, Hashim MA, Hussain MA, Bahadori L, Low CTJ (2014) Prospects of applying ionic liquids and deep eutectic solvents for renewable energy storage by means of redox flow batteries. Renew Sustain Energy Rev 30:254–270. doi:10.1016/j.rser.2013.10.004
27. Hayyan A, Hashim MA, Hayyan M, Mjalli FS, AlNashef IM (2014) A new processing route for cleaner production of biodiesel fuel using a choline chloride based deep eutectic solvent. J Clean Prod 65:246–251. doi:10.1016/j.jclepro.2013.08.031
28. Ali E, Hadj-Kali MK, Mulyono S, Alnashef I, Fakeeha A, Mjalli F, Hayyan A (2014) Solubility of CO_2 in deep eutectic solvents: experiments and modelling using the Peng–Robinson equation of state. Chem Eng Res Design 92(10):1898–1906. doi:10.1016/j.cherd.2014.02.004
29. Mandroyan A, Mourad-Mahmoud M, Doche M-L, Hihn J-Y (2014) Effects of ultrasound and temperature on copper electro reduction in deep eutectic solvents (DES). Ultrason Sonochem 21(6):2010–2019. doi:10.1016/j.ultsonch.2014.02.019
30. Wu S-H, Caparanga AR, Leron RB, Li M-H (2012) Vapor pressure of aqueous choline chloride-based deep eutectic solvents (ethaline, glyceline, maline and reline) at 30–70°C. Thermochim Acta 544:1–5. doi:10.1016/j.tca.2012.05.031
31. Hayyan M, Hashim MA, Hayyan A, Al-Saadi MA, AlNashef IM, Mirghani MES, Saheed OK (2013) Are deep eutectic solvents benign or toxic? Chemosphere 90(7):2193–2195. doi:10.1016/j.chemosphere.2012.11.004
32. Leron RB, Li M-H (2013) Solubility of carbon dioxide in a choline chloride–ethylene glycol based deep eutectic solvent. Thermochim Acta 551:14–19. doi:10.1016/j.tca.2012.09.041
33. Paiva A, Craveiro R, Aroso I, Martins M, Reis RL, Duarte ARC (2014) Natural deep eutectic solvents—solvents for the 21st century. ACS Sustainable Chem Eng 2(5):1063–1071. doi:10.1021/sc500096j
34. Dai Y, van Spronsen J, Witkamp G-J, Verpoorte R, Choi YH (2013) Natural deep eutectic solvents as new potential media for green technology. Anal Chim Acta 766:61–68. doi:10.1016/j.aca.2012.12.019
35. Abbott AP, Azam M, Ryder KS, Saleem S (2013) In situ electrochemical digital holographic microscopy; a study of metal electrodeposition in deep eutectic solvents. Anal Chem 85(14):6653–6660. doi:10.1021/ac400262c
36. Dai Y, van Spronsen J, Witkamp G-J, Verpoorte R, Choi YH (2013) Ionic liquids and deep eutectic solvents in natural products research: mixtures of solids as extraction solvents. J Nat Prod 76(11):2162–2173. doi:10.1021/np400051w
37. Hammons JA, Muselle T, Ustarroz J, Tzedaki M, Raes M, Hubin A, Terryn H (2013) Stability, assembly, and particle/solvent interactions of Pd nanoparticles electrodeposited from a deep eutectic solvent. J Phys Chem C 117(27):14381–14389. doi:10.1021/jp403739y
38. Chen R, Wu F, Xu B, Li L, Qiu X, Chen S (2007) Binary complex electrolytes based on LiX [X=N(SO2CF3)2−, CF3SO3−, ClO4−]-acetamide for electric double layer capacitors. J Electrochem Soc 154(7):A703–A708. doi:10.1149/1.2737350
39. Hu Y, Li H, Huang X, Chen L (2004) Novel room temperature molten salt electrolyte based on LiTFSI and acetamide for lithium batteries. Electrochem Commun 6(1):28–32. doi:10.1016/j.elecom.2003.10.009

40. Boisset A, Jacquemin J, Anouti M (2013) Physical properties of a new deep eutectic solvent based on lithium bis[(trifluoromethyl)sulfonyl]imide and N-methylacetamide as superionic suitable electrolyte for lithium ion batteries and electric double layer capacitors. Electrochim Acta 102:120–126. doi:10.1016/j.electacta.2013.03.150
41. Boisset A, Menne S, Jacquemin J, Balducci A, Anouti M (2013) Deep eutectic solvents based on N-methylacetamide and a lithium salt as suitable electrolytes for lithium-ion batteries. Phys Chem Chem Phys 15(46):20054–20063. doi:10.1039/c3cp53406e
42. Podolean I, Hardacre C, Goodrich P, Brun N, Backov R, Coman SM, Parvulescu VI (2013) Chiral supported ionic liquid phase (CSILP) catalysts for greener asymmetric hydrogenation processes. Catal Today 200:63–73. doi:10.1016/j.cattod.2012.06.020
43. Suzuki Y, Wakatsuki J, Tsubaki M, Sato M (2013) Imidazolium-based chiral ionic liquids: synthesis and application. Tetrahedron 69(46):9690–9700. doi:10.1016/j.tet.2013.09.017
44. Fukumoto K, Yoshizawa M, Ohno H (2005) Room temperature ionic liquids from 20 natural amino acids. J Am Chem Soc 127(8):2398–2399. doi:10.1021/ja043451i
45. Ohno H, Fukumoto K (2007) Amino acid ionic liquids. Acc Chem Res 40(11):1122–1129. doi:10.1021/ar700053z
46. Zhu J-F, He L, Zhang L, Huang M, Tao G-H (2011) Experimental and theoretical enthalpies of formation of glycine-based sulfate/bisulfate amino acid ionic liquids. J Phys Chem B 116(1):113–119. doi:10.1021/jp209649h
47. Khatri PK, Thakre GD, Jain SL (2013) Tribological performance evaluation of task-specific ionic liquids derived from amino acids. Ind Eng Chem Res 52(45):15829–15837. doi:10.1021/ie402141v
48. Sowmiah S, Srinivasadesikan V, Tseng M-C, Chu Y-H (2009) On the chemical stabilities of ionic liquids. Molecules 14(9):3780–3813
49. Rogers RD, Voth GA (2007) Ionic liquids. Acc Chem Res 40(11):1077–1078. doi:10.1021/ar700221n
50. Hough WL, Smiglak M, Rodriguez H, Swatloski RP, Spear SK, Daly DT, Pernak J, Grisel JE, Carliss RD, Soutullo MD, Davis JJH, Rogers RD (2007) The third evolution of ionic liquids: active pharmaceutical ingredients. New J Chem 31(8):1429–1436. doi:10.1039/b706677p
51. Yue C, Fang D, Liu L, Yi T-F (2011) Synthesis and application of task-specific ionic liquids used as catalysts and/or solvents in organic unit reactions. J Mol Liq 163(3):99–121. doi:10.1016/j.molliq.2011.09.001
52. Peric B, Sierra J, Martí E, Cruañas R, Garau MA (2014) A comparative study of the terrestrial ecotoxicity of selected protic and aprotic ionic liquids. Chemosphere 108:418–425. doi:10.1016/j.chemosphere.2014.02.043
53. Romero A, Santos A, Tojo J, Rodríguez A (2008) Toxicity and biodegradability of imidazolium ionic liquids. J Hazard Mater 151(1):268–273. doi:10.1016/j.jhazmat.2007.10.079
54. Frade RFM, Simeonov S, Rosatella AA, Siopa F, Afonso CAM (2013) Toxicological evaluation of magnetic ionic liquids in human cell lines. Chemosphere 92(1):100–105. doi:10.1016/j.chemosphere.2013.02.047
55. Cvjetko Bubalo M, Radošević K, Radojčić Redovniković I, Halambek J, Gaurina Srček V (2014) A brief overview of the potential environmental hazards of ionic liquids. Ecotoxicol Environ Saf 99:1–12. doi:10.1016/j.ecoenv.2013.10.019
56. Gouveia W, Jorge TF, Martins S, Meireles M, Carolino M, Cruz C, Almeida TV, Araújo MEM (2014) Toxicity of ionic liquids prepared from biomaterials. Chemosphere 104:51–56. doi:10.1016/j.chemosphere.2013.10.055
57. Huang L, Zhai M, Peng J, Xu L, Li J, Wei G (2008) Synthesis of room temperature ionic liquids from carboxymethylated chitosan. Carbohydr Polym 71(4):690–693. doi:10.1016/j.carbpol.2007.06.020
58. Dash M, Chiellini F, Ottenbrite RM, Chiellini E (2011) Chitosan—a versatile semi-synthetic polymer in biomedical applications. Prog Polym Sci 36(8):981–1014. doi:10.1016/j.progpolymsci.2011.02.001
59. Lu X, Zhang Q, Zhang L, Li J (2006) Direct electron transfer of horseradish peroxidase and its biosensor based on chitosan and room temperature ionic liquid. Electrochem Commun 8(5):874–878. doi:10.1016/j.elecom.2006.03.026

60. Singh MP, Singh RK, Chandra S (2014) Ionic liquids confined in porous matrices: physicochemical properties and applications. Prog Mater Sci 64:73–120. doi:10.1016/j.pmatsci.2014.03.001
61. Attri P, Lee S-H, Hwang SW, Kim JIL, Jang W, Kim YB, Park JH, Kwon G-C, Choi EH, Kim IT (2014) Effect of temperature on the interactions between low bandgap polymer and ionic liquids. Thermochim Acta 579:15–21. doi:10.1016/j.tca.2013.12.022
62. Wellens S, Vander Hoogerstraete T, Möller C, Thijs B, Luyten J, Binnemans K (2014) Dissolution of metal oxides in an acid-saturated ionic liquid solution and investigation of the back-extraction behaviour to the aqueous phase. Hydrometall 144–145:27–33. doi:10.1016/j.hydromet.2014.01.015
63. Noda A, Susan MABH, Kudo K, Mitsushima S, Hayamizu K, Watanabe M (2003) Brønsted acid–base ionic liquids as proton-conducting nonaqueous electrolytes. J Phys Chem B 107(17):4024–4033. doi:10.1021/jp022347p
64. Belieres J-P, Angell CA (2007) Protic ionic liquids: preparation, characterization, and proton free energy level representation. J Phys Chem B 111(18):4926–4937. doi:10.1021/jp067589u
65. Fortunato GG, Mancini PM, Bravo MV, Adam CG (2010) New solvents designed on the basis of the molecular-microscopic properties of binary mixtures of the type (protic molecular solvent+1-butyl-3-methylimidazolium-based ionic liquid). J Phys Chem B 114(36):11804–11819. doi:10.1021/jp103677q
66. Kavitha T, Attri P, Venkatesu P, Devi RSR, Hofman T (2012) Influence of alkyl chain length and temperature on thermophysical properties of ammonium-based ionic liquids with Molecular solvent. J Phys Chem B 116(15):4561–4574. doi:10.1021/jp3015386
67. Goodenough JB, Park K-S (2005) The li-ion rechargeable battery: a perspective. J Am Chem Soc 135(4):1167–1176. doi:10.1021/ja3091438
68. Jin L, Nairn KM, Forsyth CM, Seeber AJ, MacFarlane DR, Howlett PC, Forsyth M, Pringle JM (2005) Structure and transport properties of a plastic crystal ion conductor: diethyl(methyl)(isobutyl)phosphonium hexafluorophosphate. J Am Chem Soc 134(23):9688–9697. doi:10.1021/ja301175v
69. Ohno H, Yoshizawa M (2005) Preparation and properties of polymerized ionic liquids as film electrolytes. In: Rogers R et al (eds) Ionic liquids IIIB: fundamentals, progress, challenges, and opportunities, vol 902, ACS Symposium Series. American Chemical Society, Washington, DC, pp 159–170. doi:10.1021/bk-2005-0902.ch013
70. Anouti M, Jacquemin J, Lemordant D (2010) Transport properties of protic ionic liquids, pure and in aqueous solutions: effects of the anion and cation structure. Fluid Phase Equilib 297(1):13–22. doi:10.1016/j.fluid.2010.05.019
71. Pires J, Timperman L, Jacquemin J, Balducci A, Anouti M (2013) Density, conductivity, viscosity, and excess properties of (pyrrolidinium nitrate-based protic ionic liquidand propylene carbonate) binary mixture. J Chem Thermodyn 59:10–19. doi:10.1016/j.jct.2012.11.020
72. Xu Y (2013) Volumetric, viscosity, and electrical conductivity properties of aqueous solutions of two n-butylammonium-based protic ionic liquids at several temperatures. J Chem Thermodyn 64:126–133. doi:10.1016/j.jct.2013.05.019
73. Yang Q, Yu K, Xing H, Su B, Bao Z, Yang Y, Ren Q (2013) The effect of molecular solvents on the viscosity, conductivity and ionicity of mixtures containing chloride anion-based ionic liquid. J Ind Eng Chem 19(5):1708–1714. doi:10.1016/j.jiec.2013.02.010
74. Couadou E, Jacquemin J, Galiano H, Hardacre C, Anouti M (2013) A comparative study on the thermophysical properties for two bis[(trifluoromethyl)sulfonyl]imide-based ionic liquids containing the trimethyl-sulfonium or the trimethyl-ammonium cation in molecular solvents. J Phys Chem B 117(5):1389–1402. doi:10.1021/jp308139r
75. Anouti M, Jacquemin J, Porion P (2012) Transport properties investigation of aqueous protic ionic liquid solutions through conductivity, viscosity, and NMR self-diffusion measurements. J Phys Chem B 116(14):4228–4238. doi:10.1021/jp3010844
76. Yushchenko DA, Shvadchak VV, Klymchenko AS, Duportail G, Pivovarenko VG, Mély Y (2007) Modulation of excited-state intramolecular proton transfer by viscosity in protic media. J Phys Chem A 111(42):10435–10438. doi:10.1021/jp074726u

77. Chen S, Vijayaraghavan R, MacFarlane DR, Izgorodina EI (2013) Ab initio prediction of proton NMR chemical shifts in imidazolium ionic liquids. J Phys Chem B 117(11):3186–3197. doi:10.1021/jp310267x
78. Smith JA, Webber GB, Warr GG, Atkin R (2013) Rheology of protic Ionic liquids and their mixtures. J Phys Chem B 117(44):13930–13935. doi:10.1021/jp407715e
79. Han H-B, Nie J, Liu K, Li W-K, Feng W-F, Armand M, Matsumoto H, Zhou Z-B (2010) Ionic liquids and plastic crystals based on tertiary sulfonium and bis(fluorosulfonyl)imide. Electrochim Acta 55(3):1221–1226. doi:10.1016/j.electacta.2009.10.019
80. Anouti M, Caillon-Caravanier M, Dridi Y, Galiano H, Lemordant D (2008) Synthesis and characterization of new pyrrolidinium based protic ionic liquids. Good and superionic liquids. J Phys Chem B 112(42):13335–13343. doi:10.1021/jp805992b
81. Abbott AP, Capper G, Davies DL, McKenzie KJ, Obi SU (2006) Solubility of metal oxides in deep eutectic solvents based on choline chloride. J Chem Eng Data 51(4):1280–1282. doi:10.1021/je060038c
82. Guo W, Hou Y, Ren S, Tian S, Wu W (2013) Formation of deep eutectic solvents by phenols and choline chloride and their physical properties. J Chem Eng Data 58(4):866–872. doi:10.1021/je300997v
83. Takano Y, Nakamura H (2006) Quantum mechanical study of the proton transfer via a peptide bond in the novel proton translocation pathway of cytochrome c oxidase. Chem Phys Lett 430(1–3):149–155. doi:10.1016/j.cplett.2006.08.138
84. Martinus N, Crawford D, Sinclair D, Vincent CA (1977) The extended Jones—Dole equation. Electrochim Acta 22(10):1183–1187. doi:10.1016/0013-4686(77)80059-5
85. Szymczyk K (2014) Comparative study of the physicochemical properties of aqueous solutions of the hydrocarbon and fluorocarbon surfactants and their ternary mixtures. Chem Phys 433:42–47. doi:10.1016/j.chemphys.2014.02.001
86. Jones G, Dole M (1930) The electrical conductance of aqueous solutions of barium chloride as a function of the concentration. J Am Chem Soc 52(6):2245–2256. doi:10.1021/ja01369a009
87. Dole M (1984) Debye's contribution to the theory of the viscosity of strong electrolytes. J Phys Chem 88(26):6468–6469. doi:10.1021/j150670a003
88. Hammadi A, Champeney DC (1998) Ion-solvent interactions of some alkali halides in glycerol from density and viscosity data. J Chem Eng Data 43(6):1004–1008. doi:10.1021/je9801260
89. Anouti M, Porion P, Brigouleix C, Galiano H, Lemordant D (2010) Transport properties in two pyrrolidinium-based protic ionic liquids as determined by conductivity, viscosity and NMR self-diffusion measurements. Fluid Phase Equilib 299(2):229–237. doi:10.1016/j.fluid.2010.09.035
90. Schreiner C, Zugmann S, Hartl R, Gores HJ (2009) Fractional Walden rule for ionic liquids: examples from recent measurements and a critique of the so-called ideal KCl line for the Walden plot†. J Chem Eng Data 55(5):1784–1788. doi:10.1021/je900878j
91. Longinotti MP, Corti HR (2009) Fractional Walden rule for electrolytes in supercooled disaccharide aqueous solutions. J Phys Chem B 113(16):5500–5507. doi:10.1021/jp810253s
92. García A, Torres-González LC, Padmasree KP, Benavides-Garcia MG, Sánchez EM (2013) Conductivity and viscosity properties of associated ionic liquids phosphonium orthoborates. J Mol Liq 178:57–62. doi:10.1016/j.molliq.2012.11.007
93. Porion P, Dougassa YR, Tessier C, El Ouatani L, Jacquemin J, Anouti M (2013) Comparative study on transport properties for LiFAP and LiPF6 in alkyl-carbonates as electrolytes through conductivity, viscosity and NMR self-diffusion measurements. Electrochim Acta 114:95–104. doi:10.1016/j.electacta.2013.10.015
94. Kao H-M, Chang P-C, Chao S-W, Lee C-H (2006) 7Li NMR, ionic conductivity and self-diffusion coefficients of lithium ion and solvent of plasticized organic–inorganic hybrid electrolyte based on PPG-PEG-PPG diamine and alkoxysilanes. Electrochim Acta 52(3):1015–1027. doi:10.1016/j.electacta.2006.06.042

95. Blanchard JW, Belières J-P, Alam TM, Yarger JL, Holland GP (2011) NMR determination of the diffusion mechanisms in triethylamine-based protic ionic liquids. J Phys Chem Lett 2(9):1077–1081. doi:10.1021/jz200357j
96. Reddy GSM, Jayaramudu J, Varaprasad K, Sadiku R, Jailani SA, Aderibigbe BA (2014) Chapter 9—Nanostructured liquid crystals. In: Thomas S, Shanks R, Chandrasekharakurup S (eds) Nanostructured polymer blends. William Andrew, Oxford, pp 299–324. doi:10.1016/B978-1-4557-3159-6.00009-2
97. Marcus Y (2013) Surface tension and cohesive energy density of molten salts. Thermochim Acta 571:77–81. doi:10.1016/j.tca.2013.08.018
98. Marcus Y (2010) The cohesive energy of molten salts and its density. J Chem Thermodyn 42(1):60–64. doi:10.1016/j.jct.2009.07.004
99. Frost DS, Nofen EM, Dai LL (2014) Particle self-assembly at ionic liquid-based interfaces. Adv Colloid Interface Sci 206:92–105. doi:10.1016/j.cis.2013.09.004
100. Weingärtner H (2013) NMR studies of ionic liquids: structure and dynamics. Curr Opin Colloid Interface Sci 18(3):183–189. doi:10.1016/j.cocis.2013.04.001
101. Men Y, Kuzmicz D, Yuan J (2014) Poly(ionic liquid) colloidal particles. Curr Opin Colloid Interface Sci 19(2):76–83. doi:10.1016/j.cocis.2014.03.012
102. Sakai H, Saitoh T, Endo T, Tsuchiya K, Sakai K, Abe M (2009) Phytosterol ethoxylates in room-temperature ionic liquids: excellent interfacial properties and gel formation. Langmuir 25(5):2601–2603. doi:10.1021/la900139d
103. Li P, Wang W, Du Z, Wang G, Li E, Li X (2014) Adsorption and aggregation behavior of surface active trisiloxane room-temperature ionic liquids. Colloids Surf A Physicochem Eng Asp 450:52–58. doi:10.1016/j.colsurfa.2014.03.007
104. Du Z, Li E, Cao Y, Li X, Wang G (2014) Synthesis of trisiloxane-tailed surface active ionic liquids and their aggregation behavior in aqueous solution. Colloids Surf A Physicochem Eng Asp 441:744–751. doi:10.1016/j.colsurfa.2013.10.004
105. Iida M, Kawakami S, Syouno E, Er H, Taguchi E (2011) Properties of ionic liquids containing silver(I) or protic alkylethylenediamine cations with a bis(trifluoromethanesulfonyl) amide anion. J Colloid Interface Sci 356(2):630–638. doi:10.1016/j.jcis.2011.01.070
106. Meot-Ner M (2012) Update 1 of: strong ionic hydrogen bonds. Chem Rev 112(10):PR22–PR103. doi:10.1021/cr200430n
107. Greaves TL, Kennedy DF, Mudie ST, Drummond CJ (2010) Diversity observed in the nanostructure of protic ionic liquids. J Phys Chem B 114(31):10022–10031. doi:10.1021/jp103863z
108. Fernández-Castro B, Méndez-Morales T, Carrete J, Fazer E, Cabeza O, Rodríguez JR, Turmine M, Varela LM (2011) Surfactant self-assembly nanostructures in protic ionic liquids. J Phys Chem B 115(25):8145–8154. doi:10.1021/jp203204c
109. Smith J, Webber GB, Warr GG, Atkin R (2014) Silica particle stability and settling in protic ionic liquids. Langmuir 30(6):1506–1513. doi:10.1021/la403978b
110. Scholten JD, Leal BC, Dupont J (2011) Transition metal nanoparticle catalysis in ionic liquids. ACS Catal 2(1):184–200. doi:10.1021/cs200525e
111. Anouti M, Mirghani A, Jacquemin J, Timperman L, Galiano H (2013) Tunable gold nanoparticles shape and size in reductive and structuring media containing protic ionic liquids. Ionics 19(12):1783–1790. doi:10.1007/s11581-013-0915-0
112. Anouti M, Jacquemin J (2014) Structuring reductive media containing protic ionic liquids and their application to the formation of metallic nanoparticles. Colloids Surf A Physicochem Eng Asp 445:1–11. doi:10.1016/j.colsurfa.2013.12.064
113. Vollmer C, Janiak C (2011) Naked metal nanoparticles from metal carbonyls in ionic liquids: easy synthesis and stabilization. Coord Chem Rev 255(17–18):2039–2057. doi:10.1016/j.ccr.2011.03.005
114. Suarez PAZ, Consorti CS, de Souza RF, Dupont J, Gonçalves RS (2002) Electrochemical behavior of vitreous glass carbon and platinum electrodes in the ionic liquid 1-n-butyl-3-methylimidazolium trifluoroacetate. J Braz Chem Soc 13(1):106–109

115. Zein El Abedin S, Borissenko N, Endres F (2004) Electrodeposition of nanoscale silicon in a room temperature ionic liquid. Electrochem Commun 6(5):510–514. doi:10.1016/j.elecom.2004.03.013
116. Matsumoto K, Inoue K, Nakahara K, Yuge R, Noguchi T, Utsugi K (2013) Suppression of aluminum corrosion by using high concentration LiTFSI electrolyte. J Power Sources 231:234–238. doi:10.1016/j.jpowsour.2012.12.028
117. Kühnel R-S, Lübke M, Winter M, Passerini S, Balducci A (2012) Suppression of aluminum current collector corrosion in ionic liquid containing electrolytes. J Power Sources 214:178–184. doi:10.1016/j.jpowsour.2012.04.054
118. Cho E, Mun J, Chae OB, Kwon OM, Kim H-T, Ryu JH, Kim YG, Oh SM (2012) Corrosion/passivation of aluminum current collector in bis(fluorosulfonyl)imide-based ionic liquid for lithium-ion batteries. Electrochem Commun 22:1–3. doi:10.1016/j.elecom.2012.05.018
119. Garcia B, Armand M (2004) Aluminium corrosion in room temperature molten salt. J Power Sources 132(1–2):206–208. doi:10.1016/j.jpowsour.2003.12.046
120. Timperman L, Skowron P, Boisset A, Galiano H, Lemordant D, Frackowiak E, Beguin F, Anouti M (2012) Triethylammonium bis(tetrafluoromethylsulfonyl)amide protic ionic liquid as an electrolyte for electrical double-layer capacitors. Phys Chem Chem Phys 14(22):8199–8207
121. Skyllas-Kazacos M (2010) 10-electro-chemical energy storage technologies for wind energy systems. In: Kaldellis JK (ed) Stand-alone and hybrid wind energy systems. Woodhead Publishing, Cambridge, pp 323–365. doi:10.1533/9781845699628.2.323
122. Koohi-Kamali S, Tyagi VV, Rahim NA, Panwar NL, Mokhlis H (2013) Emergence of energy storage technologies as the solution for reliable operation of smart power systems: a review. Renew Sustain Energy Rev 25:135–165. doi:10.1016/j.rser.2013.03.056
123. Yang Z, Zhang J, Kintner-Meyer MCW, Lu X, Choi D, Lemmon JP, Liu J (2011) Electrochemical energy storage for green grid. Chem Rev 111(5):3577–3613. doi:10.1021/cr100290v
124. Nishi Y (2014) 2–Past, present and future of lithium-ion batteries: can new technologies open up new horizons? In: Pistoia G (ed) Lithium-ion batteries. Elsevier, Amsterdam, pp 21–39. doi:10.1016/B978-0-444-59513-3.00002-9
125. Orecchini F, Santiangeli A, Dell'Era A (2014) 10–EVs and HEVs using lithium-ion batteries. In: Pistoia G (ed) Lithium-ion batteries. Elsevier, Amsterdam, pp 205–248. doi:10.1016/B978-0-444-59513-3.00010-8
126. Vezzini A (2014) 15–Lithium-ion battery management. In: Pistoia G (ed) Lithium-ion batteries. Elsevier, Amsterdam, pp 345–360. doi:10.1016/B978-0-444-59513-3.00015-7
127. Yoshino A (2014) 1–Development of the lithium-ion battery and recent technological trends. In: Pistoia G (ed) Lithium-ion batteries. Elsevier, Amsterdam, pp 1–20. doi:10.1016/B978-0-444-59513-3.00001-7
128. Chu A, Braatz P (2002) Comparison of commercial supercapacitors and high-power lithium-ion batteries for power-assist applications in hybrid electric vehicles: I. initial characterization. J Power Sources 112(1):236–246. doi:10.1016/S0378-7753(02)00364-6
129. Thounthong P, Raël S, Davat B (2006) Control strategy of fuel cell/supercapacitors hybrid power sources for electric vehicle. J Power Sources 158(1):806–814. doi:10.1016/j.jpowsour.2005.09.014
130. Karden E, Ploumen S, Fricke B, Miller T, Snyder K (2007) Energy storage devices for future hybrid electric vehicles. J Power Sources 168(1):2–11. doi:10.1016/j.jpowsour.2006.10.090
131. Yu Z, Zinger D, Bose A (2011) An innovative optimal power allocation strategy for fuel cell, battery and supercapacitor hybrid electric vehicle. J Power Sources 196(4):2351–2359. doi:10.1016/j.jpowsour.2010.09.057
132. Trovão JP, Pereirinha PG, Jorge HM, Antunes CH (2013) A multi-level energy management system for multi-source electric vehicles—an integrated rule-based meta-heuristic approach. App Energy 105:304–318. doi:10.1016/j.apenergy.2012.12.081
133. Hannan MA, Azidin FA, Mohamed A (2014) Hybrid electric vehicles and their challenges: a review. Renew Sustain Energy Rev 29:135–150. doi:10.1016/j.rser.2013.08.097

134. Inagaki M, Kang F, Toyoda M, Konno H (2014) Chapter 11–carbon materials for electrochemical capacitors. In: Inagaki M, Kang F, Toyoda M, Konno H (eds) Advanced materials science and engineering of carbon. Butterworth-Heinemann, Boston, pp 237–265. doi:10.1016/B978-0-12-407789-8.00011-9
135. Anouti M, Timperman L (2013) A pyrrolidinium nitrate protic ionic liquid-based electrolyte for very low-temperature electrical double-layer capacitors. Phys Chem Chem Phys 15(17):6539–6548
136. Mayrand-Provencher L, Rochefort D (2009) Influence of the conductivity and viscosity of protic ionic liquids electrolytes on the pseudocapacitance of RuO_2 electrodes. J Phys Chem C 113(4):1632–1639. doi:10.1021/jp8084149
137. Mysyk R, Raymundo-Piñero E, Anouti M, Lemordant D, Béguin F (2010) Pseudocapacitance of nanoporous carbons in pyrrolidinium-based protic ionic liquids. Electrochem Commun 12(3):414–417. doi:10.1016/j.elecom.2010.01.007
138. Brandt A, Pires J, Anouti M, Balducci A (2013) An investigation about the cycling stability of supercapacitors containing protic ionic liquids as electrolyte components. Electrochim Acta 108:226–231. doi:10.1016/j.electacta.2013.06.118
139. Timperman L, Galiano H, Lemordant D, Anouti M (2011) Phosphonium-based protic ionic liquid as electrolyte for carbon-based supercapacitors. Electrochem Commun 13(10):1112–1115. doi:10.1016/j.elecom.2011.07.010
140. Kalume A, George L, Cunningham N, Reid SA (2013) Concerted and sequential pathways of proton-coupled electron transfer in hydrogen halide elimination. Chem Phys Lett 556:35–38. doi:10.1016/j.cplett.2012.11.053
141. Gagliardi CJ, Westlake BC, Kent CA, Paul JJ, Papanikolas JM, Meyer TJ (2010) Integrating proton coupled electron transfer (PCET) and excited states. Coord Chem Rev 254(21–22): 2459–2471. doi:10.1016/j.ccr.2010.03.001
142. Costentin C, Robert M, Savéant J-M (2006) Electrochemical concerted proton and electron transfers. Potential-dependent rate constant, reorganization factors, proton tunneling and isotope effects. J Electroanal Chem 588(2):197–206. doi:10.1016/j.jelechem.2005.12.027
143. Richey FW, Dyatkin B, Gogotsi Y, Elabd YA (2013) Ion dynamics in porous carbon electrodes in supercapacitors using in situ infrared spectroelectrochemistry. J Am Chem Soc 135(34):12818–12826. doi:10.1021/ja406120e
144. Zaidi W, Boisset A, Jacquemin J, Timperman L, Anouti M (2014) Deep eutectic solvents based on N-methylacetamide and a lithium salt as electrolytes at elevated temperature for activated carbon-based supercapacitors. J Phys Chem C 118(8):4033–4042. doi:10.1021/jp412552v
145. Yamada Y, Usui K, Chiang CH, Kikuchi K, Furukawa K, Yamada A (2014) General observation of lithium intercalation into graphite in ethylene-carbonate-free superconcentrated electrolytes. ACS Appl Mater Interfaces 6(14):10892–10899. doi:10.1021/am5001163
146. Scrosati B, Garche J (2010) Lithium batteries: status, prospects and future. J Power Sources 195(9):2419–2430. doi:10.1016/j.jpowsour.2009.11.048
147. Catenacci M, Verdolini E, Bosetti V, Fiorese G (2013) Going electric: expert survey on the future of battery technologies for electric vehicles. Energy Policy 61:403–413. doi:10.1016/j.enpol.2013.06.078
148. Wadia C, Albertus P, Srinivasan V (2011) Resource constraints on the battery energy storage potential for grid and transportation applications. J Power Sources 196(3):1593–1598. doi:10.1016/j.jpowsour.2010.08.056
149. Kushnir D, Sandén BA (2012) The time dimension and lithium resource constraints for electric vehicles. Res Policy 37(1):93–103. doi:10.1016/j.resourpol.2011.11.003
150. Pollet BG, Staffell I, Shang JL (2012) Current status of hybrid, battery and fuel cell electric vehicles: From electrochemistry to market prospects. Electrochim Acta 84:235–249. doi:10.1016/j.electacta.2012.03.172
151. Menne S, Pires J, Anouti M, Balducci A (2013) Protic ionic liquids as electrolytes for lithium-ion batteries. Electrochem Commun 31:39–41. doi:10.1016/j.elecom.2013.02.026

Chapter 8
Task-Specific Ionic Liquids for Electrochemical Applications

Hua Zhao

8.1 Introduction

Ionic liquids (ILs) have been widely examined as solvents, additives, or catalysts in a variety of applications including organic catalysis [1–5], inorganic and materials synthesis [6], biocatalysis [7–12], electrochemistry [13], pharmaceutical chemistry [14], polymerization [15, 16], and engineering fluids [17–19]. The physicochemical properties of ILs are tunable through a careful selection of a variety of different cations and anions and a rational combination of ion pairs. More interestingly, cations and/or anions can be further functionalized to form so-called task-specific ionic liquids (TSILs). The notion of TSILs carrying specific functionality tailored for particular applications has proven visionary, and a variety of different functional groups have been grafted onto ILs [20–22]. In particular, there is an exponential growth in constructing new TSILs as electrolytes for electrochemical applications. This chapter focuses on a critical discussion of some major types of TSILs and their electrochemical properties and applications.

8.2 Ether/Thioether- and Hydroxyl/Thiol-Functionalized ILs

8.2.1 General Physical Properties

Our recent review has captured some key physical properties of ether- and alcohol-functionalized ILs [22]. The inclusion of alkyloxy or alkyloxyalkyl groups in ILs often lowers the melting points (T_m) [23–26]. Additionally, the lowering of the

H. Zhao (✉)
Department of Chemistry and Forensic Science, Savannah State University, Savannah, GA 31404, USA
e-mail: huazhao98@gmail.com; zhaoh@savannahstate.edu

A.A.J. Torriero (ed.), *Electrochemistry in Ionic Liquids*,
DOI 10.1007/978-3-319-13485-7_8

glass transition point, T_g, is observed in ILs with no T_m [23, 27, 28]. The likely explanation for the observed T_m and T_g depression due to the ether group incorporation is that the high chain flexibility predominates over the high polarity of the ether group [23]. The complete disappearance of crystallinity and the decrease in T_m of ILs are caused by both the higher rotational freedom of ether chain and the reduced lattice energy of ILs.

The inclusion of ether groups in IL cations often results in a reduction in viscosity. Molecular dynamic simulations indicate that the low viscosity stems from the less effective assembly between more flexible ether chains, which leads to less aggregation/interaction of the chains of neighboring cations rather than any change on ionic pairing [29]. Studies by Raman-induced Kerr spectroscopy [30, 31] show that the larger ionic volume of ether substitution induces a weaker interionic interaction; however, the ether group itself causes a stronger interionic interaction but is of more flexible nature having a faster reorientation than the alkyl group. Based on the density data, the Lee group [32] suggested that the highly flexible alkoxy chains increase the total free volume, which boosts the transport of adjacent molecules and leads to a lower viscosity. On the other hand, the grafting of hydroxyl groups to the cation typically increases the IL viscosity [33]. The hydroxyl groups likely contribute to an intermolecular hydrogen-bond network leading to a higher viscosity [34–36].

Short-chain-ether-functionalized ILs are usually less thermally stable than their corresponding alkyl-substituted analogues [23, 24, 27, 37, 38], most likely due to the weakened cation–anion electrostatic interaction freeing the anions to act as nucleophiles. However, imidazolium ILs containing very long ether chain (7–16 oxygen atoms) have shown a higher thermal stability than alkyl-substituted salts, e.g., [mPEG16-Me-Im]I (containing 16 oxygen atoms) has a higher decomposition temperature (T_d) of 284 °C as compared to that of [C_2mim]I (264 °C) [39]. This was explained by the enhanced intermolecular interactions in the ILs carrying a long ether chain: despite the reduction in electrostatic interaction between the cation and anion due to the shielding of ether chain, strong van der Waals interactions are expected to exist between the polar ether chains. Additionally, due to the wrapping of long ether chain around the cation [40], the nucleophilic attack on the imidazolium ring leading to degradation may be hindered.

A higher solvent polarity is generally observed for ether-/hydroxyl-functionalized imidazolium ILs than those of non-functionalized ILs based on solvatochromic measurements [33, 41, 42] and dielectric constants [43]. The addition of hydroxyl or alkoxy group to the cation remarkably increases the hydrophilicity of ILs [44–46]. Imidazolium ILs form an extended hydrogen-bonded network in both solid and liquid states, with a main contributor being the C-2 hydrogen of the imidazolium ring [47, 48]. The strength of these H-bonds is more dependent on the type of anions than the oxygen functionalization of side chain [27] even though multiple oxygen atoms in the IL side chain can serve as H-bond acceptors for solute molecules.

8.2.2 *Electrochemical Properties*

The ionic conductivity of an IL is mainly controlled by its viscosity (known as the Walden plot), density, formula weight, and size of the cation and anion [49]. Due to the lower viscosities of ether-functionalized ILs, these liquids tend to have higher specific conductivities at 25 °C than alkyl-substituted ILs, such as [$CH_3OCH_2CH_2$-Et_2-MeN][$N(Tf)_2$] (2.6 mS/cm) vs. [Bu-Et_2-MeN][$N(Tf)_2$] (1.6 mS/cm) and [$CH_3OCH_2CH_2$-Et_2-MeN][CF_3BF_3] (3.0 mS/cm) vs. [Bu-Et_2-MeN][CF_3BF_3] (2.1 mS/cm) [23]. Another excellent example was illustrated by Donato et al. [50] as: [$Me(OCH_2CH_2)_3$-mim][BF_4] vs. [C_{10}mim][BF_4] (0.874 mS/cm vs. 0.337 mS/cm at 25 °C), and [$MeOCH_2CH_2$-mim][BF_4] vs. [C_4mim][BF_4] (0.950 mS/cm vs. 0.00124 mS/cm at 25 °C). The same conductivity increase trend was observed for ether-functionalized ILs based on imidazolium cations [38], pyrrolidinium, piperidinium, oxazolidinium, or morpholinium cations [24], phosphonium and ammonium cations [25, 51, 52], as well as guanidinium cations [53, 54]. Belhocine et al. [55] prepared a series of seven-member ring azepanium-based ILs (Scheme 8.1) and found the incorporation of alkoxyalkyl group results in a reduction in viscosities and an increase in conductivities. In addition, cyclic voltammetry indicated very wide electrochemical windows (5.0–6.5 V) of azepanium ILs, implying their high potential as electrolytes. Although [C_2mim]I showed a high ionic conductivity at temperatures above 50 °C, long ether chain-grafted ILs exhibited an even higher conductivity at room temperature than that of [C_2mim]I, which is attributed to the lower viscosity and better solvation of I^- anions in ether-derived ILs. In addition, these ether-functionalized ILs displayed higher resistance against cathodic reduction than alkyl-substituted ILs due to the protective wrapping of long ether side chain around the cation [39]. Therefore, these ether-functionalized imidazolium iodides are potentially ideal electrolytes for electrochemical devices such as dye-sensitized solar cells (DSSCs). However, when introducing two, three, or four ether groups to the quaternary ammonium salts, lower conductivities (and similar/higher viscosities) were observed over ammonium salts carrying no or one ether group [56, 57]. But if the incorporation of more than one alkoxy group to the IL cation leads to a viscosity reduction, an increase in conductivity is expected [52]. The reasonable explanation to this could be the stronger van der Waals interactions for larger

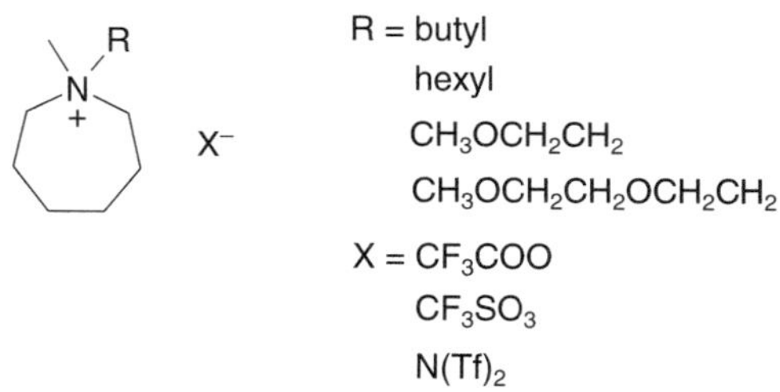

Scheme 8.1 Azepanium ILs

cations and lower cation symmetry. On the other hand, the ether functionalization and a lower viscosity do not always lead to higher conductivity. For example, methoxymethyl ethyl dimethylammonium tetrafluoroborate ($[CH_3OCH_2$-Et-$Me_2N]$ $[BF_4]$) is considered a poor ion conductor based on the Walden plot because this IL has an order of magnitude lower conductivity than ideal [58].

The electrochemical stability is also affected by the incorporation of ether groups. Grafting one or multiple short-chain ether groups in quaternary ammonium, cyclic quaternary ammonium, or sulfonium ILs generally leads to reduced electrochemical stability, but with the electrochemical windows still being sufficiently high (usually >4.0 V) for practical uses as electrolytes [23–26, 28, 51, 57, 59]. For example, $[Me(OCH_2CH_2)_3$-mim$][BF_4]$ has a smaller electrochemical window than $[C_{10}mim]$ $[BF_4]$ (1.23 V vs. 2.00 V at 25 °C); $[MeOCH_2CH_2$-mim$][BF_4]$ also has a smaller electrochemical window than $[C_4mim][BF_4]$ (1.01 V vs. 1.66 V at 25 °C) [50]. Chen et al. [52] argued that the incorporation of a second alkoxy group to the quaternary ammonium cation caused no negative impact on the electrochemical stability of the resulting ILs. The electrochemical potential windows of quaternary ammonium salts tend to be wider than that of the imidazolium type [51]. The electrochemical stability of guanidinium ILs was not affected greatly by the incorporation of ether groups [53, 54]. Liao et al. [60] observed that the grafting of an ether group (–$CH_2OCH_2CH_2CH_2CH_3$) at the C-2 position of 1-propyl-3-methylimidazolium could improve its reduction stability and achieve a very low solid electrolyte interfacial resistance. Furthermore, Jin et al. [61] synthesized 26 ether-functionalized trialkyl-imidazolium ILs at the C-2 position carrying $[N(Tf)_2]^-$ anions and found most of them have viscosities lower than 90 mPa s at 25 °C, conductivities as high as 3.02 mS/cm, a better thermal stability than ILs with alkoxy substituents at the *N* position, as well as wide electrochemical windows (>4.2 V). Chai et al. [62] prepared a series of ether-functionalized trialkylimidazolium ILs (methyl or ethyl group at the C-2 position) carrying $[N(Tf)_2]^-$ anions and reported that 26 ILs have viscosities lower than 100 mPa s (with the lowest value being 54.4 mPa s) at room temperature and these ILs have wide electrochemical windows greater than 4.3 V. Interestingly, when studying the effect of water in two ether-functionalized imidazolium ILs on cyclic voltammetry measurements using a platinum electrode, Migliorini et al. [63] observed diminished cathodic currents in $[Me(OCH_2CH_2)_3$-mim$][CH_3SO_3]$ with as little as 0.19 mM of water, and extremely reduced charge transfer processes (for both cathodic and anodic) in $[MeOCH_2CH_2$-mim$][CH_3SO_3]$ containing 0.19–1.1 mM water. It was suspected that a new species was produced between imidazolium ILs and water, which blocked the active sites of the platinum electrocatalyst. Despite its unknown identity, this species was speculated to be formed by the hydrolysis of the imidazolium cation with the release of protons. Measurements of pH confirmed the increase of proton concentration in these ILs with the addition of water. Liu et al. [64] prepared a number of ether-functionalized ILs based on ammonium cations and carborane anion [*nido*-7,8-$C_2B_9H_{12}]^-$ and found these ILs have low conductivities (the highest value being 233 μS/cm) due to high viscosity, and low electrochemical stability due to relatively high reactivity of the anion at the anode at voltages greater than 1 V. Lang and Kohl [65] suggested that the oxygen position in ether-functionalized ILs also influences the electrochemical stability, for example, [(EtOMe)EtMe$_2$N]Cl has a much lower reductive stability (1 mA/cm^2 at −0.6 V vs. Ag/AgCl) than [(MeOEt)EtMe$_2$N]Cl (−1.78 V).

Yeon et al. [66] reported that some hydroxyl-functionalized imidazolium and morpholinium ILs (paired with $[BF_4]^-$, $[PF_6]^-$, or $[N(Tf)_2]^-$) also have some superb electrochemical properties: [HOCH$_2$CH$_2$-mim]$^+$ type ILs have wide electrochemical windows in the range of 5.4–6.4 V and ionic conductivities ranging from 2.1 to 4.6 mS/cm at 25 °C; hydroxyl-functionalized morpholinium ILs have large electrochemical windows (5.2–6.2 V) as well but much poorer conductivities (0.00066–0.087 mS/cm at 25 °C) due to the high viscosities of these liquids. Pinkert et al. [34] reported that hydroxyl-containing protic ammonium ILs of acetate or formate have conductivity values ranging from 4.8×10^{-4} to 2.3 S/m at 5–75 °C, which suggests that their Walden products (molar conductivity × viscosity) are comparable to aprotic ILs. Interestingly, Wu et al. [36] indicated that hydroxyl-functionalized pyrrolidinium and piperidinium ILs [with $[N(Tf)_2]^-$ as anions] exhibit wider potential windows than those of corresponding ether-functionalized ILs.

8.2.3 Electrochemical Applications

In the early 1980s, ether-functionalized ILs were synthesized and investigated as alternative ambient-temperature electrolytes, such as methoxyethyl ethyl dimethylammonium tetrafluoroborate ([CH$_3$OCH$_2$CH$_2$-Et-Me$_2$N][BF$_4$], $T_m = 13$ °C and $T_g = -98$ °C) [67], and methoxymethyl ethyl dimethylammonium tetrafluoroborate ([CH$_3$OCH$_2$-Et-Me$_2$N][BF$_4$], $T_m = -16$ °C and $T_g = -115$ °C) [68]. In the 1990s, much attention was focused on the incorporation of cationic or anionic units on PEGs to produce the liquid state of the ion conductive polymers (cation-tethered PEG, anion-tethered PEG, or their mixtures), and many of these studies have been reviewed elsewhere [13, 69]. To demonstrate this concept, the Ohno group [70, 71] attached sulfonamide, ethylimidazolium, or sulfonate groups to the ends of PEG oligomers (MW ranging from 150 to 2,000) resulting in viscous liquids (Scheme 8.2). However, their mixtures exhibited drastically decreased T_g and increased ionic conductivity. For example, for PEG-150 with sulfonamide groups, T_g decreased from 3.2 to −61.6 °C and its ionic conductivity increased about 10^5 times to 0.0612 mS/cm at 50 °C after mixing with PEG-150 with appended ethylimidazolium groups. Similarly, Dickinson et al. [72] described the preparation of viscous molten salts containing either polyether-tailed 2-sulfobenzoate or polyether-tailed triethylammonium as counterions of ruthenium hexamine, metal trisbipyridines, metal trisphenanthrolines, and ionic forms of aluminum quinolate, anthraquinone, phthalocyanine, and porphyrins and measured their ionic conductivities and cyclic voltammetry property.

The most common application of oxygen-functionalized ILs is as *electrolytes for lithium batteries* due to their low viscosities, high conductivities, and reasonably high electrochemical stability [53, 56, 57, 73, 74]. The Quartarone group [75] prepared composite electrolytes based on PVdF-HFP [poly(vinylidene fluoride-co-hexafluoropropene)] gelled with a 0.41 mol/kg solution of Li[N(Tf)$_2$] in *N*-methoxyethyl-*N*-methylpyrrolidinium bis(trifluoromethanesulfonyl)imide; when

Scheme 8.2 PEG-derived liquid ionic polymers

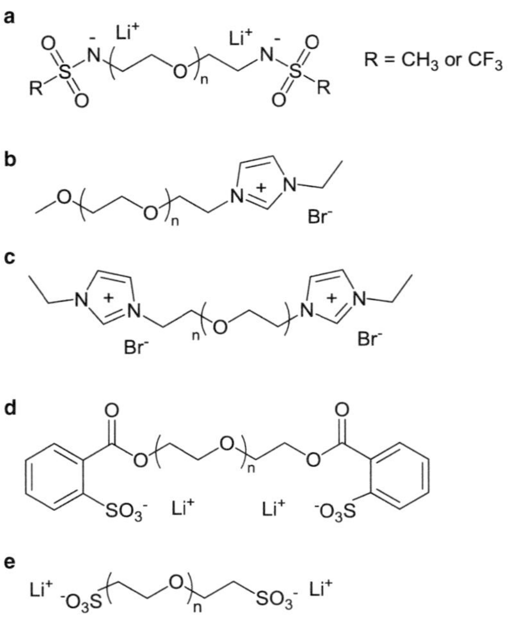

using a mesoporous silica as the filler (vs. HiSil™-based membrane), a higher lithium transference number, a more stable electrochemical interface, and better cycling performances were observed. Consequently, the Li/$LiFePO_4$ cells with this composite film showed satisfactory charge/discharge capacity, discharge efficiency close to unity, as well as low capacity losses at medium C-rates for 180 cycles. Alternatively, the Gerbaldi group [76] revealed a different approach to prepare new electrolyte membranes for Li-based cells: in situ single-step free radical photo-polymerization of dimethacrylic oligomers with up to 60 wt% *N*-methoxyethyl-*N*-methylpyrrolidinium bis(trifluoromethanesulfonyl)imide. This method combines the advantages of polymer electrolytes swollen by conventional organic liquid electrolytes with the non-flammability, high thermal, and electrochemical stability properties of ILs. Accordingly, a satisfactory ionic conductivity (0.1–1 mS/cm) and electrochemical properties are obtained over a wide temperature range. They further studied the polymer electrolyte membranes in lab-scale cells with $LiFePO_4$ as the cathode and Li metal as the anode, resulting in good charge/discharge capacities, Coulombic efficiency close to unity, and low capacity loss at medium C-rates. Han et al. [77] synthesized eight ammonium-based ILs carrying multiple ethoxyethyl groups and observed these ILs have low viscosities (79–135 mPa s at 25 °C) and high conductivities (1.0–2.8 mS/cm at 25 °C). In addition, these ILs could dissolve a high amount of lithium salt (>1.60 mol/kg), and the resulting binary electrolytes display a Li^+ ion transference number (t_{Li^+}) of 0.6–0.7. These functionalized ILs are also electrochemically stable (electrochemical windows ranging 5.42–5.73 V), affording the Li deposition/stripping at room temperature. These favorable properties indicate the high potential of these ILs as electrolytes for Li (or Li-ion) batteries. Chai et al.

[62, 78] synthesized several ether-functionalized pyrazolium ILs containing bis(trifluoromethylsulfonyl)imide anions and found these ILs have conductivities in the range of 1.6–3.31 mS/cm at 25 °C and a wide electrochemical window of 4.4–4.6 V; therefore, these ILs could be used as electrolytes for lithium battery, where a high chemical stability against lithium metal was observed for these electrolytes, and a good capacity and cycle property at the current rate of 0.1 C were obtained for Li/$LiFeO_4$ cells constructing with IL electrolytes without additives. Ferrari et al. [79] prepared two ether-functionalized ILs (i.e., [$MeOCH_2CH_2$-mim][$N(Tf)_2$] and [$Me(OCH_2CH_2)_2$-mim][$N(Tf)_2$]) and found both ILs are thermally stable and have lower viscosities than alkyl-substituted [C_4mim][$N(Tf)_2$]; more notably, both ILs have electrochemical stability windows wider than 5 V. When the new ILs containing Li[$N(Tf)_2$] were mixed with ethylene carbonate/diethyl carbonate (EC/DEC, 1:1 v/v) as electrolytes in Li/$LiFePO_4$ cells, the capacity retention higher than 92 % was achieved at C rate lower than 0.2 C. However, in such systems, the electrochemical stability windows were still under 4 V and the cycling behavior of Li cells was not improved. Nakamoto et al. [80] found that *N*,*N*-diethyl-*N*-methyl-*N*-(2-methoxyethyl)ammonium bis(trifluoromethanesulfonyl)amide exhibits a high stability against electrochemical oxidation over 4 V vs. Li/Li^+ for battery applications and a high O_2 redox reversibility; therefore, it could be a promising electrolyte for nonaqueous lithium-oxygen (Li-O_2) batteries.

In addition, ether-functionalized ILs have shown their unique roles as electrolytes in *electric double-layer capacitors (EDLCs) and dye-sensitized photoelectrochemical solar cells (DSSCs)*. EDLCs do not rely on chemical reactions for charging and discharging electricity as do rechargeable batteries, but physically store energy via the electric double layer. Generally, EDLCs have a longer lifetime and higher power density than rechargeable batteries. In addition to their high performance, EDLCs are environmentally benign devices because heavy metal materials are not required. The purpose of using ILs in EDLCs is to further improve their energy density and high-temperature stability. Sato et al. [81] noticed that an ether-grafted IL, [$CH_3OCH_2CH_2$-Me-Et_2N][BF_4], has a wide potential window (6.0 V) and a high ionic conductivity (4.8 mS cm^{-1} at 25 °C). Subsequently, an EDLC was constructed by a pair of activated carbon electrodes with this IL as the electrolyte. At a working voltage of ~2.5 V, the EDLC exhibited a higher capacity above room temperature and a better charge–discharge cycle durability at 100 °C than a conventional EDLC using an organic liquid electrolyte such as [Et_4N][BF_4] in propylene carbonate. Higashiya et al. [82] prepared oxygen-containing spirobipyrrolidinium salts (Scheme 8.3) for potential applications as electrolytes in EDLCs. This group observed that introducing an oxygen atom (and alkyl groups to break the symmetry) to the spirobipyrrolidinium ring led to a reduction in melting points and viscosities of the resulting ILs. Meanwhile, these new salts retained a high electrochemical stability: the potential window ranged from 5.9 to 7.3 V (0.65 M in acetonitrile), and the conductivity (2.0 M in acetonitrile) varied from 48 to 55 mS cm^{-1}.

Pandey et al. [83] designed new EDLC cells by using multiwalled carbon nanotube (MWCNT) electrodes and polymer electrolyte films comprising of PEG (MW ~ 6×10^5), Li(CF_3SO_3) or Mg(CF_3SO_3)$_2$, and [C_2mim][CF_3SO_3].

Scheme 8.3 Modified ether-functionalized spirobipyrrolidinium salts

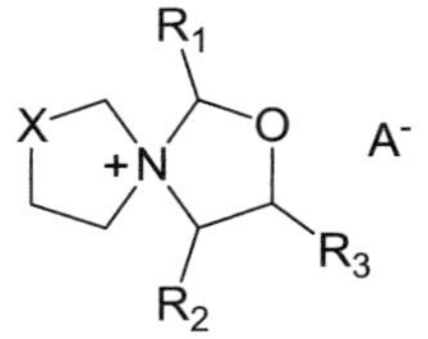

X= CH_2 or O R_1 = H, Me or Et

R_2 = H or Me R_3 = H or Me

A = BF_4, $BF_3C_2F_5$, Tf_2N

Although PEG chains were not covalently attached to the IL cation, the polymer electrolytes showed ionic conductivities of ~0.1 mS/cm at room temperature and the electrochemical stability window of ~4 V. These new EDLC cells with Mg-and Li-based polymer electrolytes exhibited specific capacitance values of 2.6–3.0 and 1.7–2.1 F/(g of MWCNT), respectively, and the Mg-based polymer electrolyte performed slightly better than the Li-type.

An ether-functionalized IL, 3-[2-(2-methoxyethoxy)ethyl]-1-methyl-imidazolium iodide, was used as the *liquid electrolyte in a DSSC*. However, the efficiency of liquid electrolyte was found inferior to those of sol–gel electrolytes [84]. The Lin group [85] attached different PEG methyl ether chains (PEG-150, 350, 550, and 750) to 1-methylimidazole to form ILs as electrolytes for DSSCs. They found that a longer PEG chain resulted in an increased apparent diffusion coefficient of triiodide, an improved dye regeneration rate, lower charge transfer resistance of triiodide reduction, and a higher ionic conductivity despite the higher solvent viscosity. Yang et al. [86] used PEG-1000-functionalized 1-methylimidazolium and ether-functionalized iodide mixed with 1-alkyl-3-methylimidazolium iodides as electrolytes for DSSCs. These new electrolytes showed high ionic conductivities and a large apparent diffusion coefficient for triiodide, particularly with the further addition of *N*-methylimidazole. Zheng et al. [87] constructed dye-sensitized solar cells using pyrazolium iodide as iodide source and ether-functionalized pyrazolium ILs as solvents and achieved a maximum of 1.7 % efficiency, photo-voltage of 0.557 V, and photocurrent density of 6.3 mA/cm^2 using the electrolyte concentration of 0.1 M I_2 and 1 M 1-(2-methoxyethyl)-2-ethylpyrazolium iodide in 1-(2-methoxyethyl)-2-ethylpyrazolium bis(trifluoromethylsulfonyl)imide without additives at 100 mW/cm^2.

The functionalized ILs have also been used in various *electrodes*. Gao et al. [88] added 1-(3-chloro-2-hydroxy-propyl)-pyridinium acetate into carbon paste as a binder to prepare so-called carbon IL electrode, which was used for the voltammetric determination of adenosine-5′-monophosphate (5′-AMP). The new electrode enhanced the oxidation peak currents, resulting in a strong oxidation peak of 5′-AMP with the oxidation peak potential of 1.38 V at pH 4.0 Britton–Robinson buffer. The same group [89] also mixed 1-(3-chloro-2-hydroxypropyl)-3-methylimidazolium tetrafluoroborate in the carbon paste to construct a carbon paste electrode, which was applied in the sensitive detection of adenosine-5′-diphosphate (ADP). They found the new electrode exhibits notable redox catalysis effect on the oxidation of ADP with a high selectivity and a low detection limit of 3.23 μmol/L. Yu et al. [90] synthesized imidazolium-based ILs functionalized by different groups such as hydroxyl, nitrile, carboxyl, and thiol and further employed these ILs in the construction of chitosan/IL composites, which was used as flexible templates for the preparation of Au/Pt nanostructures deposited in situ on the surface of composites via electrodeposition. The resulting chitosan/IL–Au/Pt biosensors for H_2O_2 reduction exhibit a high sensitivity and selectivity, low detection limit, and short response time; the functional groups of ILs influence the electrocatalytic activity of nanocomposites. The Detellier group [91] incorporated hydroxyl-functionalized ILs of different sizes in the interlayer spaces of kaolinite to prepare new nanohybrid materials based on clay minerals. The interlayer distances of the grafted materials could be controlled by the encapsulation of different ILs; as a result, the voltammperometric detection of anions with different sizes (thiocyanate, sulfite, and ferricyanide ions) could be accomplished based on the anion exchange capacity of the new nanomaterials by using glassy carbon electrodes coated with a thin film of the modified kaolinite.

KCl-type salt bridges have limited uses for samples with low ionic strength; however, newer salt bridges constructed from hydrophobic ILs are capable of overcoming this restriction [92]. Specifically, an ether-functionalized phosphonium IL, tributyl(2-methoxyethyl)phosphonium bis(pentafluoroethanesulfonyl)amide, exhibited stable liquid junction potentials when in contact with aqueous solutions with low ionic strength (as low as 1 μM), and thus is a foreseeable electrolyte for salt bridges in accurate determination of pH and other single ion activities [93]. In addition, this IL has a low solubility in water (0.2 mM at room temperature) and is more chemically stable in alkaline solutions than tetraalkylphosphonium salts.

8.2.4 Thiol/Thioether Functionality (Also See Review [94])

Lee et al. [95] demonstrated the formation of self-assembled monolayers (SAMs) of thiol-grafted imidazolium ILs on a gold surface and observed the strong influence of IL anions on the surface hydrophobicity and wettability. Furthermore, the same group [96] reported that SAM surfaces coated with these ILs on a gold electrode exhibited a selective electron transfer toward redox-probe molecules: that is, the electron transfer was observed in the presence of $Fe(CN)_6^{3-}$ but not in the case of $Ru(NH_3)_6^{3+}$. Similarly, Rong et al. [97] immobilized a thiol-functionalized imidazolium IL on the surface of a gold electrode via Au–S bonds and observed more intense and stable electrochemiluminescence (ECL) for the luminol-O_2 system at the Au–IL and water interface. Itoh et al. [98] reduced a disulfide-functionalized imidazolium salt to thiol with in situ formation and stabilization of Au nanoparticles (NPs) (see Scheme 8.4). The thiol-IL-stabilized Au NPs were water soluble and stable against Brønsted acids (such as HCl, HBr, and HBF_4). However, when either HI or HPF_6 was added, agglomeration of nanoparticles along with a color change of the solution was observed, signifying potential applications in optical sensors for anions.

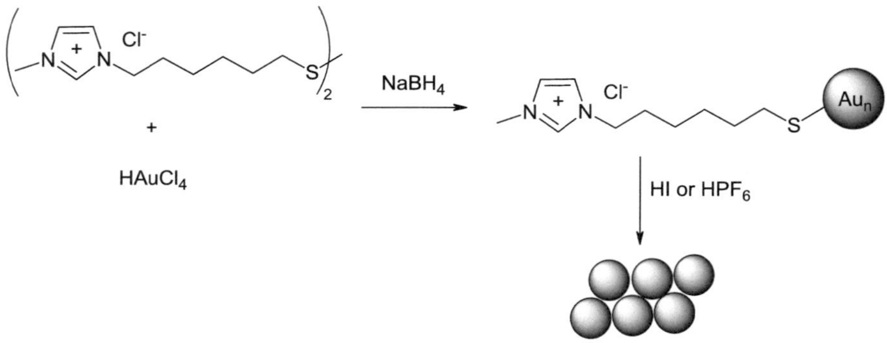

Scheme 8.4 Au NP stabilization using a thiol-functionalized IL

Torriero et al. [99] synthesized thioether-functionalized ILs based on imidazolium, tetrazolium, and benzolium cations and various anions (i.e., $[N(Tf)_2]^-$, $[CF_3SO_3]^-$, $[R_2PO_4]^-$, Cl^-, and $[PF_6]^-$) and found the potential windows (<3.94 V) of these ILs are narrower than common ILs due to the presence of thioether groups. A pyrrolidinium IL incorporating a thioether-functionalized chain was prepared, and its polarity (E_T^N) and three Kamlet–Taft parameters were determined [100]. This study suggested that the thioether-containing IL has a lower polarity and α value (H-bond acidity), but a higher π^* value (dipolarity/polarizability) than alkyl- and cyano-substituted ILs. Lesniewski et al. [101] constructed a gold–carbon nanoparticulate film electrode from oppositely charged conductive particles using a layer-by-layer approach and observed that the use of a thiol-functionalized imidazolium salt as nanoparticle stabilizer (instead of amino groups) required no protonation during the assembling process.

8.3 Carboxyl-Functionalized ILs

Zhao et al. [102] suggested that the incorporation of an ester group onto an IL cation increases the solvent's H-bond acidity due to the weakly acidic α-protons next to the ester group. Methylcarboxyl or cyano group substituted imidazolium ILs were investigated as electrolytes in lithium batteries [103]. Imidazolium, pyridinium, pyrrolidinium, piperidinium, morpholinium, and quaternary ammonium salts of $[N(Tf)_2]^-$ functionalized with carboxyl groups were used in the selective dissolution of metal oxides and hydroxides [104]. Activity coefficients at infinite dilution of various organic compounds in two ester-functionalized ILs (1-methacryloyloxyhexyl-3-methylimidazolium bromide and 1-acryloyloxypropyl-3-methylimidazolium bromide) were measured by the inverse gas chromatography (GC) method [105]. Lu et al. [106] constructed a new electrochemical sensor by printing indium tin oxide (ITO) on electrodes (working, counter, and pseudo-reference electrode), followed by coating a thin layer of [H-betaine]$[N(Tf)_2]$ (Scheme 8.5) on the ITO electrode. The thin IL layer serves as both an electrolyte and a medium for selectively dissolving

Scheme 8.5 Structure of protonated betaine bis(trifluoromethylsulfonyl) imide

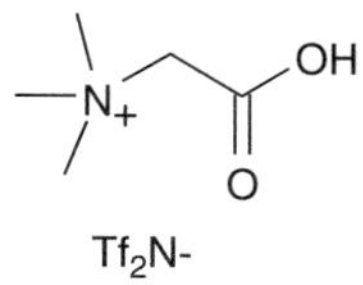

Scheme 8.6 Propylene carbonate-functionalized piperidinium RTIL

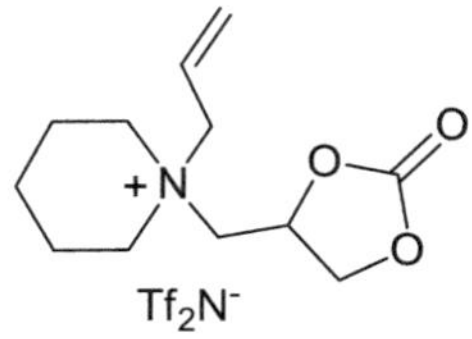

Scheme 8.7 Functionalized pyrrolidium-based ILs

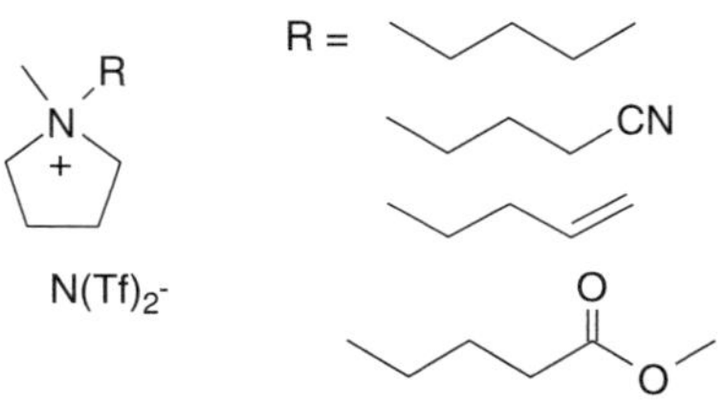

metal oxides and metal salts. As a result, this group achieved accurate quantifications of ng/L PbO, CdO, and CuO in connection with anodic stripping voltammetry by solubilizing ng/L of Bi_2O_3 in an electrolyte.

Wang et al. [107] covalently attached a carboxylic acid-functionalized IL, 1-(methylcarboxylic acid)-3-octylimidazolium bis(trifluoromethylsulfonyl)imide, to an oxidized boron-doped diamond (BDD) surface via an esterification reaction; the IL-modified electrode was found capable of selective electron transfer toward positive and negative redox species: $Fe(CN)_6^{4-}$ was detected on the IL-modified BDD interface, but surface-adsorbed $Ru(NH_3)_6^{3+}$ was not detected. Tsuda et al. [108] prepared a piperidinium IL ($T_g = -24.3$ °C) functionalized with propylene carbonate (see Scheme 8.6) as potential electrolyte for energy-storage devices (ESD) because of its high thermal stability and lithium metal deposition ability, although this IL exhibited a high viscosity (~29,000 mPa s at 40 °C) and low conductivity (0.007 mS/cm at 40 °C). Deng et al. [109] suggest that the incorporation of an ester functionality to the alkyl chain of pyridinium or ammonium cations ($[N(Tf)_2]^-$ as anions) increases the viscosity by fivefold for pyridinium ILs and twofold for ammonium ILs at 20 °C.

Baek et al. [110] prepared pyrrolidium-based ILs carrying different substituents including butyl, butyronitrile, pentenyl, and methyl butyrate (Scheme 8.7). The nitrile- and ester-functionalized ILs show higher viscosities and lower conductivities than the other two. Their potential windows (vs. Li/Li^+) fall in the range of 4.19–4.97 V and are primarily dependent on the oxidation power of the functional group on cations. It was also found that the peak currents at the reduction on the graphite electrode were 100 times stronger in butyl- and pentenyl-substituted ILs than in nitrile- and butyrate-functionalized ILs.

8.4 Amine-Functionalized ILs

The Niu group [111, 112] prepared IL-functionalized single-walled carbon nanotubes (SWNTs) by the amidation between carboxylic acid-containing SWNT and an amine-terminated IL (1-propylamine-3-methylimidazolium bromide), followed by coating the modified SWNT on the surface of glassy carbon electrode (GCE). Glucose oxidase was immobilized on the new GCE by an anion exchange process. They further observed a high bioactivity of immobilized glucose oxidase in the bionanocomposite, as well as a linear amperometric response within 0–20 mM glucose during the electrochemical detection of glucose. Following this study, Gao and Zheng [113] designed a glucose biosensor using an amine-functionalized IL through several steps (Scheme 8.8): (a) the formation of IL-SWNT by the amidation between 1-ethylamine-2,3-dimethylimidazolium bromide and SWNT, and the coating of IL-SWNT on the surface of GCE; (b) electrochemical deposition of gold nanoparticle (GNP) onto IL-SWNT to produce GNP-IL-SWNT nanocomposite; (c) the coating of 1-ethylamine-2,3-dimethylimidazolium bromide on the surface of GNP to prepare IL-GNP-IL-SWNT nanocomposite; (d) the immobilization of glucose oxidase on the new nanocomposite via anion exchange. This glucose oxidase-modified electrode exhibited a superb electrocatalytic activity to the reduction of glucose with a detection limit of 0.8 μmol/L (S/N = 3). The IL layers on the SWNT and GNP are believed to facilitate the electron transfer between glucose oxidase and the electrode.

Also following the Niu's study [112], another group [114] immobilized cytochrome *c* onto SWNTs, which was covalently modified by 1-propylamine-3-methylimidazolium bromide and then coated on GCE, for the electrochemical detection of hydrogen peroxide. The new biosensor showed a wide linear response

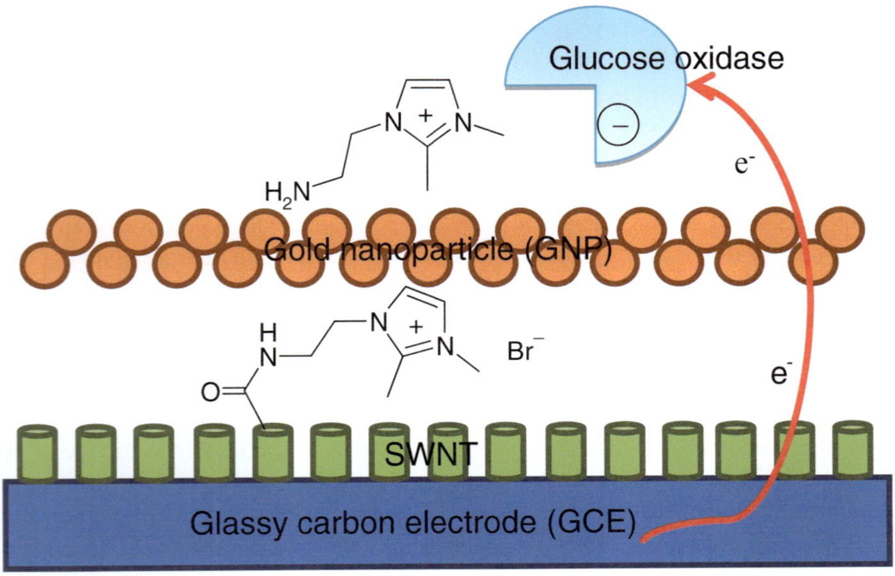

Scheme 8.8 Illustration of glucose oxidase-modified electrode using amine-functionalized IL

range of almost four orders of magnitude of H_2O_2 (4.0×10^{-8} M to 1.0×10^{-4} M) as well as a low detection limit of 1.3×10^{-8} M (based on S/N=3). In addition, the biosensor also has a high selectivity, satisfactory reproducibility, and long-term stability. Li et al. [115] modified graphene nanosheets by 1,3-di(4-amino-1-pyridinium) propane tetrafluoroborate via a covalent attachment of amino groups and epoxy groups in an alkaline solution. The functionalized nanosheets were fabricated onto glassy carbon electrode along with hemoglobin via cross-linking with chitosan and glutaraldehyde. This biosensor enables a direct electron transfer and bioelectrocatalytic reaction of immobilized hemoglobin, leading to a pair of reversible redox peaks of hemoglobin and bioelectrocatalytic detection of H_2O_2 with a high stability, reproducibility, and selectivity.

8.5 Nitrile (or Cyano)-Functionalized ILs

Egashira et al. [116] prepared a nitrile-functionalized IL named $[CTMA][N(Tf)_2]$ (CTMA = *N,N,N,N*-cyanomethyl trimethyl ammonium) and found its conductivity is in the order of 10^{-4} S cm^{-1} at 35 °C (its m.p.), which is lower than that of alkyl-substituted IL. Their further examinations on the lithium deposition/dissolution in 0.2 M $Li[N(Tf)_2]/[CTMA][N(Tf)_2]$ electrolytes suggest a better cycle behavior than using the Li electrolyte based on a non-functionalized IL. A likely explanation is the formation of a protective film on the lithium surface. The Deng group [117] produced a number of nitrile-functionalized imidazolium, pyridinium, and quaternary ammonium ILs containing different anions (Cl^-, $[BF_4]^-$, $[PF_6]^-$, $[N(CN)_2]^-$, and $[N(Tf)_2]^-$); their data indicated that the incorporation of cyano group leads to increased viscosities, decreased conductivities, lower toxicity, and higher electrochemical windows (as high as 5.8 V). The likely reason is that the inclusion of cyano group induces conformational changes of the imidazolium groups and the enhancement in hydrogen-bonding interactions. As suggested earlier by Baek et al. [110] in Scheme 8.7, the nitrile- and ester-functionalized pyrrolidium-based ILs show higher viscosities and lower conductivities than alkyl and pentenyl-substituted analogues; the nitrile-type IL has the widest potential window (4.97 V). Siriwardana et al. [118] synthesized nitrile-functionalized methimazole-type ILs by a two-step route (Scheme 8.9): alkylation of

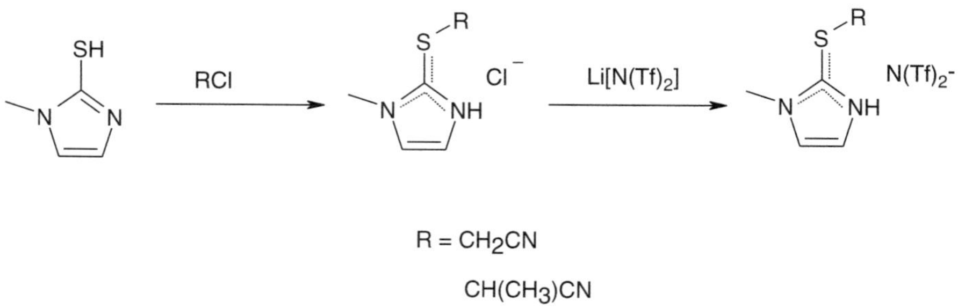

Scheme 8.9 Preparation of nitrile-functionalized methimazole-type ILs

2-mercapto-1-methylimidazole with 2-chloroacetonitrile or 2-chloropropionitrile, and anion metathesis with Li[N(Tf)$_2$]. The resulting ILs have lower potential windows (2.94–3.09 V, glassy carbon electrode) than [C$_2$mim][N(Tf)$_2$] (4.51 V).

8.6 Zwitterionic ILs

Nguyen et al. [119] synthesized several imidazolium-based zwitterionic salts functionalized by a propylsulfonate group at C-3 position and an ester group at C-1 or C-2 position. In particular, grafting an ester group at the C-2 position of the imidazolium ring substantially improved the electrochemical stability of zwitterionic salts, and when these zwitterionic salts were used as additives for lithium battery electrolytes, their cycling performance was also improved compared with those containing a zwitterionic salt with an ester group at C-1 and a hydrogen or methyl group at C-2. For example, cycle performance tests suggested that the initial cell capacity was mostly retained for up to 100 cycles at 0.5 C when 2.5 wt% of 2-(acetoxymethyl)-1-butylimidazolium-3-propylsulfonate was mixed with the model electrolyte comprising ethylene carbonate (EC), dimethyl carbonate (DMC), and ethylmethyl carbonate (EMC) at 1:1:1 (v/v/v) ratio. Rondla et al. [120] observed the liquid crystal behavior of SO_3^--functionalized imidazolium zwitterionic salts (Scheme 8.10), which tend to align perpendicularly on a glass surface due to the hydrogen-bonding interaction between the glass surface and hydrophilic SO_3^- group as well as the high local symmetry surrounding the S atom. Adding lithium salt to these zwitterionic salts decreases the melting temperature while increases the clearing temperatures, resulting in room-temperature liquid crystals and high anisotropic lithium ion conductivity. These room-temperature liquid crystals could become target ion carriers in the electrochemical applications.

8.7 Polymerized ILs

ILs containing alkene groups could be polymerized and turned into valuable polyelectrolytes. Earlier studies on the subject have been reviewed [121]; herein we provide several recent examples. The Niu group [122] constructed a new H_2O_2-detecting electrolyte by fabricating multilayer ultrathin films of polyelectrolyte-functionalized

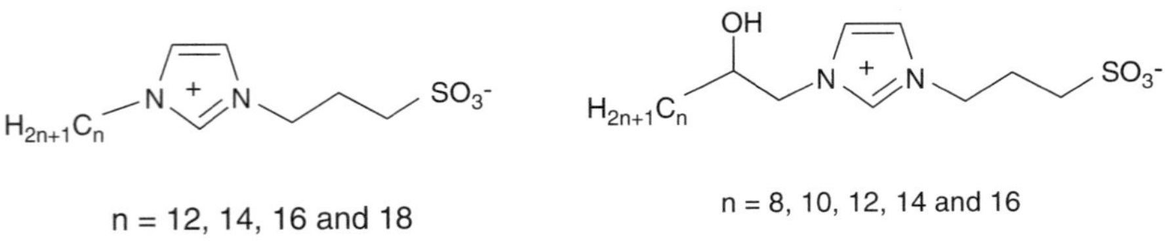

Scheme 8.10 SO_3^--functionalized imidazolium zwitterionic salts

Scheme 8.11 Polyelectrolyte-functionalized IL

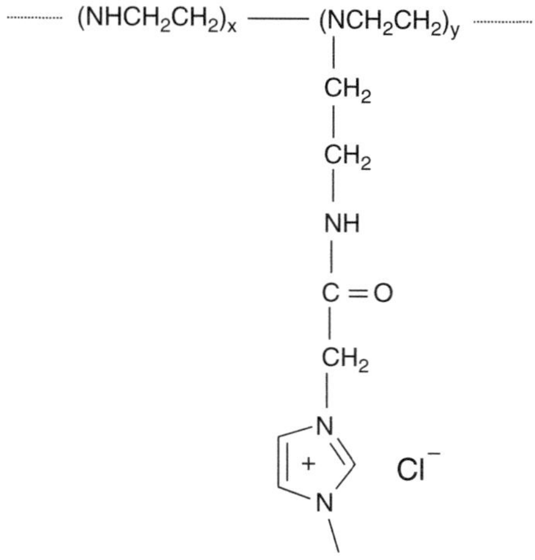

Scheme 8.12 IL monomer for synthesizing poly(ionic liquid)

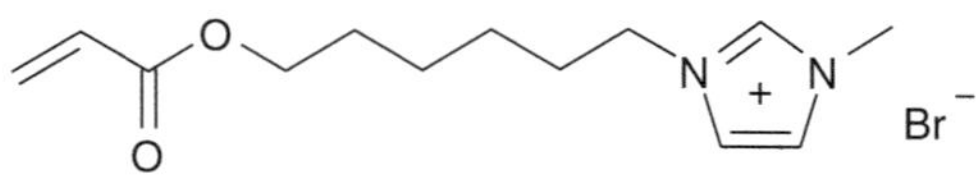

IL (Scheme 8.11) and Prussian blue (PB) nanoparticles onto indium-tin oxide conductive glass using the electrostatic layer-by-layer assembly technique. The new electrolyte exhibits an excellent reproducibility, high stability, and sensitivity (up to 10^{-7} M) for the amperometric measurement of H_2O_2. The Guo group [123] polymerized 3-ethyl-1-vinylimidazolium tetrafluoroborate on ordered mesoporous carbons (OMCs), which became the support material for the deposition of ultrafine Pt nanoparticles via self-assembly; the functionalized OMCs were used to construct a new H_2O_2 sensor with a linear range of 0.1 μM–3.2 mM. In addition, this new H_2O_2 sensor has a high sensitivity of 24.43 μA mM^{-1} and a low detection limit of 0.08 μM. To accurately determine epinephrine (an important catecholamine neurotransmitter), Wu et al. [124] designed an epinephrine sensor by fabricating an ITO electrode with a chitosan film through the electrodeposition of multiwalled carbon nanotubes coated with a polymerized IL (POIL), poly(3-ethyl-1-vinylimidazolium tetrafluoroborate), in the presence of epinephrine as the template. The new sensor shows a linear response to epinephrine in the range of 0.2 μM to 0.67 mM, and a low detection limit of 60 nM (at an S/N of 3). In addition, the electrode exhibits a high reusability and good reproducibility and stability.

Jia et al. [125] constructed ionic polymer-coated nanofibrous membranes through grafting poly(ionic liquid) brushes (Scheme 8.12) onto the electrospun SiO_2 nanofiber surface using atom transfer radical polymerization (ATRP) method. They further demonstrated the potential applications of functionalized membranes as an anion-directed molecular gating system and also a membrane with electroactivity via simple counteranion exchange of the attached poly(ionic liquid) brushes.

Scheme 8.13 Polymerized IL block copolymers

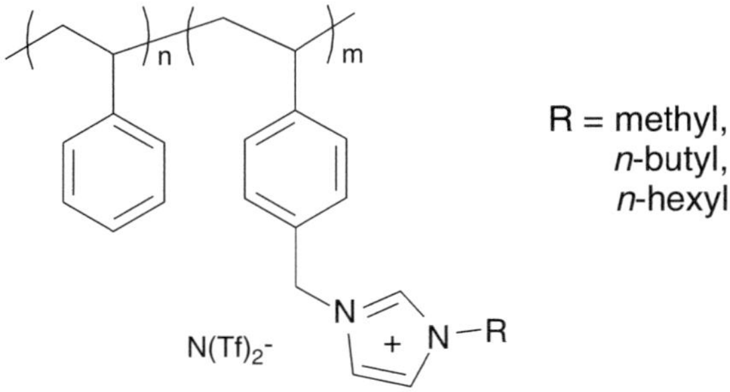

Weber et al. [126] synthesized and characterized poly(styrene-*b*-4-vinylbenzylalkylimidazolium bis(trifluoromethanesulfonyl)imide) (alkyl = CH_3, *n*-C_4H_9, *n*-C_6H_{13}) diblock copolymers (2.7–17.0 mol% POIL) (Scheme 8.13). They further obtained comparable values of conductivity (~0.1 mS cm^{-1} at 150 °C) for lamellar samples of copolymers with similar compositions but different imidazolium alkyl groups; however, the ionic conductivity of copolymers could be related to the POIL composition and the corresponding morphology (0.01 mS cm^{-1} for 8.6 mol% POIL and 0.1 mS cm^{-1} for 17.0 mol% POIL at 150 °C).

Díaz et al. [127] designed Polymer Electrolyte Membrane Fuel Cells (PEMFC) using two different types of membranes based on protic ILs: (a) Nafion membranes impregnated with 1-methyl-3-(4-sulfobutyl)imidazolium bis(trifluoromethylsulfonyl) imide ([HSO_3-Bu-mim][$N(Tf)_2$]) or 1-butyl-3-(4-sulfobutyl)imidazolium trifluoromethanesulfonate ([HSO_3-Bu-bim][CF_3SO_3]), (b) membranes based on the polymerization of 1-(4-sulfobutyl)-3-vinylimidazolium trifluoromethanesulfonate ([HSO_3-Bu-vim][CF_3SO_3]). Their measurements suggested that Nafion membranes impregnated with [HSO_3-Bu-bim][CF_3SO_3] gave the current density of 217 mA/cm^2 and maximum power of 44 mW/cm^2 under anhydrous conditions at 25 °C and [HSO_3-Bu-vim][CF_3SO_3] polymerized electrolytes produced the current density of 154 mA/cm^2 and maximum power of 33 mW/cm^2 at the same condition.

8.8 Other Cation-Functionalized ILs

Szot et al. [128] constructed an electrode based on indium-tin oxide with a controlled amount of conductive nanoparticles by combining the sol containing trimethoxysilyl-functionalized IL (Scheme 8.14) and carbon nanoparticles modified with phenyl sulfonate anionic groups. They found that the new electrode could be used as a reversible sponge for $Fe(CN)_6^{3-}$ anions because of the presence of imidazolium functional groups. In addition, 2,2′-azino-bis(3-ethylbenzothiazoline-6-sulfonate) anion as an enzyme mediator could be irreversibly accumulated on the electrode; the modified electrode shows the catalytic property in O_2-saturated aqueous solution containing laccase. The same group [129] further designed a second electrode by replacing sol with a methanolic suspension of silicate submicroparticles

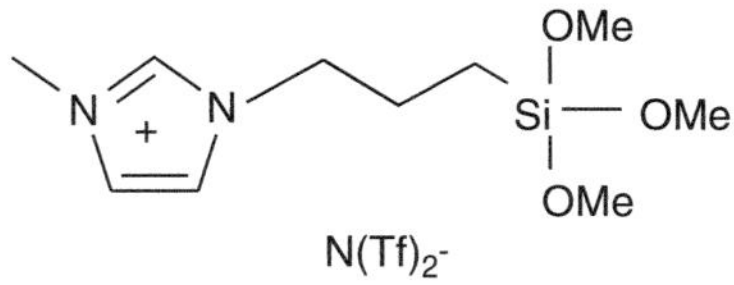

Scheme 8.14 1-Methyl-3-(3-trimethoxysilylpropyl) imidazolium-based IL

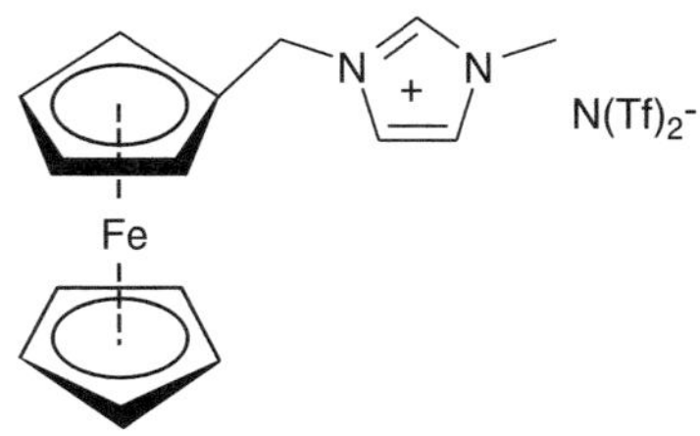

Scheme 8.15 1-Ferrocenylmethyl-3-methylimidazolium bis(trifluoromethanesulfonyl)amide

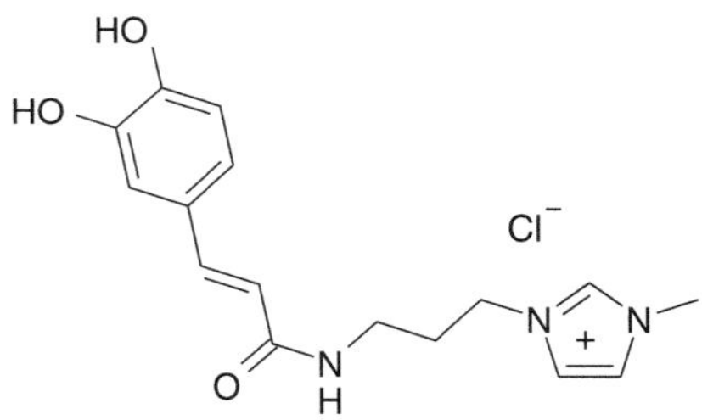

Scheme 8.16 Catechol-functionalized imidazolium chloride

with appended imidazolium functional groups. This group found that following the adsorption of bilirubin oxidase, electrodes constructed from sol and carbon nanoparticles show modest bioelectrocatalytic activity toward dioxygen reduction at pH 4.8.

Weaver et al. [130] synthesized several ferrocene-functionalized phosphonium ILs carrying PF_6^- or OTs^- anions and found the apparent diffusion coefficients for electron transport and for the IL counterions are similar to one another. These ferrocenated phosphonium ILs show similar electron transport properties as those redox ILs based on the combinations of redox groupings with imidazolium and with short-chain polyether functionalities. Forgie et al. [131] prepared a ferrocenyl-substituted imidazolium IL (Scheme 8.15). In its neat form, this IL has a high viscosity and low conductivity. However, after mixing with ethylene carbonate–diethyl carbonate solvent, the electrolytic solutions exhibit reasonable viscosity and conductivity and could be used an additive in lithium battery electrolytes in standard coin cells with a metallic Li anode and a $Li_4Ti_5O_{12}$ cathode.

The Cheng group [132] designed a catechol-functionalized imidazolium salt (Scheme 8.16) and coated it on screen-printed carbon electrode (SPCE) to construct an electrochemical sensing probe for dihydronicotinamide adenine dinucleotide (NADH). The new probe has a high electrocatalytic activity at 0.32 V for the oxidation of NADH in pH 7.0 buffer solutions and a low detection limit (S/N=3) of 0.4 μM. Weng et al. [133] synthesized sulfone-functionalized imidazolium ILs (Scheme 8.17) and found that these ILs are highly compatible with Li/$Li_{1.1}(Ni_{1/3}Co_{1/3}Mn_{1/3})_{0.9}O_2$ (NCM) cathode material in Li battery.

Scheme 8.17 Sulfone-functionalized imidazolium ILs

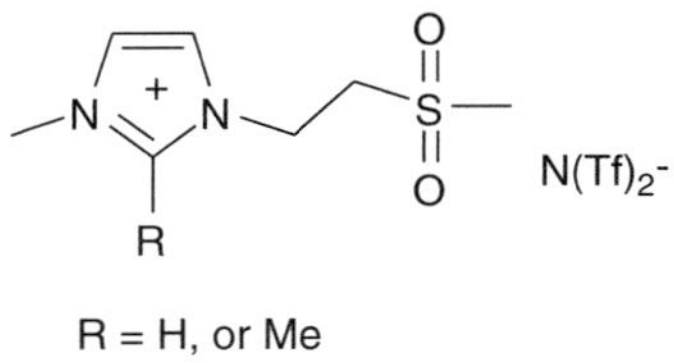

Scheme 8.18 Iodoarene-functionalized imidazolium IL

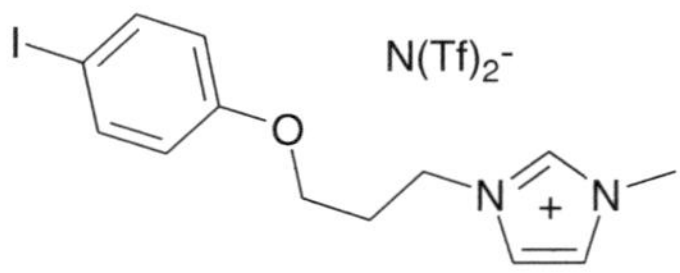

The Fuchigami group [134] prepared a new iodoarene-functionalized imidazolium IL (Scheme 8.18) and used it as a mediator for indirect electrochemical fluorination of organic compounds in a HF-based IL. The viscous IL-type mediator limits its mobility, which allows the efficient indirect electrochemical fluorination in a simple undivided cell (vs. conventional mediatory systems).

8.9 Anion-Functionalized ILs

As discussed earlier, most ether/hydroxyl-functionalized ILs are cation-modified; however, there are also examples where the functionality is incorporated into the anion's structure. For example, ionic melts formed by pairing polyether-tailed 2-sulfobenzoate with ruthenium hexamine, metal trisbipyridines, or porphyrin derivatives allow direct microelectrode voltammetry in the undiluted material [72]. Activity coefficients at infinite dilution for 1-butyl-3-methyl-imidazolium 2-(2-methoxyethoxy) ethyl sulfate were measured and these data were analyzed for the separation of hexane/benzene mixtures [135]. Zhuang et al. [136] synthesized a metal-containing IL, $[(C_{10}H_{21})_2$-benzimidazolium$]_2[Cd_2Cl_6]$ (m.p. 80 °C) and used it as a binder and electrocatalyst in constructing a bulk-modified carbon paste electrode. The new electrode was shown to be highly effective for the reduction of hydrogen peroxide and bromate at a lower detection limit. The same group [137] further constructed a new carbon paste electrode with another metal containing IL, $[(C_4H_9)_2$-benzimidazolium$][FeCl_4]$ (m.p. 70 °C), achieving excellent electrocatalytic activities toward nitrite and bromate reduction with a detection low limit and high sensitivity (0.01 μM and 100.58 μA μM^{-1} for nitrite and 0.01 μM and 83.11 μA μM^{-1} for bromate detection, respectively). Qiao et al. [138] prepared tungstate-functionalized imidazolium ILs $[Rmim]_2[WO_4]$ ($R{=}C_6H_{13}$ or $C_{12}H_{25}$) and found these new ILs have higher viscosities and lower conductivities than those carrying anions of PF_6^- and $CH_3SO_3^-$. Cyclic voltammograms suggested these functionalized ILs have electrochemical potential windows of about 4 V. Su and Cheng [139] deposited $[C_4mim]_3[Fe(CN)_6]$ onto a screen-printed carbon electrode, followed by the addition of a second layer of poly(3-(aminopropyl)trimethoxysilane) sol–gel. The new electrode shows a highly stable redox couple, and in strongly acidic solutions, the reduction of nitrite could be achieved with a low detection limit and high sensitivity also without the interference of dissolved oxygen and common salts.

1-Methyl-3-octyl-imidazolium diethyleneglycolmonomethylether sulfate (ECOENG 48 M) was studied as the solvent for the Suzuki cross-coupling of phenylboronic acid and 4-bromoacetophenone [140]. 1-Ethyl-3-methylimidazolium 2-(2-methoxyethoxy) ethylsulfate was used as a solvent for the extraction of linalool from citrus essential oil [141]. Klein et al. [142] prepared quaternary ammonium salts (including choline salts) of an oligoether carboxylate (2,5,8,11-tetraoxatridecan-13-oate) (Scheme 8.19) with a T_g of −60 °C and viscosities in the range of 260–840 mPa s at 25 °C; these organic liquids exhibit significantly lower viscosities, higher conductivities, and higher polarities than alkali oligoether carboxylates. Enomoto et al. [143] synthesized several room-temperature ILs carrying fluorosulfate anions, i.e., $[C_2mim][SO_3F]$, $[C_2mim][(FO_2SOH)OSO_2F]$, and $[C_2mim][(FO_2SOH)_2O_2SOF]$ and found that they have low viscosities (46.6 mPa s, 9.2 mPa s, and 5.1 mPa s at 25 °C, respectively) and high conductivities (10.8 mS cm^{-1}, 30.8 mS cm^{-1}, and 43.2 mS cm^{-1} at 25 °C, respectively) but different electrochemical windows (4.3 V, 2.0 V, and 2.2 V for glassy carbon electrode). The lower electrochemical stability of the latter two ILs is due to H_2 evolution at the cathode limits.

Liu et al. [144] synthesized a series of new asymmetric ether-functionalized imide anions, (fluorosulfonyl) (polyfluorooxaalkanesulfonyl)imides (Scheme 8.20) pairing with different oniums, such as 1,3-dialkylimidazolium, quaternary alkyl (or ether-functionalized alkyl)ammonium, *N*-alkyl-*N*-methyl-pyrrolidinium, and -piperidinium. Scheme 8.20a-based ILs usually exhibit higher thermal stabilities,

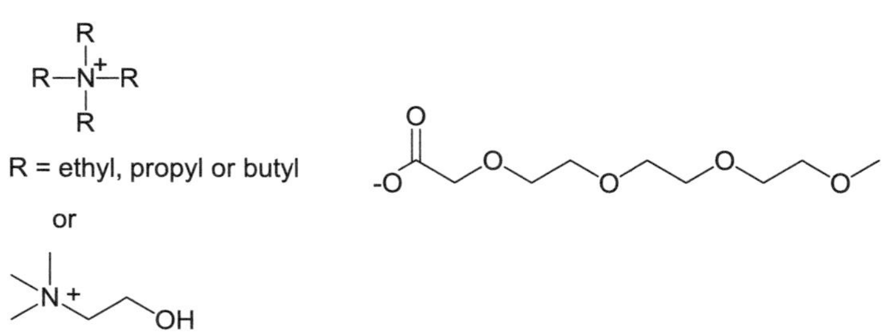

Scheme 8.19 Quaternary ammonium salts of oligoether carboxylate

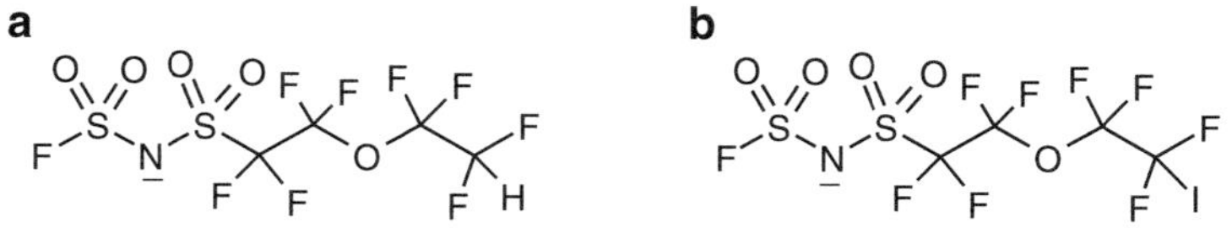

Scheme 8.20 Functionalized anions—(fluorosulfonyl)(polyfluorooxaalkanesulfonyl)imides

lower viscosities (as low as 56 mPa s at 25 °C), higher conductivities (as high as 3.44 mS/cm at 25 °C), wider electrochemical windows (up to 6.35 V), and superior capability for enabling reversible Li deposition/striping than Scheme 8.20b-based analogues. Zheng et al. [145] measured the densities, viscosities, and electrical conductivities of Lewis acidic 1-butyl- and 1-hydrogen-3-methylimidazolium chloroaluminate ILs and suggested their wide applications as electrolytes and catalysts.

It is also known that ILs with cyano-containing anions often have low viscosities (some even below 20 mPa s at room temperature) and high conductivities, such as dicyanamide $[N(CN)_2]^-$ [146, 147], tricyanomethanide $[C(CN)_3]^-$ or TCM^- [148, 149], tetracyanoborate $[B(CN)_4]^-$ [150, 151], and thiocyanate $[SCN]^-$ [152]. Weingarth et al. [153] further established that the electrochemical stability of these ILs follows a decreasing order of $[C_2mim][B(CN)_4] > [C_2mim][N(CN)_2] > [C_2mim][C(CN)_3]$.

8.10 Deep Eutectic Solvents

Recently, the Abbott group [154–156] demonstrated that a mixture of solid organic salt and a complexing agent forms a liquid at temperatures below 100 °C, so-called "deep eutectic solvents" (DES), or "deep eutectic ILs" (DEILs). A typical example is mixing choline chloride (m.p. = 302 °C, 2-hydroxyethyl-trimethylammonium chloride) and urea (m.p. = 133 °C) at a 1:2 molar ratio resulting in a room-temperature liquid (T_f = 12 °C) [154]. In addition to their applications in chemical and enzymatic reactions [22, 157], choline-based eutectic solvents are widely used in electrodeposition and electropolishing of metals [158–160], which has been reviewed for several types of eutectics [161–163]. A recent example by Mele et al. [164] illustrates the preparation of nanowires via electrodeposition in DES through a two-step strategy: (a) potentiostatic electrodeposition of an Au–Mn alloy in eutectic choline chloride/urea (1:2); (b) electrochemically controlled oxidation of Au–Mn deposits, forming simultaneously the nanoporous gold support and the functional Mn-oxide decorating the nanopores. The specific capacity of nanoporous gold-supported nanostructured Mn-oxide was found higher than that of unsupported electrodeposited Mn-oxide nanowires based on the cyclic voltammetry measurements.

8.11 Summary

ILs with a variety of different functionalities have been developed for electrochemical applications. These ILs can be customized to have a wide range of viscosities, conductivities and electrochemical stability by incorporating different take-specific functional groups and making various ion pairs. In addition, the non-volatility and high chemical/thermal stability of ILs make them ideal and versatile electrolytes for many electrochemical applications. A new trend is to combine the task-specific ILs and nanomaterials for constructing novel sensors. There is still a need for further understanding of structure/functionality–property–application relationship, especially more data on physicochemical properties of task-specific ILs.

Acknowledgments The author acknowledges the supports provided by the Henry Dreyfus Teacher-Scholar Award (2012), ACS-PRF (54875-UR9), NIH MBRS-RISE grant (1R25GM096956), and NIH NIBIB contract award (HHSN268201200011C).

References

1. Seddon KR (1997) Ionic liquids for clean technology. J Chem Technol Biotechnol 68(4): 351–356
2. Welton T (1999) Room-temperature ionic liquids–solvents for synthesis and catalysis. Chem Rev 99:2071–2083
3. Zhao H, Malhotra SV (2002) Applications of ionic liquids in organic synthesis. Aldrichimica Acta 35(3):75–83
4. Wasserscheid P, Welton T (2008) Ionic liquids in synthesis, 2nd edn. Wiley, Weinheim
5. Hallett JP, Welton T (2011) Room-temperature ionic liquids: solvents for synthesis and catalysis. 2. Chem Rev 111:3508–3576
6. Endres F, Welton T (2003) Inorganic synthesis. In: Wasserscheid P, Welton T (eds) Ionic liquids in synthesis. Wiley, Weinheim, pp 289–318
7. Kragl U, Eckstein M, Kaftzik N (2002) Enzyme catalysis in ionic liquids. Curr Opin Biotechnol 13:565–571
8. Park S, Kazlauskas RJ (2003) Biocatalysis in ionic liquids—advantages beyond green technology. Curr Opin Biotechnol 14(4):432–437
9. van Rantwijk F, Sheldon RA (2007) Biocatalysis in ionic liquids. Chem Rev 107(6):2757–2785
10. Yang Z, Pan W (2005) Ionic liquids: green solvents for nonaqueous biocatalysis. Enzym Microb Technol 37(1):19–28
11. Zhao H (2010) Methods for stabilizing and activating enzymes in ionic liquids—a review. J Chem Tech Biotechnol 85(7):891–907
12. Moniruzzaman M, Nakashima K, Kamiya N, Goto M (2010) Recent advances of enzymatic reactions in ionic liquids. Biochem Eng J 48(3):295–314
13. Ohno H (2011) Electrochemical aspects of ionic liquids, 2nd edn. Wiley, Hoboken, NJ
14. Hough WL, Rogers RD (2007) Ionic liquids then and now: from solvents to materials to active pharmaceutical ingredients. Bull Chem Soc Jpn 80(12):2262–2269
15. Kubisa P (2004) Application of ionic liquids as solvents for polymerization processes. Prog Polymer Sci 29(1):3–12
16. Carmichael AJ, Haddleton DM (2003) Polymer synthesis in ionic liquids. In: Wasserscheid P, Welton T (eds) Ionic liquids in synthesis. Wiley, Weinheim, pp 319–335
17. Brennecke JF, Maginn EJ (2003) Purification of gas with liquid ionic compounds. US 6,579,343
18. Zhao H, Xia S, Ma P (2005) Use of ionic liquids as 'green' solvents for extractions. J Chem Technol Biotechnol 80(10):1089–1096
19. Zhao H (2006) Innovative applications of ionic liquids as 'green' engineering liquids. Chem Eng Commun 193(12):1660–1677
20. Fei Z, Geldbach TJ, Zhao D, Dyson PJ (2006) From dysfunction to bis-function: on the design and applications of functionalised ionic liquids. Chem Eur J 12(8):2122–2130
21. Yue C, Fang D, Liu L, Yi T-F (2011) Synthesis and application of task-specific ionic liquids used as catalysts and/or solvents in organic unit reactions. J Mol Liq 163(3):99–121
22. Tang S, Baker GA, Zhao H (2012) Ether- and alcohol-functionalized task-specific ionic liquids: attractive properties and applications. Chem Soc Rev 41(10):4030–4066. doi:10.1039/C2CS15362A
23. Zhou Z-B, Matsumoto H, Tatsumi K (2005) Low-melting, low-viscous, hydrophobic ionic liquids: aliphatic quaternary ammonium salts with perfluoroalkyltrifluoroborates. Chem Eur J 11(2):752–766

24. Zhou Z-B, Matsumoto H, Tatsumi K (2006) Cyclic quaternary ammonium ionic liquids with perfluoroalkyltrifluoroborates: synthesis, characterization, and properties. Chem Eur J 12(8): 2196–2212
25. Tsunashima K, Sugiya M (2007) Physical and electrochemical properties of low-viscosity phosphonium ionic liquids as potential electrolytes. Electrochem Commun 9(9):2353–2358
26. Zhou Z-B, Matsumoto H, Tatsumi K (2004) Low-melting, low-viscous, hydrophobic ionic liquids: *N*-alkyl(alkyl ether)-*N*-methylpyrrolidinium perfluoroethyltrifluoroborate. Chem Lett 33(12):1636–1637
27. Schrekker HS, Silva DO, Gelesky MA, Stracke MP, Schrekker CML, Gonçalves RS, Dupont J (2008) Preparation, cation-anion interactions and physicochemical properties of ether-functionalized imidazolium ionic liquids. J Braz Chem Soc 19(3):426–433
28. Matsumoto H, Sakaebe H, Tatsumi K (2005) Preparation of room temperature ionic liquids based on aliphatic onium cations and asymmetric amide anions and their electrochemical properties as a lithium battery electrolyte. J Power Sources 146(1–2):45–50
29. Siqueira LJA, Ribeiro MCC (2009) Alkoxy chain effect on the viscosity of a quaternary ammonium ionic liquid: molecular dynamics simulations. J Phys Chem B 113(4): 1074–1079
30. Shirota H, Fukazawa H, Fujisawa T, Wishart JF (2010) Heavy atom substitution effects in non-aromatic ionic liquids: ultrafast dynamics and physical properties. J Phys Chem B 114(29):9400–9412. doi:10.1021/jp1021104
31. Shirota H, Funston AM, Wishart JF, Castner EWJ (2005) Ultrafast dynamics of pyrrolidinium cation ionic liquids. J Chem Phys 122(18):184512
32. Chen ZJ, Xue T, Lee J-M (2012) What causes the low viscosity of ether-functionalized ionic liquids? Its dependence on the increase of free volume. RSC Adv 2:10564–10574. doi:10.1039/c2ra21772d
33. Jin H, O'Hare B, Dong J, Arzhantsev S, Baker GA, Wishart JF, Benesi AJ, Maroncelli M (2008) Physical properties of ionic liquids consisting of the 1-butyl-3-methylimidazolium cation with various anions and the bis(trifluoromethylsulfonyl)imide anion with various cations. J Phys Chem B 112(1):81–92
34. Pinkert A, Ang KL, Marsh KN, Pang S (2011) Density, viscosity and electrical conductivity of protic alkanolammonium ionic liquids. Phys Chem Chem Phys 13:5136–5143
35. O'Brien RA, Mirjafari A, Jajam V, Capley EN, Stenson AC, West KN, Davis JHJ (2011) Functionalized ionic liquids with highly polar polyhydroxylated appendages and their rapid synthesis via thiol-ene click chemistry. Tetrahedron Lett 52(40):5173–5175
36. Wu T-Y, Su S-G, Lin K-F, Lin Y-C, Wang HP, Lin M-W, Gung S-T, Sun I-W (2011) Voltammetric and physicochemical characterization of hydroxyl- and ether-functionalized onium bis(trifluoromethanesulfonyl)imide ionic liquids. Electrochim Acta 56(21):7278–7287
37. Zhou Z-B, Matsumoto H, Tatsumi K (2004) Low-melting, low-viscous, hydrophobic ionic liquids: 1-alkyl(alkyl ether)-3-methylimidazolium perfluoroalkyltrifluoroborate. Chem Eur J 10(24):6581–6591
38. Monteiro MJ, Camilo FF, Ribeiro MCC, Torresi RM (2010) Ether-bond-containing ionic liquids and the relevance of the ether bond position to transport properties. J Phys Chem B 114(39):12488–12494
39. Ganapatibhotla LVNR, Zheng J, Roy D, Krishnan S (2010) PEGylated imidazolium ionic liquid electrolytes: thermophysical and electrochemical properties. Chem Mater 22(23): 6347–6360
40. Luo S, Zhang S, Wang Y, Xia A, Zhang G, Du X, Xu D (2010) Complexes of ionic liquids with poly(ethylene glycol)s. J Org Chem 75(6):1888–1891
41. Dzyuba SV, Bartsch RA (2002) Expanding the polarity range of ionic liquids. Tetrahedron Lett 43(26):4657–4659
42. Chiappe C, Pomelli CS, Rajamani S (2011) Influence of structural variations in cationic and anionic moieties on the polarity of ionic liquids. J Phys Chem B 115:9653–9661
43. Huang M-M, Jiang Y, Sasisanker P, Driver GW, Weingartner H (2011) Static relative dielectric permittivities of ionic liquids at 25 °C. J Chem Eng Data 56(4):1494–1499

44. Holbrey JD, Turner MB, Reichert WM, Rogers RD (2003) New ionic liquids containing an appended hydroxyl functionality from the atom-efficient, one-pot reaction of 1-methylimidazole and acid with propylene oxide. Green Chem 5:731–736
45. Schrekker HS, Stracke MP, Schrekker CML, Dupont J (2007) Ether-functionalized imidazolium hexafluorophosphate ionic liquids for improved water miscibilities. Ind Eng Chem Res 46(22):7389–7392
46. Leicunaite J, Klimenkovs I, Kviesis J, Zacs D, Kreišmanis JP (2010) Liquid chromatography and characterization of ether-functionalized imidazolium ionic liquids on mixed-mode reversed-phase/cation exchange stationary phase. C R Chim 13(10):1335–1340
47. Dupont J (2004) On the solid, liquid and solution structural organization of imidazolium ionic liquids. J Braz Chem Soc 15(3):341–350
48. Dupont J, Spencer J (2004) On the noninnocent nature of 1,3-dialkylimidazolium ionic liquids. Angew Chem Int Ed 43(40):5296–5297
49. Bonhote P, Dias A-P, Michel A, Papageorgiou N, Kalyanasundaram K, Gratzel M (1996) Hydrophobic, highly conductive ambient-temperature molten salts. Inorg Chem 35(5):1168–1178
50. Donato RK, Migliorini MV, Benvegnú MA, Dupont J, Gonçalves RS, Schrekker HS (2007) The electrochemical properties of a platinum electrode in functionalized room temperature imidazolium ionic liquids. J Solid State Electrochem 11:1481–1487
51. Matsumoto H, Yanagida M, Tanimoto K, Nomura M, Kitagawa Y, Miyazaki Y (2000) Highly conductive room temperature molten salts based on small trimethylalkylammonium cations and bis(trifluoromethylsulfonyl)imide. Chem Lett 29(8):922–923
52. Chen Z, Liu S, Li Z, Zhang Q, Deng Y (2011) Dialkoxy functionalized quaternary ammonium ionic liquids as potential electrolytes and cellulose solvents. New J Chem 35:1596–1606
53. Fang S, Yang L, Wang J, Li M, Tachibana K, Kamijima K (2009) Ionic liquids based on functionalized guanidinium cations and TFSI anion as potential electrolytes. Electrochim Acta 54(17):4269–4273
54. Fang S, Yang L, Wei C, Jiang C, Tachibana K, Kamijima K (2009) Ionic liquids based on guanidinium cations and TFSI anion as potential electrolytes. Electrochim Acta 54(6):1752–1756
55. Belhocine T, Forsyth SA, Gunaratne HQN, Nieuwenhuyzen M, Nockemann P, Puga AV, Seddon KR, Srinivasan G, Whiston K (2011) Azepanium ionic liquids. Green Chem 13(11):3137–3155. doi:10.1039/C1GC15189D
56. Fang S, Zhang Z, Jin Y, Yang L, Hirano S, Tachibana K, Katayama S (2011) New functionalized ionic liquids based on pyrrolidinium and piperidinium cations with two ether groups as electrolytes for lithium battery. J Power Sources 196(13):5637–5644
57. Fang S, Jin Y, Yang L, Hirano S, Kazuhiro Tachibana K, Katayama S (2011) Functionalized ionic liquids based on quaternary ammonium cations with three or four ether groups as new electrolytes for lithium battery. Electrochim Acta 56(12):4663–4671
58. Xu W, Cooper EI, Angel CA (2003) Ionic liquids: ion mobilities, glass temperatures, and fragilities. J Phys Chem B 107:6170–6178
59. Han H-B, Nie J, Liu K, Li W-K, Feng W-F, Armand M, Matsumoto H, Zhou Z-B (2010) Ionic liquids and plastic crystals based on tertiary sulfonium and bis(fluorosulfonyl)imide. Electrochim Acta 55(3):1221–1226
60. Liao C, Shao N, Han KS, Sun X-G, Jiang D-E, Hagaman EW, Dai S (2011) Physicochemical properties of imidazolium-derived ionic liquids with different C-2 substitutions. Phys Chem Chem Phys 13:21503–21510
61. Jin Y, Fang S, Zhang Z, Zhang J, Yang L, Hirano S (2013) C-2 Functionalized trialkylimidazolium ionic liquids with alkoxymethyl group: synthesis, characterization, and properties. Ind Eng Chem Res 52(22):7297–7306. doi:10.1021/ie400371q
62. Chai M, Jin Y, Fang S, Yang L, Hirano S, Tachibana K (2012) Ether-functionalized pyrazolium ionic liquids as new electrolytes for lithium battery. Electrochim Acta 66:67–74. doi:10.1016/j.electacta.2012.01.059
63. Migliorini MV, Donato RK, Benvegnú MA, Dupont J, Gonçalves RS, Schrekker HS (2008) Imidazolium ionic liquid–water mixtures: the formation of a new species that inhibits the electrocatalytical charge transfer processes on a platinum surface. Catal Commun 9(6):971–975

64. Liu S, Chen Z, Zhang Q, Zhang S, Li Z, Shi F, Ma X, Deng Y (2011) Carborane-derivatized low-melting salts with ether-functionalized cations—preparation and properties. Eur J Inorg Chem 2011(12):1910–1920. doi:10.1002/ejic.201001351
65. Lang CM, Kohl PA (2007) Investigation of ether-substituted quaternary ammonium ionic liquids. J Electrochem Soc 154(6):F106–F110. doi:10.1149/1.2721778
66. Yeon S-H, Kim K-S, Choi S, Lee H, Kim HS, Kim H (2005) Physical and electrochemical properties of 1-(2-hydroxyethyl)-3-methyl imidazolium and *N*-(2-hydroxyethyl)-*N*-methyl morpholinium ionic liquids. Electrochim Acta 50(27):5399–5407
67. Cooper EI, Angell CA (1983) Versatile organic iodide melts and glasses with high mole fractions of LiI: glass transition temperatures and electrical conductivities. Solid State Ion 9–10(1):617–622
68. Cooper EI, Angell CA (1986) Ambient temperature plastic crystal fast ion conductors (PLICFICS). Solid State Ion 18–19(1):570–576
69. Ohno H (2006) Functional design of ionic liquids. Bull Chem Soc Jpn 79(11):1665–1680
70. Nakai Y, Ito K, Ohno H (1998) Ion conduction in molten salts prepared by terminal-charged PEO derivatives. Solid State Ion 113–115:199–204
71. Ohno H, Nakai Y, Ito K (1998) Ionic conductivity of molten salts formed by polyether/salt hybrids. Chem Lett 27(1):15–16
72. Dickinson E, Williams ME, Hendrickson SM, Masui H, Murray RW (1999) Hybrid redox polyether melts based on polyether-tailed counterions. J Am Chem Soc 121(4):613–616
73. Fang S, Tang Y, Tai X, Yang L, Tachibana K, Kamijima K (2011) One ether-functionalized guanidinium ionic liquid as new electrolyte for lithium battery. J Power Sources 196(3): 1433–1441
74. Jin Y, Fang S, Yang L, Hirano S, Tachibana K (2011) Functionalized ionic liquids based on guanidinium cations with two ether groups as new electrolytes for lithium battery. J Power Sources 196(24):10658–10666
75. Ferrari S, Quartarone E, Mustarelli P, Magistris A, Fagnoni M, Protti S, Gerbaldi C, Spinella A (2010) Lithium ion conducting PVdF-HFP composite gel electrolytes based on *N*-methoxyethyl-*N*-methylpyrrolidinium bis(trifluoromethanesulfonyl)-imide ionic liquid. J Power Sources 195(2):559–566. doi:10.1016/j.jpowsour.2009.08.015
76. Gerbaldi C, Nair JR, Ferrari S, Chiappone A, Meligrana G, Zanarini S, Mustarelli P, Penazzi N, Bongiovanni R (2012) New electrolyte membranes for Li-based cells: methacrylic polymers encompassing pyrrolidinium-based ionic liquid by single step photo-polymerisation. J Membr Sci 423–424:459–467. doi:10.1016/j.memsci.2012.08.057
77. Han H-B, Liu K, Feng S-W, Zhou S-S, Feng W-F, Nie J, Li H, Huang X-J, Matsumoto H, Armand M, Zhou Z-B (2010) Ionic liquid electrolytes based on multi-methoxyethyl substituted ammoniums and perfluorinated sulfonimides: preparation, characterization, and properties. Electrochim Acta 55(23):7134–7144. doi:10.1016/j.electacta.2010.06.063
78. Chai M, Jin Y, Fang S, Yang L, Hirano S, Tachibana K (2012) Low-viscosity ether-functionalized pyrazolium ionic liquids as new electrolytes for lithium battery. J Power Sources 216:323–329
79. Ferrari S, Quartarone E, Tomasi C, Ravelli D, Protti S, Fagnoni M, Mustarelli P (2013) Alkoxy substituted imidazolium-based ionic liquids as electrolytes for lithium batteries. J Power Sources 235:142–147. doi:10.1016/j.jpowsour.2013.01.149
80. Nakamoto H, Suzuki Y, Shiotsuki T, Mizuno F, Higashi S, Takechi K, Asaoka T, Nishikoori H, Iba H (2013) Ether-functionalized ionic liquid electrolytes for lithium-air batteries. J Power Sources 243:19–23
81. Sato T, Masuda G, Takagi K (2004) Electrochemical properties of novel ionic liquids for electric double layer capacitor applications. Electrochim Acta 49(21):3603–3611
82. Higashiya S, Devarajan TS, Rane-Fondacaro MV, Dangler C, Snyder J, Haldar P (2009) Synthesis of oxygen-containing spirobipyrrolidinium salts for high conductivity room temperature ionic liquids. Helv Chim Acta 92(8):1600–1609
83. Pandey GP, Kumar Y, Hashmi SA (2011) Ionic liquid incorporated PEO based polymer electrolyte for electrical double layer capacitors: a comparative study with lithium and magnesium systems. Solid State Ion 190(1):93–98

84. Stathatos E, Lianos P, Jovanovski V, Orel B (2005) Dye-sensitized photoelectrochemical solar cells based on nanocomposite organic—inorganic materials. J Photochem Photobiol A 169(1):57–61
85. Wang M, Xiao X, Zhou X, Li X, Lin Y (2007) Investigation of PEO-imidazole ionic liquid oligomer electrolytes for dye-sensitized solar cells. Sol Energy Mater Sol Cells 91(9): 785–790
86. Yang H, Liu J, Lin Y, Zhang J, Zhou X (2011) PEO-imidazole ionic liquid-based electrolyte and the influence of NMBI on dye-sensitized solar cells. Electrochim Acta 56(18): 6271–6276
87. Zheng Y, Huang Q, Fang S, Yang L, Gan Y (2013) Ether-functionalized pyrazolium ionic liquids as electrolytes for dye sensitized solar cells. Int J Electrochem Sci 8:9558–9567
88. Gao H, Xi M, Qi X, Lu M, Zhan T, Sun W (2012) Application of a hydroxyl functionalized ionic liquid modified electrode for the sensitive detection of adenosine-5′-monophosphate. J Electroanal Chem 664:88–93. doi:10.1016/j.jelechem.2011.10.018
89. Guo Y, Hu S, Qi X, Xiang J, Sun W (2012) Voltammetric detection of adenosine-5′-diphosphate with a carbon paste electrode modified by a hydroxyl functionalized imidazolium-based ionic liquid. Acta Chim Slov 59:129–135
90. Yu Y, Sun Q, Liu X, Wu H, Zhou T, Shi G (2011) Size-controllable gold–platinum alloy nanoparticles on nine functionalized ionic-liquid surfaces and their application as electrocatalysts for hydrogen peroxide reduction. Chem Eur J 17(40):11314–11323. doi:10.1002/chem.201100010
91. Dedzo GK, Letaief S, Detellier C (2012) Kaolinite–ionic liquid nanohybrid materials as electrochemical sensors for size-selective detection of anions. J Mater Chem 22:20593–20601. doi:10.1039/C2JM34772E
92. Fujino Y, Kakiuchi T (2011) Ionic liquid salt bridge based on *N*-alkyl-*N*-methylpyrrolidinium bis(pentafluoroethanesulfonyl)amide for low ionic strength aqueous solutions. J Electroanal Chem 651(1):61–66
93. Sakaida H, Kitazumi Y, Kakiuchi T (2010) Ionic liquid salt bridge based on tributyl(2-methoxyethyl)phosphonium bis(pentafluoroethanesulfonyl)amide for stable liquid junction potentials in highly diluted aqueous electrolyte solutions. Talanta 83(2):663–666
94. Opallo M, Lesniewski A (2011) A review on electrodes modified with ionic liquids. J Electroanal Chem 656(1–2):2–16
95. Lee BS, Chi YS, Lee JK, Choi IS, Song CE, Namgoong SK, Lee S-G (2004) Imidazolium ion-terminated self-assembled monolayers on Au: effects of counteranions on surface wettability. J Am Chem Soc 126(2):480–481
96. Chi YS, Hwang S, Lee BS, Kwak J, Choi IS, Lee S-G (2005) Anion exchange-promoted $Ru^{3+/2+}$ redox switch in self-assembled monolayers of imidazolium ions on a gold electrode. Langmuir 21(10):4268–4271
97. Rong J, Chi Y, Zhang Y, Chen L, Chen G (2010) Enhanced electrochemiluminescence of luminol-O_2 system at gold–hydrophobic ionic liquidwater interface. Electrochem Commun 12(2):270–273
98. Itoh H, Naka K, Chujo Y (2004) Synthesis of gold nanoparticles modified with ionic liquid based on the imidazolium cation. J Am Chem Soc 126(10):3026–3027
99. Torriero AAJ, Siriwardana AI, Bond AM, Burgar IM, Dunlop NF, Deacon GB, MacFarlane DR (2009) Physical and electrochemical properties of thioether-functionalized ionic liquids. J Phys Chem B 113(32):11222–11231. doi:10.1021/jp9046769
100. Lee J-M, Prausnitz JM (2010) Polarity and hydrogen-bond-donor strength for some ionic liquids: effect of alkyl chain length on the pyrrolidinium cation. Chem Phys Lett 492(1–3): 55–59
101. Lesniewski A, Paszewski M, Opallo M (2010) Gold–carbon three dimensional film electrode prepared from oppositely charged conductive nanoparticles by layer-by-layer approach. Electrochem Commun 12(3):435–437
102. Zhao Q, Eichhorn J, Pitner WR, Anderson JL (2009) Using the solvation parameter model to characterize functionalized ionic liquids containing the tris(pentafluoroethyl)trifluorophosphate (FAP) anion. Anal Bioanal Chem 395(1):225–234

103. Egashira M, Todo H, Yoshimoto N, Morita M, Yamaki J-I (2007) Functionalized imidazolium ionic liquids as electrolyte components of lithium batteries. J Power Sources 174(2): 560–564
104. Nockemann P, Thijs B, Parac-Vogt TN, Van Hecke K, Van Meervelt L, Tinant B, Hartenbach I, Schleid T, Ngan VT, Nguyen MT, Binnemans K (2008) Carboxyl-functionalized task-specific ionic liquids for solubilizing metal oxides. Inorg Chem 47(21):9987–9999
105. Mutelet F, Jaubert J-N, Rogalski M, Harmand J, Sindt M, Mieloszynski J-L (2008) Activity coefficients at infinite dilution of organic compounds in 1-(meth)acryloyloxyalkyl-3-methylimidazolium bromide using inverse gas chromatography. J Phys Chem B 112(12):3773–3785
106. Lu D, Shomali N, Shen A (2010) Task specific ionic liquid for direct electrochemistry of metal oxides. Electrochem Commun 12(9):1214–1217. doi:10.1016/j.elecom.2010.06.022
107. Wang M, Schneider A, Niedziólka-Jönsson J, Marcon L, Ghodbane S, Steinmüller-Nethl D, Li M, Boukherroub R, Szunerits S (2010) Covalent modification of boron-doped diamond electrodes with an imidazolium-based ionic liquid. Electrochim Acta 55(5):1582–1587
108. Tsuda T, Kondo K, Tomioka T, Takahashi Y, Matsumoto H, Kuwabata S, Hussey CL (2011) Design, synthesis, and electrochemistry of room-temperature ionic liquids functionalized with propylene carbonate. Angew Chem Int Ed 50:1310–1313
109. Deng Y, Husson P, Delort A-M, Besse-Hoggan P, Sancelme M, Gomes MFC (2011) Influence of an oxygen functionalization on the physicochemical properties of ionic liquids: density, viscosity, and carbon dioxide solubility as a function of temperature. J Chem Eng Data 56(11):4194–4202
110. Baek B, Lee S, Jung C (2011) Pyrrolidinium cation-based ionic liquids with different functional groups: butyl, butyronitrile, pentenyl, and methyl butyrate. Int J Electrochem Sci 6:6220–6234
111. Zhang Y, Shen Y, Yuan J, Han D, Wang Z, Zhang Q, Niu L (2006) Design and synthesis of multifunctional materials based on an ionic-liquid backbone. Angew Chem Int Ed 45(35):5867–5870
112. Zhang Y, Shen Y, Han D, Wang Z, Song J, Li F, Niu L (2007) Carbon nanotubes and glucose oxidase bionanocomposite bridged by ionic liquid-like unit: Preparation and electrochemical properties. Biosens Bioelectron 23(3):438–443. doi:10.1016/j.bios.2007.06.010
113. Gao R, Zheng J (2009) Amine-terminated ionic liquid functionalized carbon nanotube-gold nanoparticles for investigating the direct electron transfer of glucose oxidase. Electrochem Commun 11(3):608–611. doi:10.1016/j.elecom.2008.12.060
114. Liu X, Bu C, Nan Z, Zheng L, Qiu Y, Lu X (2013) Enzymes immobilized on amine-terminated ionic liquid-functionalized carbon nanotube for hydrogen peroxide determination. Talanta 105:63–68
115. Li R, Liu C, Ma M, Wang Z, Zhan G, Li B, Wang X, Fang H, Zhang H, Li C (2013) Synthesis of 1,3-di(4-amino-1-pyridinium)propane ionic liquid functionalized graphene nanosheets and its application in direct electrochemistry of hemoglobin. Electrochim Acta 95:71–79
116. Egashira M, Okada S, Yamaki J, Dri DA, Bonadies F, Scrosati B (2004) The preparation of quaternary ammonium-based ionic liquid containing a cyano group and its properties in a lithium battery electrolyte. J Power Sources 138(1–2):240–244. doi:10.1016/j.jpowsour.2004.06.022
117. Zhang Q, Li Z, Zhang J, Zhang S, Zhu L, Yang J, Zhang X, Deng Y (2007) Physicochemical properties of nitrile-functionalized ionic liquids. J Phys Chem B 111(11):2864–2872. doi:10.1021/jp067327s
118. Siriwardana AI, Torriero AAJ, Reyna-Gonzalez JM, Burgar IM, Dunlop NF, Bond AM, Deacon GB, MacFarlane DR (2010) Nitrile functionalized methimazole-based ionic liquids. J Org Chem 75:8376–8382. doi:10.1021/jo101449q
119. Nguyen DQ, Bae HW, Jeon EH, Lee JS, Cheong M, Kim H, Kim HS, Lee H (2008) Zwitterionic imidazolium compounds with high cathodic stability as additives for lithium battery electrolytes. J Power Sources 183(1):303–309. doi:10.1016/j.jpowsour.2008.04.047
120. Rondla R, Lin JCY, Yang CT, Lin IJB (2013) Strong tendency of homeotropic alignment and anisotropic lithium ion conductivity of sulfonate functionalized zwitterionic imidazolium ionic liquid crystals. Langmuir 29(37):11779–11785. doi:10.1021/la402336n

121. Ohno H, Yoshizawa-Fujita M (2011) Ion conductive polymers. In: Ohno H (ed) Electrochemical aspects of ionic liquids, 2nd edn. Wiley, Hoboken, NJ, pp 419–431
122. Li F, Shan C, Bu X, Shen Y, Yang G, Niu L (2008) Fabrication and electrochemical characterization of electrostatic assembly of polyelectrolyte-functionalized ionic liquid and Prussian blue ultrathin films. J Electroanal Chem 616(1–2):1–6. doi:10.1016/j.jelechem.2007.12.020
123. Bo X, Bai J, Qi B, Guo L (2011) Ultra-fine Pt nanoparticles supported on ionic liquid polymer-functionalized ordered mesoporous carbons for nonenzymatic hydrogen peroxide detection. Biosens Bioelectron 28(1):77–83. doi:10.1016/j.bios.2011.07.001
124. Wu Y, Feng X, Zhou S, Shi H, Wu H, Zhao S, Song W (2013) Sensing epinephrine with an ITO electrode modified with an imprinted chitosan film containing multi-walled carbon nanotubes and a polymerized ionic liquid. Microchim Acta 180:1325–1332. doi:10.1007/s00604-013-1063-y
125. Jia W, Wu Y, Huang J, An Q, Xu D, Wu Y, Li F, Li G (2010) Poly(ionic liquid) brush coated electrospun membrane: a useful platform for the development of functionalized membrane systems. J Mater Chem 20:8617–8623. doi:10.1039/c0jm01179g
126. Weber RL, Ye Y, Schmitt AL, Banik SM, Elabd YA, Mahanthappa MK (2011) Effect of nanoscale morphology on the conductivity of polymerized ionic liquid block copolymers. Macromolecules 44(14):5727–5735. doi:10.1021/ma201067h
127. Díaz M, Ortiz A, Vilas M, Tojo E, Ortiz I (2013) Performance of PEMFC with new polyvinyl-ionic liquids based membranes as electrolytes. Int J Hydrog Energy. doi:10.1016/j.ijhydene.2013.04.155
128. Szot K, Lesniewski A, Niedziolka J, Jönsson M, Rizzi C, Gaillon L, Marken F, Rogalski J, Opallo M (2008) Sol–gel processed ionic liquid—hydrophilic carbon nanoparticles multilayer film electrode prepared by layer-by-layer method. J Electroanal Chem 623(2):170–176. doi:10.1016/j.jelechem.2008.07.023
129. Szot K, Lynch RP, Lesniewski A, Majewska E, Sirieix-Plenet J, Gaillon L, Opallo M (2011) The effect of linker of electrodes prepared from sol–gel ionic liquid precursor and carbon nanoparticles on dioxygen electroreduction bioelectrocatalysis. Electrochim Acta 56(28):10306–10312. doi:10.1016/j.electacta.2011.03.139
130. Weaver JEF, Breadner D, Deng F, Ramjee B, Ragogna PJ, Murray RW (2011) Electrochemistry of ferrocene-functionalized phosphonium ionic liquids. J Phys Chem C 115:19379–19385
131. Forgie JC, El Khakani S, MacNeil DD, Rochefort D (2013) Electrochemical characterisation of a lithium-ion battery electrolyte based on mixtures of carbonates with a ferrocene-functionalised imidazolium electroactive ionic liquid. Phys Chem Chem Phys 15:7713–7721. doi:10.1039/C3CP50560J
132. Wang S-M, Cheng H-H, Lai K-F, Cheng S-H (2012) Surface redox-mediated dihydronicotinamide adenine dinucleotide probes based on ionic liquids covalently bound with catechol functionality. Electrochim Acta 77:330–338. doi:10.1016/j.electacta.2012.06.008
133. Weng W, Zhang Z, Schlueter JA, Amine K (2013) Synthesis and electrochemical property of sulfone-functionalized imidazolium ionic liquid electrolytes. Electrochim Acta 92:392–396
134. Sawamura T, Kuribayashi S, Inagi S, Fuchigami T (2010) Use of task-specific ionic liquid for selective electrocatalytic fluorination. Org Lett 12(3):644–646. doi:10.1021/ol9028836
135. Letcher TM, Domanska U, Marciniak M, Marciniak A (2005) Activity coefficients at infinite dilution measurements for organic solutes in the ionic liquid 1-butyl-3-methyl-imidazolium 2-(2-methoxyethoxy) ethyl sulfate using g.l.c. at T=(298.15, 303.15, and 308.15) K. J Chem Thermodyn 37(6):587–593
136. Zhuang R, Jian F, Wang K (2009) A new binuclear Cd(II)-containing ionic liquid: preparation and electrocatalytic activities. J Organomet Chem 694(22):3614–3618. doi:10.1016/j.jorganchem.2009.07.012
137. Wang KF, Zhang L, Zhuang RR, Jian FF (2011) An iron(III)-containing ionic liquid: characterization, magnetic property and electrocatalysis. Transit Met Chem 36:785–791. doi:10.1007/s11243-011-9520-0
138. Qiao Y, Hu J, Li H, Hua L, Hu Y, Feng B, Hou Z (2010) Physicochemical properties of tungstate-based room-temperature ionic liquids. J Electrochem Soc 157(9):F124–F129. doi:10.1149/1.3462993

139. Su W-Y, Cheng S-H (2012) Functionalized ionic liquid and sol-gel composite film for the electrochemical reduction and determination of nitrite without interference from dissolved oxygen. Int J Electrochem Sci 7:9058–9073
140. Wong H, Han S, Livingston AG (2006) The effect of ionic liquids on product yield and catalyst stability. Chem Eng Sci 61(4):1338–1341
141. Francisco M, Lago S, Soto A, Arce A (2010) Essential oil deterpenation by solvent extraction using 1-ethyl-3-methylimidazolium 2-(2-methoxyethoxy) ethylsulfate ionic liquid. Fluid Phase Equilib 296(2):149–153
142. Klein R, Zech O, Maurer E, Kellermeier M, Kunz W (2011) Oligoether carboxylates: task-specific room-temperature ionic liquids. J Phys Chem B 115(29):8961–8969
143. Enomoto T, Matsumoto K, Hagiwara R (2011) Properties of fluorosulfate-based ionic liquids and geometries of $(FO_2SOH)OSO_2F^-$ and $(FO_2SOH)_2O_2SOF^-$. Dalton Trans 40:12491–12499. doi:10.1039/C1DT11014D
144. Liu C, Xu F, Feng S, Zheng L, Zhang H, Feng W, Huang X, Armand M, Nie J, Zhou Z (2013) New hydrophobic ionic liquids based on (fluorosulfonyl)(polyfluorooxaalkanesulfonyl)imides with various oniums. Electrochim Acta 99:262–272. doi:10.1016/j.electacta.2013.02.095
145. Zheng Y, Dong K, Wang Q, Zhang J, Lu X (2013) Density, viscosity, and conductivity of Lewis acidic 1-butyl- and 1-hydrogen-3-methylimidazolium chloroaluminate ionic liquids. J Chem Eng Data 58(1):32–42. doi:10.1021/je3004904
146. MacFarlane DR, Golding J, Forsyth S, Forsyth M, Deacon GB (2001) Low viscosity ionic liquids based on organic salts of the dicyanamide anion. Chem Commun 16:1430–1431
147. Fendt S, Padmanabhan S, Blanch HW, Prausnitz JM (2011) Viscosities of acetate or chloride-based ionic liquids and some of their mixtures with water or other common solvents. J Chem Eng Data 56(1):31–34
148. Wang P, Zakeeruddin SM, Grätzel M, Kantlehner W, Mezger J, Stoyanov EV, Scherr O (2004) Novel room temperature ionic liquids of hexaalkyl substituted guanidinium salts for dye-sensitized solar cells. Appl Phys A Mater Sci Process 79(1):73–77
149. Wang P, Wenger B, Humphry-Baker R, Moser J-E, Teuscher J, Kantlehner W, Mezger J, Stoyanov EV, Zakeeruddin SM, Grätzel M (2005) Charge separation and efficient light energy conversion in sensitized mesoscopic solar cells based on binary ionic liquids. J Am Chem Soc 127(18):6850–6856
150. Kuang D, Wang P, Ito S, Zakeeruddin SM, Grätzel M (2006) Stable mesoscopic dye-sensitized solar cells based on tetracyanoborate ionic liquid electrolyte. J Am Chem Soc 128(24):7732–7733
151. Kuang D, Klein C, Zhang Z, Ito S, Moser J-E, Zakeeruddin SM, Grätzel M (2007) Stable, high-efficiency ionic-liquid-based mesoscopic dye-sensitized solar cells. Small 3(12):2094–2102
152. Wang P, Zakeeruddin SM, Humphry-Baker R, Grätzel M (2004) A binary ionic liquid electrolyte to achieve>=7 % power conversion efficiencies in dye-sensitized solar cells. Chem Mater 16(14):2694–2696
153. Weingarth D, Czekaj I, Fei Z, Foelske-Schmitz A, Dyson PJ, Wokaun A, Kötz R (2012) Electrochemical stability of imidazolium based ionic liquids containing cyano groups in the anion: a cyclic voltammetry, XPS and DFT study. J Electrochem Soc 159(7):H611–H615. doi:10.1149/2.001207jes
154. Abbott AP, Capper G, Davies DL, Rasheed RK, Tambyrajah V (2003) Novel solvent properties of choline chloride/urea mixtures. Chem Commun (1):70–71
155. Abbott AP, Boothby D, Capper G, Davies DL, Rasheed RK (2004) Deep eutectic solvents formed between choline chloride and carboxylic acids: versatile alternatives to ionic liquids. J Am Chem Soc 126:9142–9147
156. Abbott AP, Capper G, Gray S (2006) Design of improved deep eutectic solvents using hole theory. ChemPhysChem 7:803–806
157. Zhao H, Baker GA (2013) Ionic liquids and deep eutectic solvents for biodiesel synthesis: a review. J Chem Tech Biotechnol 88(1):3–12. doi:10.1002/jctb.3935
158. Abbott AP, Capper G, Swain BG, Wheeler DA (2005) Electropolishing of stainless steel in an ionic liquid. Trans Inst Metal Finishing 83(1):51–53

159. Abbott AP, Capper G, McKenzie KJ, Ryder KS (2007) Electrodeposition of zinc–tin alloys from deep eutectic solvents based on choline chloride. J Electroanal Chem 599(2):288–294
160. Haerens K, Matthijs E, Chmielarz A, Van der Bruggen B (2009) The use of ionic liquids based on choline chloride for metal deposition: a green alternative? J Environ Manage 90(11):3245–3252
161. Abbott AP, McKenzie KJ (2006) Application of ionic liquids to the electrodeposition of metals. Phys Chem Chem Phys 8:4265–4279
162. Abbott AP, Ryder KS, König U (2008) Electrofinishing of metals using eutectic based ionic liquids. Trans Inst Metal Finishing 86(4):196–204
163. Endres F, MacFarlane DR, Abbott AP (2008) Electrodeposition from ionic liquids. Wiley, Weinheim
164. Mele C, Catalano M, Taurino A, Bozzini B (2013) Electrochemical fabrication of nanoporous gold-supported manganese oxide nanowires based on electrodeposition from eutectic urea/choline chloride ionic liquid. Electrochim Acta 87:918–924. doi:10.1016/j.electacta.2012.09.038

Chapter 9
Polymeric Ion Gels: Preparation Methods, Characterization, and Applications

Maitane Salsamendi, Laurent Rubatat, and David Mecerreyes

9.1 Introduction

Ion gels or ionogels are a relatively new type of gels where the liquid phase, percolating throughout the solid phase, is an ionic liquid (IL). Ion gels are unique since they combine some properties and overcome others of well-known gels such as hydrogels, where the liquid phase is water, or organogels, where the liquid phase is an organic solvent. The properties given by the ionic liquids such as the high ionic conductivity and a wide electrochemical stability window are preserved in the ion gels, making them suitable and very attractive soft solid materials for electrochemists [1–5]. Additionally, the negligible vapor pressure, the nonflammability, and the thermal stability of the liquid phase of ionic liquids will permit a use of ion gels at elevated temperatures, up to 400 °C in some reported cases. Ion gels are formed by an ionic liquid component and a gelator also named solid phase which can be polymeric or nonpolymeric, organic or inorganic. There are two clear ion gel families which depend on the chemical nature of the solid phase, inorganic or organic, i.e., polymeric. Whereas many works and recent reviews have been dedicated to ion gels where the solid phase is based on inorganic sol–gel materials, here, we will focus on the use of polymers to develop ion gels.

M. Salsamendi (✉)
POLYMAT, University of the Basque Country UPV/EHU,
Avda. Tolosa 72, 20018 San Sebastian, Spain
e-mail: maitane.salsamendi@ehu.es

L. Rubatat
EPCP, UMR IPREM 5256, Université de Pau et des Pays de l'Adour, Pau, France

D. Mecerreyes
POLYMAT, University of the Basque Country UPV/EHU,
Avda. Tolosa 72, 20018 San Sebastian, Spain

Ikerbasque, Basque Foundation for Science, 48011 Bilbao, Spain

A.A.J. Torriero (ed.), *Electrochemistry in Ionic Liquids*,
DOI 10.1007/978-3-319-13485-7_9

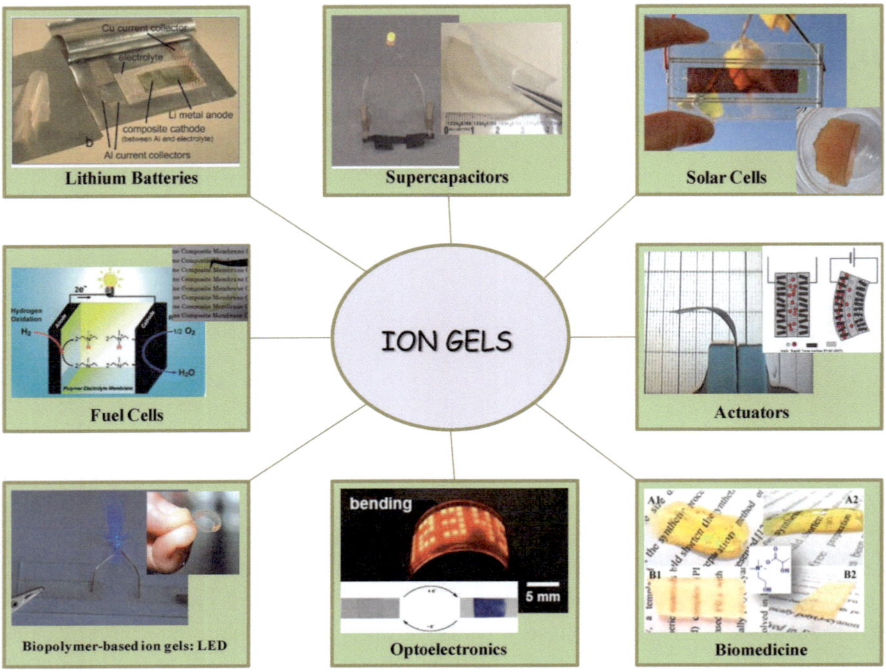

Fig. 9.1 Schematic representation of the applications of ion gels

In this chapter, a general overview of ion gels and their applications, mainly focused on electrochemical technologies, will be presented. The chapter will start with a general definition of ion gels and their preparation methods. Then we will describe the properties and advantages of ion gels, and finally their various applications. It is worth to stress that both ionic liquids and polymers offer unlimited chemical versatility; therefore, ion gels are a unique platform to design original materials [1]. Interestingly, the solid-like behavior enlarges the broad array of applications of ionic liquids. Applications of ion gels are illustrated in Fig. 9.1, which include electrolytic membranes for batteries, fuel cells, solar cells, actuators, electrochemical sensors, biosensors, supercapacitors, separation and catalytic membranes, or drug release.

9.2 Gels: Basic Definition and Classifications

Thomas Graham used the term "gel" in 1861 for the first time, and nowadays it is a common household item which is easily identified by the simple "inversion test": if the material is able to support its own weight without falling out when a pot is turned upside down, it is considered as a gel [6, 7]. However, as Dorothy Jordan Lloyd noted in 1926, gels are "…easier to recognize than to define." The generally accepted definition is that given by Flory in 1974, which is the next: A gel is a two-component, colloidal dispersion with a continuous structure with macroscopic dimensions that is permanent on the timescale of the experiment and is solid-like in its rheological behavior [8, 9].

Even though gels are predominantly liquid in composition (typically 99 % by weight, the other 1 % is the gelator), it remains a solid, i.e., a jelly-like material that does not flow [10]. By changing the nature or the concentration of the gelator (gelling component) it is possible to modify and tune the physical properties of the material. This way, a myriad of materials with different characteristics have been recognized to form gels. In a general way, gels can be classified in two major groups: (1) the physical gels, when the internal three-dimensional network is cross-linked through weak and reversible interactions (e.g., hydrogen bonds, hydrophobic interactions, van der Waals forces, etc.), and (2) the chemical gels, when the cross-links result from covalent bonding. The term supramolecular gel, which is intensively used in recent years, refers to some of the so-called physical gels where the gelating agents make use of noncovalent interactions to form the gel.

Depending on the fluid component used to form the gel, they were classified as *organogels* or *hydrogels* [11]. When water is the liquid component, the gel is named hydrogel [12], whereas the organogel term is used when the liquid is an organic solvent [13]. More recently, ion gels were proposed as an additional type of gel where the liquid phase is an ionic liquid, immobilized in a three-dimensional network presenting a solid-like behavior. Ion gels have some of the properties of hydrogels since they are ionic in nature and they usually have some inorganic components in the structure (ionic liquid's anions). However, they also have some properties similar to organogels due to the chemical nature of the ionic liquid cation. Furthermore, ion gels present some unique properties such as the low vapor pressure and thermal stability of the liquid phase and wide electrochemical applications [2]. Due to these facts, they have been classified as a new type of gel which complements the terms hydrogels and organogels.

9.3 Preparation Methods of Polymeric Ion Gels

In recent years, many studies have been carried out in order to fabricate ion gels with polymers. Preparation methods of ion gels can be classified into three main techniques as described in Fig. 9.2. These preparation methods include (i) swelling of a polymer in an ionic liquid, (ii) by a solution cast or solvent mediated method, or (iii) by in situ polymerization and cross-linking of monomers inside ionic liquids [2, 14].

Swelling a Polymer in an Ionic Liquid Two main preparation approaches are classified here. First, ion gels can be prepared by immersing a given polymer membrane (cross-linked or not) in an ionic liquid which is able to penetrate the membrane and swell the polymer without dissolving it. This synthetic method is limited by the choice of the polymer and the ionic liquid used, and the composition of the ion gel is determined by the maximum swelling capacity of the polymer [14–16]. In a typical example, Doyle et al. prepared Nafion-based ion gel membranes by using two different ionic liquids as swelling agents, 1-butyl-3-methylimidazolium trifluoromethanesulfonate [C_4mim][CF_3SO_3] and 1-butyl-3-methylimidazolium tetrafluoroborate [C_4mim][BF_4] [17].

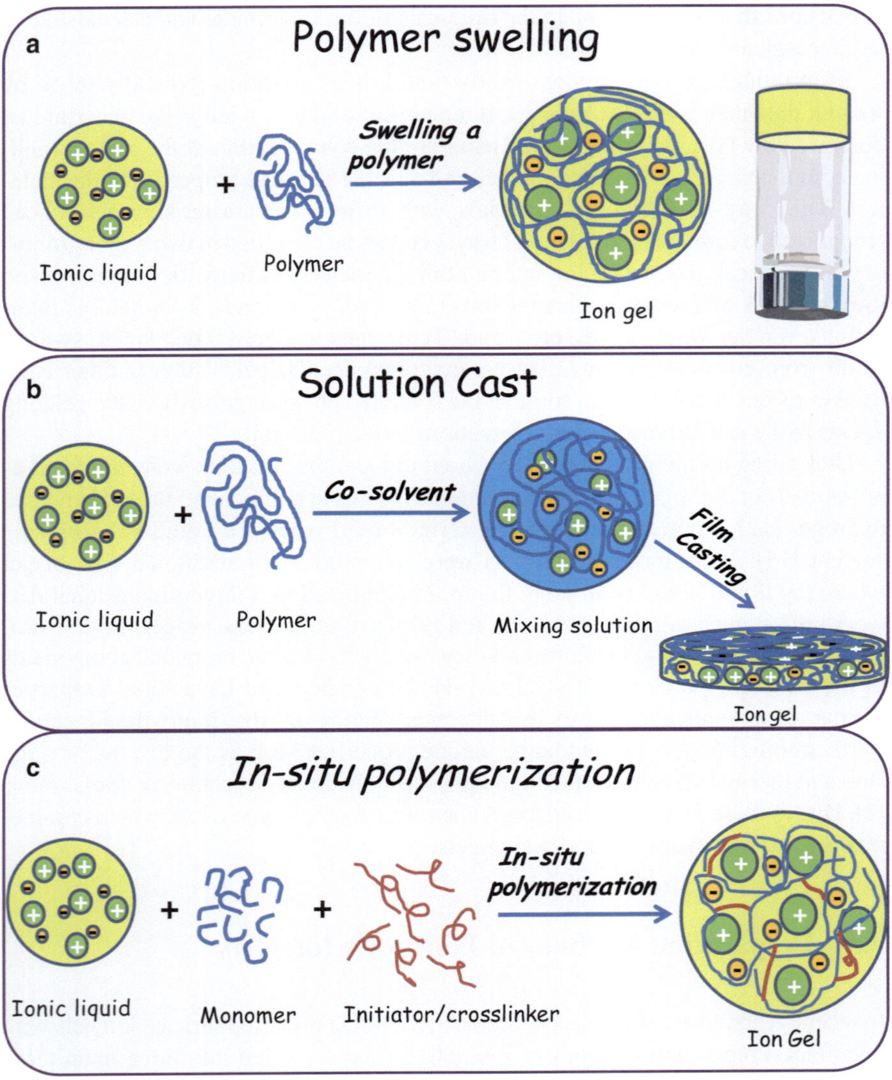

Fig. 9.2 Schematic representation of the different methods for ion gels' preparation

In the second approach, a polymer (usually a biopolymer such as gelatin, agarose, or a block copolymer) is dissolved in an ionic liquid at a given temperature. Then the polymer solution in ionic liquid is cooled down and an ion gel is formed. Due to this preparation procedure, these types of ion gels present thermo-reversible transitions. In a preparation example, Barreriros and coworkers reported the dissolution of gelatin in ionic liquids at temperatures higher than 35 °C to prepare ion jelly materials upon cooling [18]. A similar approach was used by Singh et al. using the biopolymer Agarose [19]. On the other hand, Lodge and coworkers pioneered the development of ion gels using block copolymers as gelators. Thus, poly(*N*-isopropyl acrylamide-*b*-ethylene oxide-*b*-*N*-isorpopyl acrylamide) PNIPAm-PEO-PNIPAm triblock copolymer was mixed with 1-ethyl-3-methylimidazolium bis(trifluoromethylsulfonyl)imide [C_2mim][$N(Tf)_2$] ionic liquid [20]. 10 % of the block copolymer was directly dissolved in the ionic liquid. The mixture was viscous liquid at room temperature; however, after cooling down to ~ 15 °C it became a transparent ion gel. Interestingly, a reversible gel transition was observed upon cooling–heating cycles between 5 and 50 °C. Similarly, Lodge and coworkers reported thermo-reversible ion gels based on 20 wt% of poly(styrene-b-ethylene oxide-b-styrene) and 80 wt% of [C_2mim][$N(Tf)_2$] [21].

Solution Cast Method This preparation method of ion gels consists on the mixture of a given polymer and the IL together with a cosolvent to form a homogeneous dissolution, which is subsequently casted and dried to remove the solvent [22–26]. By this method, different types of polymers and ionic liquids can be mixed in different proportions, controlling and adjusting the properties of the final ion gels [14]. However, the limitation of this method is that the weight percent of ionic liquid is usually limited and some incompatibility problems between ionic liquids and polymers could appear. In a typical example, Shalu et al. studied the preparation of poly(vinylidene fluoride) PVdF-HFP copolymer-based ion gels by increasing the amount of ionic liquid [C_4mim][BF_4] [27]. Briefly, a desired amount of PVdF-HFP was dissolved in acetone under stirring at 50 °C. Afterward, different amounts of ionic liquid (from 0 to 90 wt%) were added and stirred for 2–4 h at the same temperature until the solution of PVdF-HFP and ionic liquid became viscous. After casting it into Petri dishes and removing the solvent, rubbery films of the ion gel were obtained. The maximum conductivity value obtained was 1.6×10^{-2} S cm^{-1} for the membrane containing 90 % of ionic liquid at room temperature. The textural changes of the PVdF-HFP/SN network and phase changes from crystalline to amorphous nature provided more free volume for ease of ion transport. However, in most cases, incompatibility and segregation of the ionic liquid from the ion gel with time are observed. To avoid this ionic liquid leaching problem, Marcilla et al. proposed the use of polymeric ionic liquids as a very compatible polymer matrix to prepare ion gels [28].

Polymerization–Cross-linking of Monomers in Ionic Liquids By this method, the in situ one-step synthesis of ion gels is carried out by mixing all reactants (monomer, initiator, cross-linking agent) in an ionic liquid. The polymerization can be carried out thermally or using UV light in the presence of a photoinitiator. The compatibility between the polymer matrix and the ionic liquid used as a solvent is crucial to obtain ion gels since the loss of miscibility during polymerization may be a limitation in gel forming [2]. In this regard, vinyl monomers have been intensively studied since many of them show high compatibility with different ionic liquids [29–31]. This way, transparent, mechanically strong, and highly conductive polymer electrolyte films have been obtained. Watanabe and coworkers reported for the first time the in situ radical polymerization of a vinyl monomer 2-hydroxyethyl methacrylate (HEMA) in an ionic liquid 1-butylpyridinium tetrafluoroborate [C_4py][BF_4] leading to a conductivity of 10^{-3} S cm^{-1} [30]. Later, the same group has studied the polymerization of methyl methacrylate in 1-ethyl-3-methyl imidazolium bis(trifluoromethane sulfonyl)imide [C_2mim][$N(Tf)_2$] [31]. The monomer was chosen according to the compatibility of the vinyl monomer and the ionic liquid used, since acrylonitrile, styrene, HEMA, and acrylamide were incompatible after being polymerized.

This method is very versatile and can be applied to many different monomer/ionic liquid combinations [32]. For instance, Isik et al. recently reported for the first time ion gels which were fully composed of low toxicity and biocompatible ionic liquids [33]. Ion gels containing 40 % of cholinium lactate methacrylic monomer and 60 % of cholinium lactate ionic liquid were synthesized by photopolymerization. The effect of additional ethyleneglycol dimethacrylate (4 %) as a cross-linker was explored and found to be responsible for the reduction in the transparency and the improvement of the ion gel's integrity. As a very different example, cyclic carbonate network polymers were synthesized using dual functional cyclic carbonate methacrylate (CCMA) monomer [34]. Due to the high compatibility between the ionic liquid and the cyclic carbonate network, transparent, flexible, and stable ion gels were obtained, which contained high ionic conductivity (up to 10^{-3} S cm^{-1} at ambient temperature).

Hybrid ion gels, which are constituted by a combination of organic and inorganic materials synthesized by a sol–gel method, have attracted the interest of many researchers due to the mechanical strength that inorganic materials can provide to the gel. Although this chapter is mainly focused on organic materials, some examples should be mentioned. In 2010, Vioux and coworkers reported some work on hybrid sol–gel ion gels [35, 36]. In this type of ion gels, the three-dimensional network is formed by the condensation of alkoxysilanes such as tetraethoxy silane (TEOS) (instead of monomer polymerization). For instance, they carried out the synthesis of homogeneous and transparent PMMA-silica ion gels, fabricated from a copolymer based on MMA and methacryloxypropyltrimethoxysilane (MPTMS), then the ionic liquid [C_4mim][$N(Tf)_2$] was added, and finally a solution of tetraethoxy silane (TEOS) was incorporated to the mixture. Gelation of the hybrid ion gel occurred within 1 h and the aging continued for 14 days. In this system, the silica nanofiller covalently bonded to the polymer chains, which provided mechanical strength to the ion gel, while the ionic liquid was responsible for the flexibility and ionic conductivity of the material. High ionic liquid loadings have been described, close to 90 wt%, which contained the same ionic conductivity as the pure ionic liquid, and no leaching was observed. More recently, Noor et al. suggested that the mechanism of ionic liquid immobilization in the silica network is based on the hydrogen bonds formed between the anion of the ionic liquid (BF_4^-) and the hydroxyl group of the silica [37]. As a result, they obtained solid-state electrolytes with high loading of ionic liquid (up to 97 %).

9.4 Properties of Ion Gels

Ion gels are soft-solid materials formed by two components: (i) an ionic liquid and a (ii) polymeric cross-linking component. The intimate contact between the ionic liquid and the solid-like network provides an intrinsic hybrid character to the ion gel. Ion gels usually present the unique properties of the ionic liquid, such as thermal stability, negligible vapor pressure, nonflammability, electrochemical and chemical stability, or high ionic conductivity [38, 14]. Furthermore, the polymer network is responsible for the solid-like behavior. The properties of the ion gel can be infinitely tuned by changing the chemical structure of the polymer network, the ratios, cross-linking degree, or modifying the type of ionic liquid. The main properties that characterize ion gels are the ionic conductivity and their mechanical and rheological properties as described below.

9.4.1 *Ionic Conductivity*

The good ionic conductivity is one of the most interesting properties of ion gels when being used as electrolytic membranes in several electrochemical devices. The ionic conductivity of the ion gel can be easily controlled by adjusting the content of the ionic liquid and the type of polymer in the membrane. Indeed, by increasing the amount of ionic liquid the ionic conductivity can be enhanced. This was very well illustrated by Marcilla et al., who reported several ion gels with conductivities varying from 10^{-7} to 10^{-2} S cm^{-1} depending on the poly(ionic liquid) to ionic liquid ratio [39]. However, an excess of ionic liquid can cause poor mechanical stability and ionic liquid leaching, especially when the ionic liquid and the polymer are not totally compatible [14, 37]. Yang and coworkers have prepared ion gel membranes of PVdF-HFP with different concentrations of 1-butyl-4-methylpyridinium bis(trifluoromethanesulfonyl)imide [C_4mβpy][N$(Tf)_2$] through solution-casting method [40]. The highest ionic conductivity obtained was 2.01×10^{-4} S cm^{-1} with 33.3 wt% of ionic liquid (See Fig. 9.3).

The ionic conductivity of the gel membranes is usually lower than that of the pure ionic liquid, confirming the molecular level blending of the solid network and the ionic liquid. Pandey and Hashmi observed that the ionic conductivity of 1-ethyl-3-methylimidazolium tris(pentafluoroethyl) trifluorophosphate [C_2mim][FAP] was reduced from 3.6×10^{-3} S cm^{-1} to 1.7×10^{-3} S cm^{-1} after blending it in 20 wt% of PVdF-HFP [41]. However, the addition of salt LiPF6 in [C_2mim][FAP]/PVdF-HFP provided a slight enhancement in the conductivity value (2.6×10^{-3} S cm^{-1}), probably due to high Li^+ ion mobility in the polymer electrolyte. On the other hand, the potential window of the ion gel was slightly reduced after the addition of salt from ~4.5 V (from −2.2 to 2.3 V) to ~4 V (from −2.0 to 2.0 V), but still it was sufficient to use them in supercapacitor applications.

9.4.2 *Mechanical and Rheological Properties*

The local interactions between ILs and the solid phase of the gel have been addressed from different perspectives as functions of the systems in the case of the polymeric ion gels, the main observable effect is the plasticizing effect of the IL on the

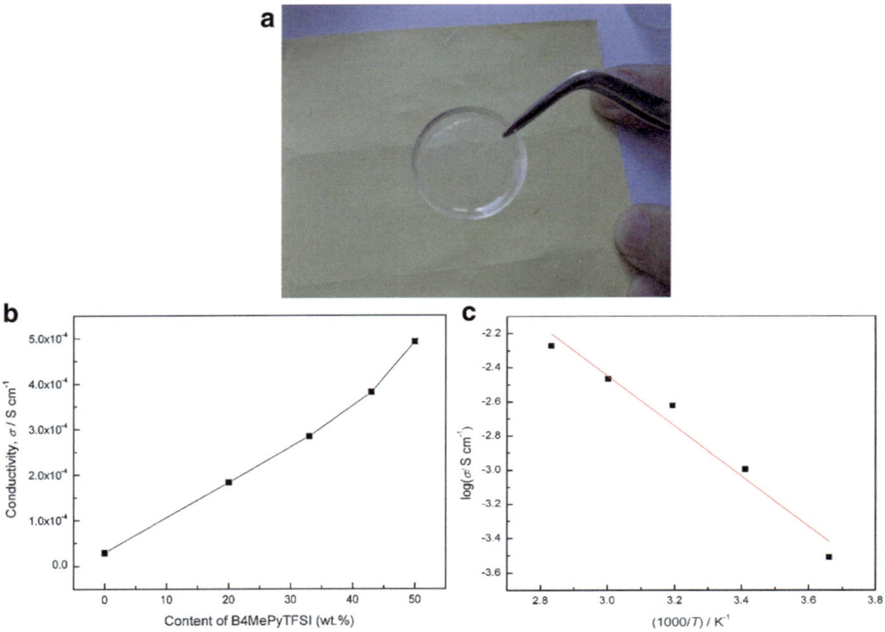

Fig. 9.3 (**a**) Photograph of the as-prepared ionic liquid gel polymer electrolyte. (**b**) Conductivity of the ion gel with various weight percentages of $[C_4m\beta py][N(Tf)_2]$ systems. (**c**) Temperature dependence of the ionic conductivity for $[C_4m\beta py][N(Tf)_2]/Li[N(Tf)_2]$/PVdF-HFP polymer electrolytes containing 33.3 wt% $[C_4m\beta py][N(Tf)_2]$ from 0 to 80 °C [40]

polymer leading usually to a glass transition lower than the polymer one and higher than the IL one [42, 29]. In a typical example, Jiang et al. have studied the decrease in glass-transition temperature (Tg) of pure PMMA from 107.5 to 23.5 °C with an increase of the mass fraction of the ionic liquid 1-butyl-3-methylimidazolium hexafluorophosphate $[C_2mim][PF6]$ from 0/10 to 5/5 of $[C_4mim][PF6]/[MMA]$ [29].

On the other hand, the mechanical and rheological properties of ion gels are important factors to be considered according to the targeted applications. Nagasawa et al. studied the properties of the ion gels consisting of several ionic liquids and the N,N'-(trans-cyclohexane-1,4-diyl)-dibenzamide-based polymer with different anions (CDBA6-X), bis(trifluoromethanesulfonyl)amide (CDBA6-$N(Tf)_2$) or bis(fluorosulfonyl)imide (CDBA6-FSI) among others [43]. The rheological properties are summarized in Fig. 9.4, demonstrating the quasi-solid behavior of the ion gel since (storage moduli) G' is greater than (loss moduli) G'' for the entire range of frequencies (Fig. 9.4a). In Fig. 9.4b, a strain-induced collapse of the gel state above the critical strain region with a dramatic decrease of the G' value is shown. However, G' increased when the concentration of CDBA6-$N(Tf)_2$ increased, reaching values of 1×10^5 Pa at 50 g/L. Regarding the mechanical properties of those ion gels, despite the fact that noncovalent networks were obtained (physical gel), they exhibited very good mechanical strength.

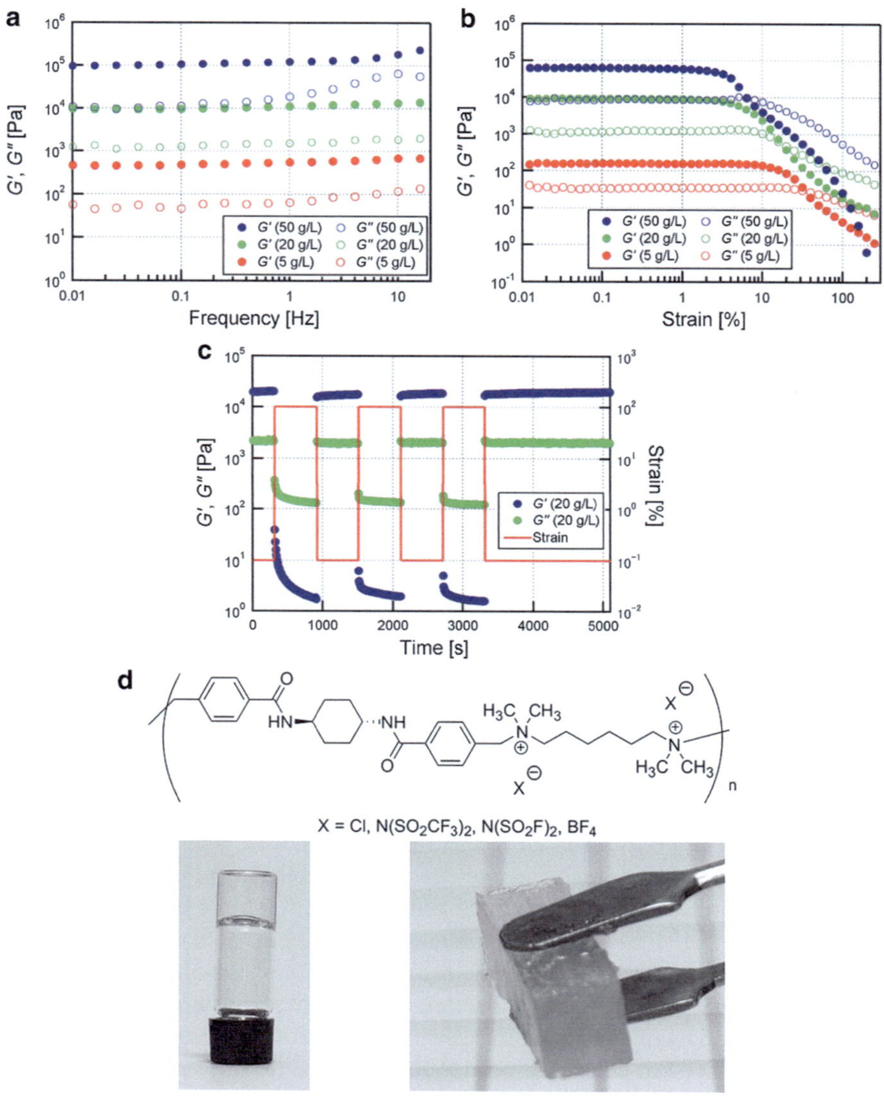

50 g/L CDBA6-TFSI/[C_2mim][N(Tf)$_2$] ion gel of 6 x 6 x 18 mm in size.

Fig. 9.4 Rheological data of the CDBA6-N(Tf)$_2$/[C_2mim][N(Tf)$_2$] and CDBA6-FSI/[C_2mim][FSI] ionogels. (**a**) Frequency sweeps of three samples of the ionogels with different concentrations. (**b**) Strain amplitude sweeps (strains at 0.1 Hz). (**c**) Step strain measurement (0.1 % ↔ 100 %) of the CDBA6-FSI/[C_2mim][FSI] ionogel (20 g/L). Step strain measurement was a continuous one (three cycles). (**d**) The chemical structure and pictures of the ion gel [43]

Highly stretchable and transparent ion gels have been reported such as the ones presented by Chen, Suo, and coworkers polymerizing acrylic acid in 1-ethyl-3-methylimidazolium ethylsulfate [C_2mim][$EtSO_4$] [44]. The ion gel exhibited a Young's modulus of 3.1 ± 0.6 kPa, fracture strength of 7.4 ± 0.4 kPa, and a fracture stretch of 4.6 ± 0.3. Furthermore, the relatively small hysteresis was observed in the cyclic stretch–stress curves after hundreds of cycles of loading and unloading. In the initial cycles, the Young's modulus decreased and became stable after 150 cycles. Therefore, good mechanical reversibility was observed. More stretchable solid electrolytes were reported by Wallace and coworkers [45]. These ion gels, which were PMMA networks with polyethylene glycol diacrylate cross-linker swollen by 1-ethyl-3-methylimidazolium bis(trifluoromethanesulnofyl)imide [C_2mim][$N(Tf)_2$], were elastically stretched more than 500 % of their original length and they exhibited larger than 1.5 MPa of tensile breaking strength.

On the other hand, confinement effects can be observed and are intensively studied exclusively on inorganic ion gels, where the confinement of the IL within the nanopores of the gel solid phase can alter significantly the physicochemical properties of the IL, such as phase transitions (appearance and disappearance of transition) and molecular dynamics as well. Nevertheless, the dynamical effects appear to be localized on the ion–wall interface preserving the liquid behavior of the IL until extreme confinement. The significative confinement effects, including enhancement in luminescence emission, observed in inorganic ion gel are reported in the review [2].

9.4.3 Morphology

The association of block copolymers with ionic liquids opens an unlimited number of combinations to be explored and exploited. ABA triblock copolymers have been already successfully used to immobilize a large amount of ionic liquids leading to unique ion gel with controlled morphologies, containing only few percent by volume of polymers. Indeed, due to the selective solubility of the blocks in the ionic liquid (only block B is soluble) the macromolecules can self-assemble to provide a highly elastic solid with typical mesh sizes in the range of 10–100 nm [1]. Using an appropriate nonsoluble block, temperature can be used as an external stimulus to provoke the solubility and then induce a reversible solid to liquid transition which can be of interest for material processing [46].

T. Russell, J. Watkins, and coworkers have reported a study on cross-linked ABA block copolymer combined with ionic liquids to obtain self-assembled ionic gels presenting decoupled ionic conductivity and mechanical strength [47]. Indeed, poly(propylene oxide)–poly(ethylene oxide)–poly-(propylene oxide) (PPO–PEO–PPO) block copolymers with cross-linkable end groups were blended with [C_4mim][PF_6] ionic liquid. Interestingly, the chosen IL selectively solubilized the PEO block, promoting the copolymer microphase separation (which were otherwise disordered or weakly ordered). According to the small-angle scattering results, different morphologies were achieved: cylindrical aggregates of PPO arrayed in a hexagonal pattern in the PEO/IL matrix and, at higher IL concentration, a body-centered cubic (BCC) morphology of PPO spheres in a PEO/IL matrix are observed. The authors conclude that the microphase separation does not affect the ionic conductivity, but it can be used to increase the mechanical stability by isolating phase-selective cross-linkers from the ion-conducting phase.

Block copolymers can be also associated with a lower amount of ILs leading to membrane-like materials such as reported by R. Segalman and coworkers using poly(styrene-block-2-vinylpyridine) blended with the ionic liquid imidazolium bis(trifluoromethane)sulfonimide [Im][$N(Tf)_2$] [48]. In this system, polystyrene (PS) is not soluble in the ILs. As demonstrated by small-angle X-ray scattering, the composite materials exhibit a variety of controlled morphologies, including lamellae, hexagonally close-packed PS cylinders, body-centered cubic PS spheres, face-centered cubic PS spheres, or disordered PS micelles, which allows us to propose lyotropic phase diagrams [48]. It has been shown that the nanostructure does not significantly impact the conductivity of the membranes, possibly due to the continuity of the conducting phase [25].

Recently, Lodge and coworkers have used a triblock terpolymer, poly[isoprene-b-(styrene-co-norbornenylethylstyrene)-b-ethylene oxide] (INSO), to blend with 1-ethyl-3-methylimidazolium bis(trifluoromethylsulfonyl)imide [C_2mim][$N(Tf)_2$] [49]. The structural characterization revealed the continuous morphologies trapped by the chemical cross-linking of the PS blocks presenting both the conductive and the high modulus domains. The conductivity measurement revealed that the ion conductivity in the PEO domains was much lower than the value measured with PEO homopolymer; this point was attributed to the potential conducting pathways' dead ends or to the presence of isolated conducting domains resulting from the solvent-casting procedure.

9.5 Applications

Most of the applications of ion gels are found as solid electrolytes in electrochemical devices. Ion gels combine two properties such as the ionic conductivity provided by the ionic liquid and the solid-like behavior which gives access to all-solid devices facilitating the preparation and security of most electrochemical devices. Indeed, the main drawback of liquid electrolytes (aqueous, organic, or ionic liquid-based) in electrochemical devices is the need of additional encapsulation, which can be overcome by the use of ion gels-based electrolytic membranes. Those ion gel solid electrolytes can be then used in batteries, fuel cells, solar cells, electrochemical sensors, biosensor, actuators, or supercapacitors. Although this chapter will mainly focus on electrolytic membranes, other promising applications of ion gels such as separation membranes, catalytic membranes, or drug release in biomedicine will be also mentioned.

9.5.1 Batteries

A battery is an energy-storage device that transforms chemical energy into electrical energy and supply electrical energy via terminals/contacts [50, 51]. In the last decades, the demand for energy-storage and portable devices has increased all over the world; therefore, the production of batteries grew considerably [52]. The lithium-ion secondary (rechargeable) batteries have become the most studied candidates among other types of batteries. They play an important role in modern society since they are the power system for sustainable vehicles, portable computers, cameras, mobile phones, medical instrumentations, as well as electrochemical storage systems in renewable energy plants [53]. However, cycling performances, costs, operational temperature, and safety are still major issues to be improved or resolved. In this perspective, all-solid batteries can be a major progress with ion gels as promising candidate for that purpose.

Electrolytic membranes based on ion gels which use poly(vinylidene fluoride)-hexafluoro propylene (PVdF-HFP) as copolymer matrix have been intensively studied [54, 55, 22, 56]. Hofmann et al. have reported an ionic liquid-based polymer gel electrolyte for lithium ion batteries prepared by combining a conducting salt (Lithium bis(trifluoromethylsulfonyl)azanide), the PVDF-HFP, different types of ionic liquids, and additives. The conductivity of the gel electrolytes varies with the ionic liquid used and values up to 1–2 mS/cm were obtained (See Fig. 9.5) [57]. Li Libo et al. have studied the increment of conductivity in the electrolytic membrane when increasing the concentration of a given ionic liquid [40, 58]. Furthermore, Fernicola et al. have prepared membranes with 60, 70, and 80 % of lithium salt–ionic liquid (Li salt–IL) mixture to give conductivity values in the order of 10^{-4} S cm^{-1} and an electrochemical stability window in the range of 1.8–4.5 V vs Li^{+}/Li [52]. The Li salt–IL mixture acts as a plasticizer for the polymer matrix, suppressing the crystallinity of the electrolytic membrane. The ionic conductivity increased with the Li salt–IL content and the temperature, reaching 10^{-2} S cm^{-1} at 110 °C. Unexpectedly, a significant enhancement in the Li transference number was observed when passing from the liquid to the membrane electrolyte system.

Recently, electrospun fibrous membrane of P(VdF-HFP) has been prepared and electrochemically activated by incorporating lithium (bis(trifluoromethylsulfonyl) imide) complexed with different ionic liquids 1-ethyl-3-methylimidazolium (bis(trifluoromethylsulfonyl)imide) $[C_2mim][N(Tf)_2]$, 1-propyl-3-methylimidazolium (bis(trifluoromethylsulfonyl)imide) $[C_3mim][N(Tf)_2]$, or 1-butyl-3-methylimidazolium (bis(trifluoromethylsulfonyl)imide) $[C_4mim][N(Tf)_2]$ [59, 60]. Interconnected fibrous structure of the membrane presented fibers with large surface area and high porosity. Due to this unique structure of the membrane, high ionic conductivity ($>2.4\times10^{-3}$ S cm^{-1}) was achieved at 25 °C. The excellent electrochemical characteristics of these fibrous ion gels allow the fabrication of safer lithium batteries with improved performance.

Another approach to obtain ion gels-based membranes for lithium batteries is to use polymeric ionic liquid (PIL) as polymer host and incorporate ionic liquid and

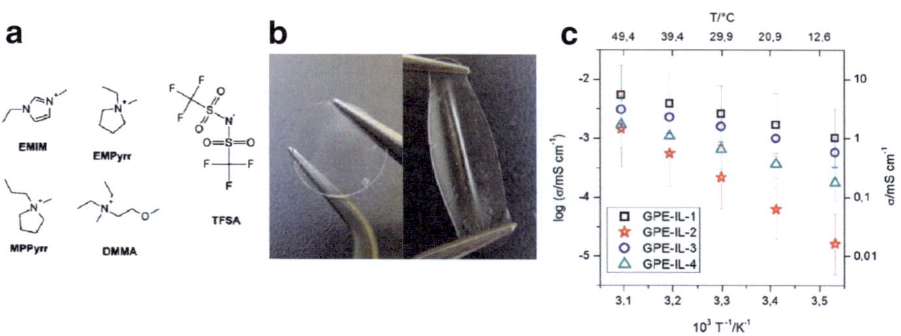

d

Sample	PVdF-HFP (wt%)	ionic liquid	Li-TFSA(mol/kg)	mp of pure IL (K)	σ/mS cm-1 (20°C)	ECW (V±0.2V)
GPE-IL-1	10	EMIM-TFSA	0.5	257.1	1.17±0.6	5.1
GPE-IL-2	10	EMPyrr-TFSA	0.5	361.2	0.066±0.019	6.1
GPE-IL-3	10	MPPyrr-TFSA	0.5	285.2	1.04±0.18	5.8
GPE-IL-4	10	DMMA-TFSA	0.5	n.a.	0.37±0.18	5.3
	0	EMIM-TFSA	0.5	257.1	3.2±0.4	5.3
	0	EMPyrr-TFSA	0.5	361.2	0.072±0.010	6.2
	0	MPPyrr-TFSA	0.5	285.2	0.91±0.07	5.7
	0	DMMA-TFSA	0.5	n.a.	0.70±0.05	5.5

Fig. 9.5 (**a**) Structures of the ionic liquids. (**b**) Photograph of gel polymer electrolyte with a diameter of 13 mm and a thickness of 200 μm. (**c**) Conductivity Arrhenius plots of four gel polymer membranes. The data are obtained from impedance spectroscopy at 10–50 °C. (**d**) Composition, conductivity data, and ECW of gel polymer electrolytes with varied ionic liquids. *mp* melting point, *ECW* electrochemical window; electrodes, stainless steel/stainless steel; first cycle; $v = 25$ mV/s; $T = 50$ °C; potential range, E vs Stainless Steel = ± 4 V [57]

lithium salt [61–63]. The use of PILs as matrices in ionic liquid-based polymer electrolytes presents some advantages comparing with other polymer hosts: (1) Good chemical affinity between PILs and ILs affording stable polymer electrolytes. (2) Electrochemical stability and the ionic character of PILs make them good candidates as electrolyte hosts. (3) As usual in ion gel-based electrolytes, PIL-based membranes are easy to process. Appetecchi et al. have reported the synthesis of electrolytic membrane using poly(diallyldimethylammonium)-bis(trifluoromethanesulfonyl)imide as a polymer host, *N*-alkyl-*N*-methylpyrrolidinium-bis(trifluoromethanesulfonyl)imide as ionic liquid, and Li[N(Tf)$_2$] as a lithium salt [39]. The electrolytic membrane obtained has presented good mechanical properties with high chemical stability even after prolonged storage times in contact with lithium anode, and it was thermally stable up to 300 °C. These membranes exhibited room temperature ionic conductivity above 10^{-4} S cm^{-1} with an electrochemical stability window up to 5.0 V vs Li$^+$/Li, time-stable interfacial resistance values, and good lithium stripping/plating performance.

Early work on polymer electrolytes indicated that polyethylene oxide (PEO) is a promising candidate as a host for lithium salt in relation to its ability to solvate Lithium cation [64–66]. This is attributed to its flexible ethylene oxide segments and ether oxygen atoms, which have a strong donor character and solvate lithium cations. However, the major drawback of PEO-based polymer electrolytes is the ionic conductivity at ambient temperatures. The introduction of room temperature ionic liquids, such as *N*-alkyl-*N*-methylpyrrolidinium bis(trifluoromethanesulfonyl)-imide, into the polymer electrolyte (PEO-Li[N(Tf)$_2$]) resulted in an increase of more than one order of magnitude in ionic conductivity at room temperature, which was attributed to the plasticizing effect of the ionic liquid. For instance, in the case of N-butyl-N-methyl-pyrrolidinium to lithium cation [C$_4$mPyr]$^+$/Li$^+$ ratio of 3.22, the ionic conductivity reaches 2×10^{-4} S cm^{-1} at 40 °C.

Moreover, other polymers have also been studied in the synthesis of ion gel-based membranes. Stepniak has reported membranes obtained by in situ photopolymerization of a mixture of ethoxylated bisphenol A diacrylate (bisAEA4) and 0.4 M solution of Li[N(Tf)$_2$] in N-methyl-N-propylpiperidinium bis(trifluoromethanesulfonyl) imide [C$_3$mpip][N(Tf)$_2$) [67]. It is worth stressing that the photopolymerization technique is a suitable method for synthesizing porous gel polymer electrolytes, due mainly to the ease and rapidity of processing. Stable, porous, and flexible membranes were characterized to give 0.64 mS cm^{-1} of ionic conductivity at 25 °C and good electrochemical stability (4.8 V). Charge–discharge test of the lithium-ion battery using $Li_4Ti_5O_{12}$ (LTO) as anode with ionic liquid gel polymer electrolyte (IL-GPE) exhibited a good specific capacity of 124 mAh g^{-1} and 110 mAh g^{-1} at C/20 and C/5, respectively. Furthermore, gel polymer membranes of polyacrylonitrile and poly(methyl methacrylate) (PAN/PMMA) incorporating room-temperature ionic liquids have been reported [68, 69]. The ionic conductivity of the membrane with [C$_4$mPyr][N(Tf)$_2$] is 3.6×10^{-3} S cm^{-1} at room temperature, and the capacity of $LiFePO_4$ in the cell Li/GPE/ $LiFePO_4$ is 139, 134, 120, and 101 mAh g^{-1} at the charge/discharge rate of 0.1, 0.2, 0.5, and 1C, respectively, remaining unchanged after ten cycles for each rate.

9.5.2 Fuel Cells

A fuel cell is an electrochemical conversion device that has a continuous supply of fuel such as hydrogen, natural gas, or methanol and an oxidant such as oxygen, air, or hydrogen peroxide [51, 70]. The most promising commercial applications of fuel cells appear to be as a stationary power for central and dispersed power stations and as mobile power for portable electronics devices and automobiles. Fuel cells offer a rather clean electrical power generation; they are quiet in operation and can be more efficient in the conversion of energy in a fuel into power than gasoline engines or utility thermal power plant.

Proton exchange membrane fuel cells (PEMFCs) are considered to be promising candidates; however, several factors, such as cost, reliability, good proton conductivity, durability, and the ability to operate at temperatures higher than 120 °C, are major challenges in the commercialization of PEMFCs [71]. Operating above 120 °C offers significant advantages such as reduced CO poisoning at the anode, increased energy efficiency, improved heat rejection, improved quality of waste heat, and easy water management. In this regard, the polymer electrolyte membrane, which acts as a charge carrier for protons, a separator for the reactant gases, and an electronic insulator that prevents electrons from passing through, is a key component to be developed [72, 73]. Water management in the PEM is crucial and challenging since polymers usually require water to promote proton mobility between both electrodes, and above 100 °C, this water is evaporating, reducing the activity of the fuel cells. Thus, there is a researching interest to develop proton-conducting electrolytes that do not require hydration for proton conduction.

Ion gels are good candidates to be used as electrolytic membranes in PEMFCs, where the ionic conductivity of the membrane depends on the concentration of the ionic liquid, on the conductivity of neat ionic liquid, and on the polymer's nature. The good miscibility of the polymer and ionic liquid is crucial while preparing membranes with sufficient ionic conductivity for PEMFC applications. Ionic liquids fall into two general classes: aprotic (with applications in lithium batteries, dye-sensitized solar cells, thermo-electrochemical cells) and protic, which are less studied in general due to the reactivity of the active proton, but this feature makes them of significant interest to be used in fuel cell electrolytes [74]. However, the good affinity between the polymer and the ionic liquid plays the most significant role, together with the nature of the polymer, in ionic conductivity. As an example, Schauer et al. have studied membranes with two relatively similar ionic liquids, 1-butyl-3-methylimidazolium trifluoromethanesulfonate [C_4mim][CF_3SO_3] and 1-ethylimidazolium trifluoromethanesulfonate [C_2mim][CF_3SO_3], respectively, aprotic and protic [73]. Both ionic liquids had very similar affinities to given polymers, and similar ionic conductivities were measured in the ionic liquid–polymer membranes.

Different polymers have been studied in combination with ionic liquids to synthesize gel polymer electrolytic membranes for PEMFCs. Perfluorosulfonated acid (PFSA)-based polymers, such as Nafion, are the most studied ones for this purpose. These polymers are mechanically, thermally, and chemically stable and have very high proton conductivity [75]. However, its conductivity drops at temperatures higher than 100 °C due to water evaporation, which has an important role in proton conduction. Díaz et al. prepared membranes based on Nafion with protic ionic liquids, such as 1-butyl-3-(4-sulfobutyl)imidazolium trifluoromethanesulfonate [HSO_3-bmim][CF_3SO_3] or 1-methyl-3-(4-sulfobutyl)imidazolium bis(trifluoromethanesulfonyl)imide [HSO_3-bmim][N(Tf)$_2$], for their use as electrolytes in PEMFCs under anhydrous conditions [76]. These membranes achieve current densities of 217 mA/cm^2 and a maximum power of 44 mW/cm^2 under anhydrous conditions at 25 °C. Mališ et al. investigated the influence of humidity and the temperature on the PVdF-HFP and Nafion-based membranes' conductivity after mixing the polymers with [C_4mim][CF_3SO_3] or [C_2mim][CF_3SO_3] [77]. All membranes showed relatively high ionic conductivity, which increases with increasing temperature. Under dry conditions, the conductivity of the Nafion membrane itself drops to zero, while in combination with ionic liquids no conductivity drop was observed. The conductivity stability has to be attributed mainly to the ionic liquid effect. On the other hand, in a humidified atmosphere the conductivity of the Nafion-based membrane is higher than in dry conditions, and a protic ionic liquid increases the content of water in the membrane. In this regard, the

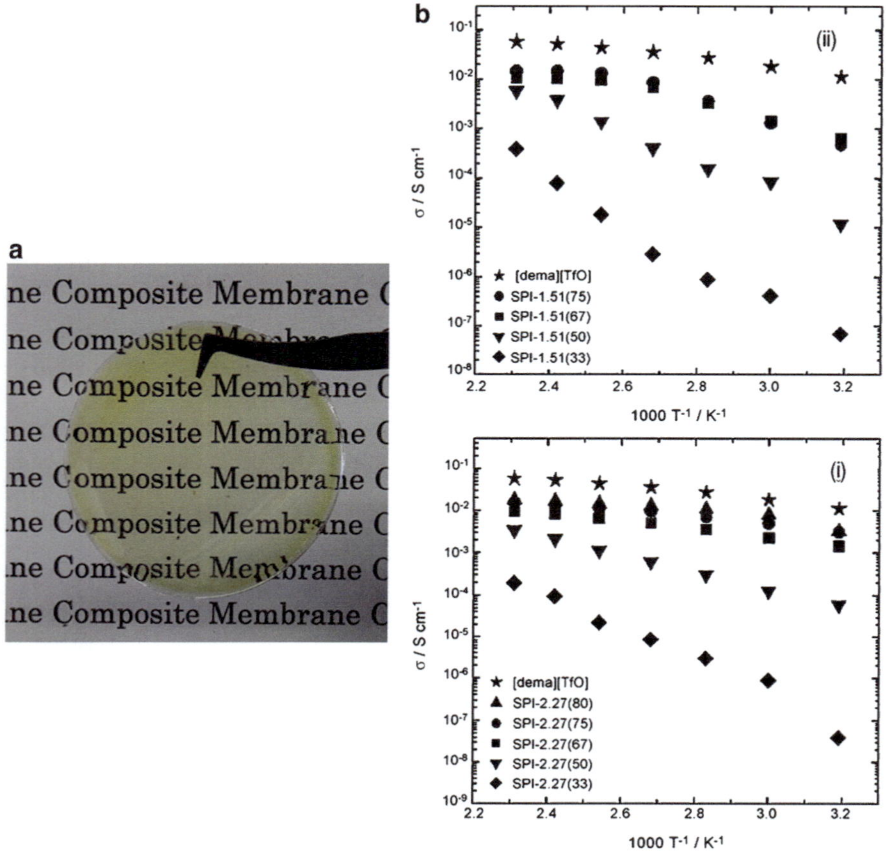

Fig. 9.6 (**a**) A photograph of SPI-2.27 composite membrane. (**b**) Arrhenius plots of ionic conductivities for [dema][CF_3SO_3] and SPI/[dema][CF_3SO_3] composite membranes based on (i) SPI-1.51 (IEC = 1.51 meq g^{-1}) and (ii) SPI-2.27 (IEC = 2.27 meq g^{-1}) [78]

lowest conductivity was achieved for the combination of aprotic ionic liquid, [C_4mim] [CF_3SO_3], and a hydrophobic fluoroelastomer, PVdF-HFP.

Moreover, sulfonated polyimides (SPIs) have been used to prepare polymer electrolyte membranes for nonhumidified fuel cells by the addition of a protic ionic liquid, diethylmethylammonium trifluoromethanesulfonate [dema][CF_3SO_3] (See Fig. 9.6) [78, 79]. The uniform, tough, and transparent membranes have good thermal stability (>300 °C) and ionic conductivity (>10^{-2} S cm^{-1} with an ionic liquid content higher than 67 %) under anhydrous conditions. A current density higher than 240 mA cm^{-2} is achieved with a maximum power density of 100 mW cm^{-2} at 80 °C when using a membrane in the H_2/O_2 fuel cell operation under dry conditions.

Many other polymers, such as poly(ether ether ketone) (PEEK), polybenzimidazole derivative containing benzofuranone moieties (PBI-O-Ph), or poly(vinyl alcohol) (PVA), have been used in order to prepare electrolytic membranes for fuel cells in combination with protic or aprotic ionic liquids [80–86]. Jothi et al. have synthesized a polymer membrane based on sulfonated poly(ether ether ketone) (SPEEK) and halogen-free ionic liquids containing imidazole cations of 1-ethyl-3-methylimidazolium, $[C_2mim]^+$, and anions of diethyl phosphate [87]. The composite membranes had good conductivity in the range of 1.25–3.16 mS cm^{-1} at 145 °C under anhydrous environment, and the fuel cell including SPEEK/50 % ionic liquid membrane gave a maximum power density of 203 mW/cm^2 at the same temperature.

Alkaline polymer electrolyte fuel cells (APEFCs): two of the major limitations of PEMFCs are the use of expensive platinum-based catalysts and also the cost of PEM [88, 89]. Therefore, alkaline polymer electrolyte fuel cells have received a great attention, which presents several advantages, such as lower activation loss in alkaline medium, facile cathode kinetics permitting the use of abundant low-cost metal catalysts (Ni, Fe, Ag, etc.), and decreased oxidant reduction overpotential at high pH, thereby increasing the efficiency. The anion exchange membrane, which acts as an electrolyte to transport anions from cathode to anode, is one of the key components of alkaline fuel cells. Lin et al. have reported the photocopolymerization of ionic liquid monomer, 1-vinyl-3-methylimidazolium iodide [Vmim][I], with styrene and acrylonitrile in the presence of 2–10 % cross-linker divinylbenzene to obtain a membrane which was converted to their OH^- forms by exchanging the iodide with hydroxide anions [90]. The membranes had hydroxide ion conductivities above 10^{-2} S cm^{-1} at room temperature and excellent chemical stabilities in high pH media at 60 °C, fulfilling the basic conductivity and chemical stability requirements [91].

9.5.3 *Solar Cells*

A solar cell is an energy production device that converts the energy of light into electricity. Dye-sensitized solar cells (DSSC) are proposed as an alternative to costly Si-solar cells, in which the electrolyte plays an important role alongside the porous TiO_2 electrode. Many studies have been reported with the objective of getting solar cells using inorganic gelators-based ion gels with promising results. Briefly, it can be found that silica, TiO_2, Al_2O_3, and ZnO nanoparticles or carbon nanotube (CNT) layers in combination with ionic liquids give ion gels for solar cells high efficiency as well as excellent stability [92–95]. Moreover, a combination of polymers with inorganic particles has also been reported as a candidate material for ion gels. Forsyth and coworkers studied poly(hydroxyethyl methacrylate-co-tetraethylene glycol diacrylate) poly(HEMA-co-TEGDA) photopolymerized in I_2-containing ionic liquid as a gel electrolyte for DSSCs [96]. TiO_2 nanoparticles were used as a photoinitiator and a cogelator in gel forming. The results from DSSCs assembled using the ionic liquid-based polymer gel electrolyte showed energy conversion efficiency of 3.9 % at 1 sun (AM 1.5) and 5.0 % at 0.39 sun illumination. However, this chapter will focus on organic ion gel electrolytes.

The most studied polymer electrolytes for solar cells are based on PEO or PVdF. An electrolytic membrane based on PEO and an ionic liquid 1-methyl-3-propylimidazolium iodide (PMII) has been reported as a novel candidate for DSSC since it has good ionic conductivity (3.56×10^{-4} S cm^{-1} when 200 wt% of ionic liquid) as well as iodide anion which is beneficial for redox couple formation (I^-/I_3) [97]. More recently, M. Forsyth and coworkers studied PEO-PVdF-based polymer electrolytes in combination with alkyl imidazolium iodide-functionalized silica (SiO_2-imI) to assemble a DSSC. The ionic conductivity of 1.07×10^{-4} S cm^{-1} and a power conversion efficiency of 3.83 % were obtained under air mass 1.5 solar spectrum illumination at 100 mW cm^{-2} with 8 wt% of SiO_2-imI composite polymer electrolyte [98]. Soni et al. have investigated polymer gel electrolytes for DSSCs [99]. Indeed, PEO/PPO-based amphiphilic block copolymers were combined with 1-butyl-3-methylimidazolium iodide [C_4mim][I] and iodide to synthesize a highly ordered polymer gel to be used in DSSCs and a 2.1 % solar to electricity conversion efficiency was achieved using a specific ruthenium-based dye D907.

Beyond the conventional polymers, polymeric ionic liquids-based gel polymer electrolytes have been also reported for solar cell applications. Azaceta et al. used poly(1-vinyl-3-alkylimidazolium) iodide polymers mixed with 1-methyl-3-propylimidazolium iodide to formulate polymer electrolytes, which showed ionic conductivities in the range of 10^{-3} to 10^{-7} S cm^{-1} at room temperature [100]. As expected, this ionic conductivity depends on the ionic liquid content. Finally, they prepared solar cells with an area of 1 cm^2 using mentioned polymer gel electrolytes to yield a maximum light-to-electricity conversion efficiency of 3.73 %. Continuing with poly(ionic liquid)-based gel electrolytes, poly(1-butyl-3-vinylimidazolium) bromide [PBvim][Br] and poly(1-butyl-3-vinylimidazolium) bis(trifluoromethanesulfonyl)imide [PBvim][$N(Tf)_2$] were dissolved at room temperature in ionic liquids to form gel electrolytes without using any volatile organic solvent [101]. Figure 9.7a shows the *J–V* curves of the fabricated DSSCs with mentioned poly(ionic liquid)s and poly(styrene-co-acrylonitrile) to compare. The cells based on [PBvim][$N(Tf)_2$] gel electrolyte showed a maximum efficiency of 4.4 % under simulated air mass 1.5 solar spectrum illumination at 100 mW cm^{-2} as well as higher long-term stability. Furthermore, bis-imidazolium-based poly(ionic liquid) electrolytes for DSSCs were reported [102]. Indeed, poly(1-butyl-3-vinylimidazolium-3-hexyl)imidazolium bis(trifluoromethanesulfonyl)imide [PBvim][C_6im][$N(Tf)_2$]$_2$ electrolytes have been prepared and used for the fabrication of organic solvent-free quasi-solid-state DSSCs to obtain an efficiency of 5.92 % under the simulated air mass 1.5 solar spectrum illumination at 100 mW cm^{-2}, which is higher than that of monocationic polymer gel electrolyte.

9.5.4 Supercapacitors

Electrochemical capacitors, also referred as supercapacitors or ultracapacitors, store energy using either ion adsorption (electrochemical double-layer capacitors EDLCs) or fast surface redox reactions (pseudocapacitors). In recent years, they have

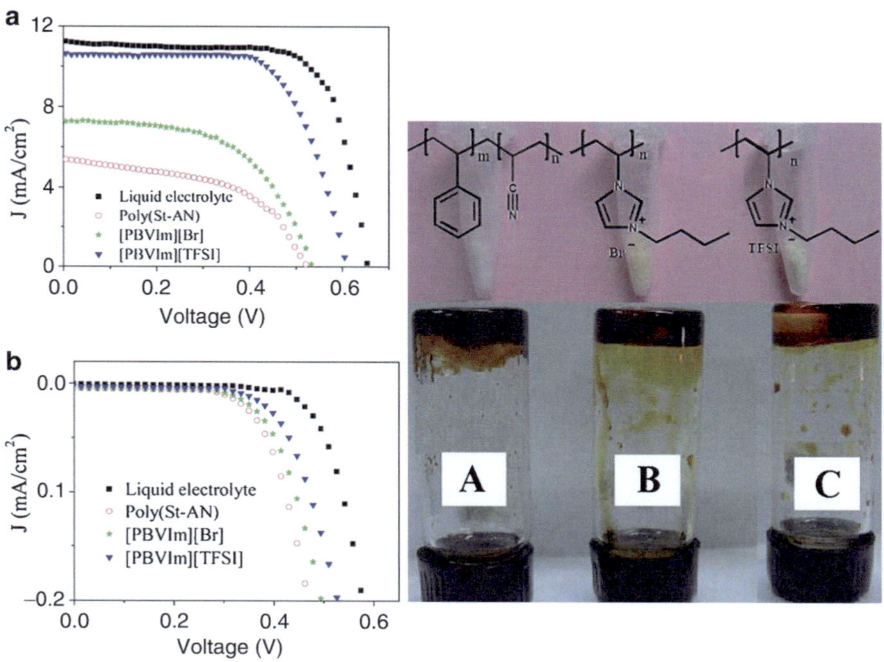

Fig. 9.7 The J–V curves of DSSCs containing liquid electrolytes, Poly(St-AN), [PBvim][Br], and [PBvim][$N(Tf)_2$]-based electrolytes (polymer contents: 25 wt%) (**a**) Under simulated AM 1.5 solar spectrum irradiation at 100 mW cm^{-2}. (**b**) In the dark [101]

received much attention, mainly due to their high-power density, long life cycle (~10^5 to 10^6 charge–discharge cycles), and fast charging (few seconds) [51, 103–106]. However, one of the main drawbacks of current capacitors is the use of aqueous or organic electrolytes, and with that the supercapacitors can show narrow cell voltage (limited to ~ 1.2 V) or high flammability and toxicity [41]. Ion gels are suitable electrolytes to avoid these inconveniences as they have wider electrochemical potential window and thermal stability.

In order to obtain ionic liquid-based gel polymer electrolytes, several polymer or copolymer matrices have been used, such as poly(vinylidene fluoride-hexafluoropropylene) PVdF-HFP, poly(methyl methacrylate) PMMA, polystyrene-poty(ethylene oxide) PS-PEO, poly(vinyl alcohol) PVA, or polyacrylonitrile PAN [41, 45, 107–109]. As an example, recently Pandey and Hashmi prepared an ionic liquid gel polymer electrolyte based on 1-ethyl-3-methylimidazolium tris(pentafluoroethyl) trifluorophosphate [C_2mim][FAP] and PVdF-HFP copolymer that exhibits an excellent thermal stability, a wide electrochemical window of ~ 4.4 V (from −2.2 to 2.2 V), and a high ionic conductivity of ~ 2×10^{-3} S cm^{-1} at room temperature [103]. Solid-state supercapacitors based on multiwalled carbon nanotube (MWCNT) electrodes and ion gel electrolytes have been fabricated; a good cycling performance with ~ 82 % capacitance retention after 10,000 charge–discharge

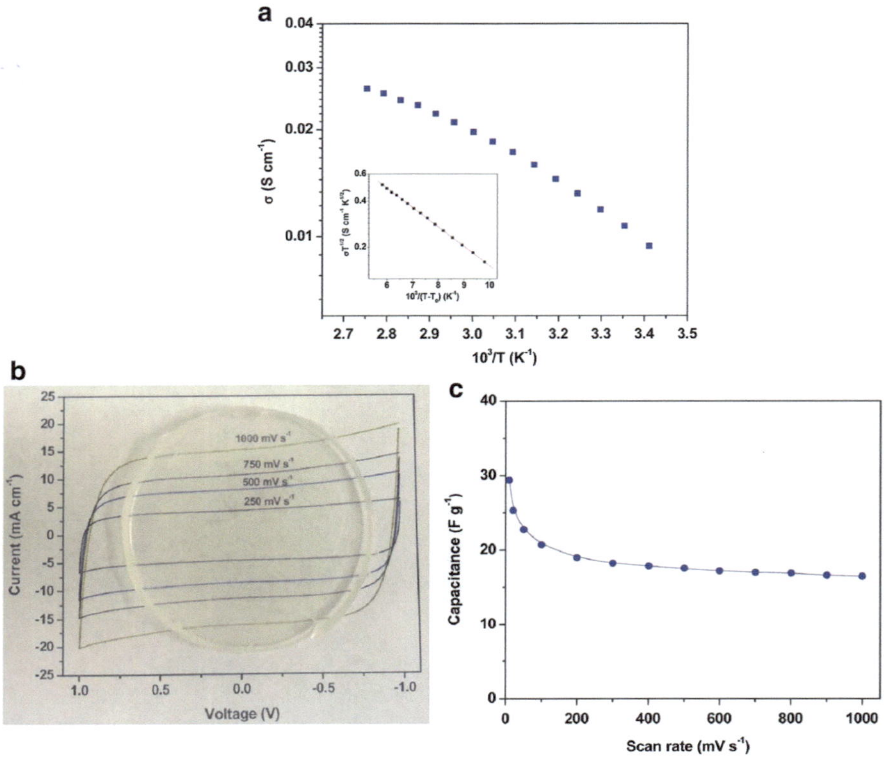

Fig. 9.8 (**a**) Temperature dependence of ionic conductivity of the ionic liquid gel polymer electrolyte film. Inset: The corresponding log (σ $T^{1/2}$) vs 1/(T – T_0) plot. (**b**) Cyclic voltammetry curves of EDLC cells at different scan rates. On the graph, there is a picture of a transparent and free-standing gel polymer electrolyte film. (**c**) Variation in the capacitance values of the EDLC cells [110]

cycles was achieved. Furthermore, the modified MWCNT electrodes show a specific capacitance of ~76 F g^{-1} and the supercapacitors show specific energy and specific power of ~ 17.2 Wh kg^{-1} and ~ 18.9 kW kg^{-1}, respectively. The synthesis of 1-ethyl-3-methylimidazolium tetracyanoborate [C_2mim][$B(CN)_4$] and PVdF-HFP copolymer-based ion gel used as an electrolyte in EDLCs has been also reported [110]. The ion gel showed excellent thermal stability up to 310 °C, an electrochemical window of ~ 3.8 V, and a high ionic conductivity of ~ 9×10^{-3} S cm^{-1} at room temperature (See Fig. 9.8). The EDLCs show single electrode-specific capacitance values of ~ 34.4 F g^{-1} and the specific energy and specific power of the capacitor are ~ 3.5 Wh kg^{-1} and ~ 4.2 kW kg^{-1}, respectively.

Besides the most common lithium ion-conducting gel polymer membranes, magnesium ion-conducting systems have also been reported to be used in electrochemical power storage devices, including EDLCs [111]. 1-ethyl-3-methylimidazolium trifluoromethanesulfonate [C_2mim][CF_3SO_3] and PVdF-HFP copolymer-based electrolytes were prepared to be used in an EDLC which contains activated charcoal

(AC) electrodes. By the incorporation of magnesium salt and ethylene carbonate–propylene carbonate cosolvent to the electrolyte, an important improvement was achieved on the properties of EDLCs. Indeed, 136 F g^{-1} of the optimal capacitance value of AC was measured, which corresponds to the specific energy and power density of ~ 18.8 Wh Kg^{-1} and ~6.67 kW kg^{-1}, respectively. On the other hand, by the addition of graphene nanosheets as an ionic conducting promoter to the common ionic liquid gel polymer electrolytes, an improved electrochemical behavior can be obtained: smaller internal resistance, higher specific capacitance, cycling stability, and better rate performance, as well as good thermal stability up to 400 °C [112].

Ion liquid-based gel polymer electrolytes in combination with electronically conducting polymer-based electrodes, such as PEDOT, have been studied to prepare supercapacitors [113, 114]. For instance, Pandey et al. have characterized devices with specific PEDOT electrode capacitance of ~ 154.5 F g^{-1}, maximum specific energy of 6.5 Wh kg^{-1}, and specific power of 11.3 kW kg^{-1}.

Ion gels play an important role to obtain all-solid-state flexible, stretchable, and lightweight supercapacitors, which are very interesting energy-storage devices for many applications [115, 45]. Kang et al. have fabricated flexible supercapacitors using bacterial nanocellulose papers coated with carbon nanotubes as electrodes and an ion gel based on PS-PEO-PS triblock copolymer with an ionic liquid 1-ethyl-3-methylimidazolium bis(trifluoromethylsulfonyl)imide, as electrolyte [109]. While the hydrophobic PS unit of the membrane ensured good adhesion to the carbon nanotube electrode via van der Waals interactions, the PEO block retained a hydrophilic ionic liquid in the ion gel system. A flexible energy-storage device was successfully fabricated and used to light a yellow light-emitting diode.

9.5.5 Actuators

In recent years, polymer actuators have attracted much attention for applications in robotics or medical devices due to their soft motion, simple structure, flexibility, and lightweight. In particular, ionic electroactive polymer actuators (polymer–metal composite actuators, conductive polymer actuators, and carbon nanotube actuators), which are driven by migration or diffusion of ions, exhibit a large deformation with low voltage application (<5 V) [116–118]. However, the main drawback arises when working in dry conditions, especially due to solvent evaporation. In this respect, ion gels have been studied as promising electrolytic membranes for this type of actuators. Fukushima, Aida, and coworkers were pioneers to report bucky-gel actuators fabricated through layer-by-layer casting. The actuators were ready to work without the need of further encapsulation [119–122]. In a typical example, the ion gel is composed of an ionic liquid, such as $[C_2mim][N(Tf)_2]$, and PVdF-HFP copolymers. The gel is then placed between two bucky-gel-electrode layers containing single-walled carbon nanotubes (SWCNT). They observed that the curvature of the actuator is proportional to the voltage applied and the thickness of the device; i.e., at a given voltage, thinner actuator's curvature is larger than the thicker ones.

Apart from conventional PVdF-HFP copolymer, other polymers were used to prepare actuator ion gels. For instance, Watanabe and coworkers have fabricated polymer actuators using ion gel electrolytes that contained 1-ethyl-3-methylimidazolium bis(trifluoromethanesulfonyl)amide [C_2mim][N(Tf)$_2$] and polystyrene–block-poly(methyl methacrylate)–block-polystyrene triblock copolymer (PS-*b*-PMMA-*b*-PS) synthesized by successive atom-transfer radical polymerizations (ATRPs) [38]. PS is not compatible with [C_2mim][N(Tf)$_2$], while PMMA phase can be dissolved in the ionic liquid to form an ion conduction pathway. This leads naturally to the self-assembly of PS-*b*-PMMA-*b*-PS copolymers in the ionic liquid giving them gel properties. Indeed, the affinity and repulsion of the polymer blocks to [C_2mim][N(Tf)$_2$] forced the formation of PS sphere domains that served as physical cross-linking points, as observed by the atomic force microscope (Fig. 9.9a). This ion gel electrolyte exhibits high ionic conductivity (>10^{-3} S cm^{-1}) at room temperature and was used for a trilaminar ionic polymer actuator sandwiched between two carbon composite electrodes. After applying low voltages to the actuator, less than 3 V, it is softly bent toward the anodic side. The same authors have studied other polymers, such as sulfonated polyimides (SPIs) or polyether-segmented polyurethaneurea (PEUU), in ion gel electrolyte formation for polymer actuators [123]. SPIs provide strength to the electrolyte due to the rigid aromatic backbone, and the counter-cations make it compatible with ionic liquids. As a result,

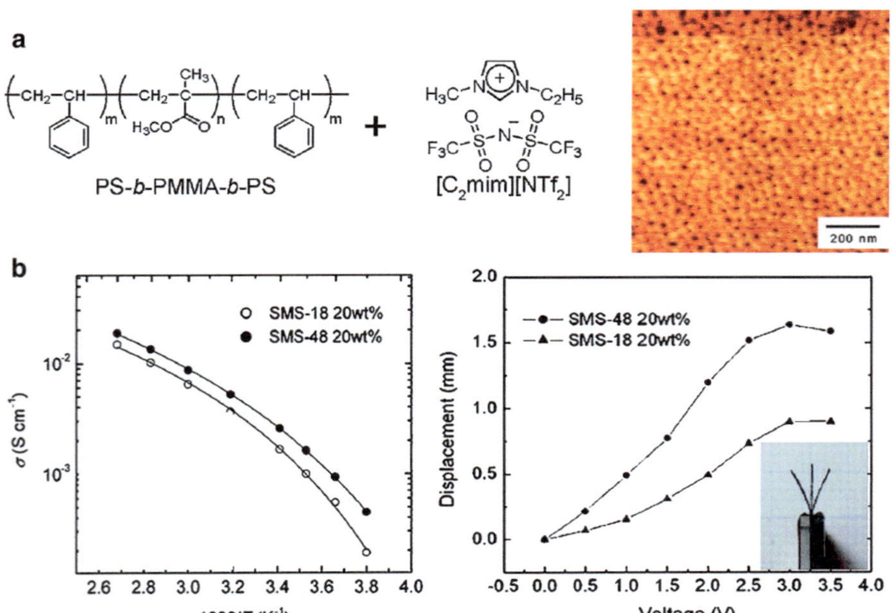

Fig. 9.9 (**a**) Chemical composition of triblock ion gel electrolyte and the membrane picture obtained by AFM; (**b**) Arrhenius plots of ionic conductivity and displacement of ion gel actuators for the ion gels containing 20 % of [C_2mim][N(Tf)$_2$] but different block lengths [38]

the SPI/IL exhibited favorable mechanical properties (elastic modulus >10^7 Pa) together with a high ionic conductivity (10^{-3} S cm^{-1}). The best trilaminar actuator obtained using SPI/IL ion gel electrolyte, reported in the literature, exhibits a relatively large displacement and high-speed response, keeping 80 % of the initial displacement even after more than 5,000 cycles [124].

9.5.6 *Emerging Applications: Electrochemical Sensors, Optoelectronic Devices, Bioelectronics, Gas Separation Membranes, Catalytic Membranes, Drug Releases*

Ion gels are finding more and more applications in different areas than the ones described before. In this section, we describe some of the most interesting emerging application areas of ion gels.

Electrochemical Sensors, Optoelectronic Devices, and Bioelectronics In the field of *sensors*, membranes based on poly(vinyl chloride) (PVC), ionic liquid 1-butyl-3-methylimidazolium hexafluorophosphate [C_4mim][PF_6], and polyazacycloalkane as ionophore were synthesized by Coll et al. and used in ion-selective electrodes as they exhibited a selective response to the hydrophilic anion sulfate [125]. Later on, Shvedene et al. have fabricated more polymer membranes for selective electrodes containing PMMA or PVC and ionic liquids [126]. The ionic liquid works as both plasticizer and ion-responsive media, and the electrodes obtained demonstrated good and stable response to ions (cations and anions). Those sensors were successfully designed to detect detergents. On the other hand, and mainly due to a biocompatibility of ionic liquids with biomolecules and enzymes, ion gels have been studied for biosensors applications [127, 128]. For instance, a film based on Nafion and a hydrophobic ionic liquid [C_4mim][PF_6] that can be used to immobilize and entrap horseradish peroxidase (HRP) and still retain its biological activity has been reported [129]. Therefore, it can be used for biosensors and biocatalysis applications as well.

Recently Owens and coworkers, who are pioneers in using ion gels for *bioelectronics*, have reported ion gels-assisted electrodes for applications in long-term cutaneous recordings [130]. They fabricated electrodes with gold and PEDOT:PSS, which were coated with ion gels. The ion gel decreased impedance at the interface with human skin, and they maintained a steady value over several days, making them suitable for a noninvasive, long-term monitoring of human electrophysiological activity (Fig. 9.10).

Electrochromic devices based on ion gels have been reported by Diamond and coworkers [131, 132]. The phosphonium ionic liquid was functionalized by a viologen molecule in order to achieve an electrochromic material, which was at the same time part of the electrolyte in an ion gel. The coloration efficiency of the device was 10.72 cm^2 C^{-1} and of relatively fast switching kinetics (221 s to obtain 95 % of absorbance) when compared with other electrochromic ionic liquids (up to 18,000 s).

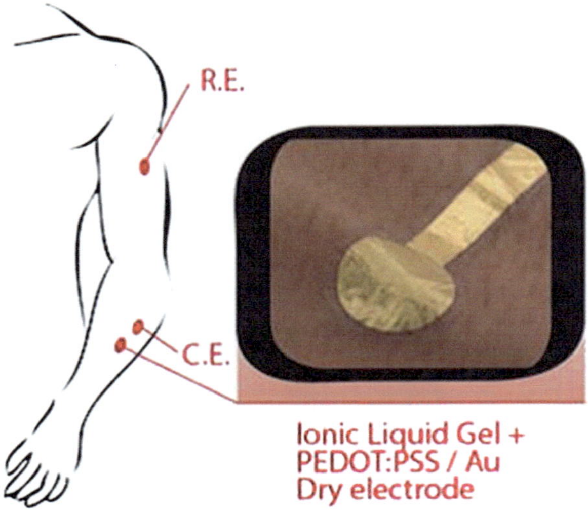

Fig. 9.10 Schematic representation of the electrode positions on an arm. The working electrode (Ionic liquid gel + PEDOT:PSS/Au dry electrode) and counter electrode (C.E.) were placed on the forearm, 5 cm away from each other. The reference electrode (R.E.) was placed on the arms, 30 cm away from the W.E. [130]

Recently Lodge, Frisbie, and coworkers have used block copolymer-based ion gel to prepare flexible, low-voltage, emissive *displays* on plastic [133]. The patterned ion gels contain redox-active luminophores and triblock copolymer of polystyrene–block-poly(methyl methacrylate)–block-polystyrene, where the solvophobic poly styrene end blocks associate form micelles in 1-ethyl-3-methylimidazolium bis(trifluoromethylsulfonyl)imide $[C_2mim][N(Tf)_2]$.

Gas Separation Membranes The electrolytic properties of ion gels and their applications in electrochemical devices are already widely explained; however, the ability of ionic liquids to dissolve a wide range of species, gases among others, make them suitable also for other types of applications, such as separation membranes. CO_2, CH_4, C_2H_6, C_2H_4, O_2, and N_2 are some of the gases that can be dissolved in ionic liquids [134]. CO_2 is one of the most common and well-known global warming driver, which can be captured by ionic liquid-based membranes [135]. For instance, PVdF and $[C_4mim][BF_4]$-based polymeric membranes have been studied to remove acid gases such as H_2S and CO_2 from crude natural gas, in which the permeability coefficient was increased when the ionic liquid content increased [136, 137].

Catalytic Membranes Ion gel membranes have been reported to be used as catalytic membranes or as a reaction media in biocatalysis. Liu et al. have reported the encapsulation of horseradish peroxidase (HRP) into an ionic liquid $[C_4mim][BF_4]$-based silica sol–gel matrix [138]. The specific activity of HRP enhanced dramatically, about

30-fold greater than that in silica gel without ionic liquid, and an excellent thermal stability of HRP was obtained after immobilizing it in ionic liquid-based matrix. It was suggested that the ionic liquid was responsible for the stabilization and protection of the immobilized HRP.

Ion Gels for Drug Release Ion gels that are able to lose gradually the ionic liquid are promising candidates to be used for drug release. Therefore, these ion gels operate in contrast to those described previously, i.e., ionic liquid leaching is required and controlled [2]. It has been demonstrated that ionic liquids can dissolve compounds with pharmaceutical activity that are insoluble or poorly soluble in the common aqueous or organic solvents, making them a promising alternative for drug solubility [139–141].

9.6 Summary and Outlook

The aim of this chapter was to give a general overview of ion gels and more in detail the use of polymer-based ion gels in electrochemical applications. Therefore, the first part of the chapter provides the definition and properties of ion gels, as well as preparation methods, such as swelling a polymer in an ionic liquid, solution casting, or direct polymerization of monomers in an ionic liquid.

Within the last 10 years, there has been a growing interest for polymer-based ion gels, probably due to the association of unique properties and their promising use in many applications, especially in electrochemical devices. Ion gels have been reported in this chapter revealing that through the polymer chemical structure and the ionic liquid fraction, (i) first the morphology of the gel can be controlled and (ii) second the mechanical properties and the ionic conductivity can be tuned independently. Without doubt, the richness, in terms of physicochemical properties, of ionic liquids and polymers makes this type of ion gels very attractive for future researches. Considering the mentioned properties of ion gels, the applications have been reported in the second part of the chapter, which are mainly focused on energy electrochemical devices. However, it is worth to mention the new properties such as biocompatibility of some ionic liquids that can be used to fabricate ion gels with applications in sensors or bioelectronics.

The intimate relationship between ion transport and the three-dimensional polymeric network has been addressed for specific polymer/ionic liquids, but is still not fully understood in detail and should definitely be mapped as a function of composite components. This again opens up a vast field of research, extremely challenging in terms of control of the morphologies and measure of the local ion dynamics toward the investigation of the ions' confinement effects in organic ion gels. Indeed, the use of multifunctional polymers such as block copolymers can provide additional functionalities to the ion gel (electron conductor, photoresponsive, etc.) or selective affinity with the ionic liquids, leading to the association of new properties or achieving original controlled morphologies.

References

1. Lodge TP (2008) A unique platform for materials design. Science 321(5885):50–51. doi:10.1126/science.1159652
2. Le Bideau J, Viau L, Vioux A (2011) Ionogels, ionic liquid based hybrid materials. Chem Soc Rev 40(2):907–925. doi:10.1039/C0CS00059K
3. MacFarlane DR, Forsyth M, Howlett PC, Pringle JM, Sun J, Annat G, Neil W, Izgorodina EI (2007) Ionic liquids in electrochemical devices and processes: managing interfacial electrochemistry. Acc Chem Res 40(11):1165–1173. doi:10.1021/ar7000952
4. Welton T (1999) Room-temperature ionic liquids. Solvents for synthesis and catalysis. Chem Rev 99(8):2071–2084. doi:10.1021/cr980032t
5. Wasserscheid P, Welton T (2008) Ionic liquids in synthesis, 2nd edn. Wiley, Weinheim
6. Piepenbrock M-OM, Lloyd GO, Clarke N, Steed JW (2009) Metal- and anion-binding supramolecular gels. Chem Rev 110(4):1960–2004. doi:10.1021/cr9003067
7. Weiss RG, Terech P (2006) Molecular gels. Materials with self-assembled fibrillar networks. Springer, Netherlands
8. Flory PJ (1974) Introductory lecture. Faraday Discuss Chem Soc 57:7–18. doi:10.1039/DC9745700007
9. Jordon Lloyd D (1926) Colloid chemistry. The Chemical Catalogue Co., New York, NY
10. Steed JW (2011) Supramolecular gel chemistry: developments over the last decade. Chem Commun 47(5):1379–1383. doi:10.1039/C0CC03293J
11. Israelachvili JN (1991) Intermolecular and surface forces. Academic, New York, NY
12. Terech P, Weiss RG (1997) Low molecular mass gelators of organic liquids and the properties of their gels. Chem Rev 97(8):3133–3160. doi:10.1021/cr9700282
13. Estroff LA, Hamilton AD (2004) Water gelation by small organic molecules. Chem Rev 104(3):1201–1218. doi:10.1021/cr0302049
14. Ye Y-S, Rick J, Hwang B-J (2013) Ionic liquid polymer electrolytes. J Mater Chem A 1(8):2719–2743. doi:10.1039/C2TA00126H
15. Izák P, Hovorka Š, Bartovský T, Bartovská L, Crespo JG (2007) Swelling of polymeric membranes in room temperature ionic liquids. J Membr Sci 296(1–2):131–138. doi:10.1016/j.memsci.2007.03.022
16. Xing C, Zheng X, Xu L, Jia J, Ren J, Li Y (2014) Toward an optically transparent, antielectrostatic, and robust polymer composite: morphology and properties of polycarbonate/ionic liquid composites. Ind Eng Chem Res 53(11):4304–4311. doi:10.1021/ie404096b
17. Doyle M, Choi SK, Proulx G (2000) High-temperature proton conducting membranes based on perfluorinated ionomer membrane-ionic liquid composites. J Electrochem Soc 147(1):34–37. doi:10.1149/1.1393153
18. Vidinha P, Lourenco NMT, Pinheiro C, Bras AR, Carvalho T, Santos-Silva T, Mukhopadhyay A, Romao MJ, Parola J, Dionisio M, Cabral JMS, Afonso CAM, Barreiros S (2008) Ion jelly: a tailor-made conducting material for smart electrochemical devices. Chem Commun 30(44):5842–5844. doi:10.1039/B811647D
19. Singh T, Trivedi TJ, Kumar A (2010) Dissolution, regeneration and ion-gel formation of agarose in room-temperature ionic liquids. Green Chem 12(6):1029–1035. doi:10.1039/B927589D
20. He Y, Lodge TP (2007) A thermoreversible ion gel by triblock copolymer self-assembly in an ionic liquid. Chem Commun 14(26):2732–2734. doi:10.1039/B704490A
21. Lee KH, Zhang S, Gu Y, Lodge TP, Frisbie CD (2013) Transfer printing of thermoreversible ion gels for flexible electronics. ACS Appl Mater Interfaces 5(19):9522–9527. doi:10.1021/am402200n
22. Liu L, Yang P, Li L, Cui Y, An M (2012) Application of bis(trifluoromethanesulfonyl)imide lithium–N-methyl-N-butylpiperidinium-bis(trifluoromethanesulfonyl)imide–poly(vinylidene difluoride-co-hexafluoropropylene) ionic liquid gel polymer electrolytes in Li/$LiFePO_4$ batteries at different temperatures. Electrochim Acta 85:49–56. doi:10.1016/j.electacta.2012.08.066

23. Pandey GP, Hashmi SA (2009) Experimental investigations of an ionic-liquid-based, magnesium ion conducting, polymer gel electrolyte. J Power Sources 187(2):627–634. doi:10.1016/j.jpowsour.2008.10.112
24. He Y, Boswell PG, Bühlmann P, Lodge TP (2006) Ion gels by self-assembly of a triblock copolymer in an ionic Liquid†. J Phys Chem B 111(18):4645–4652. doi:10.1021/jp064574n
25. Hoarfrost ML, Segalman RA (2011) Ionic conductivity of nanostructured block copolymer/ionic liquid membranes. Macromolecules 44(13):5281–5288. doi:10.1021/ma200060g
26. Suleman M, Kumar Y, Hashmi SA (2013) Structural and electrochemical properties of succinonitrile-based gel polymer electrolytes: role of ionic liquid addition. J Phys Chem B 117(24):7436–7443. doi:10.1021/jp312358x
27. Shalu CSK, Singh RK, Chandra S (2012) Thermal stability, complexing behavior, and ionic transport of polymeric gel membranes based on polymer PVdF-HFP and ionic liquid, [BMIM][BF_4]. J Phys Chem B 117(3):897–906. doi:10.1021/jp307694q
28. Marcilla R, Alcaide F, Sardon H, Pomposo JA, Pozo-Gonzalo C, Mecerreyes D (2006) Tailor-made polymer electrolytes based upon ionic liquids and their application in all-plastic electrochromic devices. Electrochem Commun 8(3):482–488. doi:10.1016/j.elecom.2006.01.013
29. Jiang J, Gao D, Li Z, Su G (2006) Gel polymer electrolytes prepared by in situ polymerization of vinyl monomers in room-temperature ionic liquids. React Funct Polym 66(10):1141–1148. doi:10.1016/j.reactfunctpolym.2006.02.004
30. Noda A, Watanabe M (2000) Highly conductive polymer electrolytes prepared by in situ polymerization of vinyl monomers in room temperature molten salts. Electrochim Acta 45(8–9):1265–1270. doi:10.1016/S0013-4686(99)00330-8
31. Susan MABH, Kaneko T, Noda A, Watanabe M (2005) Ion gels prepared by in situ radical polymerization of vinyl monomers in an ionic liquid and their characterization as polymer electrolytes. J Am Chem Soc 127(13):4976–4983. doi:10.1021/ja045155b
32. Nakajima H, Ohno H (2005) Preparation of thermally stable polymer electrolytes from imidazolium-type ionic liquid derivatives. Polymer 46(25):11499–11504. doi:10.1016/j.polymer.2005.10.005
33. Isik M, Gracia R, Kollnus LC, Tomé LC, Marrucho IM, Mecerreyes D (2013) Cholinium-based poly(ionic liquid)s: synthesis, characterization, and application as biocompatible ion gels and cellulose coatings. ACS Macro Lett 2(11):975–979. doi:10.1021/mz400451g
34. Jana S, Parthiban A, Chai CLL (2010) Transparent, flexible and highly conductive ion gels from ionic liquid compatible cyclic carbonate network. Chem Commun 46(9):1488–1490. doi:10.1039/B921517D
35. Vioux A, Viau L, Volland S, Le Bideau J (2010) Use of ionic liquids in sol-gel; ionogels and applications. C R Chim 13(1–2):242–255. doi:10.1016/j.crci.2009.07.002
36. Gayet F, Viau L, Leroux F, Monge S, Robin J-J, Vioux A (2010) Polymer nanocomposite ionogels, high-performance electrolyte membranes. J Mater Chem 20(42):9456–9462. doi:10.1039/C000033G
37. Noor SAM, Bayley PM, Forsyth M, MacFarlane DR (2013) Ionogels based on ionic liquids as potential highly conductive solid state electrolytes. Electrochim Acta 91:219–226. doi:10.1016/j.electacta.2012.11.113
38. Imaizumi S, Kokubo H, Watanabe M (2011) Polymer actuators using ion-gel electrolytes prepared by self-assembly of ABA-triblock copolymers. Macromolecules 45(1):401–409. doi:10.1021/ma2022138
39. Appetecchi GB, Kim GT, Montanino M, Carewska M, Marcilla R, Mecerreyes D, De Meatza I (2010) Ternary polymer electrolytes containing pyrrolidinium-based polymeric ionic liquids for lithium batteries. J Power Sources 195(11):3668–3675. doi:10.1016/j.jpowsour.2009.11.146
40. Yang P, Liu L, Li L, Hou J, Xu Y, Ren X, An M, Li N (2014) Gel polymer electrolyte based on polyvinylidenefluoride-co-hexafluoropropylene and ionic liquid for lithium ion battery. Electrochim Acta 115:454–460. doi:10.1016/j.electacta.2013.10.202
41. Pandey GP, Hashmi SA (2013) Solid-state supercapacitors with ionic liquid based gel polymer electrolyte: effect of lithium salt addition. J Power Sources 243:211–218. doi:10.1016/j.jpowsour.2013.05.183

42. Winterton N (2006) Solubilization of polymers by ionic liquids. J Mater Chem 16(44):4281–4293. doi:10.1039/B610143G
43. Ji N, Matsumoto H, Yoshida M (2012) Highly efficient and specific gelation of ionic liquids by polymeric electrolytes to form ionogels with substantially high gel–sol transition temperatures and rheological properties like self-standing ability and rapid recovery. ACS Macro Lett 1(9):1108–1112. doi:10.1021/mz3002808
44. Chen B, Lu JJ, Yang CH, Yang JH, Zhou J, Chen YM, Suo Z (2014) Highly stretchable and transparent ionogels as nonvolatile conductors for dielectric elastomer transducers. ACS Appl Mater Interfaces. doi:10.1021/am501130t
45. Saricilar S, Antiohos D, Shu K, Whitten PG, Wagner K, Wang C, Wallace GG (2013) High strain stretchable solid electrolytes. Electrochem Commun 32:47–50. doi:10.1016/j.elecom.2013.03.035
46. He Y, Lodge TP (2007) Thermoreversible ion gels with tunable melting temperatures from triblock and pentablock copolymers. Macromolecules 41(1):167–174. doi:10.1021/ma702014z
47. Miranda DF, Versek C, Tuominen MT, Russell TP, Watkins JJ (2013) Cross-linked block copolymer/ionic liquid self-assembled blends for polymer gel electrolytes with high ionic conductivity and mechanical strength. Macromolecules 46(23):9313–9323. doi:10.1021/ma401302r
48. Virgili JM, Hoarfrost ML, Segalman RA (2010) Effect of an ionic liquid solvent on the phase behavior of block copolymers. Macromolecules 43(12):5417–5423. doi:10.1021/ma902804e
49. McIntosh LD, Kubo T, Lodge TP (2014) Morphology, modulus, and conductivity of a triblock terpolymer/ionic liquid electrolyte membrane. Macromolecules 47(3):1090–1098. doi:10.1021/ma4022373
50. Roberts AD, Li X, Zhang H (2014) Porous carbon spheres and monoliths: morphology control, pore size tuning and their applications as Li-ion battery anode materials. Chem Soc Rev. doi:10.1039/C4CS00071D
51. Winter M, Brodd RJ (2004) What are batteries, fuel cells, and supercapacitors? Chem Rev 104(10):4245–4270. doi:10.1021/cr020730k
52. Fernicola A, Weise FC, Greenbaum SG, Kagimoto J, Scrosati B, Soleto A (2009) Lithium-ion-conducting electrolytes: from an ionic liquid to the polymer membrane. J Electrochem Soc 156(7):A514–A520. doi:10.1149/1.3122885
53. Guerfi A, Dontigny M, Kobayashi Y, Vijh A, Zaghib K (2009) Investigations on some electrochemical aspects of lithium-ion ionic liquid/gel polymer battery systems. J Solid State Electrochem 13(7):1003–1014. doi:10.1007/s10008-008-0697-x
54. Fergus JW (2010) Ceramic and polymeric solid electrolytes for lithium-ion batteries. J Power Sources 195(15):4554–4569. doi:10.1016/j.jpowsour.2010.01.076
55. Li M, Yang L, Fang S, Dong S, Jin Y, S-i H, Tachibana K (2011) Li/$LiFePO_4$ batteries with gel polymer electrolytes incorporating a guanidinium-based ionic liquid cycled at room temperature and 50 °C. J Power Sources 196(15):6502–6506. doi:10.1016/j.jpowsour.2011.03.071
56. Lewandowski A, Swiderska-Mocek A, Waliszewski L (2013) Li+ conducting polymer electrolyte based on ionic liquid for lithium and lithium-ion batteries. Electrochim Acta 92:404–411. doi:10.1016/j.electacta.2013.01.028
57. Hofmann A, Schulz M, Hanemann T (2013) Gel electrolytes based on ionic liquids for advanced lithium polymer batteries. Electrochim Acta 89:823–831. doi:10.1016/j.electacta.2012.10.144
58. Li L, Wang J, Yang P, Guo S, Wang H, Yang X, Ma X, Yang S, Wu B (2013) Preparation and characterization of gel polymer electrolytes containing N-butyl-N-methylpyrrolidinium bis(trifluoromethanesulfonyl) imide ionic liquid for lithium ion batteries. Electrochim Acta 88:147–156. doi:10.1016/j.electacta.2012.10.018
59. Kim J-K, Niedzicki L, Scheers J, Shin C-R, Lim D-H, Wieczorek W, Johansson P, Ahn J-H, Matic A, Jacobsson P (2013) Characterization of N-butyl-N-methyl-pyrrolidinium bis(trifluoromethanesulfonyl)imide-based polymer electrolytes for high safety lithium batteries. J Power Sources 224:93–98. doi:10.1016/j.jpowsour.2012.09.029

60. Raghavan P, Zhao X, Choi H, Lim D-H, Kim J-K, Matic A, Jacobsson P, Nah C, Ahn J-H (2014) Electrochemical characterization of poly(vinylidene fluoride-co-hexafluoro propylene) based electrospun gel polymer electrolytes incorporating room temperature ionic liquids as green electrolytes for lithium batteries. Solid State Ion 262:77–82. doi:10.1016/j.ssi.2013.10.044
61. Pont A-L, Marcilla R, De Meatza I, Grande H, Mecerreyes D (2009) Pyrrolidinium-based polymeric ionic liquids as mechanically and electrochemically stable polymer electrolytes. J Power Sources 188(2):558–563. doi:10.1016/j.jpowsour.2008.11.115
62. Li M, Wang L, Yang B, Du T, Zhang Y (2014) Facile preparation of polymer electrolytes based on the polymerized ionic liquid poly((4-vinylbenzyl)trimethylammonium bis(trifluoro methanesulfonylimide)) for lithium secondary batteries. Electrochim Acta 123:296–302. doi:10.1016/j.electacta.2013.12.179
63. Li M, Yang B, Wang L, Zhang Y, Zhang Z, Fang S, Zhang Z (2013) New polymerized ionic liquid (PIL) gel electrolyte membranes based on tetraalkylammonium cations for lithium ion batteries. J Membr Sci 447:222–227. doi:10.1016/j.memsci.2013.07.007
64. Shin JH, Henderson WA, Appetecchi GB, Alessandrini F, Passerini S (2005) Recent developments in the ENEA lithium metal battery project. Electrochim Acta 50(19):3859–3865. doi:10.1016/j.electacta.2005.02.049
65. Wetjen M, Kim G-T, Joost M, Winter M, Passerini S (2013) Temperature dependence of electrochemical properties of cross-linked poly(ethylene oxide)–lithium bis(trifluoromethanesulfonyl)imide–N-butyl-N-methylpyrrolidinium bis(trifluoromethanesulfonyl)imide solid polymer electrolytes for lithium batteries. Electrochim Acta 87:779–787. doi:10.1016/j.electacta.2012.09.034
66. Liao C, Sun X-G, Dai S (2013) Crosslinked gel polymer electrolytes based on polyethylene glycol methacrylate and ionic liquid for lithium ion battery applications. Electrochim Acta 87:889–894. doi:10.1016/j.electacta.2012.10.027
67. Stepniak I (2014) Compatibility of poly(bisAEA4)-LiTFSI–MPPipTFSI ionic liquid gel polymer electrolyte with $Li_4Ti_5O_{12}$ lithium ion battery anode. J Power Sources 247:112–116. doi:10.1016/j.jpowsour.2013.08.080
68. Liao Y, Sun C, Hu S, Li W (2013) Anti-thermal shrinkage nanoparticles/polymer and ionic liquid based gel polymer electrolyte for lithium ion battery. Electrochim Acta 89:461–468. doi:10.1016/j.electacta.2012.11.095
69. Rao M, Geng X, Liao Y, Hu S, Li W (2012) Preparation and performance of gel polymer electrolyte based on electrospun polymer membrane and ionic liquid for lithium ion battery. J Membr Sci 399–400:37–42. doi:10.1016/j.memsci.2012.01.021
70. Weber AZ, Newman J (2004) Modeling transport in polymer-electrolyte fuel cells. Chem Rev 104(10):4679–4726. doi:10.1021/cr020729l
71. Devanathan R (2008) Recent developments in proton exchange membranes for fuel cells. Energy Environ Sci 1(1):101–119. doi:10.1039/B808149M
72. Hasegawa T, Béléké AB, Mizuhata M (2013) Membrane modification by liquid phase deposition using small amount of TiO_2 for high-temperature operation of polymer electrolyte fuel cells. J Power Sources 233:148–156. doi:10.1016/j.jpowsour.2013.01.118
73. Schauer J, Sikora A, Plíšková M, Mališ J, Mazúr P, Paidar M, Bouzek K (2011) Ion-conductive polymer membranes containing 1-butyl-3-methylimidazolium trifluoromethanesulfonate and 1-ethylimidazolium trifluoromethanesulfonate. J Membr Sci 367(1–2):332–339. doi:10.1016/j.memsci.2010.11.018
74. Rana UA, Forsyth M, MacFarlane DR, Pringle JM (2012) Toward protic ionic liquid and organic ionic plastic crystal electrolytes for fuel cells. Electrochim Acta 84:213–222. doi:10.1016/j.electacta.2012.03.058
75. Kreuer KD (2001) On the development of proton conducting polymer membranes for hydrogen and methanol fuel cells. J Membr Sci 185(1):29–39. doi:10.1016/S0376-7388(00)00632-3
76. Díaz M, Ortiz A, Vilas M, Tojo E, Ortiz I (2014) Performance of PEMFC with new polyvinyl-ionic liquids based membranes as electrolytes. Int J Hydrog Energy 39(8):3970–3977. doi:10.1016/j.ijhydene.2013.04.155

77. Mališ J, Mazúr P, Schauer J, Paidar M, Bouzek K (2013) Polymer-supported 1-butyl-3-methylimidazolium trifluoromethanesulfonate and 1-ethylimidazolium trifluoromethanesulfonate as electrolytes for the high temperature PEM-type fuel cell. Int J Hydrog Energy 38(11):4697–4704. doi:10.1016/j.ijhydene.2013.01.126
78. Lee S-Y, Yasuda T, Watanabe M (2010) Fabrication of protic ionic liquid/sulfonated polyimide composite membranes for non-humidified fuel cells. J Power Sources 195(18):5909–5914. doi:10.1016/j.jpowsour.2009.11.045
79. Yasuda T, S-i N, Honda Y, Kinugawa K, Lee S-Y, Watanabe M (2012) Effects of polymer structure on properties of sulfonated polyimide/protic ionic liquid composite membranes for nonhumidified fuel cell applications. ACS Appl Mater Interfaces 4(3):1783–1790. doi:10.1021/am300031k
80. Fernicola A, Navarra M, Panero S (2008) Aprotic ionic liquids as electrolyte components in protonic membranes. J Appl Electrochem 38(7):993–996. doi:10.1007/s10800-008-9514-6
81. Liew C-W, Ramesh S, Arof AK (2014) A novel approach on ionic liquid-based poly(vinyl alcohol) proton conductive polymer electrolytes for fuel cell applications. Int J Hydrog Energy 39(6):2917–2928. doi:10.1016/j.ijhydene.2013.07.092
82. Dahi A, Fatyeyeva K, Langevin D, Chappey C, Rogalsky SP, Tarasyuk OP, Marais S (2014) Polyimide/ionic liquid composite membranes for fuel cells operating at high temperatures. Electrochim Acta 130:830–840. doi:10.1016/j.electacta.2014.03.071
83. Di Noto V, Negro E, Sanchez J-Y, Iojoiu C (2010) Structure-relaxation interplay of a new nanostructured membrane based on tetraethylammonium trifluoromethanesulfonate ionic liquid and neutralized nafion 117 for high-temperature fuel cells. J Am Chem Soc 132(7):2183–2195. doi:10.1021/ja906975z
84. Lee S-Y, Ogawa A, Kanno M, Nakamoto H, Yasuda T, Watanabe M (2010) Nonhumidified intermediate temperature fuel cells using protic ionic liquids. J Am Chem Soc 132(28):9764–9773. doi:10.1021/ja102367x
85. Eguizábal A, Lemus J, Pina MP (2013) On the incorporation of protic ionic liquids imbibed in large pore zeolites to polybenzimidazole membranes for high temperature proton exchange membrane fuel cells. J Power Sources 222:483–492. doi:10.1016/j.jpowsour.2012.07.094
86. Jheng L-c, Hsu SL-c, Lin B-y, Hsu Y-l (2014) Quaternized polybenzimidazoles with imidazolium cation moieties for anion exchange membrane fuel cells. J Membr Sci 460:160–170. doi:10.1016/j.memsci.2014.02.043
87. Jothi PR, Dharmalingam S (2014) An efficient proton conducting electrolyte membrane for high temperature fuel cell in aqueous-free medium. J Membr Sci 450:389–396. doi:10.1016/j.memsci.2013.09.034
88. Gopi KH, Peera SG, Bhat SD, Sridhar P, Pitchumani S (2014) Preparation and characterization of quaternary ammonium functionalized poly(2,6-dimethyl-1,4-phenylene oxide) as anion exchange membrane for alkaline polymer electrolyte fuel cells. Int J Hydrog Energy 39(6):2659–2668. doi:10.1016/j.ijhydene.2013.12.009
89. Pan J, Chen C, Li Y, Wang L, Tan L, Li G, Tang X, Xiao L, Lu J, Zhuang L (2014) Constructing ionic highway in alkaline polymer electrolytes. Energy Environ Sci 7(1):354–360. doi:10.1039/C3EE43275K
90. Lin B, Qiu L, Lu J, Yan F (2010) Cross-linked alkaline ionic liquid-based polymer electrolytes for alkaline fuel cell applications. Chem Mater 22(24):6718–6725. doi:10.1021/cm102957g
91. Yuan J, Mecerreyes D, Antonietti M (2013) Poly(ionic liquid)s: an update. Prog Polym Sci 38(7):1009–1036. doi:10.1016/j.progpolymsci.2013.04.002
92. Kato T, Kado T, Tanaka S, Okazaki A, Hayase S (2006) Quasi-solid dye-sensitized solar cells containing nanoparticles modified with ionic liquid-type molecules. J Electrochem Soc 153(3):A626–A630. doi:10.1149/1.2161578
93. Kato T, Hayase S (2007) Quasi-solid dye sensitized solar cell with straight ion paths: proposal of hybrid electrolytes for ionic liquid-type electrolytes. J Electrochem Soc 154(1):B117–B121. doi:10.1149/1.2393008

94. Kuang D, Brillet J, Chen P, Takata M, Uchida S, Miura H, Sumioka K, Zakeeruddin SM, Grätzel M (2008) Application of highly ordered TiO_2 nanotube arrays in flexible dye-sensitized solar cells. ACS Nano 2(6):1113–1116. doi:10.1021/nn800174y
95. Bai Y, Cao Y, Zhang J, Wang M, Li R, Wang P, Zakeeruddin SM, Gratzel M (2008) High-performance dye-sensitized solar cells based on solvent-free electrolytes produced from eutectic melts. Nat Mater 7(8):626–630, http://www.nature.com/nmat/journal/v7/n8/suppinfo/nmat2224_S1.html
96. Winther-Jensen O, Armel V, Forsyth M, MacFarlane DR (2010) In situ photopolymerization of a gel ionic liquid electrolyte in the presence of iodine and its use in dye sensitized solar cells. Macromol Rapid Commun 31(5):479–483. doi:10.1002/marc.200900701
97. Singh PK, Kim K-W, Park N-G, Rhee H-W (2008) Mesoporous nanocrystalline TiO_2 electrode with ionic liquid-based solid polymer electrolyte for dye-sensitized solar cell application. Synth Met 158(14):590–593. doi:10.1016/j.synthmet.2008.04.003
98. Hu M, Sun J, Rong Y, Yang Y, Liu L, Li X, Forsyth M, MacFarlane DR, Han H (2014) Enhancement of monobasal solid-state dye-sensitized solar cells with polymer electrolyte assembling imidazolium iodide-functionalized silica nanoparticles. J Power Sources 248:283–288. doi:10.1016/j.jpowsour.2013.09.107
99. Soni SS, Fadadu KB, Vekariya RL, Debgupta J, Patel KD, Gibaud A, Aswal VK (2014) Effect of self-assembly on triiodide diffusion in water based polymer gel electrolytes: an application in dye solar cell. J Colloid Interface Sci 425:110–117. doi:10.1016/j.jcis.2014.03.047
100. Azaceta E, Marcilla R, Sanchez-Diaz A, Palomares E, Mecerreyes D (2010) Synthesis and characterization of poly(1-vinyl-3-alkylimidazolium) iodide polymers for quasi-solid electrolytes in dye sensitized solar cells. Electrochim Acta 56(1):42–46. doi:10.1016/j.electacta.2009.01.058
101. Zhao J, Shen X, Yan F, Qiu L, Lee S, Sun B (2011) Solvent-free ionic liquid/poly(ionic liquid) electrolytes for quasi-solid-state dye-sensitized solar cells. J Mater Chem 21(20):7326–7330. doi:10.1039/C1JM10346F
102. Chen X, Zhao J, Zhang J, Qiu L, Xu D, Zhang H, Han X, Sun B, Fu G, Zhang Y, Yan F (2012) Bis-imidazolium based poly(ionic liquid) electrolytes for quasi-solid-state dye-sensitized solar cells. J Mater Chem 22(34):18018–18024. doi:10.1039/C2JM33273F
103. Pandey GP, Hashmi SA (2013) Performance of solid-state supercapacitors with ionic liquid 1-ethyl-3-methylimidazolium tris(pentafluoroethyl) trifluorophosphate based gel polymer electrolyte and modified MWCNT electrodes. Electrochim Acta 105:333–341. doi:10.1016/j.electacta.2013.05.018
104. Simon P, Gogotsi Y (2008) Materials for electrochemical capacitors. Nat Mater 7(11):845–854
105. Kötz R, Carlen M (2000) Principles and applications of electrochemical capacitors. Electrochim Acta 45(15–16):2483–2498. doi:10.1016/S0013-4686(00)00354-6
106. Wang G, Zhang L, Zhang J (2012) A review of electrode materials for electrochemical supercapacitors. Chem Soc Rev 41(2):797–828. doi:10.1039/C1CS15060J
107. Liew C-W, Ramesh S, Arof AK (2014) Good prospect of ionic liquid based-poly(vinyl alcohol) polymer electrolytes for supercapacitors with excellent electrical, electrochemical and thermal properties. Int J Hydrog Energy 39(6):2953–2963. doi:10.1016/j.ijhydene.2013.06.061
108. Tamilarasan P, Ramaprabhu S (2013) Graphene based all-solid-state supercapacitors with ionic liquid incorporated polyacrylonitrile electrolyte. Energy 51:374–381. doi:10.1016/j.energy.2012.11.037
109. Kang YJ, Chun S-J, Lee S-S, Kim B-Y, Kim JH, Chung H, Lee S-Y, Kim W (2012) All-solid-state flexible supercapacitors fabricated with bacterial nanocellulose papers, carbon nanotubes, and triblock-copolymer ion gels. ACS Nano 6(7):6400–6406. doi:10.1021/nn301971r
110. Pandey GP, Hashmi SA (2013) Ionic liquid 1-ethyl-3-methylimidazolium tetracyanoborate-based gel polymer electrolyte for electrochemical capacitors. J Mater Chem A 1(10):3372–3378. doi:10.1039/C2TA01347A

111. Pandey GP, Hashmi SA, Kumar Y (2010) Performance studies of activated charcoal based electrical double layer capacitors with ionic liquid gel polymer electrolytes. Energy Fuels 24(12):6644–6652. doi:10.1021/ef1010447
112. Shi M-J, Kou S-Z, Shen B-S, Lang J-W, Yang Z, Yan X-B (2014) Improving the performance of all-solid-state supercapacitors by modifying ionic liquid gel electrolytes with graphene nanosheets prepared by arc-discharge. Chin Chem Lett 25(6):859–864. doi:10.1016/j.cclet.2014.04.010
113. Pandey GP, Rastogi AC, Westgate CR (2014) All-solid-state supercapacitors with poly(3,4-ethylenedioxythiophene)-coated carbon fiber paper electrodes and ionic liquid gel polymer electrolyte. J Power Sources 245:857–865. doi:10.1016/j.jpowsour.2013.07.017
114. Sellam HSA (2013) High rate performance of flexible pseudocapacitors fabricated using ionic-liquid-based proton conducting polymer electrolyte with poly(3, 4-ethylenedioxythiophene): poly(styrene sulfonate) and its hydrous ruthenium oxide composite electrodes. ACS Appl Mater Interfaces 5(9):3875–3883. doi:10.1021/am4005557
115. Hu L, Pasta M, Mantia FL, Cui L, Jeong S, Deshazer HD, Choi JW, Han SM, Cui Y (2010) Stretchable, porous, and conductive energy textiles. Nano Lett 10(2):708–714. doi:10.1021/nl903949m
116. Smela E (2003) Conjugated polymer actuators for biomedical applications. Adv Mater 15(6):481–494. doi:10.1002/adma.200390113
117. Shahinpoor M (2003) Ionic polymer–conductor composites as biomimetic sensors, robotic actuators and artificial muscles—a review. Electrochim Acta 48(14–16):2343–2353. doi:10.1016/S0013-4686(03)00224-X
118. Baughman RH, Cui C, Zakhidov AA, Iqbal Z, Barisci JN, Spinks GM, Wallace GG, Mazzoldi A, De Rossi D, Rinzler AG, Jaschinski O, Roth S, Kertesz M (1999) Carbon nanotube actuators. Science 284(5418):1340–1344. doi:10.1126/science.284.5418.1340
119. Fukushima T, Asaka K, Kosaka A, Aida T (2005) Fully plastic actuator through layer-by-layer casting with ionic-liquid-based bucky gel. Angew Chem Int Ed 44(16):2410–2413. doi:10.1002/anie.200462318
120. Mukai K, Asaka K, Kiyohara K, Sugino T, Takeuchi I, Fukushima T, Aida T (2008) High performance fully plastic actuator based on ionic-liquid-based bucky gel. Electrochim Acta 53(17):5555–5562. doi:10.1016/j.electacta.2008.02.113
121. Takeuchi I, Asaka K, Kiyohara K, Sugino T, Mukai K, Randriamahazaka H (2010) Electrochemical impedance spectroscopy and electromechanical behavior of bucky-gel actuators containing ionic liquids. J Phys Chem C 114(34):14627–14634. doi:10.1021/jp1018185
122. Mukai K, Asaka K, Sugino T, Kiyohara K, Takeuchi I, Terasawa N, Futaba DN, Hata K, Fukushima T, Aida T (2009) Highly conductive sheets from millimeter-long single-walled carbon nanotubes and ionic liquids: application to fast-moving, low-voltage electromechanical actuators operable in air. Adv Mater 21(16):1582–1585. doi:10.1002/adma.200802817
123. Imaizumi S, Kato Y, Kokubo H, Watanabe M (2012) Driving mechanisms of ionic polymer actuators having electric double Layer capacitor structures. J Phys Chem B 116(16):5080–5089. doi:10.1021/jp301501c
124. Imaizumi S, Ohtsuki Y, Yasuda T, Kokubo H, Watanabe M (2013) Printable polymer actuators from ionic liquid, soluble polyimide, and ubiquitous carbon materials. ACS Appl Mater Interfaces 5(13):6307–6315. doi:10.1021/am401351q
125. Coll C, Labrador RH, Manez RM, Soto J, Sancenon F, Segui M-J, Sanchez E (2005) Ionic liquids promote selective responses towards the highly hydrophilic anion sulfate in PVC membrane ion-selective electrodes. Chem Commun 24:3033–3035. doi:10.1039/B503154K
126. Shvedene NV, Chernyshov DV, Khrenova MG, Formanovsky AA, Baulin VE, Pletnev IV (2006) Ionic liquids plasticize and bring ion-sensing ability to polymer membranes of selective electrodes. Electroanalysis 18(13–14):1416–1421. doi:10.1002/elan.200603537
127. Wei D, Ivaska A (2008) Applications of ionic liquids in electrochemical sensors. Anal Chim Acta 607(2):126–135. doi:10.1016/j.aca.2007.12.011
128. Xiong HY, Chen T, Zhang XH, Wang SF (2007) High performance and stability of a hemoglobin-biosensor based on an ionic liquid as nonaqueous media for hydrogen peroxide monitoring. Electrochem Commun 9(11):2671–2675. doi:10.1016/j.elecom.2007.08.020

129. Chen H, Wang Y, Liu Y, Wang Y, Qi L, Dong S (2007) Direct electrochemistry and electrocatalysis of horseradish peroxidase immobilized in Nafion-RTIL composite film. Electrochem Commun 9(3):469–474. doi:10.1016/j.elecom.2006.10.019
130. Leleux P, Johnson C, Strakosas X, Rivnay J, Hervé T, Owens RM, Malliaras GG (2014) Ionic liquid gel-assisted electrodes for long-term cutaneous recordings. Adv Healthcare Mater 3(9):1377–1380. doi:10.1002/adhm.201300614
131. Kavanagh A, Fraser KJ, Byrne R, Diamond D (2012) An electrochromic ionic liquid: design, characterization, and performance in a solid-state platform. ACS Appl Mater Interfaces 5(1):55–62. doi:10.1021/am3018948
132. Kavanagh A, Copperwhite R, Oubaha M, Owens J, McDonagh C, Diamond D, Byrne R (2011) Photo-patternable hybrid ionogels for electrochromic applications. J Mater Chem 21(24):8687–8693. doi:10.1039/C1JM10704F
133. Moon HC, Lodge TP, Frisbie CD (2014) Solution-processable electrochemiluminescent ion gels for flexible, low-voltage, emissive displays on plastic. J Am Chem Soc 136(9):3705–3712. doi:10.1021/ja5002899
134. Anderson JL, Dixon JK, Brennecke JF (2007) Solubility of CO_2, CH_4, C_2H_6, C_2H_4, O_2, and N_2 in 1-Hexyl-3-methylpyridinium Bis(trifluoromethylsulfonyl)imide: comparison to other ionic liquids. Acc Chem Res 40(11):1208–1216. doi:10.1021/ar7001649
135. Hasib-ur-Rahman M, Siaj M, Larachi F (2010) Ionic liquids for CO_2 capture—development and progress. Chem Eng Process Process Intensification 49(4):313–322. doi:10.1016/j.cep.2010.03.008
136. Lee S-H, Kim B-S, Lee E-W, Park Y-I, Lee J-M (2006) The removal of acid gases from crude natural gas by using novel supported liquid membranes. Desalination 200(1–3):21–22. doi:10.1016/j.desal.2006.03.227
137. Park Y-I, Kim B-S, Byun Y-H, Lee S-H, Lee E-W, Lee J-M (2009) Preparation of supported ionic liquid membranes (SILMs) for the removal of acidic gases from crude natural gas. Desalination 236(1–3):342–348. doi:10.1016/j.desal.2007.10.085
138. Liu Y, Wang M, Li J, Li Z, He P, Liu H, Li J (2005) Highly active horseradish peroxidase immobilized in 1-butyl-3-methylimidazolium tetrafluoroborate room-temperature ionic liquid based sol-gel host materials. Chem Commun 13:1778–1780. doi:10.1039/B417680D
139. Hough WL, Smiglak M, Rodriguez H, Swatloski RP, Spear SK, Daly DT, Pernak J, Grisel JE, Carliss RD, Soutullo MD, Davis JJH, Rogers RD (2007) The third evolution of ionic liquids: active pharmaceutical ingredients. New J Chem 31(8):1429–1436. doi:10.1039/B706677P
140. Vallet-Regí M, Balas F, Arcos D (2007) Mesoporous materials for drug delivery. Angew Chem Int Ed 46(40):7548–7558. doi:10.1002/anie.200604488
141. Fukumoto K, Yoshizawa M, Ohno H (2005) Room temperature ionic liquids from 20 natural amino acids. J Am Chem Soc 127(8):2398–2399. doi:10.1021/ja043451i

Chapter 10
Structural Aspect on Li Ion Solvation in Room-Temperature Ionic Liquids

Kenta Fujii, Shiro Seki, Hiroyuki Doi, and Yasuhiro Umebayashi

10.1 Introduction

As is well recognized now, room-temperature ionic liquids (ILs) possess many advantages for electrochemistry due to their unique properties such as high ionic conductivity, wide electrochemical window, good chemical and thermal stabilities, non-flammability, and non-volatility. So, ILs have been widely applied to chemical synthesis and separation fields as a green solvent and also new electrolyte for electrochemical devices such as Li secondary batteries [1–4]. The ILs generally consist of bulky organic cations such as 1-alkyl-3-methylimidazolium $[C_n mim]^+$, N-alkyl-N-methylpyrrolidinium $[C_n mPyr]^+$, quaternary ammonium, etc., and inorganic and/or organic anions such as tetrafluoroborate $[BF_4]^-$, hexafluorophosphate $[PF_6]^-$, and bis(trifluoromethylsulfonyl)imide $[N(Tf)_2]^-$. Therefore, the physicochemical properties of a solvent can be easily controlled by changing a combination of cation and anion, which is one of advantages in ionic liquid chemistry. In electrochemical applications, $[N(Tf)_2]^-$ and its analogue bis(fluorosulfonyl)amide $[FSI]^-$ have so far been widely used as the anion of ILs. These anions combined with $[C_n mPyr]^+$ or *N*-methyl-*N*-(2-methoxyethyl)ammonium $[DEME]^+$ are well utilized for Li battery applications using ILs, which is ascribed to stable formation of solid electrolyte interface (SEI) [5–10]. Furthermore, other electrochemical device applications

K. Fujii (✉)
Graduate School of Science and Engineering, Yamaguchi University,
2-16-1, Tokiwadai, Ube, Yamaguchi 755-8611, Japan
e-mail: k-fujii@yamaguchi-u.ac.jp

S. Seki
Materials Science Research Laboratory, Central Research Institute of Electric Power Industry,
2-11-1 Iwado-kita, Komae, Tokyo 201-8511, Japan

H. Doi • Y. Umebayashi
Graduate School of Science and Technology, Niigata University,
8050, Ikarashi, 2-no-cho, Nishi-ku, Niigata City 950-2181, Japan

A.A.J. Torriero (ed.), *Electrochemistry in Ionic Liquids*,
DOI 10.1007/978-3-319-13485-7_10

(field effect transfer [11, 12] and electric double layer [13]) have been reported, particularly in FSI-based ILs.

In development of ILs for electrochemical devices such as Li secondary batteries, it is important to reduce the viscosity, because a lower viscous solvent leads to a higher ionic conductivity and transport properties. Therefore, several low viscous ILs have been synthesized and proposed up until now [14–18]. Here, we focus on FSI-based ILs, which is well known as one of low viscous ILs. $[FSI]^-$ has a similar molecular structure to $[N(Tf)_2]^-$, i.e., the CF_3 groups of $[N(Tf)_2]^-$ are only substituted by F atoms. In spite of similar molecular structure between $[FSI]^-$ and $[N(Tf)_2]^-$, it is established that the FSI-based ILs show a significantly lower viscosity than the $N(Tf)_2$-based ones [5, 6, 19]. For example, the viscosity of $[C_2mim]$ [FSI] is 17 mPa s, which is appreciably lower than that of $[C_2mim][N(Tf)_2]$, 28 mPa s. In general, the addition of Li salt to solvent IL leads to the increase in the viscosity. Indeed, the viscosity of $N(Tf)_2$-based ILs significantly increases with the addition of $Li[N(Tf)_2]$ salt. Interestingly, the increase in the viscosity of FSI-based ILs with the addition of Li[FSI] salt is appreciably small relative to the corresponding $[N(Tf)_2]$ system [19], resulting in excellent performance of electrochemical devices using the FSI-based ILs. Tsuzuki et al. reported temperature dependence of the viscosity of $[C_2mim]$[FSI] and $[C_2mim][N(Tf)_2]$ with the corresponding Li salts [19]. The viscosity strongly depends on temperature in the $[N(Tf)_2]$ system, and then the relative viscosity, i.e., η(neat IL)/η(IL with 0.32 mol kg^{-1} Li salt), largely changes from 1.4 (10 °C) to 1.9 (80 °C). On the other hand, there is no or less temperature dependence of the relative viscosity for the FSI system, and the extent is very small within 0.1 in the same temperature range. The ionic conductivity and ionic diffusion have been also reported. The self-diffusion coefficients of ions (Li^+, IL cation and anion) in the FSI-based IL solutions were investigated in detail by both experimental and theoretical methods and compared with those in the $[N(Tf)_2]$-based ones at a microscopic level [19–21]. Here, note that such dynamic properties of IL systems are strongly related to the static liquid structure and Li ion solvation. We have therefore studied conformational isomerism and liquid structure in neat FSI- and $[N(Tf)_2]$-based ILs by using Raman/IR spectroscopic and X-ray/neutron scattering experiments and theoretical calculations [22–36]. Particularly, it is indispensable to elucidate solvation structure of Li ion in the ILs as the fundamental knowledge on understanding Li ion reactivity and mobility, which plays a key role for improving Li ion battery performance at the molecular level. In this chapter, we describe our recent works on Li ion solvation structure in the [FSI]- and $[N(Tf)_2]$-based ILs and discuss their difference and similarity from the structural aspects.

10.2 Liquid Structure of Neat ILs

Solvation of metal ion in electrolyte solutions strongly depends not only on the metal ion–solvent interactions but also on the solvent–solvent interactions and/or liquid structure in the bulk [37–40]. When a metal ion is introduced into a solvent, (1) the metal ion destroys the solvent structure to create isolated solvent molecules around it as the first step and then (2) the metal ion is solvated with the isolated solvent molecules to form a solvation cluster as the second step. The liquid structure in the bulk is thus effective for metal ion solvation, particularly in the former step. So, we firstly describe about liquid structure of neat ILs (mainly, [C_2mim][FSI] and [C_2mim][N$(Tf)_2$]) to deeply understand Li ion solvation in the IL systems.

For investigation of static liquid structure in the bulk ILs, it is well known that X-ray and neutron scatterings (or diffractions) are one of powerful tools to directly determine the interaction distance between ions. Furthermore, computer simulations are also useful to investigate a detailed static structure at the atomistic level, and thus often combined with X-ray/neutron experimental data. Soper et al. firstly applied such combination of experimental and simulation data to the structural investigation of ILs, i.e., the large-angle neutron diffraction with empirical potential structure refinement (EPSR) based on Monte-Carlo simulations [41, 42]. We also reported liquid structure of various IL systems such as imidazolium- [27, 32, 35], pyrrolidinium- [26, 43], and ammonium [30, 44]-based ILs studied by large-angle X-ray scattering or high-energy X-ray diffraction (LAXS and HEXRD, respectively) with the aid of molecular dynamics (MD) simulations. In general, the liquid structure and ion–ion interaction obtained from X-ray/neutron experiment or computer simulation study are discussed in terms of radial distribution function given by Furrier transformation of the structure factor, $S(q)$ where q denotes the scattering vector. Figure 10.1a shows X-ray radial distribution functions, $G^{\mathrm{exp}}(r)$s as the form of $r^2[G(r) - 1]$ for [C_2mim][FSI] [36] and [C_2mim][N$(Tf)_2$] [35] observed by HEXRD experiments. The HEXRD measurements were performed by using high-energy X-ray diffraction apparatus of BL04B2 beam line at SPring-8 (Japan Synchrotron Radiation Research Institute, JASRI, Japan) [45–47], and their details (instruments and data corrections) are described in our previous work [26, 35]. As can be seen in this figure, the experimental $r^2[G^{\mathrm{exp}}(r) - 1]$ exhibits the sharp and broad peaks at $r<3$ Å and $r>5$ Å, respectively, for both [C_2mim][FSI] (solid line, black) and [C_2mim][N$(Tf)_2$] (solid line, gray). The sharp peaks at $r<3$ Å are mainly assigned to the intramolecular interactions of cation and anion, whereas the three broad peaks at around 5, 9, and 15 Å to the intermolecular interactions of cation–cation, anion–anion, and cation–anion components [27, 35]. The peak centered at around 5 Å can be assigned to the first neighboring cation–anion interactions, which are discussed below in detail. On the other hand, two broad peaks at around 9 and 15 Å may be originated from the long-range correlations between ions (including cation–anion, cation–cation, and anion–anion) in the second and the third neighbor regions. Here, note that the peak positions of 9 and 15 Å are slightly shorter for the [C_2mim][FSI] than for the [C_2mim][N$(Tf)_2$] due to smaller anion size of [FSI]$^-$ than [N$(Tf)_2$]$^-$. In this long r-range, the peak intensities or oscillations in the $r^2[G^{\mathrm{exp}}(r) - 1]$ for the [C_2mIm][FSI] are significantly weaker than those for the [C_2mIm][N$(Tf)_2$]. This implies that the extent of the long-range ordering or structuredness is lower in the FSI system than in the [N$(Tf)_2$] one. Figure 10.1b shows X-ray weighted radial distribution functions derived from MD simulations, $G^{\mathrm{MD}}(r)$s [36]. It is clear that the observed $G^{\mathrm{exp}}(r)$s are well represented by the theoretical $G^{\mathrm{MD}}(r)$s for both systems. The total $G^{\mathrm{MD}}(r)$ can be deconvoluted into intra- and intermolecular components, i.e., $G^{\mathrm{MD}}{}_{\mathrm{intra}}(r)$ and $G^{\mathrm{MD}}{}_{\mathrm{inter}}(r)$. The $G^{\mathrm{MD}}{}_{\mathrm{inter}}(r)$ is composed of three partial $G^{\mathrm{MD}}{}_{\mathrm{inter}}(r)$s

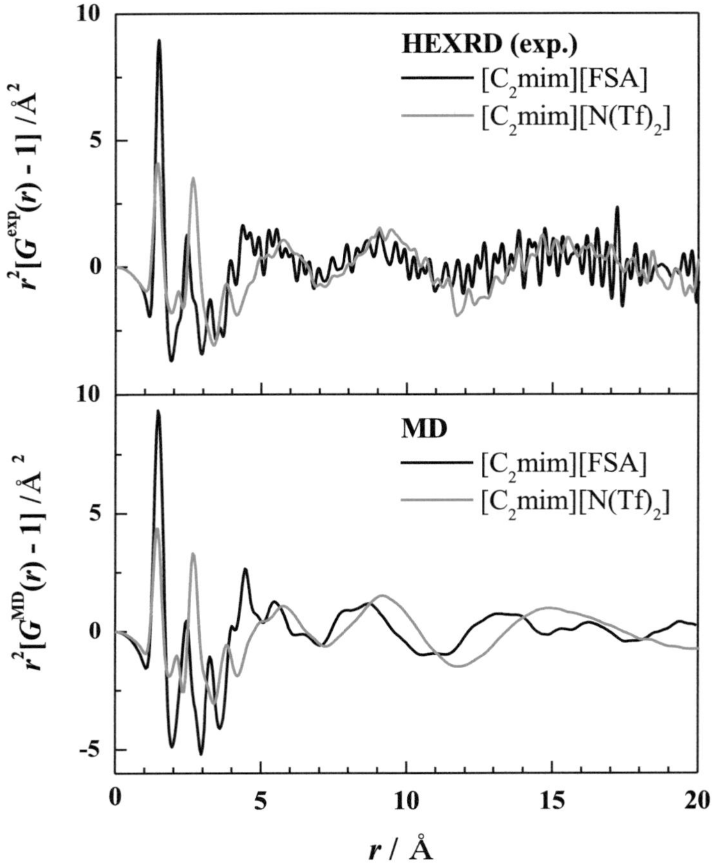

Fig. 10.1 X-ray radial distribution functions as a form of $r^2[G(r)-1]$ for neat $[C_2mim][FSI]$ (*black*) and $[C_2mim][N(Tf)_2]$ (*gray*) obtained by (**a**) HEXRD experiments and (**b**) MD simulations [36] and [62]

for cation–anion, anion–anion, and cation–cation components, which are shown in Fig. 10.2. In the partial $G^{MD}{}_{inter}(r)$s for the $[C_2mim][FSI]$ system, the cation–anion correlation in the first neighboring sphere (3–7 Å) is mainly composed of the peaks at 3.4, 4.5, and 5.6 Å. The 3.4 Å peak corresponds to the closest atom–atom interaction in this system, which is mainly ascribed to the carbon (imidazolium ring)–oxygen ([FSI]) interaction [36]. The same is applied to the $[C_2mim][N(Tf)_2]$ system, i.e., the closest carbon (imidazolium ring)–oxygen ($[N(Tf)_2]$) interaction appears at around 3.4 Å. The partial $G^{MD}{}_{inter}(r)$ for the cation–anion component also has the broad peak centered at 11 Å, which corresponds to the cation–anion interaction in the second neighbors. For the anion–anion components, the broad correlations appear at around 8 Å as the first peak. The cation–cation correlations are also found at around 8 Å in the $G^{MD}{}_{inter}(r)$, although they are significantly weak in the intensity. This is because $[C_2mim]^+$ is constructed by H, C, and N atoms with small X-ray scattering factor, *f*, unlike $[FSI]^-$ with large ones of F and S. It is thus concluded that

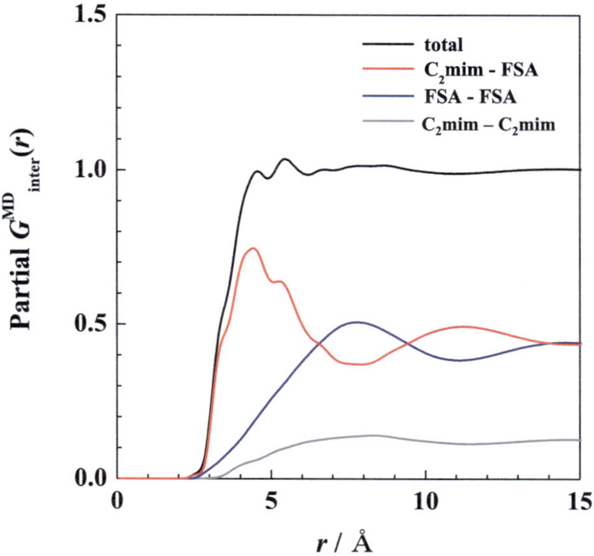

Fig. 10.2 Partial $G^{MD}{}_{inter}(r)$s for the C_2mim–FSI (*red*), FSI–FSI (*blue*) and C_2mim–C_2mim (*gray*) interactions derived from MD simulations, together with the corresponding total $G^{MD}{}_{inter}(r)$ (*black*) [36]

the anion–anion correlations mainly contribute to the 8.5 Å peak observed in the total $G^{exp}{}_{inter}(r)$, whereas the cation–cation ones less contribute to the total $G^{exp}{}_{inter}(r)$.

With regard to the first neighboring cation–anion interaction, we further analyzed the MD results to evaluate space distribution function, SDF. Figure 10.3a shows SDF for $[FSI]^-$ (center of mass) around C_2mim cation, calculated with the MD trajectories [36]. It is easily found in this figure that $[FSI]^-$s distribute above and below the imidazolium ring plane and near to the C4 and the C5 carbons. The coordination number, $n(r)$ of $[FSI]^-$s around the C_2mim cation, is estimated to be 3.8 (approximately 4). With an increase in the probability density from 1.5 to 2.4, the distribution of $[FSI]^-$ still remained above and below the imidazolium ring, whereas that near to C4 and C5 positions disappeared (Fig. 10.3b). This means that the potential surface of the interaction between $[FSI]^-$ and $[C_2mim]^+$ is deeper at above and below the imidazolium ring than at C4 and C5 positions. Here, note that the H of C2 position of the imidazolium cation is capable of forming a hydrogen bonding interaction with the anion in the ILs composed of smaller-size anions such as Cl, Br, and BF_4 [48]. In FSI system, however, no significant probability of $[FSI]^-$ is found to the direction of C2–H bond in the SDF. It is thus expected that the hydrogen bonding C2–H···O ($[FSI]^-$) cannot be formed in $[C_2mim][FSI]$ due to bulkiness of anion. For $[C_2mim][N(Tf)_2]$ system (Fig. 10.3c, d), we can see that the $[C_2mIm]^+$ is also surrounded by four anions in the first neighbor region, and the orientation is also similar to the $[C_2mim][FSI]$ system. From these results, we can propose that there is no significant difference in the closest cation–anion interactions between [FSI] and $[N(Tf)_2]$systems in terms of static structure, although the size of $[FSI]^-$ is appreciably smaller than that of $[N(Tf)_2]^-$.

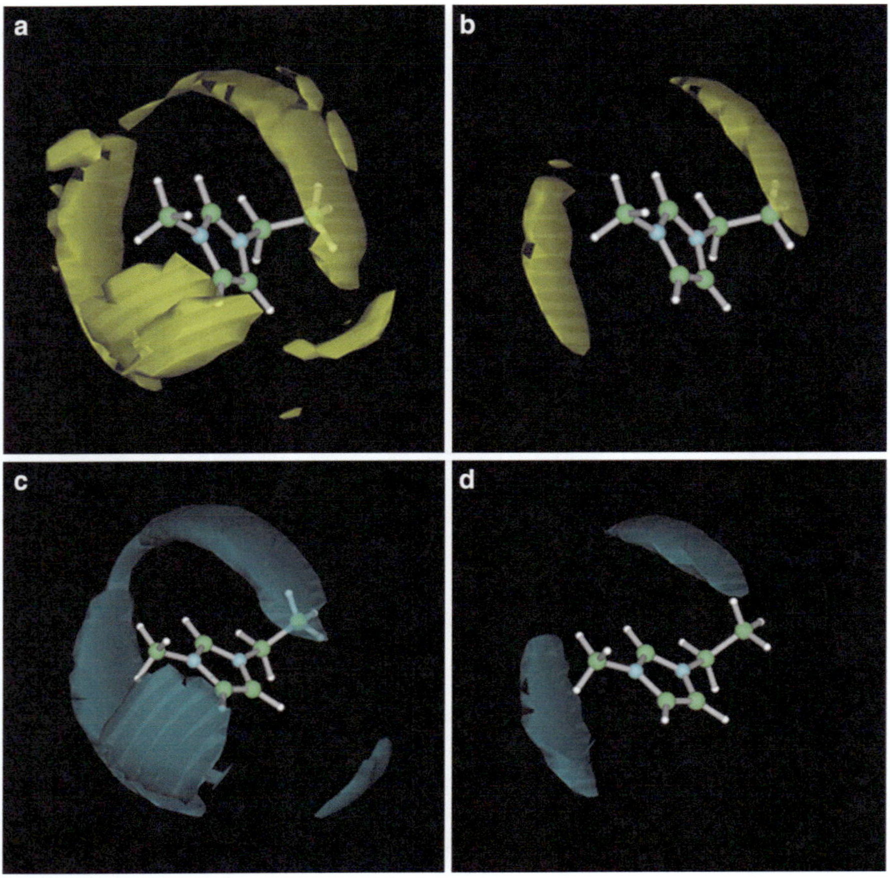

Fig. 10.3 Space distribution functions, SDFs for center of mass of $[FSI]^-$ around $[C_2mim]^+$. The colored clouds represent the isoprobability surfaces of (**a**) 1.5 and (**b**) 2.4 larger probability density relative to the bulk one. The SDFs for the $N(Tf)_2$ system with (**c**) 1.5 and (**d**) 2.4 larger probability density are also shown [36]

10.3 Li Ion Solvation: Raman Spectroscopic Study

As mentioned in Introduction, lithium batteries using ILs as an electrolyte have been studied in terms of the development of high-energy density battery with high safety. The battery performances such as current density and charge–discharge reactivity at electrolyte/electrode interface are closely related to the ionic conductivity in ILs and the redox reaction and the desolvation behavior of Li ion at the interface, respectively. Here, note that they are directly and/or indirectly influenced by the Li ion solvation, i.e., the solvation number of IL anions bound to Li ion and the strength of Li ion–IL anion interactions. Therefore, the Li ion solvation structure in ILs has been widely investigated by both experimental [49–52] and theoretical [53–55] techniques.

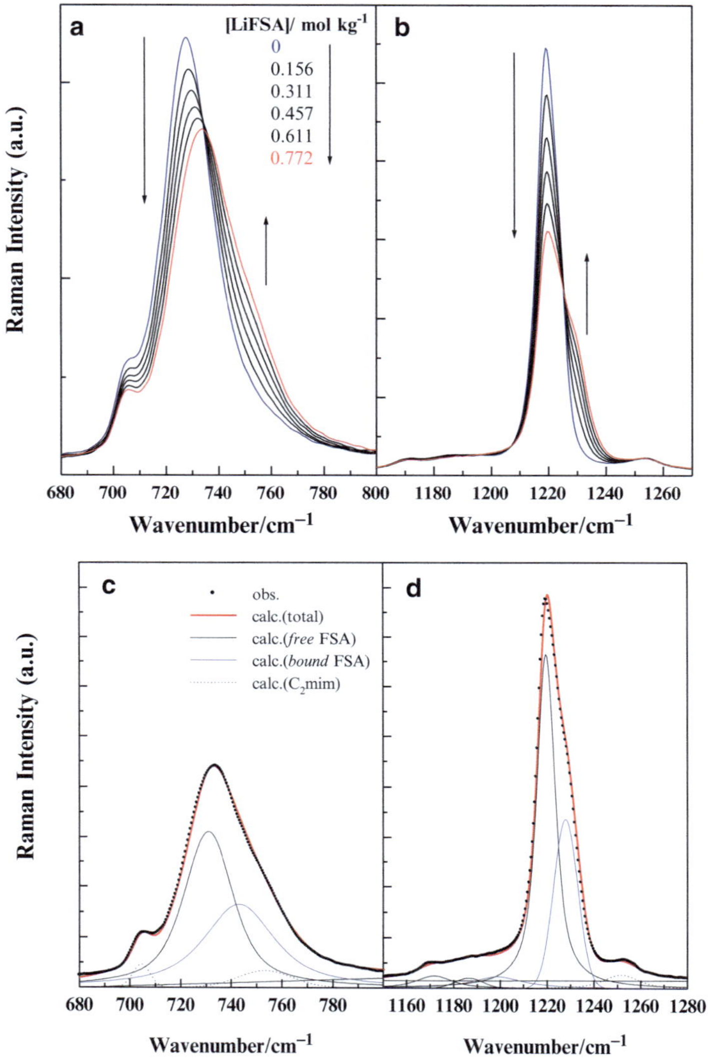

Fig. 10.4 *Top*: LiFSI salt concentration dependence of Raman spectra observed for the [C_2mim][FSI] solutions in the frequency range of (**a**) 680–800 cm^{-1} and (**b**) 1,160–1,260 cm^{-1} at 298 K. *Bottom*: Typical curve fitting results for the solutions (0.772 mol kg^{-1} LiFSI concentration) in the range of (**c**) 680–800 cm^{-1} and (**d**) 1,160–1,260 cm^{-1} [59]

We also studied the [FSI]- and [N(Tf)$_2$]-based IL solutions including Li ion by Raman/IR spectroscopy and quantum calculations to elucidate the Li ion solvation structure [56–59]. Figure 10.4a shows typical Raman spectra in the range of 680–800 cm^{-1} observed for [C_2mim][FSI] solutions containing Li[FSI] salt (0–0.772 mol kg^{-1}). At Li[FSI] = 0 mol kg^{-1} (neat IL), [C_2mim][FSI] exhibits a main peak at 731 cm^{-1}, which can be assigned to the ν_s(N–S) vibration of $[FSI]^-$ (free [FSI]) [24]. The Raman band at 731 cm^{-1} is gradually weakened with increasing Li[FSI] concentration, and then the higher-shifted band at 744 cm^{-1} is intensified. The same behavior is observed in the $[N(Tf)_2]$-based RTILs [56, 57, 60, 61], i.e., Raman band of $[N(Tf)_2]^-$ at around 740–750 cm^{-1} is weakened and shifted to the higher frequency with increasing metal ion concentration. According to our previous work [56, 57, 60, 61], the former band is assigned to the $[N(Tf)_2]^-$ in the bulk (free $N(Tf)_2$) and the latter to the $[N(Tf)_2]^-$ bound to Li^+ ion (bound $[N(Tf)_2]$). Thus, in the FSI system, it is concluded that the observed 731 and 744 cm^{-1} bands are originated from free and bound [FSI], respectively. Figure 10.4b shows Raman spectra of the ν_s(S–O) vibration of $[FSI]^-$ in the range of 1,220–1,230 cm^{-1}. The Li[FSI] concentration dependence of the ν_s(S–O) vibration is similar to that of the ν_s(N–S) vibration described above, indicating that the Raman band at lower frequency side is originated from the free FSI and that at higher one from the bound FSI. To discuss the spectral variations in detail, in general, the observed Raman spectra are deconvoluted into some single bands by using curve fitting analysis. Typical results examined in the range of 680–800 and 1,150–1,280 cm^{-1} are shown in Fig. 10.4c, d, respectively. As can be seen in these figures, both Raman spectra observed for Li[FSI]=0.177 mol kg^{-1} solution are well explained in terms of the free and bound [FSI] (solid gray and blue lines) and some $[C_2mim]$ (bashed line) components. With an increase of the salt concentration, the free [FSI] bands centered at 731 and 1,220 cm^{-1} decrease and the bound bands at 744 and 1,228 cm^{-1} intensify without appreciable change in peak frequency and half width at half maximum. By analyzing the integrated intensity of the free band extracted, the number of $[FSI]^-$s solvated to Li^+ ion can be estimated, as follows.

The integrated intensity, I_f, of free [FSI] in the bulk is given as $I_f = J_f m_f$, where J_f and m_f stand for the Raman scattering coefficient and the concentration, respectively, of free [FSI]. By taking into account a mass balance equation, $m_f = m_T - m_b = m_T - n m_{Li}$, where m_T, m_b, m_{Li} denote concentrations of total [FSI], bound [FSI], and Li^+ ion, respectively, and n is the number of $[FSI]^-$ bound to the Li^+ ion. As the result, we obtain the following equation: $I_f/m_{Li} = J_f(m_T/m_{Li} - n)$. Plots of I_f/m_{Li} values against m_T/m_{Li} give a straight line with the slope ($=J_f$) and the intercept ($=-J_f n$), which is shown in [59]. The solvation number, n, is thus obtained according to $n=-$slope/intercept and was estimated to be 3.0(1) and 2.9(1) at 731 and 1,220 cm^{-1} bands, respectively, at 298 K. This suggests that Li ion is solvated with three $[FSI]^-$s in the $[C_2mim][FSI]$ at 298 K. The solvation number of Li ion in the $[N(Tf)_2]$-based ILs was also estimated in the concentration range of $LiN[(Tf)_2] < 1$ mol dm^{-3} by the similar analysis [56–58]. According to our Raman spectroscopic study with the aid of DFT calculations, the Li ion is solvated with two $[N(Tf)_2]^-$s acting as a bidentate ligand, resulting in a tetrahedrally four-coordinated structure with O atom of two bidentate $[N(Tf)_2]^-$s. We can point out here that the n (=3) in [FSI] system significantly differs from the n (=2) in $[N(Tf)_2]$ system. To discuss a possible solvation structure of Li ion in the [FSI] system, we extended our Raman study to the temperature dependence of the n.

Figure 10.5 shows temperature dependence of the solvation number, n values for the $Li[FSI]/[C_2mim][FSI]$ system, together with those for the $Li[N(Tf)_2]/[C_2mim]$

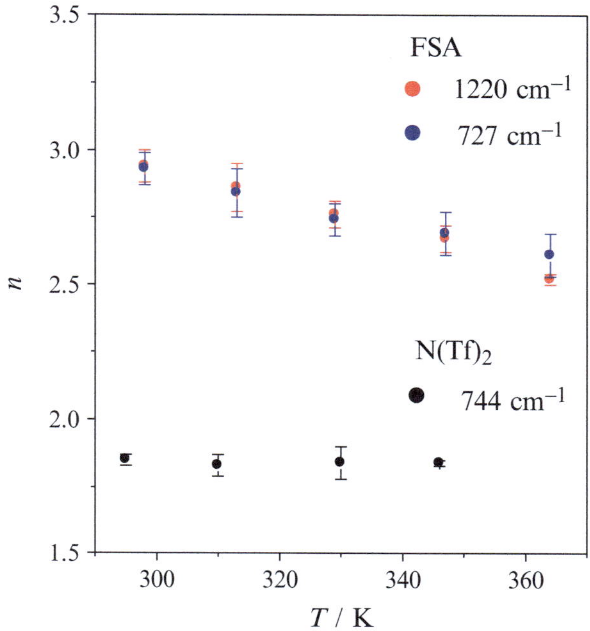

Fig. 10.5 Temperature dependence of the solvation number, *n*, of Li ion in [C_2mim][FSI] estimated by using Raman bands at 1,220 cm^{-1} (*red*) and 727 cm^{-1} (*blue*), together with that in [C_2mim][$N(Tf)_2$] (*black*) [59]

[$N(Tf)_2$] one. In the [FSI] system, the *n* gradually decreases with increasing temperature. This indicates that the solvated $[FSI]^-$ around Li ion are liberated to the bulk at high temperature to decrease the *n* from 3 (298 K) to 2.5 (370 K). On the other hand, there is no temperature dependence on the solvation number in the [$N(Tf)_2$] system, i.e., the *n* value is almost a constant at any temperature examined here. This implies that the desolvation of bidentate $[N(Tf)_2]^-$ ligand coordinated to Li ion does not occur or is very difficult in the [$N(Tf)_2$] system. From these results, we can propose that the Li ions in the [C_2mim][FSI] coexist as $[Li(FSI)_3]^{2-}$ and $[Li(FSI)_2]^-$ at high temperature. Furthermore, it is expected that the $[Li(FSI)_3]^{2-}$ complex involves both bidentate and monodentate $[FSI]^-$s that strongly and weakly coordinate to Li ion, respectively, resulting in the desolvation of weakly coordinated monodentate $[FSI]^-$ from the $[Li(FSI)_3]^{2-}$ complex at high temperature.

The temperature dependence of the *n* value can be quantitatively analyzed to give an equilibrium constant, *K*, with regard to the [FSI] desolvation process by assuming the following equilibrium equation [59].

$$\left[\mathrm{Li}\left(\mathrm{FSI}\right)_3\right]^{2-} = \left[\mathrm{Li}\left(\mathrm{FSI}\right)_2\right]^{-} + \left[\mathrm{FSI}\right]^{-}$$

$$K = \left[\mathrm{Li}\left(\mathrm{FSI}\right)_2\right]^{-} / \left[\mathrm{Li}\left(\mathrm{FSI}\right)_3\right]^{2-} = \left(3-n\right)/\left(n-2\right)$$

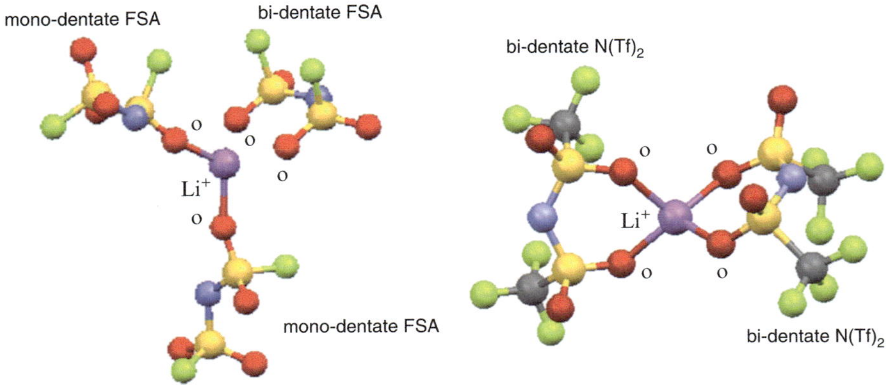

Fig. 10.6 Most stable optimized geometry of (**a**) $[Li(FSI)_3]^{2-}$ and (**b**) $[Li(N(Tf)_2)_2]^-$ complexes in gas phase [59]

The thermodynamic parameters of enthalpy, entropy, and Gibbs free energy ($\Delta H°$, $\Delta S°$, and $\Delta G°$) for the equilibrium are also estimated from the relationship of $\Delta G° = -RT\ln K$ and $\Delta G° = \Delta H° - T\Delta S°$, i.e., van't Hoff plots for the equilibrium reaction. By using the n values obtained from 1,220 cm^{-1} Raman band (filled circles colored with red in Fig. 10.5), the $\Delta H°$ and $T\Delta S°$ (298 K) are evaluated to be 35(2) kJ mol^{-1} and 29(2) kJ mol^{-1}, respectively.

According to our ab initio calculations, the possible model of $[Li(FSI)_3]^{2-}$ complex has been proposed [59], which is shown in Fig. 10.6a. In the calculation, we examined the geometry optimizations for four types of the complexes: $[Li(\text{monodentate FSI})_2(\text{bidentate FSI})_1]^{2-}$, $[Li(\text{monodentate FSI})_1(\text{bidentate FSI})_2]^{2-}$, $[Li(\text{monodentate FSI})_3]^{2-}$, and $[Li(\text{bidentate FSI})_3]^{2-}$. As the results, the most stable geometry is obtained for the $[Li(\text{monodentate FSI})_2(\text{bidentate FSI})_1]^{2-}$. As can be seen in the picture, the Li ion interacts with the oxygen atoms of $[FSI]^-$s and the coordination number is 4, which is similar to the $[Li(\text{bidentate } N(Tf)_2)_2]^-$ complex (Fig. 10.6b). The interaction energy of the Li–monodentate FSI in the $[Li(\text{monodentate FSI})_2(\text{bidentate FSI})_1]^{2-}$ complex is appreciably weaker than that of the Li–bidentate FSI. The calculated interaction energies using the geometry shown in Fig. 10.6a are −444.3 and −560.6 kJ mol^{-1} for the monodentate and bidentate $[FSI]^-$s, respectively. This supports Raman spectroscopic results described above, i.e., the monodentate $[FSI]^-$ preferentially desolvates from the first solvation shell of the Li ion.

10.4 Atomistic Solvation Structure of Li Ion in ILs: HEXRD and MD Simulation Study

Figure 10.7a shows the experimental radial distribution function as a form of $r^2[G(r) - 1]$ observed for LiFSI in $[C_2mim][FSI]$ solutions by HEXRD measurements. The peaks at around 5 (4.5 and 5.5), 9, and 15 Å can be fully assigned to the

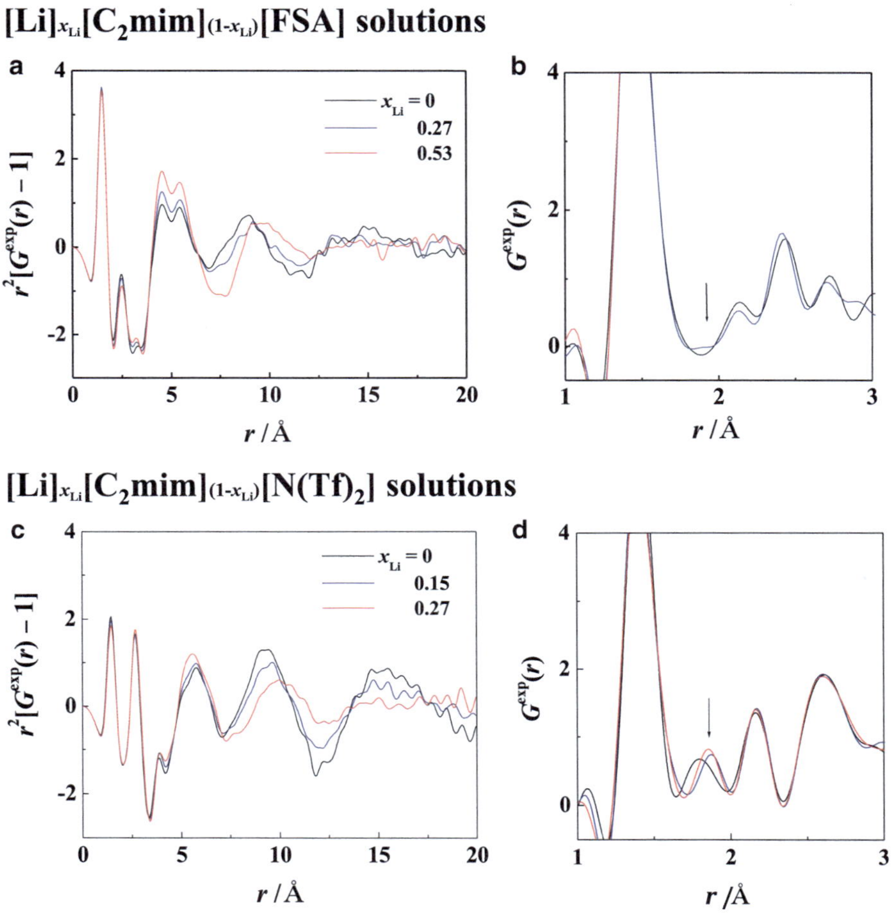

Fig. 10.7 X-ray radial distribution functions as a form of $r^2[G^{exp}(r) - 1]$ for (**a**) $[Li]x_{Li}[C_2mim](1{-}x_{Li})[FSI]$ and (**c**) $[Li]x_{Li}[C_2mim](1{-}x_{Li})[N(Tf)_2]$ solutions obtained by HEXRD experiments. (**b**) and (**d**) show the corresponding normal $G^{exp}(r)$s to easily see the closest Li^+–O (anion) interactions shown by *arrows* [59] and [62]

intermolecular interactions [26, 36, 59]. It is found that the peak intensities at 4.5 and 5.5 Å increase with increasing Li[FSI] salt concentration. They are mainly originated from the correlations in the $[Li(FSI)_3]^{2-}$ complexes and the closest $[C_2mim]^+$–$[FSI]^-$ interactions, which is discussed below in detail. Focusing on the broad peaks at around 9 Å, the peak intensity at 8 Å decreases but that at 11 Å increases with increasing salt concentration. The same is also found in the corresponding $[N(Tf)_2]$ system (Fig. 10.7c). The variations in the $[N(Tf)_2]$ system are ascribed to the long-range anion–anion interactions in the bulk and in the $[Li(N(Tf)_2)_2]^-$ complex [62]. Therefore, in the FSI system, it is expected that the formation of $[Li(FSI)_3]^{2-}$ affects the radial distribution function at both of the short and long ranges. Here, we note the closest atom–atom interactions in the $[Li(FSI)_3]^{2-}$ complex, particularly Li^+–O

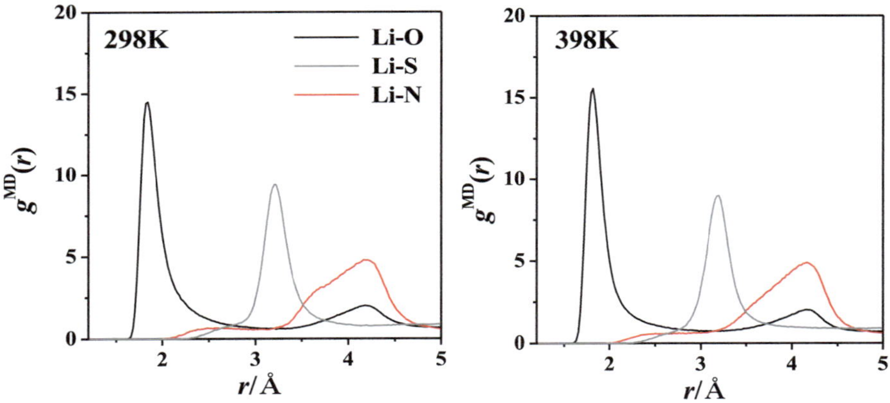

Fig. 10.8 Pair distribution functions, $g^{MD}(r)$s for the Li^+–X (O, S, and N) correlations in $[Li]x_{Li}[C_2mim](1-x_{Li})[FSI]$ ($x_{Li}=0.27$) at 298 and 398 K [59]

(FSI) interactions. Figure 10.7b, d shows the radial distribution functions, $G^{exp}(r)$s (normal form) for the [FSI] and $[N(Tf)_2]$ systems, respectively. In general, it is quite difficult to observe the Li^+–X correlations in the Li ion solvation structure in the solutions by an X-ray scattering technique due to a smaller X-ray scattering ability of Li ion. However, by using HEXRD with high *r*-resolution, we can find the Li^+–O (FSI) interactions in the $G^{exp}(r)$. The closest Li^+–O interaction in the $[Li(FSI)_3]^{2-}$ complex appears at 1.94 Å and increases in intensity with an increase of Li salt concentration. The distance of 1.94 Å is consistent with the averaged Li^+–O distance (1.97 Å) obtained by the present ab *initio* calculations for the $[Li(\text{monodentate FSI})_2(\text{bidentate FSI})_1]^{2-}$ (see Fig. 10.6). Similarly, the Li^+–O interaction is also experimentally observed in the corresponding $Li[N(Tf)_2]/[C_2mim][N(Tf)_2]$ system. The Li^+–O distance is 1.86 Å in the $[N(Tf)_2]$ system, which is slightly shorter than that in the FSI system. The difference in the Li^+–O interactions between [FSI] and $[N(Tf)_2]$ systems can be explained in terms of the solvation number *n* in the Li complexes, i.e., the larger $n=3$ in the [FSI] system than in the $[N(Tf)_2]$ one ($n=2$) leads to the longer atom—atom distance in the first coordination sphere of the Li ion.

The solvation structure at the atomistic level can be also investigated by MD simulations to give important evidence from theoretical aspect. Figure 10.8 shows the atom–atom pair correlation functions, $g^{MD}(r)$ for the Li^+–X (X=O, S, and N within [FSI] anion) derived from MD simulations. At 298 K, the Li^+–O in the $g^{MD}(r)$ is clearly found as a closest atom–atom correlation at 1.84 Å. The integral area intensity of the 1.84 Å peak that corresponds to the coordination number of O around Li^+ in the first neighboring is 3.7 (approximately 4). This suggests that Li ion is coordinated with four oxygen atoms within $[FSI]^-$s to form a tetrahedral-like structure. The Li^+–S correlation locates at 3.20 Å and the integrated value is also near to 4 (3.98). Here, note that the there are two peaks of the Li^+–N correlation at 3.72 and 4.19 Å in the $g^{MD}(r)$. The former can be assigned to the bidentate $[FSI]^-$ solvated to the Li ion, whereas the latter to the monodentate [FSI]. The integrated intensity ratio of the monodentate $[FSI]^-$ to the bidentate one is estimated to be 1.97. This means that the monodentate and bidentate $[FSI]^-$s in the first solvation shell of the Li ion coexist as mono:bi = 2:1. These results are completely consistent with those obtained from the Raman spectroscopy and ab initio calculations as described above. It is found in the $g^{MD}(r)$s at high temperature (Fig. 10.8, 398 K) that the coordination numbers of O, S, and N decrease to 3.44, 3.71, and 2.91, respectively. We thus found that the desolvation of $[FSI]^-$ in the solvation sphere of Li ion evidently occurs at high temperature, which strongly supports Raman spectroscopic study.

10.5 Conclusion

Raman spectroscopy and high-energy X-ray diffraction with the aid of theoretical calculations are quite useful to elucidate Li ion solvation structure in ionic liquid systems. The molecular structure of $[FSI]^-$ is similar to that of $[N(Tf)_2]^-$ one; however, the Li ion solvation in the FSI-based ILs is largely different from that in the $N(Tf)_2$-based ILs. The difference in the Li ion solvation affects the physicochemical properties such as viscosity and ionic conductivity for the ILs containing Li ion. We believe that structural investigations at the molecular level presented in this chapter are important and helpful to understand electrochemical applications using ILs, for example, Li battery performances.

Acknowledgments The authors greatly appreciate Dr. Seiji Tsuzuki (AIST) for his high-level ab initio calculations.

References

1. Ohno H (ed) (2005) Electrochemical aspects of ionic liquids. Wiley, Hoboken, NJ
2. Armand M, Endres F, MacFarlane DR, Ohno H, Scrosati B (2009) Ionic-liquid materials for the electrochemical challenges of the future. Nat Mater 8:621–629
3. Wasserscheid P, Welton T (eds) (2008) Ionic liquids in synthesis, vol 1 and 2, 2nd edn. Wiley, Weinheim
4. Picquet M, Tkatchenko I, Tommasi I, Wasserscheid P, Zimmermann J (2003) Ionic liquids, 3. Synthesis and utilisation of protic imidazolium salts in homogeneous catalysis. Adv Synth Catal 345:959–962
5. Matsumoto H, Sakaebe H, Tatsumi K, Kikuta M, Ishiko E, Kono M (2006) Fast cycling of Li/LiCoO2 cell with low-viscosity ionic liquids based on bis(fluorosulfonyl)imide [FSI](-). J Power Sources 160:1308
6. Ishikawa M, Sugimoto T, Kikuta M, Ishiko E, Kono M (2006) Pure ionic liquid electrolytes compatible with a graphitized carbon negative electrode in rechargeable lithium-ion batteries. J Power Sources 162:658–662
7. Seki S, Kobayashi Y, Miyashiro H, Ohno Y, Mita Y, Terada N, Charest P, Guerfi A, Zaghib K (2008) Compatibility of N-methyl-N-propylpyrrolidinium cation room-temperature ionic liquid electrolytes and graphite electrodes. J Phys Chem C 112:16708–16713
8. Sugimoto T, Atsumi Y, Kikuta M, Ishiko E, Kono M, Ishikawa M (2009) Ionic liquid electrolyte systems based on bis (fluorosulfonyl) imide for lithium-ion batteries. J Power Sources 189:802–805

9. Abouimrane A, Ding J, Davidson IJ (2009) Liquid electrolyte based on lithium bis-fluorosulfonyl imide salt: aluminum corrosion studies and lithium ion battery investigations. J Power Sources 189:693–696
10. Lux SF, Schmuck M, Appetecchi GB, Passerini S, Winter M, Balducci A (2009) Lithium insertion in graphite from ternary ionic liquid–lithium salt electrolytes: II. Evaluation of specific capacity and cycling efficiency and stability at room temperature. J Power Sources 192:606–611
11. Ono S, Seki S, Hirahara R, Tominari Y, Takeya J (2008) High-mobility, low-power, and fast-switching organic field-effect transistors with ionic liquids. Appl Phys Lett 92:103313
12. Ono S, Miwa K, Seki S, Takeya J (2009) A comparative study of organic single-crystal transistors gated with various ionic-liquid electrolytes. Appl Phys Lett 94:063301
13. Handa N, Sugimoto T, Yamagata M, Kikuta M, Kono M, Ishikawa M (2008) A neat ionic liquid electrolyte based on FSI anion for electric double layer capacitor. J Power Sources 185:1585–1588
14. Seki S, Hayamizu K, Tsuzuki S, Fujii K, Umebayashi Y, Mitsugi T, Kobayashi T, Ohno Y, Kobayashi Y, Mita Y, Miyashiro H, Ishiguro S (2009) Relationships between center atom species (N, P) and ionic conductivity, viscosity, density, self-diffusion coefficient of quaternary cation room-temperature ionic liquids. Phys Chem Chem Phys 11:3509–3514
15. Tsunashima K, Sugiya M (2007) Physical and electrochemical properties of low-viscosity phosphonium ionic liquids as potential electrolytes. Electrochem Commun 9:2353–2358
16. Zhou ZB, Matsumoto H, Tatsumi K (2006) Cyclic quaternary ammonium ionic liquids with perfluoroalkyltrifluoroborates: synthesis, characterization, and properties. Chem Eur J 12: 2196–2212
17. Matsumoto H, Sakaebe H, Tatsumi K (2005) Preparation of room temperature ionic liquids based on aliphatic onium cations and asymmetric amide anions and their electrochemical properties as a lithium battery electrolyte. J Power Sources 146:45–50
18. Zhou ZB, Matsumoto H, Tatsumi K (2005) Low-melting, low-viscous, hydrophobic ionic liquids: aliphatic quaternary ammonium salts with perfluoroalkyltrifluoroborates. Chem Eur J 11:752–766
19. Tsuzuki S, Hayamizu K, Seki S (2010) Origin of the low-viscosity of [emim][(FSO2)2N] ionic liquid and its lithium salt mixture: experimental and theoretical study of self-diffusion coefficients, conductivities, and intermolecular interactions. J Phys Chem B 114: 16329–16336
20. Hayamizu K, Tsuzuki S, Seki S, Fujii K, Suenaga M, Umebayashi Y (2010) Studies on the translational and rotational motions of ionic liquids composed of N-methyl-N-propyl-pyrrolidinium (P13) cation and bis(trifluoromethanesulfonyl)-amide and bis(fluorosulfonyl) amide anions and their binary systems including lithium salts. J Chem Phys 133:194505
21. Hayamizu K, Tsuzuki S, Seki S, Umebayashi Y (2011) Nuclear magnetic resonance studies on the rotational and translational motions of ionic liquids composed of 1-ethyl-3-methylimidazolium cation and bis(trifluoromethanesulfonyl)amide and bis(fluorosulfonyl) amide anions and their binary systems including lithium salts. J Chem Phys 135:084505
22. Fujii K, Kanzaki R, Takamuku T, Fujimori T, Umebayashi Y, Ishiguro S (2006) Conformational equilibrium of bis(trifluoromethanesulfonyl) imide anion of a room-temperature ionic liquid – Raman spectroscopic study and DFT calculations. J Phys Chem B 110:8179–8183
23. Ishiguro S, Umebayashi Y, Fujii K, Kanzaki R (2006) Solvent conformation and ion solvation: from molecular to ionic liquids. Pure Appl Chem 78:1595–1609
24. Fujii K, Seki S, Fukuda S, Kanzaki R, Takamuku T, Umebayashi Y, Ishiguro S (2007) Anion conformation of low-viscosity room-temperature ionic liquid 1-ethyl-3-methylimidazolium bis(fluorosulfonyl) imide. J Phys Chem B 111:12829
25. Fujimori T, Fujii K, Kanzaki R, Chiba K, Yamamoto H, Umebayashi Y, Ishiguro S (2007) Conformational structure of room temperature ionic liquid N-buthyl- N-methyl-pyrrolidinium bis(trifluoromethanesulfonyl) imide – Raman spectroscopic study and DFT calculations. J Mol Liquids 131(132):216–224
26. Fujii K, Seki S, Fukuda S, Takamuku T, Kohara S, Kameda Y, Umebayashi Y, Ishiguro S (2008) Liquid structure and conformation of a low-viscosity ionic liquid, N-methyl-N-

propylpyrrolidinium bis(fluorosulfonyl) imide studied by high-energy X-ray scattering. J Mol Liq 143:64–69
27. Fujii K, Soejima Y, Kyoshoin Y, Fukuda S, Kanzaki R, Umebayashi Y, Yamaguchi T, Ishiguro S, Takamuku T (2008) Liquid structure of room-temperature ionic liquid, 1-ethyl-3-methylimidazolium bis-(trifluoromethanesulfonyl) imide. J Phys Chem B 112:4329–4336
28. Lopes JNC, Shimizu K, Padua AAH, Umebayashi Y, Fukuda S, Fujii K, Ishiguro S (2008) A tale of two ions: the conformational landscapes bis(trifluoromethanesulfonyl) amide and N, N-dialkylpyrrolidinum. J Phys Chem B 112:1465–1472
29. Lopes JNC, Shimizu K, Pádua AAH, Umebayashi Y, Fukuda S, Fujii K, Ishiguro S (2008) Potential energy landscape of bis(fluorosulfonyl)amide. J Phys Chem B 112:9449–9455
30. Umebayashi Y, Chung W-L, Mitsugi T, Fukuda S, Takeuchi M, Fujii K, Takamuku T, Kanzaki R, Ishiguro S (2008) Liquid structure and the ion-ion interaction of ethylammonium nitrate ionic liquid studied by large angle X-ray scattering and molecular dynamics simulations. J Comput Chem Jpn 7:125–134
31. Fujii K, Mitsugi T, Takamuku T, Yamaguchi T, Umebayashi Y, Ishiguro S (2009) Effect of methylation at the C2 position on the liquid structure of ionic liquids revealed by large-angle X-ray scattering experiments and MD simulations. Chem Lett 38:340–341
32. Kanzaki R, Mitsugi T, Fukuda S, Fujii K, Takeuchi M, Soejima Y, Takamuku T, Yamaguchi T, Umebayashi Y, Ishiguro S (2009) Ion-ion interaction in room temperature ionic liquid 1-ethyl-3-methylimidazolium tetrafluoroborate studied by large-angle X-ray scattering experiment and molecular dynamics simulations. J Mol Liq 147:77
33. Umebayashi Y, Mitsugi T, Fujii K, Seki S, Chiba K, Yamamoto H, Lopes JNC, Pádua AAH, Takeuchi M, Ishiguro S (2009) Raman spectroscopic study, DFT calculations and MD simulations on the conformational isomerism of N-alkyl-N-methylpyrrolidinium bis-(trifluoromethanesulfonyl) amide ionic liquids. J Phys Chem B 113:4338–4346
34. Umebayashi Y, Hamano H, Tsuzuki S, Lopes JNAC, Pádua AAH, Kameda Y, Kohara S, Yamaguchi T, Fujii K, Ishiguro S (2010) Dependence of the conformational isomerism in 1-N-butyl-3-methylimidazolium ionic liquids on the nature of the halide anion. J Phys Chem B 114:11715–11724
35. Fujii K, Kanzaki R, Takamuku T, Kameda Y, Kohara S, Kanakubo M, Shibayama M, Ishiguro S, Umebayashi Y (2011) Experimental evidences for molecular origin of low-Q peak in neutron/X-ray scattering of 1-alkyl-3-methylimidazolium bis(trifluoromethanesulfonyl)amide ionic liquids. J Chem Phys 135:244502
36. Fujii K, Seki S, Ohara K, Kameda Y, Doi H, Saito S, Umebayashi Y (2014) High-energy X-ray diffraction and MD simulation study on the ion-ion interaction in 1-ethyl-3-methylimidazolium bis(fluorosulfonyl)amide. J. Solution Chem 43: 1655–1668
37. Ohtaki H, Radnai T (1993) Structure and dynamics of hydrated ions. Chem Rev 93:1157–1204
38. Burgess J (1999) Ions in solution: basic principles of chemical interactions. Horwood, Chichester
39. Fujii K, Kumai T, Takamuku T, Umebayashi Y, Ishiguro S (2006) Liquid structure and preferential solvation of metal ions in solvent mixtures of N, N-dimethylformamide and N-methylformamide. J Phys Chem A 110:1798–1804
40. Fujii K, Umebayashi Y, Kanzaki R, Kobayashi D, Matsuura R, Ishiguro S (2005) Thermodynamic aspect on metal-ion complexation in structured solvent N-methylformamide. J Solut Chem 34:739–752
41. Hardacre C, McMath SEJ, Nieuwenhuyzen M, Boeron DT, Soper AK (2003) Liquid structure of 1, 3-dimethylimidazolium salts. J Phys Condens Matter 15:S159
42. Deetlefs M, Hardacre C, Nieuwenhuyzen M, Pádua AAH, Sheppard O, Soper AK (2006) Liquid structure of the ionic liquid 1,3-dimethylimidazolium bis{(trifluoromethyl)sulfonyl} amide. J Phys Chem B 110:12055
43. Fukuda S, Takeuchi M, Fujii K, Kanzaki R, Takamuku T, Chiba K, Yamamoto H, Umebayashi Y, Ishiguro S (2008) Liquid structure of N-methyl-N-pyrrolidinium bis-(trifluoromethanesulfonyl) amide ionic liquid studied by large angle X-ray scattering and molecular dynamics simulations. J Mol Liq 143:2–7
44. Song X, Hamano H, Minofar B, Kanzaki R, Fujii K, Kameda Y, Kohara S, Watanabe M, Ishiguro S, Umebayashi Y (2012) Structural heterogeneity and unique distorted hydrogen

bonding in primary ammonium nitrate ionic liquids studied by high-energy X-ray diffraction experiments and MD simulations. J Phys Chem B 116:2801–2813
45. Kohara S, Suzuya K, Kashihara Y, Matsumoto N, Umesaki N, Sakai I (2001) A horizontal two-axis diffractometer for high-energy X-ray diffraction using synchrotron radiation on bending magnet beamline BL04B2 at SPring-8. Nucl Inst Meth A 467–468:1030
46. Isshiki M, Ohishi Y, Goto S, Takeshita K, Oshikawa T (2001) High-energy X-ray diffraction beamline: BL04B2 at SPring-8. Nucl Inst Meth A 467–468:663
47. Benmore CJ (2012) A review of high-energy X-ray diffraction from glasses and liquids. ISRN Mater Sci 1–19
48. Dong K, Zhang S, Wang D, Yao X (2006) Hydrogen bonds in imidazolium ionic liquids. J Phys Chem A 110(31):9775–9782
49. Castriota M, Caruso T, Agostino RG, Cazzanelli E, Henderson WA, Passerini S (2005) Raman investigation of the ionic liquid N-methyl-N-propylpyrrolidinium bis (trifluoromethanesulfonyl) imide and its mixture with LiN(SO2CF3)2. J Phys Chem A 109:92–96
50. Hardwick LJ, Holzapfel M, Wokaun A, Novák P (2007) Raman study of lithium coordination in EMI–TFSI additive systems as lithium–ion battery ionic liquid electrolytes. J Raman Spectrosc 38:110–112
51. Lassègues J-C, Grondin J, Talaga D (2006) Lithium solvation in bis (trifluoromethanesulfonyl) imide-based ionic liquids. Phys Chem Chem Phys 8:5629–5632
52. Lassègues J-C, Grondin J, Aupetit C, Johansson P (2008) Spectroscopic identification of the lithium ion transporting species in LiTFSI-doped ionic liquids. J Phys Chem A 113:305–314
53. Valencia H, Kohyama M, Tanaka S, Matsumoto H (2012) First-principles study of EMIM-FAFSI molecule adsorption on a Li (100) surface as a model for Li-Ion battery electrodes. J Phys Chem C 116:8493–8509
54. Borodin O, Smith GD, Henderson WA (2006) Li + cation environment, transport, and mechanical properties of the LiTFSI doped N-methyl-N-alkylpyrrolidinium + TFSI-ionic liquids. J Phys Chem B 110:16879–16886
55. Monteiro MJ, Bazito FFC, Siqueira LJA, Ribeiro MCC, Torresi RM (2008) Transport coefficients, Raman spectroscopy, and computer simulation of lithium salt solutions in an ionic liquid. J Phys Chem B 112:2102–2109
56. Umebayashi Y, Mitsugi T, Fukuda S, Fujimori T, Fujii K, Kanzaki R, Takeuchi M, Ishiguro S (2007) Lithium-ion solvation in room temperature ionic liquids involving bis-(trifluorosulfonyl) imide anion studied by Raman spectroscopy and DFT calculations. J Phys Chem B 111: 13028–13032
57. Shirai A, Fujii K, Umebayashi Y, Ishiguro S, Ikeda Y (2008) A study on solvation of lithium ion in N, N-diethyl-N- methyl-N-(2-methoxy-ethyl)ammonium bis(trifluoromethanesulfonyl) imide using Raman and multinuclear NMR spectroscopy. Anal Sci 24:1291–1296
58. Umebayashi Y, Tsuzuki S, Fujii K, Mori S, Seki S, Hayamizu K, Ishiguro S (2010) Raman spectroscopic studies and ab initio calculations on conformational isomerism of TFSI solvated to a lithium ion in ionic liquis: effects of the second solvation sphere of the lithium ion. J Phys Chem B 114:6513–6521
59. Fujii K, Hamano H, Doi H, Song X, Tsuzuki S, Hayamizu H, Seki S, Kameda Y, Dokko K, Watanabe M, Umebayashi Y (2013) Unusual Li + ion solvation structure in bis(fluorosulnonyl) amide based ionic liquid. J Phys Chem C 117:19314–19324
60. Fujii K, Nonaka T, Akimoto Y, Umebayashi Y, Ishiguro S (2008) Solvation structures of some transition metal(II) ions in a room-temperature ionic liquid, 1-ethyl-3-methylimidazolium bis(trifluoromethanesulfonyl)amide. Anal Sci 24:1377–1380
61. Umebayashi Y, Yamaguchi T, Fukuda S, Mitsugi T, Takeuchi M, Fujii K, Ishiguro S (2008) Raman spectroscopic study on alkaline metal ion solvation in 1-butyl-3-methylimidazolium bis(trifluoromethanesulfonyl)amide ionic liquid. Anal Sci 24:1297
62. Umebayashi Y, Hamano H, Seki S, Minofar B, Fujii K, Hayamizu K, Tsuzuki S, Kameda Y, Kohara S, Watanabe M (2011) Liquid structure of and Li + ion solvation in bis(trifluoromethanesulfonyl) amide based ionic liquids composed of 1-ethyl-3-methylimidazolium and N-methyl-N-propylpyrrolidinium cations. J Phys Chem B 115:12179–12191

Symbols and Abbreviations

Ionic Liquid Abbreviations

Cations

$[Hmim]^+$	1-Methylimidazolium
$[C_1mim]^+$	1,3-Dimethylimidazolium
$[C_2mim]^+$	1-Ethyl-3-methylimidazolium
$[C_2dmim]^+$	1-Ethyl-2,3-dimethylimidazolium
$[C_3mim]^+$	3-Methyl-1-propylimidazolium
$[C_3dmim]^+$	2,3-Dimethyl-1-propylimidazolium
$[C_4mim]^+$	1-Butyl-3-methylimidazolium
$[C_4dmim]^+$	1-Butyl-2,3-dimethylimidazolium
$[C_4C_4im]^+$	1,3-Dibutylimidazolium
$[C_5mim]^+$	3-Methyl-1-pentylimidazolium
$[C_6mim]^+$	1-Hexyl-3-methylimidazolium
$[C_8mim]^+$	3-Methyl-1-octylimidazolium
$[C_{10}mim]^+$	1-Decyl-3-methylimidazolium
$[dC_{18}im]^+$	1,3-Dioctadecylimidazolium
$[HOCH_2CH_2\text{-mim}]^+$	1-(2-Hydroxyl)ethyl-3-methylimidazolium
$[MeOCH_2CH_2\text{-mim}]^+$	1-(2-Methoxy)ethyl 3-methylimidazolium
$[Me(OCH_2CH_2)_2\text{-mim}]^+$	1-(2-Methoxyethoxy)ethyl)-3-methylimidazolium
$[Me(OCH_2CH_2)_3\text{-mim}]^+$	1-(2-(2-Methoxyethoxy)ethoxy)ethyl)-3-methylimidazolium
$[HSO_3\text{-Bu-mim}]^+$	1-Methyl-3-(4-sulfobutyl)imidazolium
$[HSO_3\text{-Bu-bim}]^+$	1-Butyl-3-(4-sulfobutyl)imidazolium
$[HSO_3\text{-Bu-vim}]^+$	1-(4-Sulfobutyl)-3-vinylimidazolium
$[vmim]^+$	3-Methyl-1-vinylimidazolium
$[C_4vim]^+$	3-Butyl-1-vinylimidazolium
$[C_8vim]^+$	3-Octyl-1-vinylimidazolium
$[C_{12}vim]^+$	3-Dodecyl-1-vinylimidazolium

(continued)

A.A.J. Torriero (ed.), *Electrochemistry in Ionic Liquids*,
DOI 10.1007/978-3-319-13485-7

(continued)

$[C_{18}vim]^+$	3-Octadecyl-1-vinylimidazolium
$[C_{22}vim]^+$	3-Docosyl-1-vinylimidazolium
$[C_2C_4OOC_2im]^+$	3-(4-Ethoxy-4-oxobutyl)-1-ethylimidazol-3-ium
$[C_2PhC_2im]^+$	1-Ethyl-3-phenethylimidazol-3-ium
$[C_3PhC_3im]^+$	3-(3-Phenylpropyl)-1-propyl-1H-imidazol-3-ium
$[C_3PhOC_3im]^+$	3-(4-Phenoxybutyl)-1-propyl-1H-imidazol-3-ium
$[FcC_1mim]^+$	1-(Ferrocenylmethyl)-3-methylimidazolium
$[Amim]^+$	1-Allyl-3-methylimidazolium
$[HPyr]^+$	Pyrrolidinium
$[C_1HPyr]^+$	1-Methylpyrrolidinium
$[C_3mPyr]^+$	1-Methy-1-propylpyrrolidinium
$[C_4mPyr]^+$	1-Butyl-1-methylpyrrolidinium
$[C_5mPyr]^+$	1-Methyl-1-pentylpyrrolidinium
$[C_3bPyr]^{2+}$	1,1′-(Propane-1,3-diyl)bis(4-methylpyridinium)
$[C_4bPyr]^{2+}$	1,1′-(Butane-1,4-diyl)bis(4-methylpyridinium)
$[C_3mpip]^+$	1-Methyl-1-propylpiperidinium
$[C_4mpip]^+$	1-Buthyl-1-methylpiperidinium
$[C_2C_6pip]^+$	1-Ethyl-1-hexylpiperidinium
$[HP_{4\,4\,4}]^+$	Tributylphosphonium
$[P_{1\,4\,4\,4}]^+$	Tributylmethylphosphonium
$[P_{4\,4\,4\,4}]^+$	Tetrabutylphosphonium
$[P_{6\,6\,6\,14}]^+$	Trihexyl(tetradecyl)phosphonium
$[H_3NEt]^+$	Ethylammonium
$[HN_{1\,1\,1}]^+$	Trimethylammonium
$[HN_{2\,2\,2}]^+$	Triethylammonium
$[N_{1\,1\,1\,4}]^+$	Butyltrimethylammonium
$[N_{1\,1\,1\,6}]^+$	Hexyl(trimethyl)ammonium
$[N_{2,2,1,m}]^+$	*N*,*N*-diethyl-*N*-methyl-*N*-(2-methoxyethyl)ammonium
$[N_{2\,2\,2\,4}]^+$	Butyl(trimethyl)ammonium
$[N_{2\,2\,2\,6}]^+$	Hexyl(trimethyl)ammonium
$[N_{3\,3\,3\,p}]^+$	Trimethylphenylammonium
$[N_{4\,4\,4\,4}]^+$	Tetrabutylammonium
$[N_{\,4\,4\,4\,6}]^+$	Tributyl(hexyl)ammonium
$[N_{6\,6\,6\,6}]^+$	Tetrahexylammonium
$[N_{6\,6\,6\,14}]^+$	Trihexyl(tetradecyl)ammonium
$[N_{1\,8\,8\,8}]^+$	Methyl(trioctyl)ammonium
$[N_{8\,8\,8\,8}]^+$	Tetraoctylammonium
$[N_{1\,12\,12\,12}]^+$	Tridodecylmethylammonium
$[CTMA]^+$	*N*,*N*,*N*,*N*-cyanomethyl trimethyl ammonium
$[dema]^+$	Diethylmethylammoium
$[dmea]^+$	Dimethylethylamine
$[DEME]^+$	*N*-methyl-*N*-(2-methoxyethyl)ammonium
$[NHEt(i\text{-}Pr)_2]^+$	*N*,*N*-diisopropylethylamine

(continued)

(continued)

$[H_3N(CH_2CH_2OH)]^+$	Ethanolammonium
$[H_2N(CH_2CH_2OH)_2]^+$	Bis(2-hydroxyethyl)ammonium
$[(HOCH_2CH_2)N_{1\,1\,1}]^+$	Choline
$[CH_3OCH_2CH_2\text{-Et-}Me_2N]^+$	(2-Methoxy)ethyl ethyl dimethylammonium
$[CH_3OCH_2CH_2\text{-}Et_2\text{-MeN}]^+$	(2-Methoxy)ethyl diethyl methylammonium
$[\text{Bu-}Et_2\text{-MeN}]^+$	Butyl diethyl methylammonium
$[Hpy]^+$	Pyridinium
$[C_4py]^+$	1-Butylpyridinium
$[C_4m\beta py]^+$	1-Butyl-3-methylpyridinium
$[C_8py]^+$	1-Octylpyridinium
$[C_{18}py]^+$	1-Octadecylpyridinium
$[HmimSC_4]^+$	1-Methyl-2-butylthiolonium
$[S_{1\,1\,1}]^+$	Trimethylsulfonium
$[S_{2\,2\,1}]^+$	Diethyl(methyl)sulfonium
$[S_{2\,2\,2}]^+$	Triethylsulfonium
$[CDBA6]^+$	*N,N′*-(*trans*-cyclohexane-1,4-diyl)-dibenzamide
$[\text{H-betaine}]^+$	Betainium
$[PD]^{3+}$	1,3,5-Tris(3-((ferrocenylmethyl)amino) pyridiniumyl)-2,4,6-triethylbenzene
$[(N_{111})(N_{112})BH_2]^+$	(Trimethylamine)-(dimethylethylamine)dihydroborate

Anions

$[FSI]^-$	Bis(fluorosulfonyl)imide
$[BETI]^-$	Bis(perfluoroethylsulfonyl)imide
$[N(Tf)_2]^-$	Bis(trifluoromethylsulfonyl)imide
$[C_4C_4N]^-$	Bis(nonafluorobutylsulfonyl)imide
$[NTf]^-$	Brifluoromethanesulfonate
$[N(CN)_2]^-$	Bicyanamide
$[C(CN)_3]^-$	Tricyanomethanide
$[PF_6]^-$	Hexafluorophosphate
$[NO_3]^-$	Nitrate
$[B(CN)_4]^-$	Tetracyanoborate
$[BF_4]^-$	Tetrafluoroborate
$[CF_3BF_3]^-$	Trifluoromethyltrifluoroborate
$[BSB]^-$	Bis(salicylate[2-])borate
$[TFPB]^-$	Tetrakis(3,5-trifluoromethylphenyl)borate anion
$[SCN]^-$	Thiocyanate
$[HCO_2]^-$	Formate
$[CF_3CO_2]^-$	Trifluoroacetate
$[CH_3CO_2]^-$ or $[Ac]^-$	Acetate
$[C_9H_{19}CO_2]^-$	Decanoate

(continued)

(continued)

$[CF_3SO_3]^-$	Trifluoromethanesulfonate
$[CH_3SO_3]^-$	Methanesulphonate
$[HSO_4]^-$	Hydrogen sulfate
$[MeSO_4]^-$	Methylsulphate
$[EtSO_4]^-$	Ethyl sulphate
$[C_8H_{17}SO_4]^-$	Octylsulfate
$[FAP]^-$	Trifluorotris(pentafluoroethyl)phosphate
$[R_2PO_4]^-$	Dialkylphosphate
$[Me_2PO_4]^-$	Dimethylphosphate
$[(EH)_2PO_4]^-$	Bis(2-ethylhexyl)phosphate
$[TPTP]^-$	Tris(pentafluoroethyl)trifluorophosphate
$[(i\text{-}C_8)PO_2]^-$	Bis(2,4,4-trimethylpentyl)phosphinate
$[DPP]^-$	Diphenylphosphate
$[DBP]^-$	Dibutylphosphate
$[EtPO_3H]^-$	Ethylphosphonate
$[(i\text{-}C_4)_2PS_2]^-$	Diisobutyldithiophosphinate
$[(OC_2)_2PS_2]^-$	O,O′-diethyldithiophosphate
$[OH]^-$	Hydroxide
$[I]^-$	Iodide
$[Cl]^-$	Chloride
$[CS]^-$	Camphorsulfonate
$[SiF_6]^{2-}$	Hexafluorosilicate
$[SbF_6]^-$	Hexafluoroantimonate
$[Cu_2Cl_3]^-$	Copper chloride
$[Cu_2Br_3]^-$	Copper bromide
$[AlCl_4]^-$	Tetrachloroaluminate
$[tos]^-$	Tosylate
$[Adip]^{2-}$	Adipate
$[Succ]^{2-}$	Succinate

Chemical Abbreviations

AA	Ascorbic acid
AC	Activated carbons
ACN	Acetronitrile
Au	Gold
γ-BL	Gamma butyrolactone
Bu_4N	Tetrabutylammonium
Cc^+	Cobaltocenium
Chi	Chitosan
cis-Mn	*cis*-$[Mn(CN)(CO)_2[P(oph)_3](DPM)]$ (DPM = $Ph_2PCH_2PPh_2$
cis-W	*cis*-$[W(CO)_2(DPE)_2]$ ($DPE_5Ph_2PCH_2CH_2PPh_2$)

(continued)

(continued)

CNTs	Carbon nanotubes
Cp	Cyclopentadienyl
DEC	Diethyl carbonate
DMC	Dimethyl carbonate
DMF	Dimethyl formamide
DmFc	Decamethylferrocene
DMSO	Dimethyl sulfoxide
EAN	Ethylammonium nitrate
EC	Ethylene carbonate
EMC	Ethylmethyl carbonate
Fc	Ferrocene
Fc^{+}	Ferrocenium
FTO	Fluorine-doped tin oxide
HOPG	Highly ordered pyrolytic graphite
ITO	Indium tin oxide
LTO	Lithium titanium oxide
MAc	*N*-methylacetamide
MEA	Monoethanolamine
MeCN	Acetonitrile
MWCNT	Multiwall carbon nanotubes
NADH	Nicotinamide adenine dinucleotide
NHC	*N*-heterocyclic carbenes
OMCs	Ordered mesoporous carbons
PC	Propylene carbonate
	Polycarbonate
PEDOT	Poly(3,4-ethylenedioxythiophene)
PEG	Polyethylene glycol
PS	Polystyrene
Pt	Platinum
PVA	Polyvinyl alcohol
PVF	Polyvinylferrocene
RE	Rare earth
SAPG	Stress-annealed pyrolytic graphite
$scCO_2$	Supercritical carbon dioxide
SWCNT	Single wall carbon nanotubes
trans-Mn	*trans*-$[Mn\text{-}(CN)(CO)_2\{P(oph)_3\}(Ph_2PCH_2PPh_2)]$

Abbreviations

AC	Alternating current
AFM	Atomic force microscopy
AIL	Aprotic ionic liquid

(continued)

(continued)

ANS	Analyte
ATR	Attenuated total reflectance
BDD	Boron doped diamond
bp	Boiling point
CGT	Complexed gas technology
CILEs	Carbon ionic liquid electrodes
CP	Conducting polymer
CPE	Carbon paste electrodes
CPET	Concerted proton electron transfer
CV	Cyclic voltammetry
CVD	Chemical vapor deposition
DES	Deep eutectic solvents
DLC	Diamond-like coatings
DLS	Dynamic light scattering
DRIFTS	Diffuse reflectance infrared Fourier-transform spectroscopy
DSSC	Dye-sensitized photoelectrochemical solar cells
E-ALD	Electrochemical atomic layer deposition
E-ALE	Electrochemical atomic layer epitaxy
ECC	Electrocapillary curve
ECL	Electrochemiluminescence
ECM	Electrocapillary maximum
EDL	Electrical double layer
EDLC	Electric double layer capacitors
EDR	Equivalent diffusion resistance
EDX	Energy dispersive X-ray
EE	Electrochemical mechanism with sequential electron transfer
EIS	Electrochemical impedance spectroscopy
EPT	Electron proton transfer
EQCM	Electrochemical quartz crystal microbalance
ESR	Equivalent series resistance
ESW	Electrochemical stability windows
ET	Electron transfer
EVs	Electric vehicles
EW	Electrochemical potential windows
FTIR	Fourier transform infrared spectroscopy
GC	Glassy carbon
HBD	Hydrogen bond donor
HER	Hydrogen evolution reaction
HEVs	Hybrid electric vehicles
HFCs	Hydrogen fuel cell
HOR	Hydrogen oxidation reaction
IDE	Interdigitated electrode
IL, RTIL	Ionic liquid

(continued)

(continued)

ILEs	Ionic liquid-based electrolytes
IL-G	Ionic liquid-graphene composite film
IR	Infrared spectroscopy
IRRS	Internal reference redox systems/internal reference redox scale
IS	Immittance spectroscopy
JCPDS	Joint committee on powder diffraction standards
LED	Light-emitting diode
MBE	Molecular beam epitaxy
MEP	Molecular electrostatic potential
MFT	Mean field theory
mp	Melting point
NADES	Natural deep eutectic solvents
NHE	Normal hydrogen electrode
NMR	Nuclear magnetic resonance
NP	Nanoparticle
OCP	Open circuit potential
OER	Oxygen evolution reaction
OLEs	Organic solvent-based electrolytes
OPD	Overpotential deposition
ORR	Oxygen reduction reaction
OX_{ANS}	Oxidised redox component of the ANS
OX_{IRRS}	Oxidised redox component of the IRRS
QRE	Quasi-reference (or pseudo-reference) electrodes
PDDF	Pair distance distribution function
PEMFC	Polymer electrolyte membrane fuel cells
PET	Proton electron transfer
PFIL	Polyelectrolyte-functionalized ionic liquid
PILs	Protic ionic liquids
ppb	Part per billion
ppm	Part per million
PVD	Physical vapor deposition
PZC	Potential of zero charge
RDE	Rotating disk electrode
RED_{ANS}	Reduced redox component of the ANS
RED_{IRRS}	Reduced redox component of the IRRS
RFB	Redox flow batteries
rpm	Revolutions per minute
RRDE	Rotating ring-disk electrode
RTMS	Room-temperature molten salts
SAS	Small angle scattering
SAXS	Small angle X-ray scattering
SCE	Saturated calomel electrode
SECM	Scanning electrochemical microscopy

(continued)

(continued)

SEM	Scanning electron microscopy
SFG	Sum frequency generation spectroscopy
SHG	Second harmonic generation
SPCE	Screen-printed carbon electrode
STM	Scanning tunneling microscope
STS	Scanning tunneling spectroscopy
ToF-SIMS	Time-of-flight secondary ion mass spectrometry
TSIL	Task-specific ionic liquids
UHV	Ultrahigh vacuum
UMEs	Ultramicroelectrodes
UPD	Underpotential deposition
UV	Ultraviolet
UV-Vis	Ultraviolet–visible
VOCs	Volatile organic compounds
VSFS	Vibrational sum frequency spectroscopy
WAXS	Wide angle X-ray scattering
XFEL	X-ray free electron laser
XPCS	X-ray photon correlation spectroscopy
XPS	X-ray photoelectron spectroscopy
XRD	X-ray diffraction
ZIL	Zwitterionic liquid

Roman Symbols

A	Electrode area
a	Microdisk electrode radius
a_i	Activity of ions
B	Porod scaling factor
C	Capacitance
	Coulomb
c	Bulk ionic concentration (also referred as C)
c_-	Concentration of anion
c_+	Concentration of cation
C_0	Capacitance at the minimum (PZC)
C_d	Double layer capacitance
C_{sp}	Specific capacitance
D	Diffusion coefficient
D_O	Diffusion coefficient of an oxidised redox component
$D_n(R)$	Number distribution function
$D(q)$	Effective diffusion constant
D_R	Diffusion coefficient of a reduced redox component

(continued)

(continued)

$D(\lambda)$	Detector efficiency
d	Tip-substrate distance
	Electrolyte density
	Repeat distance
d_i	Diameter of an ion
d_m	Mass fractal dimension
d_s	Surface fractal dimension
e	Charge on the electron
E	Energy
	Potential
E'	Electric field of the visible and infrared incoming beams
$E^{\ominus\prime}$	Formal potential (also referred as $E^{0\prime}$)
$E_{1/2}$	Measured or expected half-wave potential in voltammetry
$E_{1/4}$	Potential where $I/I_d = 1/4$
$E_{3/4}$	Potential where $I/I_d = 3/4$
E_a	Activation energy
E_{end}	Ending potential in a voltammetric scan
E_m	Mid-point potential
ΔE_m	Distance between the E_m of two redox processes
E_T^N	Normalized charge transfer
ΔE_p	Peak-to-peak potential separation
E_{rev}	Reversing potential in a voltammetric scan
E_{start}	Starting potential in a voltammetric scan
F	Faraday constant-the number of coulombs per mole of electrons (F = 96485 C mol^{-1})
$F(q,..)$	Form factor
f	Scattering power of a free electron (excluding anomalous effects)
f_0	Fundamental frequency
Δf	Frequency shift
G	Guinier scaling factor
$\Delta G°$	Gibbs free energy
g	Gravitational force
$g_2(q,\tau)$	Normalized second order intensity correlation function
$H°$	Enthalpy
$H(q)$	Hydrodynamic interaction
h	Hour
I	Faradaic current (also referred as i)
Io	Incident intensity flux
I_d	Current under diffusion control
I_e	Intensity scattered by a single electron
I_p	Peak current (also referred as i_p)
$I(q)$	Measured intensity from a small angle scattering experiment
i_{ss}	Steady-state mass transport limited current (also referred as I_L)
i_T	Tip current

(continued)

(continued)

J	Raman scattering coefficient
j	Current density (also referred as J)
K	Binding constant
$k^{0\prime}$	Heterogeneous charge transfer rate constant (cm^2 s)
k_B	Boltzmann constant
K_{cross}	Equilibrium constant for a simple outer-sphere second-order cross-reaction
k_{cross}	Rate constant for a simple outer-sphere second-order cross-reaction
k_d	Homogeneous rate constant for "forward" reaction (the unit depends on reaction order)
L	Dimension of any inhomogeneity
M	Convolved/semiintegrated current
M_L	Mass transport limited convolved/semiintegrated current
m	Mass transfer coefficient
Δm	Mass changes
N	Number density of scatterers
n	Number of moles of electrons
	Integer
P	Pressure
	Real space vector between two inhomogeneities
Q	Charge
Q_{inv}	Scattering invariant
q	Scattering vector
R	Resistance
	Universal gas constant (8.314 J K^{-1} mol^{-1})
	Radius
R_{ct}	Charge transfer resistance
R_g	Radius of gyration
Rg_{CO}	Radius of gyration cutoff (unified equation)
R_{HS}	Hard sphere radius
R_s	Solution resistance
R_u	Uncompensated resistance
r	Real space vector between two points
	Radial distance from the center of an electrode (cm)
r_0	Electrode radius
r_g	Radius of the sheath
S	Surface area
	Siemens
S°	Entropy
$S(q,\ldots)$	Structure factor
s,s_o	Unit vector of scattered and incoming X-rays, respectively
T	Temperature
T_g	Glass transition temperature
T_m	Melting point
Tr	Transmission

(continued)

(continued)

ts	Material thickness
V	Voltage (also referred as U)
ν	Particle volume fraction
ν_p	Volume occupied by a single point
W	Warburg element
w	Mass fraction
x	Distance from a planar electrode
x_2	Distance between the electrode surface and the inner Helmholtz plane

Greek Symbols

α	Electron transfer coefficient
$\beta(q)$	Optical transfer function
γ	Decay exponent
γ_s	Interfacial tension
$\gamma_o(r)$	Normalized auto-correlation function
ε	Dielectric constant
	Permittivity
ε_r	Relative permittivity
η	Efficiency of the process
	Viscosity
η^{-1}	Fluidity
$\overline{\eta^2}$	Square of the volume average contrast
Λ	Molar conductivity
λ	Wavelength of light (nm)
$\propto_q$	Quartz shear modulus,
μ/ρ	Mass absorption coefficient
υ	Scan rate
ρ	Scattering length density
ρ_q	Quartz density
ρ_s	Mass density of a material
$\overline{\rho}^2$	Auto-correlation of scattering length density function
$\rho^2(\lambda)$	Scattering power (includes absorption)
σ	Ionic conductivity
	Standard deviation
τ	Cell time constant
	Time delay between two measurements
ϕ	Phase
ω	Electrode angular velocity

Index for First Volume

A.A.J. Torriero (ed.), *Electrochemistry in Ionic Liquids*,
DOI 10.1007/978-3-319-13485-7

Index for Second Volume

A.A.J. Torriero (ed.), *Electrochemistry in Ionic Liquids*,
DOI 10.1007/978-3-319-13485-7

MIX
Papier aus verantwortungsvollen Quellen
Paper from responsible sources
FSC® C105338

If you have any concerns about our products,
you can contact us on
ProductSafety@springernature.com

In case Publisher is established outside the EU,
the EU authorized representative is:
Springer Nature Customer Service Center GmbH
Europaplatz 3, 69115 Heidelberg, Germany

Printed by Libri Plureos GmbH
in Hamburg, Germany